Optimal and Robust Estimation

With an Introduction to Stochastic Control Theory

SECOND EDITION

AUTOMATION AND CONTROL ENGINEERING

A Series of Reference Books and Textbooks

1. Nonlinear Control of Electric Machinery, *Darren M. Dawson, Jun Hu, and Timothy C. Burg*
2. Computational Intelligence in Control Engineering, *Robert E. King*
3. Quantitative Feedback Theory: Fundamentals and Applications, *Constantine H. Houpis and Steven J. Rasmussen*
4. Self-Learning Control of Finite Markov Chains, *A. S. Poznyak, K. Najim, and E. Gómez-Ramírez*
5. Robust Control and Filtering for Time-Delay Systems, *Magdi S. Mahmoud*
6. Classical Feedback Control: With MATLAB®, *Boris J. Lurie and Paul J. Enright*
7. Optimal Control of Singularly Perturbed Linear Systems and Applications: High-Accuracy Techniques, *Zoran Gajif and Myo-Taeg Lim*
8. Engineering System Dynamics: A Unified Graph-Centered Approach, *Forbes T. Brown*
9. Advanced Process Identification and Control, *Enso Ikonen and Kaddour Najim*
10. Modern Control Engineering, *P. N. Paraskevopoulos*
11. Sliding Mode Control in Engineering, *edited by Wilfrid Perruquetti and Jean-Pierre Barbot*
12. Actuator Saturation Control, *edited by Vikram Kapila and Karolos M. Grigoriadis*
13. Nonlinear Control Systems, *Zoran Vukiç, Ljubomir Kuljaãa, Dali Donlagiã, and Sejid Tesnjak*
14. Linear Control System Analysis & Design: Fifth Edition, *John D'Azzo, Constantine H. Houpis and Stuart Sheldon*
15. Robot Manipulator Control: Theory & Practice, Second Edition, *Frank L. Lewis, Darren M. Dawson, and Chaouki Abdallah*
16. Robust Control System Design: Advanced State Space Techniques, Second Edition, *Chia-Chi Tsui*
17. Differentially Flat Systems, *Hebertt Sira-Ramirez and Sunil Kumar Agrawal*

18. Chaos in Automatic Control, *edited by Wilfrid Perruquetti and Jean-Pierre Barbot*
19. Fuzzy Controller Design: Theory and Applications, *Zdenko Kovacic and Stjepan Bogdan*
20. Quantitative Feedback Theory: Fundamentals and Applications, Second Edition, *Constantine H. Houpis, Steven J. Rasmussen, and Mario Garcia-Sanz*
21. Neural Network Control of Nonlinear Discrete-Time Systems, *Jagannathan Sarangapani*
22. Autonomous Mobile Robots: Sensing, Control, Decision Making and Applications, *edited by Shuzhi Sam Ge and Frank L. Lewis*
23. Hard Disk Drive: Mechatronics and Control, *Abdullah Al Mamun, GuoXiao Guo, and Chao Bi*
24. Stochastic Hybrid Systems, *edited by Christos G. Cassandras and John Lygeros*
25. Wireless Ad Hoc and Sensor Networks: Protocols, Performance, and Control, *Jagannathan Sarangapani*
26. Optimal and Robust Estimation: With an Introduction to Stochastic Control Theory, Second Edition, *Frank L. Lewis, Lihua Xie, and Dan Popa*

Optimal and Robust Estimation

With an Introduction to Stochastic Control Theory

SECOND EDITION

FRANK L. LEWIS
LIHUA XIE
DAN POPA

CRC Press
Taylor & Francis Group
Boca Raton London New York

CRC Press is an imprint of the
Taylor & Francis Group, an **informa** business

CRC Press
Taylor & Francis Group
6000 Broken Sound Parkway NW, Suite 300
Boca Raton, FL 33487-2742

Printed in the United States of America on acid-free paper
10 9 8 7 6 5 4 3 2 1

International Standard Book Number-13: 978-0-8493-9008-1 (Hardcover)

Library of Congress Cataloging-in-Publication Data

Lewis, Frank L.
Optimal and robust estimation : with an introduction to stochastic control theory / Frank L. Lewis, Lihua Xie, and Dan Popa. -- 2nd ed.
p. cm. -- (Automation and control engineering series ; 25)
Rev. ed. of: Optimal estimation. New York : Wiley, c1986.
Includes bibliographical references and index.
ISBN 978-0-8493-9008-1 (alk. paper)
1. Stochastic control theory. 2. Mathematical optimization. I. Xie, Lihua. II. Popa, Dan, 1969- III. Lewis, Frank L. Optimal estimation. IV. Title.

QA402.3.L4875 2007
629.8'312--dc22 2007008437

Visit the Taylor & Francis Web site at
http://www.taylorandfrancis.com

and the CRC Press Web site at
http://www.crcpress.com

To Galina, Roma, Chris.
Frank L. Lewis

To my wife, Meiyun, and daughters, Jessica and Julie, for their constant support and enduring love.
Lihua Xie

In memory of my late wife Priya Mancharam, for her constant support and inspiration during the revision of the book.
Dan Popa

Contents

Preface xv
Authors xvii
Subject xix
Audience xxi

I Optimal Estimation **1**

1 Classical Estimation Theory **3**
1.1 Mean-Square Estimation 3
1.1.1 Mean-Square Estimation of a Random Variable X by a Constant 4
1.1.2 Mean-Square Estimation of a Random Variable X Given a Random Variable Z: General Case 5
1.1.3 The Orthogonality Principle 13
1.1.4 Linear Mean-Square Estimation of a Random Variable X Given a Random Variable Z 14
1.2 Maximum-Likelihood Estimation 19
1.2.1 Nonlinear Maximum-Likelihood Estimation 19
1.2.2 Linear Gaussian Measurements 20
1.3 The Cramér–Rao Bound 25
1.4 Recursive Estimation 28
1.4.1 Sequential Processing of Measurements 28
1.4.2 Sequential Maximum-Likelihood Estimation 31
1.4.3 Prewhitening of Data 33
1.5 Wiener Filtering 34
1.5.1 The Linear Estimation Problem 36
1.5.2 Solution of the Wiener–Hopf Equation 38
1.5.2.1 Infinite-Delay Steady-State Smoothing 38
1.5.2.2 Causal Steady-State Filtering 41
Problems 50

2 Discrete-Time Kalman Filter **59**
2.1 Deterministic State Observer 59
2.2 Linear Stochastic Systems 64
2.2.1 Propagation of Means and Covariances 65
2.2.2 Statistical Steady-State and Spectral Densities 68

2.3 The Discrete-Time Kalman Filter 70
2.3.1 Kalman Filter Formulations 71
2.4 Discrete Measurements of Continuous-Time Systems 84
2.4.1 Discretization of Continuous Stochastic Systems 85
2.4.2 Multiple Sampling Rates 98
2.4.3 Discretization of Time-Varying Systems 101
2.5 Error Dynamics and Statistical Steady State 101
2.5.1 The Error System 101
2.5.2 The Innovations Sequence 102
2.5.3 The Algebraic Riccati Equation 103
2.5.4 Time-Varying Plant 110
2.6 Frequency Domain Results 112
2.6.1 A Spectral Factorization Result 112
2.6.2 The Innovations Representation 114
2.6.3 Chang–Letov Design Procedure for the Kalman Filter 115
2.6.4 Deriving the Discrete Wiener Filter 118
2.7 Correlated Noise and Shaping Filters 123
2.7.1 Colored Process Noise 124
2.7.2 Correlated Measurement and Process Noise 127
2.7.3 Colored Measurement Noise 130
2.8 Optimal Smoothing 132
2.8.1 The Information Filter 133
2.8.2 Optimal Smoothed Estimate 136
2.8.3 Rauch–Tung–Striebel Smoother 138
Problems 140

3 Continuous-Time Kalman Filter **151**
3.1 Derivation from Discrete Kalman Filter 151
3.2 Some Examples 157
3.3 Derivation from Wiener–Hope Equation 166
3.3.1 Introduction of a Shaping Filter 167
3.3.2 A Differential Equation for the Optimal Impulse Response 168
3.3.3 A Differential Equation for the Estimate 169
3.3.4 A Differential Equation for the Error Covariance 170
3.3.5 Discussion 172
3.4 Error Dynamics and Statistical Steady State 177
3.4.1 The Error System 177
3.4.2 The Innovations Sequence 178
3.4.3 The Algebraic Riccati Equation 179
3.4.4 Time-Varying Plant 180
3.5 Frequency Domain Results 180
3.5.1 Spectral Densities for Linear Stochastic Systems 181
3.5.2 A Spectral Factorization Result 181
3.5.3 Chang–Letov Design Procedure 184

3.6 Correlated Noise and Shaping Filters 188
3.6.1 Colored Process Noise 188
3.6.2 Correlated Measurement and Process Noise 189
3.6.3 Colored Measurement Noise 190
3.7 Discrete Measurements of Continuous-Time Systems 193
3.8 Optimal Smoothing 197
3.8.1 The Information Filter 200
3.8.2 Optimal Smoothed Estimate 201
3.8.3 Rauch–Ting–Striebel Smoother 203
Problems 204

4 Kalman Filter Design and Implementation 213
4.1 Modeling Errors, Divergence, and Exponential Data Weighting 213
4.1.1 Modeling Errors 213
4.1.2 Kalman Filter Divergence 223
4.1.3 Fictitious Process Noise Injection 226
4.1.4 Exponential Data Weighting 230
4.2 Reduced-Order Filters and Decoupling 236
4.2.1 Decoupling and Parallel Processing 236
4.2.2 Reduced-Order Filters 242
4.3 Using Suboptimal Gains 249
4.4 Scalar Measurement Updating 253
Problems 254

5 Estimation for Nonlinear Systems 259
5.1 Update of the Hyperstate 259
5.1.1 Discrete Systems 259
5.1.2 Continuous Systems 263
5.2 General Update of Mean and Covariance 265
5.2.1 Time Update 266
5.2.2 Measurement Update 268
5.2.3 Linear Measurement Update 269
5.3 Extended Kalman Filter 271
5.3.1 Approximate Time Update 271
5.3.2 Approximate Measurement Update 272
5.3.3 The Extended Kalman Filter 273
5.4 Application to Adaptive Sampling 283
5.4.1 Mobile Robot Localization in Sampling 284
5.4.2 The Combined Adaptive Sampling Problem 284
5.4.3 Closed-Form Estimation for a Linear Field without Localization Uncertainty 286
5.4.4 Closed-Form Estimation for a Linear Field with Localization Uncertainty 290
5.4.5 Adaptive Sampling Using the Extended Kalman Filter 295

5.4.6 Simultaneous Localization and Sampling Using a Mobile Robot ... 297
Problems ... 305

II Robust Estimation **313**

6 Robust Kalman Filter **315**
6.1 Systems with Modeling Uncertainties ... 315
6.2 Robust Finite Horizon Kalman *a Priori* Filter ... 317
6.3 Robust Stationary Kalman *a Priori* Filter ... 321
6.4 Convergence Analysis ... 326
6.4.1 Feasibility and Convergence Analysis ... 326
6.4.2 ε–Switching Strategy ... 329
6.5 Linear Matrix Inequality Approach ... 331
6.5.1 Robust Filter for Systems with Norm-Bounded Uncertainty ... 332
6.5.2 Robust Filtering for Systems with Polytopic Uncertainty ... 335
6.6 Robust Kalman Filtering for Continuous-Time Systems ... 341
Proofs of Theorems ... 343
Problems ... 350

7 $\mathcal{H}_\infty$ Filtering of Continuous-Time Systems **353**
7.1 $\mathcal{H}_\infty$ Filtering Problem ... 353
7.1.1 Relationship with Two-Person Zero-Sum Game ... 356
7.2 Finite Horizon $\mathcal{H}_\infty$ Linear Filter ... 357
7.3 Characterization of All Finite Horizon $\mathcal{H}_\infty$ Linear Filters ... 361
7.4 Stationary $\mathcal{H}_\infty$ Filter—Riccati Equation Approach ... 365
7.4.1 Relationship between Guaranteed $\mathcal{H}_\infty$ Norm and Actual $\mathcal{H}_\infty$ Norm ... 371
7.4.2 Characterization of All Linear Time-Invariant $\mathcal{H}_\infty$ Filters ... 373
7.5 Relationship with the Kalman Filter ... 373
7.6 Convergence Analysis ... 374
7.7 $\mathcal{H}_\infty$ Filtering for a Special Class of Signal Models ... 378
7.8 Stationary $\mathcal{H}_\infty$ Filter—Linear Matrix Inequality Approach ... 382
Problems ... 383

8 $\mathcal{H}_\infty$ Filtering of Discrete-Time Systems **387**
8.1 Discrete-Time $\mathcal{H}_\infty$ Filtering Problem ... 387
8.2 $\mathcal{H}_\infty$ *a Priori* Filter ... 390
8.2.1 Finite Horizon Case ... 391
8.2.2 Stationary Case ... 397
8.3 $\mathcal{H}_\infty$ *a Posteriori* Filter ... 400
8.3.1 Finite Horizon Case ... 400
8.3.2 Stationary Case ... 405

8.4 Polynomial Approach to $\mathcal{H}_\infty$ Estimation 408
8.5 J-Spectral Factorization 410
8.6 Applications in Channel Equalization 414
Problems 419

III Optimal Stochastic Control 421

9 Stochastic Control for State Variable Systems 423
9.1 Dynamic Programming Approach 423
9.1.1 Discrete-Time Systems 424
9.1.2 Continuous-Time Systems 435
9.2 Continuous-Time Linear Quadratic Gaussian Problem 443
9.2.1 Complete State Information 445
9.2.2 Incomplete State Information and the Separation Principle 449
9.3 Discrete-Time Linear Quadratic Gaussian Problem 453
9.3.1 Complete State Information 454
9.3.2 Incomplete State Information 455
Problems 457

10 Stochastic Control for Polynomial Systems 463
10.1 Polynomial Representation of Stochastic Systems 463
10.2 Optimal Prediction 465
10.3 Minimum Variance Control 469
10.4 Polynomial Linear Quadratic Gaussian Regulator 473
Problems 481

Appendix: Review of Matrix Algebra 485
A.1 Basic Definitions and Facts 485
A.2 Partitioned Matrices 486
A.3 Quadratic Forms and Definiteness 488
A.4 Matrix Calculus 490

References 493

Index 501

Preface

This book is intended for use in a second graduate course in modern control theory. A background in both probability theory and the state variable representation of systems is assumed. One advantage of modern control theory is that it employs matrix algebra, which results in a simplification in notation and mathematical manipulations when dealing with multivariable systems. The Appendix provides a short review of matrices for those who need it.

The book is also intended as a reference. Equations of recurring usefulness are displayed in tabular form for easy access. Many examples are used to illustrate the concepts and to impart intuition, and the reader is shown how to write simple software implementations of estimators using MATLAB®. In this edition, the example problems have been revisited and updated. Sample code in MATLAB will be available from the publisher's Web site for download.

Optimal control and optimal estimation are the dual theories that provide the foundation for the modern study of systems. Optimal control can be studied in a purely deterministic context in which the unrealistic assumption is made that perfect information about nature is available. The solutions to control problems are discovered by a backward progression through time.

Optimal estimation, however, is the problem of casting into a useful form information from an inherently noisy and substantially uncooperative environment; it is, therefore, intimately concerned with probabilistic notions. Solutions to estimation problems are discovered by a forward progression through time.

Stochastic control theory uses information reconstructed from noisy measurements to control a system so that it has a desired behavior; hence, it represents a marriage of optimal estimation and deterministic optimal control.

In Part I, we discuss the estimation problem, covering in detail a system that extracts information from measured data and which is known as the Kalman filter. Kalman filters are used, for example, in communication networks and in the navigation of ships, aircraft, and space vehicles. Classical estimation theory, including the Wiener filter, is reviewed in Chapter 1, and throughout the book we point out relationships between the Kalman filter and this older body of knowledge.

In Part II, we focus our discussions on robust estimation in which the dynamics of the underlying system are not exactly known. First, we are concerned with the design of a filter that would have an optimal guaranteed performance for systems with parametric uncertainties of a norm-bounded type or polytopic type. However, when little information about the statistics of noises is given, an H_∞ filter would be more desirable than a Kalman filter. In Chapters 7 and 8, we give a detailed account of H_∞ filtering.

In Part III, we introduce stochastic control theory, treating both state variable systems and polynomial systems. For a complete foundation in modern system theory and more insight on stochastic control, a course in deterministic optimal control is required.

Frank L. Lewis—This book is dedicated to my teachers, J.B. Pearson and E.W. Kamen, whose knowledge and wisdom I have imperfectly transcribed into this written volume wherein the insight and the better ideas are completely theirs. It is also dedicated to N.B. Nichols, whose achievements have been a source of inspiration and whose friendship a source of strength. Finally, it is dedicated to Brian Stevens, a true engineer, and a close friend who has given me more insight than he realizes in our years of working together.

Lihua Xie—The book is dedicated to C.E. de Souza, M. Fu, and U. Shaked for their willingness to share their extensive knowledge and their research collaborations on robust estimation. We are also indebted to Y.C. Soh and H. Zhang for many useful discussions and S. Sun and J. Xu for their careful reading of Part II of the book. The help from J. Xu in some simulation examples of Part II is greatly appreciated. The book is also dedicated to my wife, Meiyun, and daughters, Jassica and Julie, for their constant support and enduring love.

Dan Popa wishes to dedicate his contribution to this edition in the memory of his late wife Priya Mancharam, for her constant support and inspiration during the revision of the book. We also acknowledge the collaboration work with Professors Arthur Sanderson and Harry Stephanou, and the help of Chad Helm, Vadiraj Hombal, Koushil Sreenath, and Muhammad Mysorewala on the example of adaptive sampling with robots added to Chapter 5.

Finally, we thank Muhammad Mysorewala for his patience and diligence working through most of the examples, codes, and the diagrams in the book.

Frank L. Lewis
Lihua Xie
Dan Popa

Authors

Frank L. Lewis, Fellow IEEE, Fellow U.K. Institute of Measurement and Control, PE Texas, U.K. Chartered Engineer, is distinguished scholar professor and Moncrief-O'Donnell chair at the University of Texas at Arlington's Automation and Robotics Research Institute. He obtained his Bachelor's degree in physics/EE and his MSEE degree at Rice University, his MS degree in aeronautical engineering from the University of West Florida, and his PhD at the Georgia Institute of Technology. He works in feedback control, intelligent systems, and sensor networks. He is the author of five U.S. patents, 174 journal papers, 286 conference papers, and 11 books. He received the Fulbright Research Award, NSF Research Initiation Grant, and ASEE Terman Award. He also received the Outstanding Service Award from the Dallas IEEE Section and was selected as engineer of the year by the Ft. Worth IEEE Section. He is listed in the Ft. Worth Business Press Top 200 Leaders in Manufacturing. He was appointed to the NAE Committee on Space Station in 1995. Dr. Lewis is an elected guest consulting professor at both South China University of Technology and Shanghai Jiao Tong University. He is the founding member of the Board of Governors of the Mediterranean Control Association. He helped win the IEEE Control Systems Society Best Chapter Award (as founding chairman of the DFW Chapter), the National Sigma Xi Award for Outstanding Chapter (as president of UTA Chapter), and the US SBA Tibbets Award in 1996 (as director of ARRI's SBIR Program).

Lihua Xie received his BE and ME degrees in electrical engineering from Nanjing University of Science and Technology in 1983 and 1986, respectively, and his PhD degree in electrical engineering from the University of Newcastle, Australia, in 1992. Since 1992, he has been with the School of Electrical and Electronic Engineering, Nanyang Technological University, Singapore, where he is a professor. He held teaching appointments in the Department of Automatic Control, Nanjing University of Science and Technology from 1986 to 1989. He has also held visiting appointments with the University of Newcastle, the University of Melbourne, and Hong Kong Polytechnic University.

Dr Xie's research interests include robust control and estimation, networked control systems, time delay systems, control of disk drive systems, and sensor networks. He has published over 120 journal papers and coauthored the books *H-Infinity Control and Filtering of Two-Dimensional Systems* (with C. Du), *Optimal and Robust Estimation* (with F.L. Lewis and D. Popa), and *Control and Estimation of Systems with Input/Output Delays* (with H. Zhang). He is an associate editor of the *IEEE Transactions on Automatic Control,*

Automatica, and *IEEE Transactions on Circuits and Systems II*. He is also a member of the editorial board of *IEE Proceedings on Control Theory and Applications*. He served as an associate editor of the *International Journal of Control, Automation and Systems* from 2004 to 2006 and an associate editor of the Conference Editorial Board, IEEE Control Systems Society from 2000 to 2005. He was the general chairman of the 9th International Conference on Control, Automation, Robotics and Vision. Dr. Xie is a fellow of IEEE and a fellow of Institution of Engineers, Singapore.

Dan Popa is presently an assistant professor of electrical engineering at the Automation and Robotics Research Institute at the University of Texas at Arlington. He received a BA degree in engineering, mathematics, and computer science and an MS degree in engineering, both from Dartmouth College where he was a Montgomery Scholar from 1990 to 1994. He received a PhD in EE from Rensselaer Polytechnic Institute in 1998, focusing on nonholonomic robotics. He then joined the Center for Automation Technologies at RPI, where he held the rank of senior research scientist until 2004. Dr. Popa led numerous projects sponsored by government agencies and industry in the areas of robotics, control, and microsystems assembly and packaging.

Dr. Popa has a broad experience base in control and robotics, including the design, characterization, modeling, and control of MEMS actuators; the design of parallel and multiscale assembly architectures; the deployment of mobile sensor networks; and the development of new processes for 3D wafer integration. His current research focuses on sensors, actuators, and integrated microsystems, and their deployment using robots. He is the recipient of several prestigious awards and the author of over 60 refereed papers, and a member of IEEE, ASME, SME, and IMAPS.

Subject

The book covers estimation theory and design techniques that are important in navigation, communication systems and signal processing, aerospace systems, industry, seismology and earthquake prediction, and elsewhere. Estimators seek to combine the available measurements and knowledge of system dynamics to provide the most accurate knowledge possible of desired characteristics or quantities of interest. An example is in spacecraft navigation where dynamical information about the motion of the vehicle and also measurements from navigation systems and star positioning information are available. The Kalman filter combines all this information to provide the best possible estimate of the vehicle's location. Another example is in communications, where one seeks to reconstruct the signals in the presence of noise and distortions. The book covers optimal filtering including the Kalman filter, optimal smoothing, and nonlinear estimation. Three chapters on robust estimation and H_∞ estimation show how to apply filtering for modern-day high-performance systems with disturbances and uncertainty. Both continuous-time and discrete-time systems are discussed. Extensive computer examples and case studies provide insight and design experience. MATLAB code is available on the Web site. Design equations and algorithms are collected in tables for ease of use and reference.

Audience

The book is intended for senior undergraduate or first graduate courses on estimation. The first edition has been used for both. It is also useful for researchers in academia and government labs. Industry engineers currently using the book will welcome the new material on robust and H-infinity filtering.

The material is important in navigation, communication systems and signal processing, aerospace systems, industry, seismology and earthquake prediction, biological systems, control systems, and elsewhere. As such it will market in the above IEEE societies and other corresponding societies nationally and internationally.

Part I

Optimal Estimation

1

Classical Estimation Theory

There are many ways to estimate an unknown quantity from available data. This chapter basically covers mean-square estimation and the Wiener filter. A discussion of recursive estimation is included, since it provides a good background for the Kalman filter. The techniques presented here can be described as *classical techniques* in that they represent the state of the art prior to 1960, when Kalman and Bucy developed their approach (Kalman 1960, 1963; Kalman and Bucy 1961).

1.1 Mean-Square Estimation

Suppose there is an unknown vector X and available data Z that is somehow related to X, so that knowing Z we have some information about X. Mean-square estimation implies a setting in which there is prior information about both the unknown $X \in R^n$ and the measurement $Z \in R^p$, so that both X and Z are stochastic. That is, we know the joint statistics of X and Z. If Z is given, we would like to make an *estimate* $\hat{X}(Z)$ of the value of X so that the *estimation error*

$$\tilde{X} = X - \hat{X} \tag{1.1}$$

is small. "Small" can be defined in several ways, leading to different methods of estimation.

Define the *mean-square error* as the expected value of the Euclidean norm squared of $\tilde{X}$,

$$J = \overline{\tilde{X}^T\tilde{X}} = \overline{(X - \hat{X})^T(X - \hat{X})} \tag{1.2}$$

Since $\tilde{X}$ is a random variable, the expected value (overbar) is required as shown. (We use both overbars and $E(\cdot)$ for expected value.)

The *mean-square estimation criterion* can be stated as follows. Using all available information, select an estimate $\hat{X}$ to minimize Equation 1.2.

It is worth saying a few words about the "available information." In the next subsection, we consider the case where there is no information other than the statistics of X. Next, we consider the case where a random variable Z, the measurement, provides additional information on X. There will be no assumption about how Z derives from X, so this latter result will be completely general. Finally, in the third subsection we make a useful restricting assumption on the form of the estimate. This presentation is modified from the one in Papoulis (1965).

Mean-square estimation is a special case of *Bayesian estimation*. The general Bayesian estimation problem is to find the estimate that minimizes the *Bayes risk*

$$J = E[C(\tilde{X})] = \int_{-\infty}^{\infty} C(x - \hat{X}) f_{XZ}(x, z) \mathrm{d}x \, \mathrm{d}z \tag{1.3}$$

where $C(\cdot)$ is a *cost function*. Selecting $C(\tilde{X}) = \tilde{X}^T \tilde{X}$ results in Equation 1.2. Another special case is *maximum a posteriori estimation* ($C(\tilde{X}) = 0$, if all components of $\tilde{X}$ have magnitude less than a given $\varepsilon, C(\tilde{X}) = 1$ otherwise). It can be shown that as long as the cost is symmetric and convex, and the conditional probability density function (PDF) $f_{X/Z}(x/z)$ is symmetric about its mean, then all Bayes estimates yield the same optimal estimate (Sorenson 1970a). As we shall discover, this is the conditional mean of the unknown, given the data $\overline{X/Z}$.

1.1.1 Mean-Square Estimation of a Random Variable *X* by a Constant

Suppose the measurement Z is not yet given, but we know only the *a priori statistics* of the unknown X. That is, the PDF $f_X(x)$ is known, or equivalently all moments of X are given. The problem is to choose a constant that will yield the best a *a priori estimate* $\hat{X}$ for X in the mean-square sense.

To solve this problem, work with the mean-square error (Equation 1.2)

$$\begin{aligned} J = \overline{(X - \hat{X})^T (X - \hat{X})} &= \overline{X^T X} - \hat{X}^T \bar{X} - \bar{X}^T \hat{X} + \hat{X}^T \hat{X} \\ &= \overline{X^T X} - 2\hat{X}^T \bar{X} + \hat{X}^T \hat{X} \end{aligned}$$

where we have used the fact that, since it is a constant, $\hat{X}$ does not depend on X and so can be taken out from under the expectation operator. Now, differentiating this, it is seen that

$$\frac{\partial J}{\partial \hat{X}} = -2\bar{X} + 2\hat{X} = 0$$

or

$$\hat{X}_{\mathrm{MS}} = \bar{X} \tag{1.4}$$

We have thus arrived at the possibly obvious and certainly intuitive conclusion that if only the statistics of X are known, then the best constant to estimate X is the *mean* of X.

To obtain a measure of confidence for the estimate (Equation 1.4), let us find the *error covariance*. Notice first that

$$E(\tilde{X}) = E(X - \bar{X}) = \bar{X} - \bar{X} = 0$$

so that the expected estimation error is equal to zero. This property is of sufficient importance to give it a name. Accordingly, define an estimate $\hat{X}$ to be *unbiased* if

$$E(\tilde{X}) = 0 \tag{1.5}$$

This is equivalent to

$$E(\hat{X}) = \bar{X} \tag{1.6}$$

which says that the expected value of the estimate is equal to the mean of the unknown. (In the absence of data, the optimal mean-square estimate *is* the mean of the unknown, as we have just discovered.)

We can now write for the error covariance

$$\begin{aligned} P_{\tilde{X}} &= \overline{(\tilde{X} - \bar{\tilde{X}})(\tilde{X} - \bar{\tilde{X}})^T} = \overline{\tilde{X}\tilde{X}^T} \\ &= \overline{(X - \hat{X})(X - \hat{X})^T} = \overline{(X - \bar{X})(X - \bar{X})^T} \end{aligned}$$

but this is just equal to the covariance of X. The *a priori* error covariance is therefore just

$$P_{\tilde{X}} = P_X \tag{1.7}$$

If the error covariance is small, then the associated estimate is a good one. We call the inverse of the error covariance, $P_{\tilde{X}}^{-1}$, the *information matrix.*

1.1.2 Mean-Square Estimation of a Random Variable X Given a Random Variable Z: General Case

Now, additional information is prescribed. Suppose that in addition to the statistics of X, the value of another related random variable Z is known (presumably by some measurement process). We shall make no assumptions on how Z derives from X. It is assumed that only the joint density $f_{XZ}(x, z)$ of X and Z is given.

The optimal mean-square estimate depends on the (known) measurement, so write $\hat{X}(Z)$. We call this estimate after taking the measurement the *a posteriori estimate* for X. As the value of Z changes, so does $\hat{X}$, but if Z has a fixed value of $Z = z$, then $\hat{X}(Z)$ has the fixed value $\hat{X}(z)$. This fact will be required later.

By definition, to find $\hat{X}(Z)$, write Equation 1.2 as

$$J = \int_{-\infty}^{\infty} \int [x - \hat{X}(z)]^T [x - \hat{X}(z)] f_{XZ}(x, z) \mathrm{d}x \, \mathrm{d}z$$

or, using Bayes' rule,

$$J = \int_{-\infty}^{\infty} f_Z(z) \int_{-\infty}^{\infty} [x - \hat{X}(z)]^T [x - \hat{X}(z)] f_{X/Z}(x/z) \mathrm{d}x \, \mathrm{d}z$$

Since $f_Z(z)$ and the inner integral are nonnegative, J can be minimized by fixing $Z = z$ and minimizing for every such z the value of the *conditional mean-square error*

$$J' = \int_{-\infty}^{\infty} [x - \hat{X}(z)]^T [x - \hat{X}(z)] f_{X/Z}(x/z) \mathrm{d}x \tag{1.8}$$

Now, repeat the derivation of Equation 1.4 using conditional means. Then,

$$\begin{aligned} J' &= \overline{[X - \hat{X}(Z)]^T [X - \hat{X}(Z)]/Z} \\ &= \overline{X^T X/Z} - 2\hat{X}(Z)^T \overline{X/Z} + \hat{X}(Z)^T \hat{X}(Z) \end{aligned} \tag{1.9}$$

where we have used our previous realization that $\hat{X}(Z)$ is fixed if Z is fixed, so that $\hat{X}(Z)$ comes out from under the conditional expectation operator. Differentiate to find that

$$\frac{\partial J'}{\partial \hat{X}(Z)} = -2\overline{X/Z} + 2\hat{X}(Z) = 0$$

or finally

$$\hat{X}_{\mathrm{MS}}(Z) = \overline{X/Z} \tag{1.10}$$

This is the not so obvious but equally intuitive counterpart to Equation 1.4. If a measurement $Z = z$ is given, then we should take as the *a posteriori* estimate for X the conditional mean of X, given that $Z = z$. If X and Z are independent then $f_{XZ} = f_X f_Z$ so that

$$\begin{aligned} \overline{X/Z} = \int x f_{X/Z}(x/z) \mathrm{d}x &= \int x \frac{f_{XZ}(x, z)}{f_Z(z)} \mathrm{d}x \\ &= \int x f_X(x) \mathrm{d}x = \bar{X} \end{aligned} \tag{1.11}$$

and the *a priori* and *a posteriori* estimates are the same. Then, knowledge of Z does not provide any information about X.

Equation 1.10 appears very simple, but the simplicity of its appearance belies the difficulty of its use, the computation of the conditional mean $\overline{X/Z}$ may be an intractable problem. Worse than this, to find $\overline{X/Z}$ we require f_{XZ}, which may not be known. In some cases, we may know only its first and second moments.

Equation 1.10 is the mean-square estimate in the most general case. It is often called the *conditional mean estimate of* X and, under the general conditions we have mentioned, all Bayes estimation schemes lead to this same optimal estimate.

It is sometimes convenient to express the conditional mean $\overline{X/Z}$ in terms of the PDF $f_{Z/X}(z/x)$. To do this, note that by Bayes' rule

$$\overline{X/Z} = \int_{-\infty}^{\infty} x f_{X/Z}(x/z)\,\mathrm{d}x$$

$$\overline{X/Z} = \int_{-\infty}^{\infty} x \frac{f_{X/Z}(x,z)}{f_Z(z)}\mathrm{d}x$$

$$= \frac{\int_{-\infty}^{\infty} x f_{XZ}(x,z)\mathrm{d}x}{\int_{-\infty}^{\infty} f_{XZ}(x,z)\mathrm{d}x} \tag{1.12}$$

$$\overline{X/Z} = \frac{\int_{-\infty}^{\infty} x f_{Z/X}(z/x) f_X(x)\mathrm{d}x}{\int_{-\infty}^{\infty} f_{Z/X}(z/x) f_X(x)\mathrm{d}x} \tag{1.13}$$

which is the desired result.

The conditional mean estimate is unbiased since $\bar{X} = E(\overline{X/Z})$:

$$E(\tilde{X}) = E(\overline{\tilde{X}/Z}) = E(\overline{(X - \hat{X})/Z}) = E(\overline{(X - \overline{X/Z})/Z})$$
$$= E(\overline{X/Z} - \overline{\overline{X/Z}/Z}) = E(\overline{X/Z} - \overline{X/Z}) = 0$$

(Note that taking the same expected value twice is the same as taking it once.)

A measure of confidence in an estimate is given by the error covariance. To find the error covariance associated with the conditional mean estimate, write, since $\bar{\tilde{X}} = 0$,

$$\begin{aligned} P_{\tilde{X}} &= \overline{\tilde{X}\tilde{X}^T} \\ &= \overline{(X - \hat{X})(X - \hat{X})^T} \\ &= E(\overline{(X - \overline{X/Z})(X - \overline{X/Z})^T/Z}) \\ &= E(P_{X/Z}) \end{aligned}$$

or

$$P_{\tilde{X}} = \overline{P_{X/Z}} \tag{1.14}$$

Under some conditions $P_{\tilde{X}} = P_{X/Z}$, so that the unconditional error covariance is the same as the conditional error covariance given a particular measurement Z. In general, however, Equation 1.14 must be used since $P_{X/Z}$ may depend on Z.

It is worth noting that J as given by Equation 1.2 is equal to trace $(P_{\tilde{X}})$ so that $\hat{X}_{\mathrm{MS}}$ minimizes the trace of the error covariance matrix.

The next example shows how to compute mean-square estimates. We shall use $(\bar{X}, P_X)$ to denote a random variable with mean $\bar{X}$ and covariance P_X. If X is normal we write $N(\bar{X}, P_X)$.

Example 1.1 Nonlinear Mean-Square Estimation (Melsa and Cohn 1978)
Let unknown X be distributed uniformly according to

$$f_X(x) = \begin{cases} 1, & 0 \le x \le 1 \\ 0, & \text{otherwise} \end{cases} \tag{1.15}$$

a. In the absence of other information, choose a constant to estimate X in the optimal mean-square sense (i.e., find the *a priori* estimate).
   According to Equation 1.4, in this case the best estimate is

$$\begin{aligned} \hat{X}_{\text{MS}} = \hat{X}_{ap} = \bar{X} &= \int_{-\infty}^{\infty} x f_X(x)\mathrm{d}x \\ &= \int_0^1 x\,\mathrm{d}x = \frac{1}{2} \end{aligned} \tag{1.16}$$

b. Now, suppose a Z is measured, which is related to X by

$$Z = \ln\left(\frac{1}{X}\right) + V \tag{1.17}$$

   where V is noise with exponential distribution

$$f_V(v) = \begin{cases} e^{-v}, & v \ge 0 \\ 0, & v < 0 \end{cases} \tag{1.18}$$

   and X and V are independent. Find the conditional mean estimate.

If X and V are independent, then (function of a random variable; Papoulis 1965)

$$\begin{aligned} f_{Z/X}(z/x) &= f_V\left[z - \ln\left(\frac{1}{x}\right)\right] \\ &= \begin{cases} e^{-(z-\ln(1/x))}, & z - \ln\frac{1}{x} \ge 0 \\ 0, & z - \ln\frac{1}{x} < 0 \end{cases} \\ &= \begin{cases} \frac{1}{x}e^{-z}, & x \ge e^{-z} \\ 0, & x < e^{-z} \end{cases} \end{aligned} \tag{1.19}$$

Now using Equation 1.13 we get

$$\begin{aligned} \hat{X}_{\text{MS}}(z) &= \frac{\int_{e^{-z}}^1 e^{-z}\mathrm{d}x}{\int_{e^{-z}}^1 \frac{1}{x}e^{-z}\mathrm{d}x} \\ &= \frac{xe^{-z}|_{e^{-z}}^1}{e^{-z}\ln x|_{e^{-z}}^1} \\ &= \frac{(1-e^{-z})e^{-z}}{ze^{-z}} \end{aligned}$$

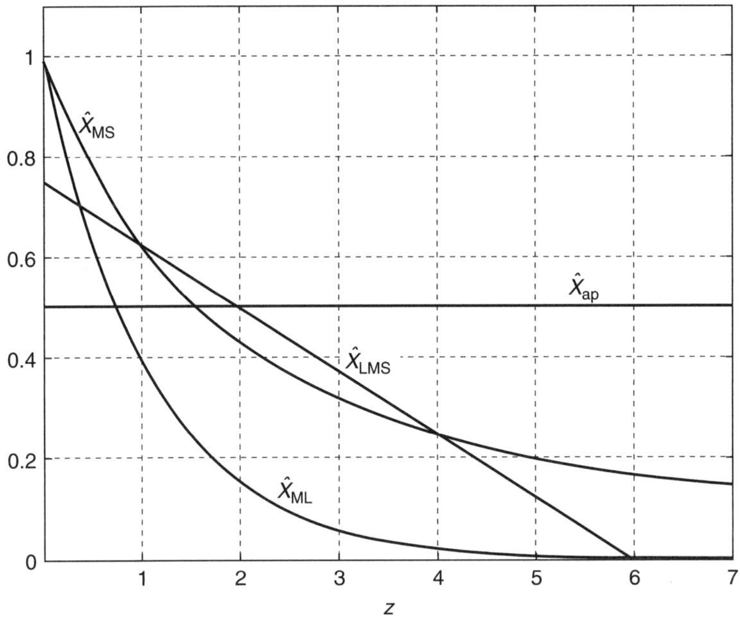

FIGURE 1.1
Nonlinear mean-square estimate.

or

$$\hat{X}_{\text{MS}} = \frac{1 - e^{-Z}}{Z} \tag{1.20}$$

This nonlinear estimate is sketched in Figure 1.1 along with the *a priori* estimate X_{ap} and some other estimates for the same problem, which we shall discuss in Examples 1.4 and 1.6. □

Example 1.2 Gaussian Conditional Mean and Covariance

This is an extremely important example, which we shall constantly use in subsequent sections.

Let $Y \in R^{n+p}$ be normal, symbolized by $N(\bar{Y}, P_Y)$, so that its PDF is

$$f_Y(y) = \frac{1}{\sqrt{(2\pi)^{n+p}|P_Y|}} e^{-(1/2)(y-\bar{Y})^T P_Y^{-1}(y-\bar{Y})} \tag{1.21}$$

Now, partition Y as $Y = [X^T \;\; Z^T]^T$, where $X \in R^n$ and $Z \in R^p$ are themselves random vectors (RV). Suppose we perform an experiment where the value z of Z is measured and that it is then desired to get some information about the unmeasured portion X of Y.

To find an expression for the conditional PDF, note first that X and Z are both normal, that is, $X \sim N(\bar{X}, P_X), Z \sim N(\bar{Z}, P_Z)$. The cross-covariance is

$P_{XZ} = E(X - \bar{X})(Z - \bar{Z})^T$. Then

$$y = \begin{bmatrix} x \\ z \end{bmatrix}, \qquad \bar{Y} = \begin{bmatrix} \bar{X} \\ \bar{Z} \end{bmatrix}, \qquad P_Y = \begin{bmatrix} P_X & P_{XZ} \\ P_{ZX} & P_Z \end{bmatrix} \tag{1.22}$$

and so (Appendix)

$$P_Y^{-1} = \begin{bmatrix} D^{-1} & -D^{-1}P_{XZ}P_Z^{-1} \\ -P_Z^{-1}P_{ZX}D^{-1} & P_Z^{-1} + P_Z^{-1}P_{ZX}D^{-1}P_{XZ}P_Z^{-1} \end{bmatrix} \tag{1.23}$$

where the Schur complement of P_X is $D = P_X - P_{XZ}P_Z^{-1}P_{ZX}$.

Noting that $f_Y(y)$ and $f_{XZ}(x, z)$ are the same PDF

$$f_{X/Z}(x/z) = \frac{f_{XZ}(x, z)}{f_Z(z)} = \frac{f_Y(y)}{f_Z(z)}$$

we can write

$$f_{X/Z}(x/z) = \frac{\sqrt{(2\pi)^p|P_Z|}}{\sqrt{(2\pi)^{n+p}|P_Y|}} \exp\left\{-\frac{1}{2}[(y - \bar{Y})^T P_Y^{-1}(y - \bar{Y}) - (z - \bar{Z})^T P_Z^{-1}(z - \bar{Z})]\right\} \tag{1.24}$$

(To write Equation 1.21 as $f_{XZ}(x, z)$, we can substitute from Equation 1.22.)

By using Equations 1.22 and 1.23, write the exponent as

$$\begin{aligned}
&\begin{bmatrix} x - \bar{X} \\ z - \bar{Z} \end{bmatrix}^T P_Y^{-1} \begin{bmatrix} x - \bar{X} \\ z - \bar{Z} \end{bmatrix} - (z - \bar{Z})^T P_Z^{-1}(z - \bar{Z}) \\
&\quad = (x - \bar{X})^T D^{-1}(x - \bar{X}) - (x - \bar{X})^T D^{-1}P_{XZ}P_Z^{-1}(z - \bar{Z}) \\
&\qquad - (z - \bar{Z})^T P_Z^{-1}P_{ZX}D^{-1}(x - \bar{X}) \\
&\qquad + (z - \bar{Z})^T (P_Z^{-1} + P_Z^{-1}P_{ZX}D^{-1}P_{XZ}P_Z^{-1})(z - \bar{Z}) \\
&\qquad - (z - \bar{Z})^T P_Z^{-1}(z - \bar{Z})
\end{aligned} \tag{1.25}$$

Define the *conditional mean* as

$$\overline{X/Z} = \bar{X} + P_{XZ}P_Z^{-1}(Z - \bar{Z}) \tag{1.26}$$

and the *conditional covariance* as

$$P_{X/Z} = P_X - P_{XZ}P_Z^{-1}P_{ZX} \tag{1.27}$$

the Schur complement of P_X in P_Y. Now, it takes only patience to show that Equation 1.25 can be written as

$$(x - \overline{X/Z = z})^T P_{X/Z}^{-1}(x - \overline{X/Z = z}) \tag{1.28}$$

Furthermore, it can be demonstrated (Appendix) that

$$|P_Y|/|P_Z| = |P_{X/Z}| \tag{1.29}$$

We have thus proven that X/Z is a normal RV with PDF

$$f_{X/Z}(x/z) = \frac{1}{\sqrt{(2\pi)^n |P_{X/Z}|}} \exp\left[-\frac{1}{2}(x - \overline{X/Z = z})^T P_{X/Z}^{-1}(x - \overline{X/Z = z})\right] \tag{1.30}$$

The conditional mean estimate in the Gaussian case is therefore given by Equation 1.26. Note that in the Gaussian case, $P_{X/Z}$ is independent of Z so that the error covariance

$$P_{\tilde{X}} = P_{X/Z} \tag{1.31}$$

is given by Equation 1.27.

From Equation 1.27 it is clear that $P_{X/Z} \leq P_X$. If in a particular case, $P_{X/Z} < P_X$, then, after measuring Z we are more certain of the value of X than we were before (i.e., the *a posteriori* RV X/Z is distributed more closely about its mean $\overline{X/Z}$ than is the *a priori* RV X about its mean $\bar{X}$). □

Example 1.3 Conditional Mean for Linear Gaussian Measurements

Here we find the Gaussian conditional mean estimate and error covariance in the case where X and Z have a particular relation.

Suppose the unknown X and the data Z are related by the *linear Gaussian measurement model*

$$Z = HX + V \tag{1.32}$$

where $X \sim N(\bar{X}, P_X)$ and $V \sim N(0, R)$ is zero mean *measurement noise*, and measurement matrix H is known and deterministic (i.e., constant). Suppose $R > 0$ and X and V are orthogonal so that $\overline{XV^T} = 0$. Let $X \in R^n$ and $V \in R^p$, so that H is a $p \times n$ matrix and $Z \in R^p$.

The measurement Z is then a normal RV, and it is not difficult to determine its mean $\bar{Z}$ and variance P_Z in terms of the known quantities $\bar{X}$, P_X, R, and H. Thus

$$\bar{Z} = \overline{HX + V} = H\bar{X} + \bar{V}$$

by linearity, or

$$\bar{Z} = H\bar{X} \tag{1.33}$$

Furthermore,

$$\begin{aligned} P_Z &= \overline{(Z - \bar{Z})(Z - \bar{Z})^T} = \overline{(HX + V - H\bar{X})(HX + V - H\bar{X})^T} \\ &= \overline{H(X - \bar{X})(X - \bar{X})^T H^T} + \overline{H(X - \bar{X})V^T} + \overline{V(X - \bar{X})^T H^T} + \overline{VV^T} \\ &= HP_X H^T + \overline{HXV^T} - H\bar{X}\bar{V}^T + \overline{VX^T H^T} - \bar{V}\bar{X}^T H^T + R \end{aligned}$$

or

$$P_Z = HP_X H^T + R \tag{1.34}$$

Therefore, Z is $N(H\bar{X}, HP_X H^T + R)$.

To find the cross-covariance of X and Z, write

$$P_{XZ} = \overline{(X-\bar{X})(Z-\bar{Z})^T} = \overline{(X-\bar{X})(HX+V-H\bar{X})^T}$$
$$= \overline{(X-\bar{X})(X-\bar{X})^T H^T} + \overline{XV^T} - \bar{X}\bar{V}^T \tag{1.35}$$

or

$$P_{ZX} = P_{XZ}^T = HP_X \tag{1.36}$$

since P_X is symmetric.

The error covariance is obtained by using the above expressions for P_Z, P_{XZ}, and P_{ZX} in Equation 1.27, which yields

$$P_{\tilde{X}} = P_{X/Z} = P_X - P_{XZ}P_Z^{-1}P_{ZX}$$
$$= P_X - P_X H^T(HP_X H^T + R)^{-1}HP_X \tag{1.37}$$

By using the matrix inversion lemma (Appendix), this can be written as

$$P_{\tilde{X}} = P_{X/Z} = (P_X^{-1} + H^T R^{-1} H)^{-1} \tag{1.38}$$

if $|P_X| \neq 0$. Notice that $P_{X/Z} \leq P_X$. If strict inequality holds, measuring RV Z decreases our uncertainty about the unknown RV X.

The conditional mean estimate is found by substituting for P_{XZ}, P_Z, and $\bar{Z}$ in Equation 1.26

$$\hat{X} = \overline{X/Z} = \bar{X} + P_{XZ}P_Z^{-1}(Z-\bar{Z})$$
$$= \bar{X} + P_X H^T(HP_X H^T + R)^{-1}(Z - H\bar{X}) \tag{1.39}$$

To find a neater expression for $\overline{X/Z}$, note that

$$P_X H^T(HP_X H^T + R)^{-1}$$
$$= P_X H^T(HP_X H^T + R)^{-1}[(HP_X H^T + R) - HP_X H^T]R^{-1}$$
$$= P_X H^T[I - (HP_X H^T + R)^{-1}(HP_X H^T)]R^{-1}$$
$$= [P_X - P_X H^T(HP_X H^T + R)^{-1}HP_X]H^T R^{-1}$$
$$= P_{X/Z}H^T R^{-1}$$

Hence,

$$\hat{X} = \overline{X/Z} = \bar{X} + P_{X/Z}H^T R^{-1}(Z - H\bar{X}) \tag{1.40}$$

The PDF $f_{X/Z}(x/z)$ is normal with conditional mean $\overline{X/Z}$ and conditional covariance $P_{X/Z}$.

It is important to note that $\overline{X/Z}$ depends on the value of Z, but that covariance $P_{X/Z}$ is independent of the value of Z. Thus, $P_{X/Z}$ can be computed *before* we take any measurements.

It is not difficult to redo this example for the case of nonorthogonal X and V. See the problems. ▯

EXERCISE 1.1 Kalman Gain

In the case of linear Gaussian measurements, define the *Kalman gain* by

$$K = P_X H^T (HP_X H^T + R)^{-1} \tag{1.41}$$

a. Express $P_{X/Z}$ and $\overline{X/Z}$ in terms of K.

b. Prove that

$$P_{X/Z} = (I - KH)P_X(I - KH)^T + KRK^T = (I - KH)P_X \tag{1.42}$$

1.1.3 The Orthogonality Principle

The following *orthogonality principle* is basic to probabilistic estimation theory. Let RVs X and Z be jointly distributed. Then, for any function $g(\cdot)$,

$$\overline{g(Z)(X - \overline{X/Z})^T} = 0 \tag{1.43}$$

That is, any function of Z is orthogonal to X once the conditional mean has been subtracted out. To show this, write

$$\begin{aligned}\overline{g(Z)(X - \overline{X/Z})^T} &= \overline{g(Z)X^T} - \overline{g(Z)\overline{X^T/Z}} \\ &= \overline{g(Z)X^T} - \overline{\overline{g(Z)X^T/Z}} = \overline{g(Z)X^T} - \overline{g(Z)X^T} = 0\end{aligned}$$

We used the fact that

$$g(Z) \cdot \overline{X^T/Z} = \overline{g(Z)X^T/Z} \tag{1.44}$$

for any function $g(\cdot)$, since $g(Z)$ is deterministic if Z is fixed.

The idea behind this can be illustrated as in Figure 1.2 where orthogonal RVs are represented as being at right angles. All functions of Z are represented as being in the direction of Z. (In the figure, interpret $\hat{X}_{\text{LMS}}$ as $\overline{X/Z}$.) The orthogonality principle says that the RV

$$\tilde{X} = X - \overline{X/Z} \tag{1.45}$$

which is the estimation error, is orthogonal to all other RVs $g(Z)$. It is emphasized that this figure is extremely heuristic, since if nonlinear functions $g(\cdot)$ are allowed, then RVs of the form $g(Z)$ do not form a subspace. The figure should only be considered as a mnemonic device.

We can quickly show that the orthogonality principle has an immediate application to estimation theory. In fact, if $g(\cdot)$ is any function, then

$$E\|X - \overline{X/Z}\| \le E\|X - g(Z)\| \tag{1.46}$$

where "$\|\cdot\|$" denotes the Euclidean norm. Thus, no other function of Z provides a "closer approximation to X" in a probabilistic sense than does $\overline{X/Z}$.

This is a partial statement of the *stochastic projection theorem*. To show it, a probabilistic formulation of Pythagoras' theorem is used: if Y and Z are

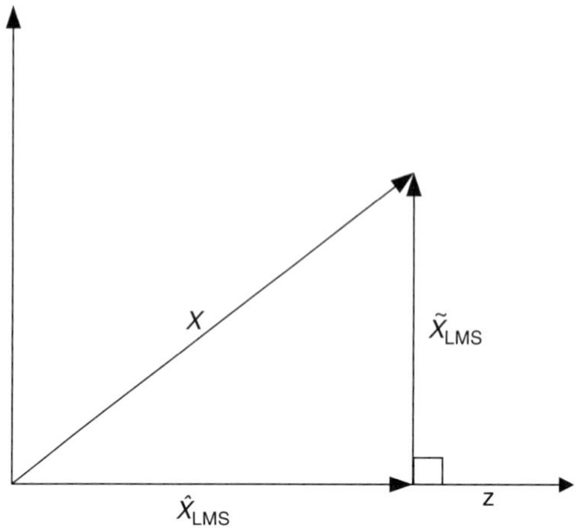

FIGURE 1.2
The orthogonality principle.

orthogonal, then

$$E\|Y+Z\|^2 = E\|Y\|^2 + E\|Z\|^2 \tag{1.47}$$

Now use Equation 1.43 to write for any $g(\cdot)$

$$\begin{aligned} E\|X-g(Z)\|^2 &= E\|X-\overline{X/Z}+\overline{X/Z}-g(Z)\|^2 \\ &= E\|X-\overline{X/Z})\|^2 + E\|\overline{X/Z}-g(Z)\|^2 \geq E\|X-\overline{X/Z}\|^2 \end{aligned}$$

with equality if and only if $g(Z)=\overline{X/Z}$.

1.1.4 Linear Mean-Square Estimation of a Random Variable *X* Given a Random Variable *Z*

No assumptions are made here on the relation between Z and X; the results to be presented are valid for any possible nonlinear relation between measurement and unknown. In the realization that Equation 1.10 is in general too much trouble to compute, except for the Gaussian case, let us restrict the admissible estimate $\hat{X}(Z)$ to those that depend linearly on Z, that is,

$$\hat{X}(Z) = AZ + b \tag{1.48}$$

for some constant matrix $A \in R^{n\times p}$ and vector $b \in R^n$. This *linear estimate restriction* should not be confused with the linear measurement model (Equation 1.32).

An additional motive for restricting an admissible estimate to those of the form in Equation 1.48 is that this form allows processing of the measurements Z by a linear system to obtain $\hat{X}$.

Since the class of admissible estimate is restricted, the *linear mean-square estimate* will not, in general, be as good in the sense of minimizing Equation 1.2 as the general mean-square estimate (Equation 1.10).

The constants A and b depend on the measurement process and on the *a priori* information available before the measurement is made. They must be chosen to minimize the mean-square error (Equation 1.2). Recall that $\text{trace}(MN) = \text{trace}(NM)$ for compatible matrices M and N and write

$$
\begin{aligned}
J &= \overline{(X - \hat{X})^T(X - \hat{X})} = \text{trace}\overline{(X - X)(X - X)^T} \\
&= \text{trace}\overline{(X - AZ - b)(X - AZ - b)^T} \\
&= \text{trace}\overline{[X - \bar{X}) - (AZ + b - \bar{X})][(X - \bar{X}) - (AZ + b - \bar{X})]^T}
\end{aligned}
$$

After some straightforward work we arrive at

$$
\begin{aligned}
J = \text{trace}[&P_X + A(P_Z + \overline{ZZ}^T)A^T + (b - \bar{X})(b - \bar{X})^T \\
&+ 2A\bar{Z}(b - \bar{X})^T - 2AP_{ZX}]
\end{aligned}
\tag{1.49}
$$

From Appendix, $(\mathrm{d}/\mathrm{d}A)\ \text{trace}\ (ABA^T) = 2AB$ and $(\mathrm{d}/\mathrm{d}A)\ \text{trace}(BAC) = B^TC^T$; so for a minimum

$$\frac{\partial J}{\partial b} = 2(b - \bar{X}) + 2A\bar{Z} = 0 \tag{1.50a}$$

and

$$\frac{\partial J}{\partial A} = 2A(P_Z + \bar{Z}\bar{Z}^T) - 2P_{XZ} + 2(b - \bar{X})\bar{Z}^T = 0 \tag{1.50b}$$

(Recall $P_{ZX}^T = P_{XZ}$.) From Equation 1.50a

$$b = \bar{X} - A\bar{Z}$$

and substituting this into Equation 1.50b

$$AP_Z - P_{XZ} = 0$$

or

$$A = P_{XZ}P_Z^{-1}$$

Using this A and b in Equation 1.48 the optimal linear mean-square estimate is determined as

$$\hat{X}_{\text{LMS}} = \bar{X} + P_{XZ}P_Z^{-1}(Z - \bar{Z}) \tag{1.51}$$

This equation is typical of the form of many estimators. We can consider $\bar{Z}$ as an estimate for the measurement Z. We also define the *residual* as

$$\tilde{Z} = Z - \bar{Z} \tag{1.52}$$

The mean $\bar{X}$ is an *a priori* estimate for X and the second term is a correction based on the residual. If we measure exactly what we expect, so that $Z = \bar{Z}$, then the *a priori* estimate $\bar{X}$ is correct and it does not need to be modified; in that case $\hat{X} = \bar{X}$. The weighting $P_{XZ}P_Z^{-1}$ by which the measurement error $(Z - \bar{Z})$ is incorporated into the estimate $\hat{X}$ is dependent on the second-order joint statistics. Thus, if P_{XZ} is smaller, then X and Z depend on each other to a lesser degree and measurement of Z yields less information about X. In fact, if X and Z are independent, then $P_{XZ} = 0$ and Z yields no information about X. In this case $\hat{X} = \bar{X}$ for any value of Z, as given by Equation 1.4. Or again, if P_Z is very large, then we have little confidence in Z and so $(Z - \bar{Z})$ is weighted less heavily into $\hat{X}$. If we have great confidence in Z, then P_Z is small and the residual has a greater role in determining $\hat{X}$.

Compare Equation 1.51 to the general mean-square estimate $\hat{X}_{\text{MS}}$ given by Equation 1.10. To find $\hat{X}_{\text{MS}}$, all the joint statistics of X and Z are required since the PDF $f_{Z/X}$ (or equivalently f_{XZ} and f_Z) is needed. To find the best linear estimate $\hat{X}_{\text{LMS}}$, however, all that is required is the *first-order statistics* (i.e., means) and *second-order statistics* (i.e., covariances).

To find the error covariance associated with $\hat{X}_{\text{LMS}}$, first verify that $\hat{X}_{\text{LMS}}$ is unbiased by writing

$$\begin{aligned}
\bar{\hat{X}}_{\text{LMS}} &= \bar{\bar{X}} + P_{XZ}P_Z^{-1}\overline{(Z - \bar{Z})} \\
&= \bar{X} + P_{XZ}P_Z^{-1}(\bar{Z} - \bar{Z}) \\
&= \bar{X}
\end{aligned}$$

Therefore, $\bar{\tilde{X}} = 0$ and

$$\begin{aligned}
P_{\tilde{X}} &= \overline{(\tilde{X} - \bar{\tilde{X}})(\tilde{X} - \bar{\tilde{X}})^T} = \overline{\tilde{X}\tilde{X}^T} \\
&= \overline{(X - \hat{X})(X - \hat{X})^T} \\
&= \overline{[(X - \bar{X}) - P_{XZ}P_Z^{-1}(Z - \bar{Z})][(X - \bar{X}) - P_{XZ}P_Z^{-1}(Z - \bar{Z})]^T} \\
&= P_X - P_{XZ}P_Z^{-1}P_{ZX} - P_{XZ}P_Z^{-1}P_{ZX} + P_{XZ}P_Z^{-1}P_ZP_Z^{-1}P_{ZX}
\end{aligned}$$

or

$$P_{\tilde{X}} = P_X - P_{XZ}P_Z^{-1}P_{ZX} \tag{1.53}$$

In this expression, P_X represents the *a priori* error covariance and the second term represents the reduction in uncertainty because of the measurement. We see again that if X and Z are independent then $P_{XZ} = 0$ and $P_{\tilde{X}} = P_X$, therefore measuring Z provides no increase in our knowledge of X.

A very important point will now be made. Equations 1.51 and 1.53 are identical on the right-hand sides to Equations 1.26 and 1.27. This means that in the Gaussian case, the optimal mean-square estimate is *linear*, since $\hat{X}_{\text{MS}} = \overline{X/Z}$ and $\hat{X}_{\text{LMS}}$ are the same.

The similarity of Equations 1.51 and 1.53 to Equations 1.26 and 1.27 implies another very important result which we shall need later. The expressions derived in Example 1.3 are the optimal *linear* mean-square estimate and error covariance for the case of linear measurements even if the measurement noise is not Gaussian.

EXERCISE 1.2 The Orthogonality Principle with Linear Estimates

If we restrict our attention to linear estimates, we have a particularly nice geometric formulation of the orthogonality principle, for we can talk in terms of vector spaces.

a. The linear mean-square estimate does not satisfy $\overline{g(Z)(X-\hat{X}_{\text{LMS}})^T}=0$ for all function $g(\cdot)$. Show that it does satisfy the orthogonality condition

$$\overline{Z(X-\hat{X}_{\text{LMS}})^T}=0 \tag{1.54}$$

Thus, $(X-\hat{X}_{\text{LMS}})$ is orthogonal to all *linear* functions of Z of the form $g(Z)=AZ$ for constant A. Since $\hat{X}_{\text{LMS}}$ is unbiased show that, $(X-\hat{X}_{\text{LMS}})$ is orthogonal to all linear functions $g(Z)=AZ+b$ for constants A and b. Show this.

b. Show that the set of all linear functions $g(Z)=AZ$ is a vector space, with an inner product of two elements defined as $\overline{Z_1^TZ_2}=\text{trace}\ \overline{Z_2Z_1^T}$.

c. In terms of these constructions we can now use Figure 1.2, where the axis labeled "Z" represents the subspace of linear functions of Z. This should be contrasted to the situation in the previous subsection, where the orthogonality principle was first discussed. There, we did not restrict $g(\cdot)$ to be linear, so we could not speak in terms of subspaces.

d. Show that if $g(\cdot)$ is any linear function, then

$$E\|X-\hat{X}_{\text{LMS}}\|\le E\|X-g(Z)\| \tag{1.55}$$

with equality only if $g(Z)=\hat{X}_{\text{LMS}}$. Thus, $\hat{X}_{\text{LMS}}$ provides a "closer approximation" in a probabilistic sense to the unknown X than any other *linear* function $g(Z)$. Does Equation 1.55 hold in general for nonlinear $g(\cdot)$?

e. The orthogonality principle can be used to derive $\hat{X}_{\text{LMS}}$. Suppose we want to derive an optimal estimate of the form $\hat{X}=AZ+b$.

First, assume that A is given. Then the best value for b is the optimal estimate for the RV $(X-AZ)$ by a constant. Write it down.

Now, to find A, show that, assuming $\hat{X}$ is unbiased, the optimal A satisfies

$$\overline{[(X-\bar{X})-A(Z-\bar{Z})](Z-\bar{Z})^T}=0 \tag{1.56}$$

Hence, find A. Is $\hat{X}$ unbiased?

Example 1.4 Linear Mean-Square Estimation

Find the optimal linear mean-square estimate for Example 1.1.

To apply Equation 1.51 we need to compute the first and second moments of X and Z. Some of the intermediate steps will be left out

$$
\begin{aligned}
\bar{X} &= \int_{-\infty}^{\infty} x f_X(x)\mathrm{d}x = \frac{1}{2} \\
f_{XZ}(x,z) &= f_{Z/X}(z/x) f_X(x) \\
&= \begin{cases} \frac{1}{x}e^{-z}; & e^{-z} \le x \le 1; \quad 0 \le z \\ 0, & \text{otherwise} \end{cases} \\
f_Z(z) &= \int_{-\infty}^{\infty} f_{XZ}(x,z)\mathrm{d}x \\
&= \begin{cases} ze^{-z}; & z \ge 0 \\ 0; & z < 0 \end{cases} \\
\bar{Z} &= \int_{-\infty}^{\infty} z f_Z(z)\mathrm{d}z = 2 \\
\overline{XZ} &= \int_{-\infty}^{\infty}\int_{-\infty}^{\infty} xz f_{XZ}(x,z)\mathrm{d}x\,\mathrm{d}z \\
&= \int_{0}^{\infty}\int_{e^{-z}}^{1} ze^{-z}\mathrm{d}x\,\mathrm{d}z = \frac{3}{4} \\
P_{XZ} &= \overline{XZ} - \bar{X}\bar{Z} = \frac{3}{4} - 1 = -\frac{1}{4} \\
\overline{Z^2} &= \int_{-\infty}^{\infty} z^2 f_Z(z)\mathrm{d}z = 6 \\
P_Z &= \overline{Z^2} - \bar{Z}^2 = 6 - 4 = 2
\end{aligned}
$$

and finally, using Equation 1.51,

$$
\begin{aligned}
\hat{X}_{\mathrm{LMS}} &= \bar{X} + P_{XZ}P_Z^{-1}(Z - \bar{Z}) \\
&= \frac{1}{2} - \frac{1}{4}\left(\frac{1}{2}\right)(Z - 2) \\
&= \frac{3}{4} - \frac{1}{8}Z
\end{aligned}
$$

Note that $\hat{X}_{\mathrm{LMS}}$ is unbiased since $\bar{\hat{X}}_{\mathrm{LMS}} = \frac{3}{4} - \frac{1}{8}\bar{Z} = \frac{1}{2}$.

To compare this to $\hat{X}_{\mathrm{MS}}$ found previously, examine Figure 1.1. Note that $\hat{X}_{\mathrm{LMS}}$ cannot generally be obtained simply by linearizing $\hat{X}_{\mathrm{MS}}$.

This example has not presented linear mean-square estimation under the most flattering of circumstances. The computation of the moments is tedious, and finding $\hat{X}_{\mathrm{LMS}}$ is more work in this case than is finding $\hat{X}_{\mathrm{MS}}$. However, it is not always possible to find the conditional mean, but given the first and second moments, $\hat{X}_{\text{LMS}}$ can always be found. Besides this, the means and variances can often be determined from the measurement process without knowing the joint PDF f_{XZ} itself. Linear mean-square estimation comes into its own as we proceed in our development. □

Example 1.5 Quadratic Mean-Square Estimation

It is desired to find the optimal estimate $\hat{X}$ that depends on Z quadratically, that is,

$$\hat{X}(Z) = CZ^2 + AZ + b \tag{1.57}$$

where, for simplicity, it is supposed that X and Z are scalars. The mean-square criterion becomes

$$J = \overline{(X - \hat{X})^T(X - \hat{X})} = \overline{(X - CZ^2 - AZ - b)^2}$$

and to find the optimal C, A, b, write

$$\begin{aligned}
\frac{\partial J}{\partial b} &= -2\overline{(X - CZ^2 - AZ - b)} = 0 \\
\frac{\partial J}{\partial A} &= -2\overline{Z(X - CZ^2 - AZ - b)} = 0 \\
\frac{\partial J}{\partial C} &= -2\overline{Z^2(X - CZ^2 - AZ - b)} = 0
\end{aligned}$$

which can be collected into the "normal equations"

$$\begin{bmatrix} 1 & \bar{Z} & \overline{Z^2} \\ \bar{Z} & \overline{Z^2} & \overline{Z^3} \\ \overline{Z^2} & \overline{Z^3} & \overline{Z^4} \end{bmatrix} \begin{bmatrix} b \\ A \\ C \end{bmatrix} = \begin{bmatrix} \bar{X} \\ \overline{XZ} \\ \overline{XZ^2} \end{bmatrix} \tag{1.58}$$

Several things are interesting about these equations. First, clearly if $C = 0$ they provide us with an alternative derivation of Equation 1.51. Second, we have seen that the optimal linear mean-square estimate depends on the joint statistics of X and Z through order 2. It is now apparent that the quadratic mean-square estimate depends on joint statistics through order 4. □

1.2 Maximum-Likelihood Estimation

1.2.1 Nonlinear Maximum-Likelihood Estimation

The connotation of "maximum likelihood" is a setting in which nothing is known *a priori* about the unknown quantity, but there is prior information on the measurement process itself. Thus, X is deterministic and Z is stochastic.

The conditional PDF of Z given the unknown X, $f_{Z/X}(z/X)$, contains information about X, and if it can be computed, then X may be estimated according to the *maximum-likelihood estimation criterion*, which can be stated as follows: *Given a measurement Z, the maximum-likelihood estimate $\hat{X}_{ML}$ is the value of X which maximizes $f_{Z/X}$, the likelihood that X resulted in the observed Z.*

The PDF $f_{Z/X}$ is a *likelihood function*. Since $\ln(\cdot)$ is a monotonically increasing function, the estimate $\hat{X}$ could equally well be chosen to maximize $\ln(f_{Z/X})$, which is often more convenient. Thus, $\ln(f_{Z/X})$ is also a likelihood function.

If the first derivatives are continuous and $f_{Z/X}(z/X)$ has its maximum interior to its domain as a function of X, then the maximum-likelihood criterion can be expressed as

$$\left.\frac{\partial f_{Z/X}(z/X)}{\partial X}\right|_{X=\hat{X}} = 0 \tag{1.59a}$$

or

$$\left.\frac{\partial \ln f_{Z/X}(z/X)}{\partial X}\right|_{X=\hat{X}} = 0 \tag{1.59b}$$

This is called the *likelihood equation.*

Note that since X is deterministic it makes no sense here to talk about $f_{X/Z}$.

Example 1.6 Nonlinear Maximum-Likelihood Estimation

Let measurement Z and unknown X be related as in Example 1.1b, where X is now deterministic (i.e., no *a priori* statistics are given on X).

The likelihood function is given by Equation 1.19 in Example 1.1 (where lowercase x's should be replaced by uppercase since X is deterministic). To maximize the likelihood function, select X as small as possible; so the maximum-likelihood estimate is

$$\hat{X}_{\text{ML}} = e^{-Z}$$

See Melsa and Cohn (1978).

Note that the estimate is given by $\hat{X}_{\text{ML}} = e^{-Z}$, while the *value* of the estimate given the data $Z = z$ is $\hat{X}_{\text{ML}} = e^{-z}$. See Figure 1.1. □

1.2.2 Linear Gaussian Measurements

For our applications, assume now that Z depends linearly on X according to

$$Z = HX + V \tag{1.60}$$

Assume the measurement noise is normal so that

$$f_V(v) = \frac{1}{\sqrt{(2\pi)^p|R|}} e^{-(1/2)v^T R^{-1} v} \tag{1.61}$$

Then the likelihood function is given by a shifted version of $f_V(v)$,

$$f_{Z/X}(z/X) = f_V(z - HX)$$
$$= \frac{1}{\sqrt{(2\pi)^p|R|}} e^{-(1/2)(z-HX)^T R^{-1}(z-HX)} \tag{1.62}$$

To maximize $f_{Z/X}$, we can equivalently maximize $\ln(f_{Z/X})$, or minimize

$$J = \frac{1}{2}(z - HX)^T R^{-1}(z - HX) \tag{1.63}$$

Differentiating,

$$\frac{\partial J}{\partial X} = H^T R^{-1}(z - HX) = 0$$

so that the optimal estimate is given by

$$\hat{X}_{\text{ML}} = (H^T R^{-1} H)^{-1} H^T R^{-1} Z \tag{1.64}$$

Evidently, to obtain the maximum-likelihood estimate, the measurement matrix H must have full rank.

Note that the mean-square performance index (Equation 1.2) depends on the *estimation error*, while Equation 1.63 depends on the *residual*, since $\hat{Z} = H\hat{X}$ is an estimate for Z so that the residual is

$$\tilde{Z} = Z - H\hat{X} \tag{1.65}$$

To find the error covariance, first note that the estimation error is

$$\tilde{X} = X - \hat{X} = X - (H^T - R^{-1}H)^{-1} H^T R^{-1}(HX + V)$$
$$= -(H^T R^{-1} H)^{-1} H^T R^{-1} V \tag{1.66}$$

(Of course we cannot actually compute $\tilde{X}$ by this equation since noise V is unknown!) Now, taking expected values

$$\bar{\tilde{X}} = -(H^T R^{-1} H)^{-1} H^T R^{-1} \bar{V} = 0 \tag{1.67}$$

so that the maximum-likelihood estimate for the linear Gaussian case is unbiased.

Now the error covariance becomes

$$P_{\tilde{X}} = \overline{(\tilde{X} - \bar{\tilde{X}})(\tilde{X} - \bar{\tilde{X}})^T} = \overline{\tilde{X}\tilde{X}^T}$$
$$= (H^T R^{-1} H)^{-1} H^T R^{-1} \overline{VV^T} R^{-1} H (H^T R^{-1} H)^{-1}$$

or

$$P_{\tilde{X}} = (H^T R^{-1} H)^{-1} \tag{1.68}$$

In terms of $P_{\tilde{X}}$ the optimal estimate can be expressed as

$$\hat{X}_{\text{ML}} = P_{\tilde{X}} H^T R^{-1} Z \tag{1.69}$$

Note that if $H = I$ then the error covariance $P_{\tilde{X}}$ is equal to R, the noise covariance, which makes sense intuitively.

An *estimator* will be taken to mean an estimate plus its error covariance. The estimator described by Equations 1.68 and 1.69 is called the *Gauss–Markov* estimator.

The absence of *a priori* information on X can be expressed as

$$\bar{X} = 0, \quad P_X \to \infty \tag{1.70}$$

which models complete ignorance. Using these values in Equations 1.38 and 1.40 we obtain Equations 1.68 and 1.69. The maximum-likelihood estimate is, therefore, in the linear Gaussian case, a limiting case of the mean-square estimate.

The error covariance $P_{\tilde{X}}$ is independent of the measurement Z, so that once we have designed an experiment described by $Z = HX + V, P_{\tilde{X}}$ can be computed off-line to see if the accuracy of the experiment is acceptable *before* we build the equipment and take any measurements. This is an important result.

Example 1.7 Experiment Design

To estimate a deterministic voltage X, there are two alternatives for designing an experiment. Two expensive meters can be used, each of which add $N(0, 2)$ Gaussian noise to X, or four inexpensive meters can be used, each of which add $N(0, 3.5)$ Gaussian noise to X. Which design would result in the more reliable estimate for X?

Design number one is described by

$$Z = \begin{bmatrix} 1 \\ 1 \end{bmatrix} X + V$$

where

$$V \sim N\left(0, \begin{bmatrix} 2 & 0 \\ 0 & 2 \end{bmatrix}\right)$$

This gives an error covariance of

$$P_{\tilde{X}} = (H^T R^{-1} H)^{-1} = \left(\begin{bmatrix} 1 & 1 \end{bmatrix} \begin{bmatrix} \frac{1}{2} & 0 \\ 0 & \frac{1}{2} \end{bmatrix} \begin{bmatrix} 1 \\ 1 \end{bmatrix}\right)^{-1} = 1$$

Design number two is described by

$$Z = \begin{bmatrix} 1 \\ 1 \\ 1 \\ 1 \end{bmatrix} X + V, \quad V \sim N(0, 3.5I)$$

where I is the identity matrix, so

$$P_{\tilde{X}} = (H^T R^{-1} H)^{-1} = \frac{7}{8}$$

Design number two using four inexpensive meters is therefore a more reliable scheme since it has a lower error covariance.

The noise covariance matrix R must be determined before maximum-likelihood methods can be used in the linear Gaussian case. The effects of inaccurately known noise covariance are explored in the problems. The accurate determination of the noise parameters is a crucial part of the estimation process.

The next examples provide further insight into maximum-likelihood estimation. ▯

Example 1.8 Independent Measurement of a Scalar Unknown

If the measurements are independent, then R is diagonal. Suppose an experiment is described by three scalar measurements (one vector measurement)

$$Z = \begin{bmatrix} 1 \\ 1 \\ 1 \end{bmatrix} X + V, \qquad V \sim N\left(0, \begin{bmatrix} \sigma_1^2 & & 0 \\ & \sigma_2^2 & \\ 0 & & \sigma_3^2 \end{bmatrix}\right)$$

with X deterministic. Then the error covariance is given by

$$P_{\tilde{X}} = (H^T R^{-1} H)^{-1} = \left[\frac{1}{\sigma_1^2} + \frac{1}{\sigma_2^2} + \frac{1}{\sigma_3^2}\right]^{-1} \tag{1.71}$$

and the estimate is

$$\begin{aligned} \hat{X}_{\text{ML}} &= P_{\tilde{X}} H^T R^{-1} Z \\ &= \frac{(Z_1/\sigma_1^2 + Z_2/\sigma_2^2 + Z_3/\sigma_3^2)}{(1/\sigma_1^2 + 1/\sigma_2^2 + 1/\sigma_3^2)} \end{aligned} \tag{1.72}$$

where $Z = [Z_1\ Z_2\ Z_3]^T$ is the measurement vector, with Z_i being scalar random variables. This is a very natural and appealing result, as can be seen by noticing a few points.

First, suppose that $\sigma_3^2 \to \infty$, indicating that we have no confidence in Z_3. Then, $1/\sigma_3^2 \to 0$ and the effects of Z_3 simply drop out, leaving a two-measurement experiment. Now, suppose that $\sigma_3^2 \to 0$, indicating that we have great confidence in Z_3. Then $P_{\tilde{X}} \to \sigma_3^2$ and $\hat{X}_{\text{ML}} \to Z_3$, so that the maximum-likelihood procedure would say to disregard Z_1 and Z_2 in this situation. What we have here, then, is an estimator that weights Z_i into $\hat{X}_{\text{ML}}$, depending on our measure of confidence σ_i^2 in Z_i.

The generalization to an arbitrary number of measurements Z_i is immediate. If all σ_i are equal to σ, then Equation 1.72 becomes

$$\hat{X}_{\text{ML}} = \frac{Z_1 + Z_2 + Z_3}{3}$$

the *sample mean*, and Equation 1.71 becomes

$$P_{\tilde{X}} = \frac{\sigma^2}{3}$$

▯

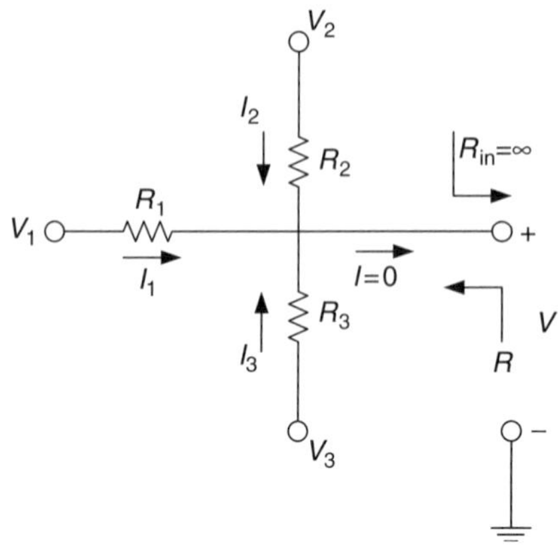

FIGURE 1.3
An electric circuit analog to maximum-likelihood estimation.

Example 1.9 An Electric Circuit Analog

Examine Figure 1.3. Let voltages V_i be applied between one end of resistance R_i and ground. The other ends of the resistances are connected to a common point at potential V. Find the voltage V if $R_{\text{in}} = \infty$.

This problem can be solved by using Kirchhoff's current law $I_1+I_2+I_3 = I$, where $R_{\text{in}} = \infty$ means that $I = 0$. Now apply Ohm's law to each I_i to obtain

$$I_1 + I_2 + I_3 = \frac{V_1 - V}{R_1} + \frac{V_2 - V}{R_2} + \frac{V_3 - V}{R_3} = 0$$

or

$$\frac{V_1}{R_1} + \frac{V_2}{R_2} + \frac{V_3}{R_3} = V\left(\frac{1}{R_1} + \frac{1}{R_2} + \frac{1}{R_3}\right)$$

The solution is

$$V = \left(\frac{1}{R_1} + \frac{1}{R_2} + \frac{1}{R_3}\right)^{-1}\left(\frac{V_1}{R_1} + \frac{V_2}{R_2} + \frac{V_3}{R_3}\right) \tag{1.73}$$

which corresponds exactly to Equation 1.72!

The "effective resistance seen by V" is

$$R = \left(\frac{1}{R_1} + \frac{1}{R_2} + \frac{1}{R_3}\right)^{-1} \tag{1.74}$$

This corresponds to the error covariance $P_{\tilde{X}}$ in Equation 1.71. In electric circuit theory, R is written as the *parallel combination* of R_i, that is, $R = R_1//R_2//R_3$, which is just a shorthand notation for "the reciprocal of the sums of the reciprocals." With this notation, Equation 1.71 can be written as

$$P_{\tilde{X}} = \sigma_1^2//\sigma_2^2//\sigma_3^2 \tag{1.75}$$

so that the error covariance is just the parallel combination of the individual noise variances. As such, it is smaller than any individual σ_i^2, so our confidence in $\hat{X}$ after the experiment is greater than our confidence in any one measurement.

Equation 1.73 is called *Millman's theorem* and it is an exact analog to maximum-likelihood estimation of a scalar unknown with independent measurements. □

1.3 The Cramér–Rao Bound

To examine the performance of the maximum-likelihood estimate, we can use the following lower bound on the variance of any unbiased estimate. If $\hat{X}$ is any unbiased estimate of a *deterministic* variable X based on measurement Z, then the covariance of the estimation error $\tilde{X} = X - \hat{X}$ is bounded by

$$P_{\tilde{X}} \geq J_F^{-1} \tag{1.76}$$

where the *Fisher information matrix* is given by

$$J_F = e\left\{\left[\frac{\partial \ln f_{Z/X}(z/X)}{\partial X}\right]\left[\frac{\partial \ln f_{Z/X}(z/X)}{\partial X}\right]^T\right\} \tag{1.77}$$

$$= -E\left[\frac{\partial^2 \ln f_{Z/X}(z/X)}{\partial X^2}\right] \tag{1.78}$$

Equality holds in Equation 1.76 if and only if

$$\frac{\partial \ln f_{Z/X}(z/X)}{\partial X} = k(X)(X - \hat{X}) \tag{1.79}$$

It is assumed that $\partial f_{Z/X}/\partial X$ and $\partial^2 f_{Z/X}/\partial X^2$ exist and are absolutely integrable.

Equation 1.79 is a limitation on the form of the derivative: it must be proportional to $\tilde{X}$ with constant of proportionality at most a function of the unknown X. An estimate $\hat{X}$ is *efficient* if it satisfies the *Cramér–Rao bound* (Equation 1.76) with equality, that is, if Equation 1.79 holds. Proof for the Cramér–Rao bound can be found in Van Trees (1968) or Sorenson (1970a,b).

In general, $\hat{X}_{\text{ML}}$ is not necessarily efficient. However, it is not hard to show that if Equation 1.79 is ever satisfied so that there exists efficient estimate, then this estimate is given by $\hat{X}_{\text{ML}}$. The Cramér–Rao bound thus demonstrates the importance of the maximum-likelihood estimate.

To gain some insight into the Fisher information matrix, consider the linear Gaussian measurement model. From Equation 1.62 we obtain

$$\frac{\partial \ln f_{Z/X}(z/X)}{\partial X} = H^T R^{-1}(z - HX) \tag{1.80}$$

and differentiating again we find

$$J_F = H^T R^{-1} H = P_{\tilde{X}}^{-1} \tag{1.81}$$

the inverse of the error covariance matrix.

According to Equation 1.71, we have for Example 1.8 that $J_F = P_{\tilde{X}}^{-1} = 1/\sigma_1^2 + 1/\sigma_2^2 + 1/\sigma_3^2$, which can be phrased in the intuitively appealing statement that in the scalar case of independent measurements, the information J_F at the end of an experiment is the *sum* of the information provided by each measurement. An electrical analog is provided by the conductance $1/R$, which adds according to Example 1.9 Equation 1.74 when resistances are put in parallel.

The maximum-likelihood estimate has several nice properties as the number of independent measurements N goes to infinity. To discuss them we need the concept of *stochastic convergence*. The sequence $Y_1, Y_2, \ldots$ of RVs *converges in probability* to RV Y if for all $\varepsilon > 0$

$$P(\|Y_N - Y\| > \varepsilon) \to 0 \quad \text{for} \quad N \to \infty \tag{1.82}$$

Under fairly general conditions, the maximum-likelihood estimate $\hat{X}_{\text{ML}}$ has the following limiting properties:

1. $\hat{X}_{\text{ML}}$ converges in probability to the correct value of X as $N \to \infty$. (Any estimate with this property is said to be *consistent*.)
2. $\hat{X}_{\text{ML}}$ becomes efficient as $N \to \infty$.
3. As $N \to \infty$, $\hat{X}_{\text{ML}}$ becomes Gaussian $N(X, P_{\tilde{X}})$.

For a *stochastic* unknown X, a result similar to the Cramér–Rao bound can be stated as follows:

If $\hat{X}$ is any estimate of a stochastic variable X based on measurement Z, then the covariance of the estimation error $\tilde{X} = X - \hat{X}$ is bounded by

$$P_{\tilde{X}} \geq L^{-1} \tag{1.83}$$

where the *information matrix* L is given by

$$L = E\left\{ \left[\frac{\partial \ln f_{XZ}(x,z)}{\partial x} \right] \left[\frac{\partial \ln f_{XZ}(x,z)}{\partial x} \right]^T \right\} \tag{1.84}$$

$$= -E\left[\frac{\partial^2 \ln f_{XZ}(x,z)}{\partial x^2} \right] \tag{1.85}$$

Equality holds in Equation 1.83 if and only if

$$\frac{\partial \ln f_{XZ}(x,z)}{\partial x} = k(x - \hat{X}) \tag{1.86}$$

where k is a constant. It is assumed that $\partial f_{XZ}/\partial x$ and $\partial^2 f_{XZ}/\partial x^2$ exist and are absolutely integrable with respect to both variables, and that

$$\lim_{x \to \infty} B(x) f_X(x) = 0 \tag{1.87a}$$

and

$$\lim_{x \to -\infty} B(x) f_X(x) = 0 \tag{1.87b}$$

where

$$B(x) \triangleq \int_{-\infty}^{\infty} (x - \hat{X}) f_{Z/X}(z/x) \mathrm{d}z \tag{1.88}$$

An estimate $\hat{X}$ for which the bound is satisfied with equality (i.e., condition Equation 1.86 on the form of f_{XZ} holds) is said to be *efficient*. The proof is similar to that of the Cramér–Rao bound, but expectations are now taken with respect to Z and X (Van Trees 1968; Sorenson 1970a,b). Several points should be noted.

1. This bound depends on the joint PDF f_{XZ} while the Cramér–Rao bound (for deterministic unknown X) depends on the likelihood function $f_{Z/X}$.
2. The Cramér–Rao bound deals with unbiased estimates $\hat{X}$; this bound substitutes requirements 1.87a or 1.87b for unbiasedness of $\hat{X}$.
3. The efficiency condition (Equation 1.86) in terms of the *constant* k should be contrasted with the efficiency condition of the Cramér–Rao bound in terms of the function $k(X)$ of X.
4. Differentiating Equation 1.86 yields the equivalent condition for efficiency

$$\frac{\partial^2 \ln f_{XZ}(x,z)}{\partial x^2} = k \tag{1.89}$$

or (since $f_Z(z)$ is independent of X)

$$\frac{\partial^2 \ln f_{X/Z}(x,z)}{\partial x^2} = k \tag{1.90}$$

Integrating twice and taking the antilog yields

$$f_{X/Z}(x/z) = e^{-(k/2)x^T x + k\hat{X}^T x + c} \tag{1.91}$$

so that the *a posteriori* PDF must be Gaussian for an efficient estimate to exist.

It is interesting to relate information matrix L to the Fisher information matrix J_F. Since $f_{XZ} = f_{Z/X} f_X$, it follows that

$$L = J_F - E\left[\frac{\partial^2 \ln f_X(x)}{\partial x^2}\right] \tag{1.92}$$

so that the stochastic description of X enters independent of the noise statistics.

We can gain a little more insight by considering the linear Gaussian measurement model. Thus, let measurement Z and unknown X be related by Equation 1.60 with $V = N(0, R)$. Then, by Equation 1.81 we know that $J_F = H^T R^{-1} H$. To find L, note that

$$f_X(x) = \frac{1}{\sqrt{(2\pi)^n |P_X|}} e^{-(1/2)(x-\bar{X})^T P_X^{-1}(x-\bar{X})} \tag{1.93}$$

so that

$$\frac{\partial \ln f_X(x)}{\partial x} = -P_X^{-1}(x - \bar{X}) \tag{1.94}$$

Then, by Equation 1.92

$$L = H^T R^{-1} H + P_X^{-1} \tag{1.95}$$

The total information available has been *increased* by *a priori* information on X. See Equation 1.38.

1.4 Recursive Estimation

1.4.1 Sequential Processing of Measurements

Up to this point, we have considered *batch processing* of the measurements in which all information is incorporated in one single step into the estimate. This leads to the necessity for inverting $p \times p$ matrices, where p is the number of measurements. The next examples demonstrate a more sensible approach when p is large.

Example 1.10 Recursive Computation of Sample Mean and Variance

Given k measurements, Z_i the sample mean is

$$\hat{X}_k = \frac{1}{k}\sum_{i=1}^{k} Z_i \tag{1.96}$$

Suppose that $\hat{X}_k$ has been computed based on measurements Z_i for $i = 1, \ldots, k$. Now, one more measurement Z_{k+1} is made. The new sample mean could be computed as

$$\hat{X}_{k+1} = \frac{1}{k+1}\sum_{i=1}^{k+1} Z_i \tag{1.97}$$

but this clearly involves duplication of some of the work we have already done.

To avoid this, $\hat{X}_{k+1}$ can be computed in terms of $\hat{X}_k$ and Z_{k+1} by proceeding as follows:

$$\begin{aligned}\hat{X}_{k+1} &= \frac{1}{k+1}\left(\frac{1}{k}\sum_{i=1}^{k} Z_i\right) + \frac{1}{k+1}Z_{k+1} \\ &= \frac{1}{k+1}\hat{X}_k + \frac{1}{k+1}Z_{k+1}\end{aligned}$$

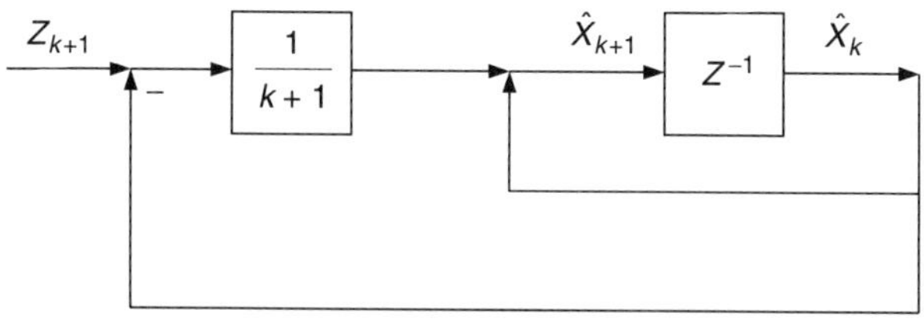

FIGURE 1.4
Linear system for recursive generation of sample mean.

To make this look like some of our previous results, write it as

$$\hat{X}_{k+1} = \hat{X}_k + \frac{1}{k+1}(Z_{k+1} - \hat{X}_k) \tag{1.98}$$

This recursive equation can be represented as a time-varying linear system, as in Figure 1.4. The quantity $\tilde{Z}_{k+1} = Z_{k+1} - \hat{X}_k$ can be interpreted as a residual. If the Z_i are measurements of a constant unknown X, then $\hat{X}_k$ is the estimate of X based on the first k measurements and Equation 1.98 is the *estimate update equation* that adds in the effects of the next measurement Z_{k+1}. Compare this equation to Equation 1.40. Conceptually, the two are identical; they both express the current estimate as a linear combination of the previous estimate and the residual. Note that as k increases, the measurements Z_{k+1} have less and less effect as more weighting is given to $\hat{X}_k$.

A recursive method can also be found to compute the sample variance. To wit, after k measurement we have

$$\sigma_k^2 = \frac{1}{k}\sum_{i=1}^{k}(Z_i - \hat{X}_k)^2 \tag{1.99}$$

Introducing one more measurement Z_{k+1}, there results

$$\begin{aligned}\sigma_{k+1}^2 &= \frac{1}{k+1}\sum_{i=1}^{k+1}(Z_i - \hat{X}_{k+1})^2 \\ &= \frac{1}{k+1}\sum_{i=1}^{k+1}[(Z_i - \hat{X}_k) + (\hat{X}_k - \hat{X}_{k+1})]^2\end{aligned}$$

After some manipulation, we achieve

$$\sigma_{k+1}^2 = \frac{k}{k+1}\sigma_k^2 + \frac{1}{k+1}(Z_{k+1} - \hat{X}_k)^2 - (\hat{X}_{k+1} - \hat{X}_k)^2 \tag{1.100}$$

□

Now, using Equation 1.98 write

$$(\hat{X}_{k+1} - \hat{X}_k)^2 = \frac{1}{(k+1)^2}(Z_{k+1} - \hat{X}_k)^2$$

so that

$$\sigma_{k+1}^2 = \frac{k}{k+1}\sigma_k^2 + \frac{k}{(k+1)^2}(Z_{k+1} - \hat{X}_k)^2$$
$$\sigma_{k+1}^2 = \sigma_k^2 + \frac{1}{k+1}\left[\frac{k}{k+1}(Z_{k+1} - \hat{X}_k)^2 - \sigma_k^2\right] \tag{1.101}$$

Schemes such as the above in which the current estimate depends on the previous estimate and the current measurement are called *recursive* or *sequential estimation* schemes. Such schemes rely on *sequential processing* of data, where the measurements are processed in stages.

Example 1.11

Suppose there are four measurements of a scalar unknown X given by $Z_1 = 10$, $Z_2 = 12$, $Z_3 = 11.5$, and $Z_4 = 11$. Find the sample mean and variance recursively.

Using Equation 1.98 several times with $\hat{X}_0 = 0$,

$$\hat{X}_1 = \hat{X}_0 + (Z_1 - \hat{X}_0) = Z_1 = 10$$
$$\hat{X}_2 = 10 + \frac{1}{2}(12 - 10) = 11$$
$$\hat{X}_3 = 11 + \frac{1}{3}(11.5 - 11) = 11\frac{1}{6}$$
$$\hat{X}_4 = 11\frac{1}{6} + \frac{1}{4}\left(11 - 11\frac{1}{6}\right) = 11\frac{1}{8}$$

Thus, after each such measurement Z_k, $\hat{X}_k$ provides an up-to-date estimate of X. These estimates are up to date in the maximum-likelihood sense, since there is no *a priori* information about X.

To find σ_k^2 recursively, let $\sigma_0^2 = 0$. Then, applying Equation 1.101,

$$\sigma_1^2 = 0$$
$$\sigma_2^2 = \frac{1}{4}(Z_2 - \hat{X}_1)^2 = 1$$
$$\sigma_3^2 = 1 + \frac{1}{3}\left[\frac{2}{3}(11.5 - 11)^2 - 1\right] = 0.722$$
$$\sigma_4^2 = 0.722 + \frac{1}{4}\left[\frac{3}{4}\left(11 - 11\frac{1}{6}\right)^2 - 0.722\right] = 0.547$$

The sample variance σ_k^2 yields a measure of confidence in $\hat{X}_k$ after each measurement Z_k. Sequence σ_k^2 is decreasing, indicating that each measurement increases the reliability of $\hat{X}_k$. □

1.4.2 Sequential Maximum-Likelihood Estimation

Let us generalize our results by deriving a sequential maximum-likelihood estimator. Let there be given a linear Gaussian measurement Z of an unknown constant X:

$$Z = HX + V \tag{1.102}$$

where $X \in R^n, Z \in R^p, V \in R^p, V \sim N(0, R)$, and $|R| \neq 0$. Recall that "maximum likelihood" implies ignorance of any statistics of X.

Write Equation 1.102 explicitly in terms of the components of vectors as

$$\begin{bmatrix} z_1 \\ z_2 \\ \vdots \\ z_p \end{bmatrix} = \begin{bmatrix} h_1^T \\ h_2^T \\ \vdots \\ h_p^T \end{bmatrix} X + \begin{bmatrix} v_1 \\ v_2 \\ \vdots \\ v_p \end{bmatrix}$$

where h_i^T is the ith row of H and suppose that $R = \text{diag}\{\sigma_i^2\}$. This assumption says that the measurements are independent and is sufficient to guarantee that a recursive estimation procedure exists. (Note that in this subsection, lowercase letters are used to define components of RV, which are themselves random variables.) Define

$$Z_k = \begin{bmatrix} z_1 \\ z_2 \\ \vdots \\ z_k \end{bmatrix}, \qquad H_k = \begin{bmatrix} h_1^T \\ h_2^T \\ \vdots \\ h_k^T \end{bmatrix}, \qquad V_k = \begin{bmatrix} v_1 \\ v_2 \\ \vdots \\ v_k \end{bmatrix}$$

as the first k components of Z, H, and V, respectively. Define $R_k = \text{diag}\{\sigma_1^2, \sigma_2^2, \ldots, \sigma_k^2\}$.

Considering the first k components of Z, we have

$$Z_k = H_k X + V_k, \qquad V_k \sim N(0, R_k)$$

Based on this information, the error covariance P_k is given by Equation 1.68, or

$$P_k = (H_k^T R_k^{-1} H_k)^{-1} \tag{1.103}$$

and the estimate $\hat{X}_k$ is given by Equation 1.69, or

$$\hat{X}_k = P_k H_k^T R_k^{-1} Z_k \tag{1.104}$$

Now, it is necessary to include the effects of measurement component Z_{k+1} to update $\hat{X}_k$ to $\hat{X}_{k+1}$. To do this, write

$$Z_{k+1} = H_{k+1} X + V_{k+1}, \qquad V_{k+1} \sim N(0, R_{k+1})$$

or

$$Z_{k+1} = \begin{bmatrix} H_k \\ h_{k+1}^T \end{bmatrix} X + V_{k+1}, \qquad V_{k+1} \sim N\left(0, \begin{bmatrix} R_k & 0 \\ 0 & \sigma_{k+1}^2 \end{bmatrix}\right)$$

The error covariance P_{k+1} based on the first $k+1$ measurement components is

$$\begin{aligned} P_{k+1} &= (H_{k+1}^T R_{k+1}^{-1} H_{k+1})^{-1} \\ &= \left(\begin{bmatrix} H_k \\ h_{k+1}^T \end{bmatrix}^T \begin{bmatrix} R_k & 0 \\ 0 & \sigma_{k+1}^2 \end{bmatrix}^{-1} \begin{bmatrix} H_k \\ h_{k+1}^T \end{bmatrix} \right)^{-1} \\ &= \left(H_k^T R_k^{-1} H_k + \frac{h_{k+1} h_{k+1}^T}{\sigma_{k+1}^2} \right)^{-1} \end{aligned}$$

or

$$P_{k+1} = \left(P_k^{-1} + \frac{h_{k+1} h_{k+1}^T}{\sigma_{k+1}^2} \right)^{-1} \tag{1.105}$$

This is the *error covariance update* equation. To find an alternative form for Equation 1.105, use the matrix-inversion lemma to write

$$P_{k+1} = P_k - P_k h_{k+1} (h_{k+1}^T P_k h_{k+1} + \sigma_{k+1}^2)^{-1} h_{k+1}^T P_k \tag{1.106}$$

To find the estimate update equation, we have

$$\begin{aligned} \hat{X}_{k+1} &= P_{k+1} H_{k+1}^T R_{k+1}^{-1} Z_{k+1} \\ &= P_{k+1} \begin{bmatrix} H_k \\ h_{k+1}^T \end{bmatrix}^T \begin{bmatrix} R_k & 0 \\ 0 & \sigma_{k+1}^2 \end{bmatrix}^{-1} \begin{bmatrix} Z_k \\ z_{k+1} \end{bmatrix} \\ &= P_{k+1} \left(H_k^T R_k^{-1} Z_k + \frac{h_{k+1}}{\sigma_{k+1}^2} z_{k+1} \right) \end{aligned}$$

By Equation 1.104, this becomes

$$\hat{X}_{k+1} = P_{k+1} \left(P_k^{-1} \hat{X}_k + \frac{h_{k+1}}{\sigma_{k+1}^2} z_{k+1} \right) \tag{1.107}$$

This is the *estimate update*. To find a more familiar form for this update, note that

$$\begin{aligned} P_{k+1} P_k^{-1} &= P_{k+1} \left(P_{k+1}^{-1} - \frac{h_{k+1} h_{k+1}^T}{\sigma_{k+1}^2} \right) \\ &= I - P_{k+1} \frac{h_{k+1} h_{k+1}^T}{\sigma_{k+1}^2} \end{aligned}$$

hence Equation 1.107 is equivalent to

$$\hat{X}_{k+1} = \hat{X}_k + P_{k+1} \frac{h_{k+1}}{\sigma_{k+1}^2} (z_{k+1} - h_{k+1}^T \hat{X}_k) \tag{1.108}$$

Note that $\tilde{z}_{k+1} = z_{k+1} - h_{k+1}^T \hat{X}_k$ is the residual.

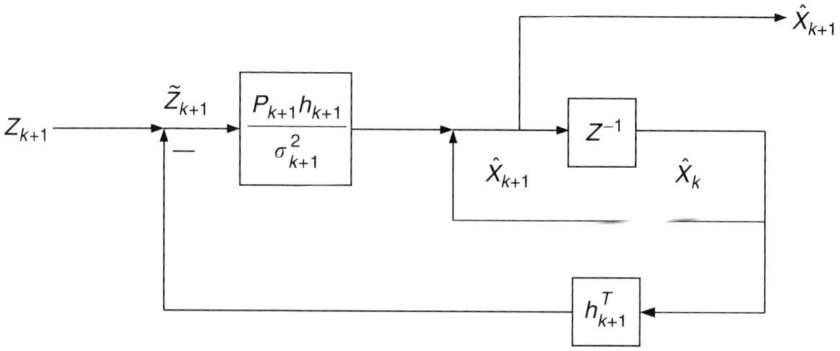

FIGURE 1.5
Linear system for sequential maximum-likelihood estimation.

The recursions (Equations 1.105 and 1.108) should be started using $P_0^{-1} = 0$, $\hat{X}_0 = 0$, which models complete ignorance of the *a priori* statistics of X.

To summarize our results then, if the measurements are independent, the maximum-likelihood (batch) estimator (Equations 1.68 and 1.69) is equivalent to the sequential maximum-likelihood estimator given by Equations 1.106 and 1.108, which requires no matrix inversions but only scalar divisions. The sequential estimator can be implemented as in Figure 1.5. Note that owing to the unity feedback, it has the structure of a summer or integrator. It is a time-varying system where σ^2_{k+1} and h_{k+1} are given directly by the structure of the measurement process and the P_{k+1} are computed off-line (i.e., before we begin to take measurements) from the knowledge of the measurement process. Compare with Figure 1.4.

If *a priori* statistics $\bar{X}$ and P_X on X are available, then we should use $\hat{X}_0 = \bar{X}, P_0 = P_X$ to initialize Equations 1.106 and 1.108, which then amount to a recursive mean-square estimator.

1.4.3 Prewhitening of Data

Now, suppose that the assumptions required to derive Equations 1.105 and 1.108 all hold except that R is nonsingular but arbitrary. For sequential processing of Z, one component at a time, we require a diagonal R, that is, the measurement components must be independent. The data can be *prefiltered* to make R diagonal as follows. Let S be any square root of R^{-1} so that $R^{-1} = S^T S$. Now premultiply Equation 1.102 by S to obtain

$$SZ = SHX + SV$$

and define new quantities by

$$\begin{aligned} Z^\omega &= SZ \\ H^\omega &= SH \\ V^\omega &= SV \end{aligned} \tag{1.109}$$

so that

$$Z^\omega = H^\omega X = V^\omega \tag{1.110}$$

The covariance of the prefiltered noise V^ω is

$$\begin{aligned} R^\omega &= \overline{SV(SV)^T} = S\overline{VV^T}S^T \\ &= SRS^T \\ &= S(S^TS)^{-1}S^T = I \end{aligned}$$

so that the components of V^ω are independent and V^ω is a white noise vector. Hence the superscripts. To implement Equation 1.109, we define a new measurement matrix H^ω and process all our measurements Z through a linear *prewhitening filter* S. We can now proceed as above to derive Equations 1.105 and 1.108 using Z^ω, H^ω, and $R^\omega = I$ instead of Z, H, and R.

1.5 Wiener Filtering

Wiener developed the ideas we are about to discuss in the early 1940s for application to antiaircraft fire control systems. His monograph in 1949 brought stochastic notions into control and communication theory, and was the basis for a great deal of subsequent work. Wiener dealt with the continuous-time case; Kolmogorov (1939) applied mean-square theory to discrete stochastic processes. See Van Trees (1968) and the excellent survey by Kailath (1974).

Recall (Papoulis 1965) that if $x(t)$ is a stationary stochastic process with autocorrelation function

$$R_X(\tau) = \overline{x(t+\tau)x^T(t)} \tag{1.111}$$

then its spectral density $\Phi_X(\omega)$ is

$$\Phi_X(s) = \int_{-\infty}^{\infty} R_X(t)e^{-st}\mathrm{d}t \tag{1.112}$$

evaluated at $s = j\omega$. In the discrete case, we have

$$R_X(k) = \overline{x_{i+k}x_i^T} \tag{1.113}$$

and the spectral density is

$$\Phi_X(z) = \sum_{k=-\infty}^{\infty} R_X(k)z^{-k} \tag{1.114}$$

evaluated at $z = e^{j\omega}$.

If $u(t)$ and $y(t)$ are the input and output of a linear system with impulse response $h(t)$ and transfer function $H(s)$, then the autocorrelation functions and spectral densities are related by

$$R_{YU}(t) = h(t) * R_U(t) \tag{1.115}$$

$$R_Y(t) = h(t) * R_U(t) * h^T(-t) \tag{1.116}$$

$$\Phi_{YU}(s) = H(s)\Phi_U(s) \tag{1.117}$$

$$\Phi_Y(s) = H(s)\Phi_U(s)H^T(-s) \tag{1.118}$$

where * represents convolution. In the discrete counterparts, $-s$ is replaced by z^{-1}.

We shall have occasion to use the *spectral factorization* theorem, which can be stated as follows (Åström 1970; Anderson and Moore 1979):

Let $|\Phi_Y(s)| \neq 0$ for almost every s (such a process is said to be of *full rank*). Then, $\Phi_Y(s)$ can be factored uniquely as

$$\Phi_Y(s) = H(s)\Omega H^T(-s) \tag{1.119}$$

where $H(s)$ is a square, real, rational transfer function with all poles in the open left-half plane; $\lim_{s\to\infty} H(s) = I; H^{-1}(s)$ has all poles in the left-half plane or on the $j\omega$ axis; and $\Omega = \Omega^T > 0$ is a real matrix.

If $|\Phi_Y(s)| \neq 0$ for all s satisfying $\text{Re}(s) = 0$, then $H^{-1}(s)$ has all poles in the open left-half plane.

For a square multivariable transfer function $H(s)$ as described above, the *zeros* of $H(s)$ are the poles of $H^{-1}(s)$. A transfer function with all poles and zeros stable is said to be of *minimum phase*. A similar result holds for discrete systems, with $-s$ replaced by z^{-1} and using the discrete version of stability (i.e., with respect to the unit circle). According to Equation 1.118 and the spectral factorization theorem, the process $y(t)$ can be regarded as the output of the stable system $H(s)$ when the input is white noise with spectral density Ω. This is the *representation theorem*.

Example 1.12 Discrete Spectral Factorization

The rational (in $\cos\omega$) spectral density

$$\Phi(\omega) = \frac{1.25 + \cos\omega}{1.0625 + 0.5\cos\omega} = \frac{(e^{j\omega} + 0.5)(e^{-j\omega} + 0.5)}{(e^{j\omega} + 0.25)(e^{-j\omega} + 0.25)}$$

can be factored in four ways:

$$H_1(z) = \frac{z + 0.5}{z + 0.25}$$

$$H_2(z) = \frac{z + 0.5}{1 + 0.25z}$$

$$H_3(z) = \frac{1 + 0.5z}{z + 0.25}$$

$$H_4(z) = \frac{1 + 0.5z}{1 + 0.25z}$$

Only $H_1(z)$ has no poles and zeros outside the unit circle. Thus, if the stable system $H_1(z)$ has a white noise input with unit spectral density, the output will have spectral density given by $\Phi(\omega)$. ◻

Example 1.13 Continuous Spectral Factorization

The rational function of ω

$$\Phi(\omega) = \frac{\omega^2 + 4}{\omega^2 + 1} = \frac{(j\omega + 2)(-j\omega + 2)}{(j\omega + 1)(-j\omega + 1)}$$

can be factored in four ways, one of which has

$$H(s) = \frac{s + 2}{s + 1}$$

the minimum phase. The system with this transfer function will have output with spectral density $\Phi(\omega)$ when driven by continuous-time white noise with unit spectral density. ◻

1.5.1 The Linear Estimation Problem

Suppose that $x(t) \in R^n$ is an unknown stochastic process. A related process $z(t) \in R^p$ is measured, and based on this data it is desired to reconstruct $x(t)$. We want to use a linear filter, in general time-varying, for this purpose; so suppose that the estimate for $x(t)$ is

$$\hat{x}(t) = \int_{t_0}^{T} h(t, \tau) z(\tau) \mathrm{d}\tau \tag{1.120}$$

The interval $[t_0, T]$ is the time interval over which the data are available. It is desired to find the optimal filter impulse response $h(t, \tau)$, which is called the time-varying *Wiener filter.*

If $T < t$, the determination of $h(t, \tau)$ is called the linear *prediction* problem; if $T = t$, it is the linear *filtering* problem; and if $T > t$, it is called linear *smoothing* or interpolation. The general problem is known as *linear estimation.*

To find the optimal linear after $h(t, \tau)$, which reconstructs unknown $x(t)$ given the data $z(\tau)$ for $t_0 \leq \tau \leq T$, minimize the mean-square error at time t

$$J(t) = E[\tilde{x}(t)^T \tilde{x}(t)] = \mathrm{trace} P_{\tilde{X}}(t) \tag{1.121}$$

where the estimation error is $\tilde{x}(t) = x(t) - \hat{x}(t)$, $f_{E[\tilde{x}(t)]} = 0$. The error covariance is

$$P_{\tilde{X}}(t) = \overline{[x(t) - \hat{x}(t)][x(t) - \hat{x}(t)]^T}$$
$$= E\left\{\left[x(t) - \int_{t_0}^{T} h(t,\tau)z(\tau)\mathrm{d}\tau\right]\left[x(t) - \int_{t_0}^{T} h(t,\tau)z(\tau)\mathrm{d}\tau\right]^T\right\} \tag{1.122}$$

According to the orthogonality principle, $J(t)$ is minimized when $\tilde{x}(t)$ is orthogonal to the data over the interval (t_0, T), that is,

$$\overline{\tilde{x}(t)z(u)^T} = E\left\{\left[x(t) - \int_{t_0}^{T} h(t,\tau)z(\tau)\mathrm{d}\tau\right]z(u)^T\right\}$$
$$= 0; \qquad t_0 < u < T \tag{1.123}$$

This results in the nonstationary *Wiener–Hopf* equation

$$\overline{x(t)z(u)^T} = \int_{t_0}^{T} h(t,\tau)\overline{z(\tau)z(u)^T}\mathrm{d}\tau; \qquad t_0 < u < T \tag{1.124}$$

In terms of correlation functions, this can be written as

$$R_{XZ}(t,u) = \int_{t_0}^{T} h(t,\tau)R_Z(\tau,u)\mathrm{d}\tau; \qquad t_0 < u < T \tag{1.125}$$

It is important to note that the optimal *linear* filter is found by knowing only the first- and second-order joint statistics of $x(t)$ and $z(t)$.

If the statistics are Gaussian, then Equation 1.125 provides the *optimal* filter for reconstructing $x(t)$ from $z(t)$. For arbitrary statistics, it provides the optimal *linear* filter. (In the Gaussian case, the optimal filter is linear.)

With impulse response given by the Wiener–Hopf equation, the error covariance is equal to (use Equation 1.122 and the orthogonality of $\tilde{x}(t)$ and $z(\tau)$)

$$P_{\tilde{X}}(t) = E\left\{\left[x(t) - \int_{t_0}^{T} h(t,\tau)z(\tau)\mathrm{d}\tau\right]x(t)^T\right\}$$

or

$$P_{\tilde{X}}(t) = R_X(t,t) - \int_{t_0}^{T} h(t,\tau)R_{ZX}(\tau,t)\mathrm{d}\tau \tag{1.126}$$

If $x(t)$ and $z(t)$ are stationary and $t_0 = -\infty$ so that there is an infinite amount of data, then the optimal filter turns out to be time-invariant, $h(t,\tau) = h(t-\tau)$. In this *steady-state* case, Equation 1.125 becomes

$$R_{XZ}(t-u) = \int_{-\infty}^{T} h(t-\tau)R_Z(\tau-u)\mathrm{d}\tau; \qquad -\infty < u < T$$

or by changing variables

$$R_{XZ}(\tau) = \int_{t-T}^{\infty} h(u)R_Z(\tau - u)\mathrm{d}u; \qquad t - T < \tau < \infty \tag{1.127}$$

This stationary version is the equation Wiener actually worked with, and it is the one he and Hopf provided an elegant frequency domain solution for in 1931 in another context. Wiener was quite delighted in 1942 to discover that the linear estimation problem depended on the solution of this equation he had studied 11 years earlier!

In the steady-state case, the error covariance is a constant given by

$$\begin{aligned} P_{\tilde{X}} &= R_X(0) - \int_{t-T}^{\infty} h(\tau)R_{ZX}(-\tau)\mathrm{d}\tau \\ &= R_X(0) - \int_{t-T}^{\infty} h(\tau)R_{XZ}^T(\tau)\mathrm{d}\tau \end{aligned} \tag{1.128}$$

1.5.2 Solution of the Wiener–Hopf Equation

1.5.2.1 Infinite-Delay Steady-State Smoothing

Let us begin by solving Equation 1.127 with $T = \infty$. This means that we must wait until all the data $z(t), -\infty \le t \le \infty$, are received before we begin processing and so it amounts to non-real-time data processing. It corresponds to the infinite-delay smoothing problem.

In this case, the Wiener-Hopf equation 1.127 becomes

$$R_{XZ}(t) = \int_{-\infty}^{\infty} h(\tau)R_Z(t - \tau)\mathrm{d}\tau, \qquad -\infty < t < \infty \tag{1.129}$$

The solution is trivial, for according to Fourier transform theory we can write this convolution in terms of spectral densities as

$$\Phi_{XZ}(\omega) = H(j\omega)\Phi_Z(\omega) \tag{1.130}$$

The optimal linear filter is therefore given by

$$H(j\omega) = \Phi_{XZ}(\omega)\Phi_Z^{-1}(\omega) \tag{1.131}$$

We can quickly find an expression for the error covariance (Equation 1.128) in terms of spectral densities, for, if we define an auxiliary function by

$$f(t) = R_X(t) - \int_{-\infty}^{\infty} h(\tau)R_{ZX}(t - \tau)\mathrm{d}\tau$$

then $P_{\tilde{X}} = f(0)$. But, using the convolution property of Fourier transforms:

$$F(\omega) = \Phi_X(\omega) - H(j\omega)\Phi_{ZX}(\omega) \tag{1.132}$$

Substituting for the optimal $H(j\omega)$ and taking the inverse transform, there results

$$P_{\tilde{X}} = \frac{1}{2\pi}\int_{-\infty}^{\infty}[\Phi_X(\omega) - \Phi_{XZ}(\omega)\Phi_Z^{-1}(\omega)\Phi_{ZX}(\omega)]\mathrm{d}\omega \tag{1.133}$$

Compare this to the time-domain expression (Equation 1.53).

Example 1.14 Infinite-Delay Smoothing for Unknown Markov Process

Suppose the unknown process $x(t)$ is first-order Markov. Then it can be modeled as being generated by the stable shaping filter

$$\dot{x}(t) = -ax(t) + w(t) \tag{1.134}$$

where $a > 0$ and the *process noise* $w(t)$ is white, with spectral density $2a\sigma_w^2$. For simplicity, we consider the scale case. In the steady-state case, $x(t)$ has spectral density (Equation 1.118) and autocorrelation (Equation 1.116) given by (prove this)

$$\Phi_X(\omega) = \frac{2a\sigma_w^2}{\omega^2 + a^2} \tag{1.135}$$

$$R_X(t) = \sigma_w^2 e^{-a|t|} \tag{1.136}$$

To find Φ_Z and Φ_{XZ}, let us assume the special case of linear measurements with additive noise. Hence, let

$$z(t) = cx(t) + v(t) \tag{1.137}$$

where $c \in R$ and the *measurement noise* $v(t)$ is white with spectral density r. Suppose here that $x(t)$ and $v(t)$ are orthogonal. Then,

$$\begin{aligned} R_{XZ}(t) &= \overline{x(\tau + t)z(\tau)} \\ &= cR_X(t) \end{aligned} \tag{1.138}$$

and

$$\begin{aligned} R_Z(t) &= \overline{z(\tau + t)z(\tau)} \\ &= \overline{[cx(\tau + t) + v(\tau + t)][cx(\tau) + v(\tau)]} \\ &= c^2 R_X(t) + r\delta(t) \end{aligned} \tag{1.139}$$

with $\delta(t)$ the Kronecker delta. Hence,

$$\Phi_{XZ}(\omega) = c\Phi_X(\omega) \tag{1.140}$$

and

$$\Phi_Z(\omega) = c^2\Phi_X(\omega) + r \tag{1.141}$$

In this linear measurement case, then the optimal linear filter (Equation 1.131) is

$$
\begin{aligned}
H(j\omega) &= \frac{\Phi_{XZ}(\omega)}{\Phi_Z(\omega)} = \frac{c\Phi_X(\omega)}{c^2\Phi_X(\omega) + r} \\
&= \frac{2ac\gamma}{\omega^2 + (a^2 + 2ac^2\gamma)}
\end{aligned} \tag{1.142}
$$

where the signal-to-noise ratio $\gamma = \sigma_w^2/r$. In this problem it is convenient to define another parameter (Van Trees 1968)

$$
\lambda = \frac{2c^2\gamma}{a} \tag{1.143}
$$

which can also be thought of as a signal-to-noise ratio. Note that $\lambda > 0$. Then,

$$
H(j\omega) = \frac{2ac\gamma}{\omega^2 + a^2(1+\lambda)} \tag{1.144}
$$

so the impulse response of the optimal linear filter is

$$
h(t) = \frac{c\gamma}{\sqrt{1+\lambda}} e^{-a\sqrt{1+\lambda}|t|} \tag{1.145}
$$

This impulse response is sketched in Figure 1.6.

The error variance (Equation 1.133) becomes

$$
\begin{aligned}
P_{\tilde{X}} &= \frac{1}{2\pi}\int_{-\infty}^{\infty} \frac{r\Phi_X(\omega)}{c^2\Phi_X(\omega) + r} \mathrm{d}\omega \\
&= \frac{1}{2\pi}\int_{-\infty}^{\infty} \frac{2a\sigma_\omega^2}{\omega^2 + (a^2 + 2ac^2\gamma)} \mathrm{d}\omega
\end{aligned}
$$

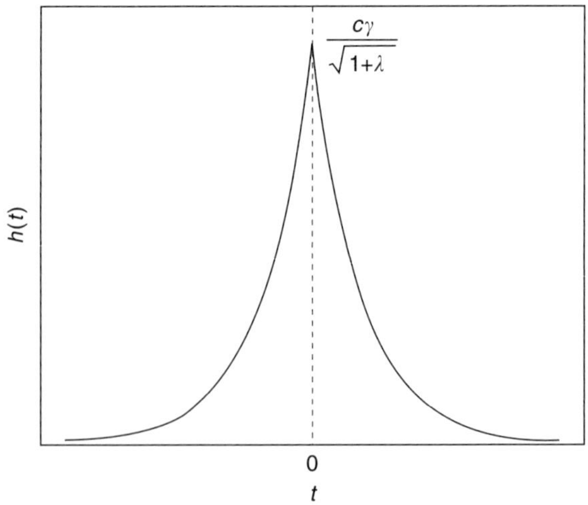

FIGURE 1.6
Optimal infinite-delay smoother.

Letting $\alpha = a\sqrt{1+\lambda}$, we get

$$P_{\tilde{X}} = \frac{2a\sigma_w^2}{\pi\alpha}\int_0^\infty \frac{\alpha}{\omega^2+\alpha^2}\mathrm{d}\omega = a\sigma_w^2/\alpha$$

or

$$P_{\tilde{X}} = \frac{\sigma_w^2}{\sqrt{1+\lambda}} \tag{1.146}$$

If $\gamma = 0$, so that the measurements provide no information, then $h(t) = 0$ so that the estimate is $\hat{x}(t) = 0$, the expected value of $x(t)$. The error variance is then

$$P_{\tilde{X}} = \sigma_w^2 = R_X(0) \tag{1.147}$$

which is independent of the measurement noise.

If $\gamma = \infty$, then the measurements are perfect. Examine Equation 1.142 to see that in this case

$$h(t) = \frac{1}{c}\delta(t) \tag{1.148}$$

so that

$$\hat{x}(t) = \frac{1}{c}z(t) \tag{1.149}$$

Then the error variance is zero.

If

$$\gamma = \frac{a}{2c^2} \tag{1.150}$$

(i.e., $\lambda = 1$), then the optimal infinite-delay smoother is

$$h(t) = \frac{c}{\sqrt{2}r}\sigma_w^2 e^{-\sqrt{2}a|t|} \tag{1.151}$$

with error variance

$$P_{\tilde{X}} = \frac{\sigma_w^2}{\sqrt{2}} = \frac{R_X(0)}{\sqrt{2}}$$

This example points out a problem with the simple-minded approach of letting $T = \infty$. Although this assumption leads to an easy solution, the resulting impulse response is noncausal and so unrealizable. □

1.5.2.2 Causal Steady-State Filtering

To find a stable causal $h(t)$ that can be used for real-time processing of the data, we present the solution of Wiener and Hopf in a modification owing to Bode and Shannon (1950). Accordingly, consider the steady-state Wiener–Hopf equation 1.127 with $T = t$, which corresponds to the filtering problem,

$$R_{XZ}(t) = \int_0^\infty h(\tau) R_Z(t-\tau)\mathrm{d}\tau; \quad 0 < t < \infty \tag{1.152}$$

Recall that the steady-state case means that $x(t)$ and $z(t)$ are stationary and a data signal $z(t)$ with infinite past history (i.e., $t_0 = -\infty$) is available.

Because of the constraint $t > 0$, the Fourier transform theorems can no longer be directly used to provide a simple solution $h(t)$ to the Wiener–Hopf equation. However, if the data signal has a rational spectral density $\Phi_Z(\omega)$, we can proceed as follows. For simplicity, we initially consider the scalar case. Our presentation is modified from that of Van Trees (1968).

Note first that if the data $z(t)$ are white then Equation 1.152 has a trivial solution, for then $R_Z(t) = r\delta(t)$ and

$$h(t) = \begin{cases} \frac{1}{r} R_{XZ}(t), & t > 0 \\ 0, & t < 0 \end{cases} \tag{1.153}$$

This will not happen except in pathological cases, but we can always use spectral factorization to *prewhiten* the data!

To wit, we know that the rational spectral density of $z(t)$ can always be factored as

$$\Phi_Z(\omega) = W^{-1}(j\omega) W^{-1}(-j\omega) \tag{1.154}$$

where $W^{-1}(s)$ is stable with zeros in the left-half plane. The representation theorem says that $z(t)$ can be considered as the output of the linear system $W^{-1}(s)$ driven by a white noise process $v(t)$. It is clear then that the "inverse" of this theorem says that $W(s)$ describes a stable linear system which results in a white noise output $v(t)$ when the input is the data $z(t)$. Thus, $W(s)$ is a *whitening filter* for $z(t)$. See Figure 1.7.

The fact that there exists a realizable whitening filter with a realizable inverse for any process with rational spectral density is called the *whitening property*. It can be considered as the "inverse" of the representation theorem.

Since $v(t)$ is white, it is uncorrelated with itself from instant to instant. The information contained in the data $z(t)$ is present also in $v(t)$, but the uncorrelatedness of $v(t)$ means that each value brings us new information, unlike the values of $z(t)$, which are correlated with each other. The information

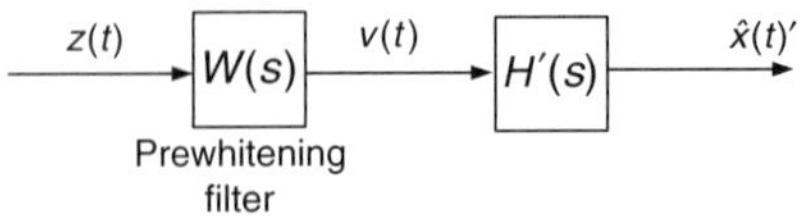

FIGURE 1.7
Solution of the Wiener–Hopf equation by prewhitening of the data.

in $z(t)$ is therefore available in the whitened process $v(t)$ in a particularly convenient form; this process is called the *innovations* or new information process. See Kailath (1981).

The prewhitening of the data $z(t)$ by spectral factorization of $\Phi_Z^{-1}(\omega)$ should be compared to the prewhitening we did in the random variable case in Section 1.4. There we factored the inverse of the noise covariance as $R^{-1} = S^T S$.

There is now only one step left in our solution of Equation 1.152. From Figure 1.7 we see that it is necessary to design an optimal filter $h'(t)$ which reconstructs $x(t)$ from the innovations $v(t)$. Thus, $h'(t)$ satisfies

$$R_{Xv}(t) = \int_0^\infty h'(\tau) R_v(t-\tau)\mathrm{d}\tau; \qquad 0 < t < \infty \tag{1.155}$$

Since $R_v(t) = \delta(t)$ (note that $v(t)$ has spectral density equal to 1), the optimal solution is given by

$$h'(t) = \begin{cases} R_{Xv}(t), & t > 0 \\ 0, & t < 0 \end{cases} \tag{1.156}$$

It is not hard to find $\Phi_{Xv}(\omega)$ in terms of $\Phi_{XZ}(\omega)$ and $W(j\omega)$ and so provide a frequency-domain solution. In fact, using a modification of Equation 1.117,

$$\Phi_{Xv}(\omega) = \Phi_{XZ}(\omega) W(-j\omega) \tag{1.157}$$

Since $h'(t)$ is causal, we retain only the portion of $R_{Xv}(t)$ for $t \geq 0$. This means that we must do a partial fraction expansion of $\Phi_{Xv}(s)$ and retain the *realizable portion* $[\Phi_{Xv}(s)]_+$, which has no poles in the open right-half plane,

$$\Phi_{Xv}(s) = [\Phi_{Xv}(s)]_+ + [\Phi_{Xv}(s)]_- \tag{1.158}$$

The second term has poles in the right-half plane and corresponds to the anticausal (i.e., $t < 0$) portion of $R_{Xv}(t)$. Any constant terms in $\Phi_{Xv}(s)$ are included in the realizable portion. In this notation,

$$H'(j\omega) = [\Phi_{XZ}(\omega) W(-j\omega)]_+ \tag{1.159}$$

The overall Wiener filter $h(t)$ that solves Equation 1.152 is given by the cascade of the whitening filter and $H'(s)$, or

$$H(j\omega) = [\Phi_{XZ}(\omega) W(-j\omega)]_+ W(j\omega) \tag{1.160}$$

where

$$\Phi_Z^{-1}(\omega) = W(-j\omega) W(j\omega) \tag{1.161}$$

This should be compared with our solution (Equation 1.131) to the infinite-delay smoothing problem. The transfer function $H(s)$ manufactured by Equation 1.160 represents a realizable system and allows real-time processing of the data $z(t)$ to find the optimal estimate $\hat{x}(t)$ of the unknown process $x(t)$.

To find an expression for the error covariance (i.e., Equation 1.128 with $T = t$), substitute Equation 1.160 into Equation 1.132 to get (scalars!)

$$\begin{aligned} F(\omega) &= \Phi_X(\omega) - [\Phi_{XZ}(\omega)W(-j\omega)]_+ W(j\omega)\Phi_{XZ}(-\omega) \\ &= \Phi_X(\omega) - [\Phi_{Xv}(\omega)]_+ \Phi_{Xv}(-\omega) \end{aligned} \tag{1.162}$$

where Equation 1.157 was used. If the cross-correlation of unknown $x(t)$ and innovations $v(t)$ is $R_{Xv}(t)$, we can retain only the causal portion by noting that

$$\Im[R_{Xv}(t)u_{-1}(t)] = [\Phi_{Xv}(\omega)]_+. \tag{1.163}$$

where $\Im[\cdot]$ denotes Fourier transform and $u_{-1}(t)$ is the unit step. Hence, Equation 1.162 can be transformed to obtain (recall the transform of $\Phi_{Xv}(-\omega)$ is $R_{Xv}(-t)$)

$$f(t) = R_X(t) - \int_{-\infty}^{\infty} R_{Xv}(\tau)u_{-1}R_{Xv}(\tau - t)\mathrm{d}\tau \tag{1.164}$$

so that $P_{\tilde{X}} = f(0)$ or

$$P_{\tilde{X}} = R_X(0) - \int_0^{\infty} R_{Xv}^2(\tau)\mathrm{d}\tau \tag{1.165}$$

We can also use these constructions to obtain a time-domain expression for the optimal error covariance obtained by using the unrealizable filter (Equation 1.131). Considering the scalar case and using superscript u for clarity, there results, from Equations 1.131, 1.132, and 1.161,

$$F^u(\omega) = \Phi_X(\omega) - \Phi_{XZ}(\omega)W(-j\omega)W(j\omega)\phi_{XZ}(-\omega) \tag{1.166}$$

Now use Equation 1.157 to write

$$F^u(\omega) = \Phi_X(\omega) - \Phi_{Xv}(\omega)\phi_{Xv}(-\omega) \tag{1.167}$$

Transforming this, we get

$$f^u(t) = R_X(t) - \int_{-\infty}^{\infty} R_{Xv}(\tau)R_{Xv}(\tau - t)\mathrm{d}\tau \tag{1.168}$$

so that

$$P_{\tilde{X}}^u = R_X(0) - \int_{-\infty}^{\infty} R_{Xv}^2(\tau)\mathrm{d}\tau \tag{1.169}$$

Therefore, the (unattainable!) error using the unrealizable filter is in general less than the error (Equation 1.165) using the realizable Wiener filter.

Example 1.15 Causal Steady-State Filter for Unknown Markov Process

Let us find the realizable Wiener filter (Equation 1.160) for Example 1.14. From that example,

$$\Phi_X(\omega) = \frac{2a\sigma_w^2}{\omega^2 + a^2} \tag{1.170}$$

$$\Phi_{XZ}(\omega) = c\Phi_X(\omega) = \frac{2ac\sigma_w^2}{\omega^2 + a^2} \tag{1.171}$$

$$\Phi_Z(\omega) = c^2\Phi_X(\omega) + r = \frac{r[\omega^2 + a^2(1+\lambda)]}{\omega^2 + a^2} \tag{1.172}$$

To find the whitening filter, factor the inverse of the spectral density of $z(t)$ to get

$$\begin{aligned}\Phi_Z^{-1}(\omega) &= W(-j\omega)W(j\omega)\\ &= \frac{(-j\omega + a)}{\sqrt{r}(-j\omega + a\sqrt{1+\lambda})} \cdot \frac{(j\omega + a)}{\sqrt{r}(j\omega + a\sqrt{1+\lambda})}\end{aligned} \tag{1.173}$$

Therefore,

$$\begin{aligned}\Phi_{Xv}(\omega) &= \Phi_{XZ}(\omega)W(-j\omega)\\ &= \frac{2ac\sigma_w^2}{\sqrt{r}} \frac{(-j\omega + a)}{(-j\omega + a\sqrt{1+\lambda})(\omega^2 + a^2)}\end{aligned} \tag{1.174}$$

To facilitate things, convert this to a function of $s = \sigma + j\omega$. Then a partial fraction expansion yields

$$\Phi_{Xv}(s) = \frac{-2ac\sigma_w^2}{\sqrt{r}} \frac{s - a}{(s - a\sqrt{1+\lambda})(s-a)(s+a)} \tag{1.175}$$

$$= \frac{2c\sigma_w^2}{\sqrt{r}(1 + \sqrt{1+\lambda})} \left(\frac{1}{s+a} + \frac{-1}{s - a\sqrt{1+\lambda}} \right) \tag{1.176}$$

The causal portion of $R_{Xv}(t)$ is given by the first term, which contains the left-half plane pole. (Note that the region of convergence of Equation 1.176 contains the imaginary axis.) Therefore, by Equation 1.160 (recall $\gamma = \sigma_w^2/r$)

$$\begin{aligned}H(s) &= [\Phi_{Xv}(s)]_+ W(s)\\ &= \frac{2c\gamma}{1 + \sqrt{1+\lambda}} \cdot \frac{1}{s + a\sqrt{1+\lambda}}\end{aligned} \tag{1.177}$$

so that

$$h(t) = \frac{2c\gamma}{1 + \sqrt{1+\lambda}} e^{-a\sqrt{1+\lambda}t} u_{-1}(t) \tag{1.178}$$

This is the impulse response of the Wiener filter for reconstructing $x(t)$ from the data $z(t)$. It is causal and stable, and should be compared to the unrealizable filter (Equation 1.145) found in Example 1.14. Note that we *cannot* obtain Equation 1.178 by taking the $t \geq 0$ portion of the unrealizable filter!

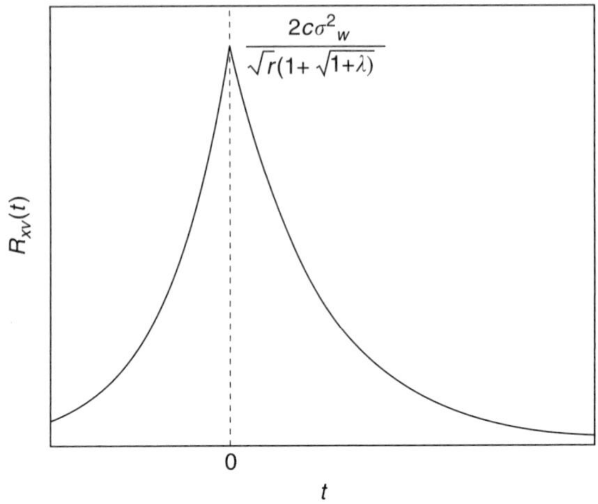

FIGURE 1.8
Cross-correlation of unknown with innovations.

From Equation 1.177 it is clear that the Wiener filter is a low-pass filter whose bandwidth depends on a and λ.

The optimal error covariance is found by using $R_{Xv}(t)$, which by Equation 1.176 is

$$R_{Xv}(t) = \begin{cases} \frac{2c\sigma_w^2}{\sqrt{r}(1+\sqrt{1+\lambda})} e^{-at}, & t \geq 0 \\ \frac{2c\sigma_w^2}{\sqrt{r}(1+\sqrt{1+\lambda})} e^{a\sqrt{1+\lambda}t}, & t < 0 \end{cases} \tag{1.179}$$

This cross-correlation is sketched in Figure 1.8. Now, by Equation 1.166,

$$P_{\tilde{X}} = \sigma_w^2 - \int_0^\infty \frac{4c^2\sigma_w^2\gamma}{(1+\sqrt{1+\lambda})^2} e^{-2at} dt \tag{1.180}$$

$$= \sigma_w^2 \frac{[(1+\sqrt{1+\lambda})^2 - \lambda]}{(1+\sqrt{1+\lambda})^2} \tag{1.181}$$

$$= \frac{2\sigma_w^2}{1+\sqrt{1+\lambda}} \tag{1.182}$$

This error covariance and $P_{\tilde{X}}^u$ given by Equation 1.146 in Example 1.14 are both sketched in Figure 1.9 as a function of λ. ▯

Example 1.16 The Wiener Filter for Linear Measurements

Here we derive expressions for the Wiener filter and error covariance for the special case of scalar linear measurements

$$z(t) = cx(t) + v(t) \tag{1.183}$$

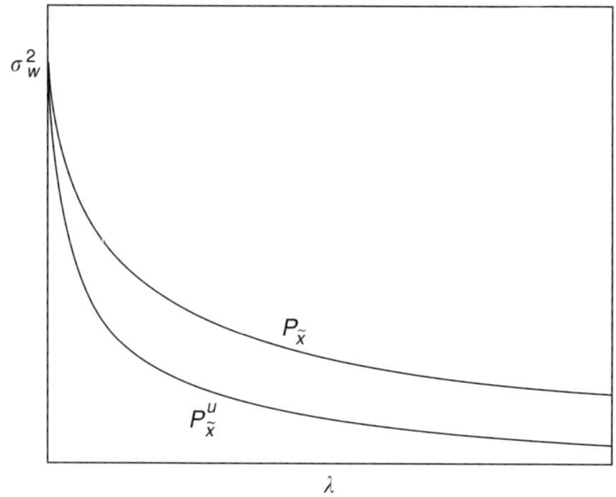

FIGURE 1.9
Error variances for realizable and unrealizable optimal linear filters.

where $c \in R$ and measurement noise $v(t)$ is white with spectral height r. We assume $v(t)$ is uncorrelated with the unknown $x(t)$.

From Example 1.14,

$$\Phi_{XZ}(\omega) = c\Phi_X(\omega) \tag{1.184}$$

and

$$\Phi_Z(\omega) = c^2\Phi_X(\omega) + r \tag{1.185}$$

To find the Wiener filter, we factor $\Phi_Z(\omega)$ into unrealizable and realizable spectral factors as

$$\Phi_Z(\omega) = W^{-1}(j\omega)W^{-1}(-j\omega) \tag{1.186}$$

where $W^{-1}(s)$ has poles in the open left-half plane (i.e., not on the $j\omega$ axis). From the spectral factorization theorem we know that $W^{-1}(s)$ can only be guaranteed to have zeros in the left-half plane, including possible zeroes on the $j\omega$ axis. Thus, if we write

$$\Phi_Z^{-1}(\omega) = W(-j\omega)W(j\omega) \tag{1.187}$$

we cannot in general guarantee that $W(-j\omega)$ has all poles in the s plane with real parts strictly greater than zero (i.e., unstable). However, it is easy to show that if $\Phi_Z(\omega)$ has the form of Equation 1.185 with $r \neq 0$, then $W(-j\omega)$ does have all poles in the open right-half plane. (We shall soon see why we want $W(-j\omega)$ to be completely unstable!) For

$$\Phi_Z^{-1}(s) = \frac{1/r}{1 + (c^2/r)\Phi_X(s)} \tag{1.188}$$

is in the exact canonical form for the root-locus theory. Hence, as the "feedback gain" c^2/r increases, the poles of $\Phi_Z^{-1}(s)$ migrate from the poles of $\Phi_X(s)$ to the zeros of $\Phi_X(s)$. Since $\Phi_X(s)$ corresponds to a spectral density, it may have zeros but not poles on the imaginary axis. It does not have poles or zeros in the right-half plane. Hence, if $r \neq 0$, the poles of $\Phi_Z^{-1}(s)$ do not occur on the $j\omega$ axis and the spectral factors given by

$$\Phi_Z^{-1}(\omega) = W(-j\omega)W(j\omega) \tag{1.189}$$

are such that $W(s)$ has all poles in the open left-half plane and $W(s^*)$ has all poles in the open right-half plane (where * denotes complex conjugate).

We can use this fact to find an easy expression for the realizable part of $W(-j\omega)$, which we will need shortly. Let

$$\Phi_X(s) = \frac{n(s)}{d(s)} \tag{1.190}$$

for polynomials $n(s)$ and $d(s)$ in s^2. If $x(t)$ is band-limited, the degree of $d(s)$ in s^2 is at least one greater than the degree of $n(s)$ (why?). Therefore, $\Phi_Z^{-1}(s)$ has a relative degree of zero since

$$\Phi_Z^{-1}(s) = \frac{(1/r)d(s)}{d(s) + (c^2/r)n(s)} \tag{1.191}$$

In Equation 1.189, a multiplier of $1/\sqrt{r}$ is assigned to each of the spectral factors, which also have relative degrees of zero. $W(-j\omega)$ thus has the form

$$W(-j\omega) = \frac{1}{\sqrt{r}} + \text{unrealizable part} \tag{1.192}$$

and so

$$[W(-j\omega)]_+ = \frac{1}{\sqrt{r}} \tag{1.193}$$

This can be used to find a very simple expression for the Wiener filter in the case of linear measurements. From Equation 1.160, the Wiener filter is

$$H(j\omega) = \frac{1}{c}[c^2\Phi_X(\omega)W(-j\omega)]_+W(j\omega) \tag{1.194}$$

(note that the "realizable part" operator is linear), or

$$\begin{aligned} H(j\omega) &= \frac{1}{c}[(c^2\Phi_X(\omega) + r)W(-j\omega) - rW(-j\omega)]_+W(j\omega) \\ &= \frac{1}{c}[W^{-1}(j\omega) - rW(-j\omega)]_+W(j\omega) \\ &= \frac{1}{c}[W^{-1}(j\omega)]_+W(j\omega) - \frac{r}{c}[W(-j\omega)]_+W(j\omega) \end{aligned} \tag{1.195}$$

Now, since $W^{-1}(s)$ is completely realizable (all poles in the open left-half plane), the first term is $1/c$. Therefore (see Equation 1.193),

$$H(j\omega) = \frac{1}{c}[1 - \sqrt{r}W(j\omega)] \tag{1.196}$$

which is the desired expression for the Wiener filter in the linear measurement case.

It is not too difficult to show (Yovits and Jackson 1955) that the error covariance is given by

$$P_{\tilde{X}} = \frac{r}{2\pi c^2} \int_{-\infty}^{\infty} \ln\left[1 + \frac{c^2\Phi_X(\omega)}{r}\right] d\omega \tag{1.197}$$

Using Equations 1.196 and 1.197, the results of Example 1.15 can be obtained at somewhat less expense!

A useful identity when using Equation 1.197 is

$$\frac{1}{2\pi} \int_{-\infty}^{\infty} \ln\left(\frac{\omega^2 + \alpha^2}{\omega^2 + \beta^2}\right) d\omega = \alpha - \beta \tag{1.198}$$

It should be noted that although the whitening filter $W(s)$ is the central concept in our derivation of the Wiener filter, only the latter is actually implemented in an application.

The Wiener filter can be generalized to multivariable processes and a counterpart also exists in the discrete case. In fact, if the unknown x_k and the data z_k are discrete vector random processes, then Equations 1.160 and 1.161 become

$$H(z) = [\Phi_{XZ}(z)W^T(z^{-1})]_+ W(z) \tag{1.199}$$

where

$$\Phi_Z^{-1}(z) = W^T(z^{-1})W(z) \tag{1.200}$$

In this case, the realizable part operator means to retain the portion with poles inside or on the unit circle. If $|\Phi_Z(z)| \neq 0$ on $|z| = 1$, then $W(z)$ is minimum phase and the realizable poles are strictly within the unit circle.

In the multivariable case with linear measurements,

$$z(t) = x(t) + v(t) \tag{1.201}$$

where $v(t) \sim (0, R)$, the results of Example 1.16 can be generalized. The Wiener filter is given in this case by the simple formula

$$H(s) = I - R^{1/2}W(s) \tag{1.202}$$

where $W(s)$ is the stable spectral factor of $\phi_Z^{-1}(s)$ and $R = R^{1/2}R^{T/2}$ with $R^{1/2}$ the square root of R (Kailath 1981). We derive the discrete counterpart to this result in Section 2.6. □

Problems

Section 1.1

1. Least-Squares Solution of Linear Equations. The connotation of *least-squares* estimation is a setting in which both the unknown X and the data Z are deterministic, so that no *a priori* information is available on either one.

Let

$$Z = HX \tag{1.203}$$

with H of full column rank. Minimize the performance index

$$J = \frac{1}{2}(Z - HX)^T R^{-1}(Z - HX) \tag{1.204}$$

where R^{-1} is a weighting matrix, to show that the (weighted) least-squares estimate is

$$\hat{X} = (H^T R^{-1} H)^{-1} H^T R^{-1} Z \tag{1.205}$$

Note that the least-squares performance index is expressed in terms of the *residual*.

2. Nonlinear LS Estimation. Let the data Z and the unknown X be related by

$$Z = h(X) + e \tag{1.206}$$

where e is a residual and $h(\cdot)$ is a (possibly nonlinear) function.

(a) Show that the estimate minimizing the least-squares error

$$J = \frac{1}{2}\langle e, e\rangle = \frac{1}{2}\langle Z - h(X), Z - h(X)\rangle \tag{1.207}$$

where $\langle \cdot, \cdot \rangle$ represents inner product is the value of X satisfying the *least-squares normal equations*

$$\left\langle \frac{\partial h}{\partial X}, h(X)\right\rangle = \left\langle \frac{\partial h}{\partial X}, Z\right\rangle \tag{1.208}$$

(b) Expand $h(X)$ in a Taylor series about a given estimate $\hat{X}$ to show that

$$\langle H, HX\rangle = \langle H, H\hat{X}\rangle + \langle H, Z - h(\hat{X})\rangle \tag{1.209}$$

where $H = \partial h/\partial X$ is the Jacobian. This provides an iterative method for solving Equation 1.208, for it shows how to determine a more accurate updated estimate for X from a previous estimate $\hat{X}$.

3. Continuous Estimate Updating. Let the inner product in Problem 2 be defined as

$$J = \frac{1}{2}\int_0^T e^T R^{-1} e \, dt \tag{1.210}$$

where $R > 0$.

(a) Write the linearized normal Equations 1.209.

(b) Note that $H = -\partial e/\partial X$ and define

$$M = \int_0^T H^T R^{-1} H \, dt \tag{1.211}$$

Using your result in part (a), let T become small to show that the estimate satisfies

$$\dot{X} = -M^{-1}\left(\frac{\partial e}{\partial X}\right)^T R^{-1} e \tag{1.212}$$

where $e = z - h(x)$.

Note that

$$\dot{M} = H^T R^{-1} H \tag{1.213}$$

Equations 1.212 and 1.213 provide a method for continuously updating the least-squares estimate X if data are taken continuously over a time interval.

4.

(a) Suppose we have three measurements Z_i of a constant nonrandom voltage X. The Z_i are corrupted by additive noises $V_1 \sim (0,1)$, $V_2 \sim (0,2)$, and $V_3 \sim (0,4)$, respectively. If we observe $Z_1 = 4V$, $Z_2 = 3V$, and $Z_3 = 5V$, find the mean-square estimate for X.

(b) Find the minimum error variance $\sigma_{\tilde{X}}^2$.

5. Estimate for Linear Function of an Unknown. If $\hat{X}$ is the linear mean-square estimate for unknown X, show that the linear mean-square estimate for the linear function $f(X) = CX + d$ is given by $\hat{f} = C\hat{X} + d$.

6. Fill in the steps in showing Equation 1.30.

7. A stochastic unknown X and the data Z are jointly distributed according to

$$f_{XZ}(x,z) = \frac{6}{7}(x+z)^2; \quad 0 \le x \le 1 \quad \text{and} \quad 0 \le z \le 1$$

(a) Show that

$$f_X(x) = \frac{6}{7}\left(x^2 + x + \frac{1}{3}\right); \quad 0 \le x \le 1$$

$$f_{X/Z}(x/z) = \frac{x^2 + 2xz + z^2}{\frac{1}{3} + z + z^2}; \quad 0 \le x \le 1$$

(b) Find the best estimate for X and its error covariance in the absence of data.

(c) Find the best mean-square estimate for X given the data Z.

(d) Find the best linear mean-square estimate for X and its error covariance given the data Z.

8. Multiplicative Noise. An unknown X is measured in the presence of multiplicative noise so that

$$Z = (1 + W)X$$

where $X \sim (0, \sigma_X^2)$ and the noise is $W \sim (0, \sigma_W^2)$. The noise and unknown are uncorrelated.

(a) Find the optimal linear estimate $\hat{X} = aZ$.

(b) Find the error variance.

9. Linear Measurements with Correlated Noise and Unknown. Repeat Example 1.3 if $\overline{XV^T} = P_{XV} \neq 0$.

10. Experiment Design. An unknown voltage S is distributed according to

$$f_S(s) = \frac{1}{2\sqrt{2\pi}} e^{-s^2/8}$$

Consider two ways to measure S.

(a) Use two expensive meters that add to S noise, which is $(0, 2)$ so that

$$\begin{aligned} Z_1 &= S + V_1, \qquad & V_1 \sim (0, 2) \\ Z_2 &= S + V_2, \qquad & V_2 \sim (0, 2) \end{aligned}$$

with V_1 and V_2 uncorrelated with each other and with S. Find linear mean-square estimate and error covariance.

(b) Use four inexpensive meters that add to S noise, which is $(0, 3)$. Find linear mean-square estimate and error covariance.

(c) For the best estimate, should we use scheme (a) or scheme (b)?

(d) Find the numerical values for $\hat{S}_{\text{LMS}}$: in (a) if $Z_1 = 1$ and $Z_2 = 2$; in (b) if $Z_1 = 1$, $Z_2 = 2$, $Z_3 = 0.5$, and $Z_4 = 1.2$.

Section 1.2

11. Inaccurately Known Noise Covariance. To examine the effects of inaccurately known noise parameters, suppose we have

$$Z = \begin{bmatrix} 1 \\ 1 \end{bmatrix} X + v, \qquad v \sim N\left(0, \begin{bmatrix} \sigma^2 & 0 \\ 0 & 1 \end{bmatrix}\right)$$

(a) Compute $\hat{X}_{\text{ML}}$ and $P_{\tilde{X}}$ in terms of σ^2 if the measurement is

$$Z = \begin{bmatrix} 5 \\ 8 \end{bmatrix}$$

Sketch $P_{\tilde{X}}$ and $\hat{X}_{\text{ML}}$ versus σ^2.

(b) Evaluate $\hat{X}_{\text{ML}}$ and $P_{\tilde{X}}$ for $\sigma^2 = \frac{1}{2}$, 1, and 2.

12. Let

$$Z = \begin{bmatrix} 1 & 2 \\ 0 & -1 \end{bmatrix} X + V, \qquad V \sim N\left(\begin{bmatrix} 0 \\ 0 \end{bmatrix}, \begin{bmatrix} 1 & 1 \\ 1 & 3 \end{bmatrix} \right)$$

$$X \sim N\left(\begin{bmatrix} 1 \\ -1 \end{bmatrix}, \begin{bmatrix} 1 & 2 \\ 2 & 5 \end{bmatrix} \right)$$

and let the measured data be

$$z = \begin{bmatrix} 2 \\ 0.5 \end{bmatrix}$$

(a) Ignore statistics and use matrix inverse to find $\hat{X}$ (i.e., $\hat{X} = H^{-1}Z$).

(b) Ignore statistics and use weighted least-squares estimation (Problem 1) to find $\hat{X}$. Use weighting matrix R equal to the covariance of V. Can you find an error covariance?

(c) Include statistics on V but not on X. Find maximum-likelihood estimate. Find $P_{\tilde{X}}$.

(d) Include statistics on X and V. Find linear mean-square estimate. Find $P_{\tilde{X}}$.

(e) What is your estimation for X before Z is measured? Find error covariance $P_{\tilde{X}}$ in this case.

13. Consider two measurements of a scalar unknown random variable X

$$\begin{aligned} Z_1 &= X + V_1, \qquad & V_1 &\sim (0, \sigma_1^2) \\ Z_2 &= X + V_2, \qquad & V_2 &\sim (0, \sigma_2^2) \end{aligned}$$

with noises V_1 and V_2 uncorrelated.

(a) X is known *a priori* to be $(\bar{X}, \sigma_X^2)$. Find the linear mean-square estimate $\hat{X}$ and error covariance $P_{\tilde{X}}$. Compare with Example 1.8.

(b) Find linear mean-square estimate if no statistics are available on X.

14. Let the data Z be related to an unknown X by

$$Z = \ln X + V \tag{1.214}$$

with scalar random variables X and V independent. Suppose X is uniformly distributed between 0 and b, and noise V is Laplacian with $a = 1$ [i.e., $f_V(v) = (a/2)e^{-a|v|}$].

(a) Find the optimal *a priori* estimate and error covariance.

(b) Find the maximum-likelihood estimate.

(c) Find the conditional mean estimate.

15. Arbitrary Linear Combination of Data. Let a constant scalar unknown X be measured with two meters corrupted by additive noise so that

$$
\begin{aligned}
Z_1 &= X + V_1 \\
Z_2 &= X + V_2
\end{aligned}
\tag{1.215}
$$

where $V_1 \sim N(0, \sigma_1^2)$, and $V_2 \sim N(0, \sigma_2^2)$, and X, V_1, and V_2 are independent. We can estimate X by an arbitrary linear combination of the data,

$$\hat{X} = aZ_1 + bZ_2 \tag{1.216}$$

(a) Show that for this estimate to be unbiased, we require

$$b = 1 - a \tag{1.217}$$

(b) In the unbiased case, find the error covariance $P_{\tilde{X}}$. For what value of a is this estimate equal to $\hat{X}_{\text{ML}}$?

(c) Is it possible to find a biased estimate (Equation 1.216) with smaller error covariance than that in (b)?

16. Optimal Combination of Two Independent Estimates. Suppose two estimates $\hat{X}_1$ and $\hat{X}_2$ of an unknown random variable X are obtained by independent means. The associated error covariances are P_1 and P_2. It is desired to find the overall optimal linear estimate of X, that is, the optimal estimate of the form

$$\hat{X} = A_1\hat{X}_1 + A_2\hat{X}_2 \tag{1.218}$$

where A_1 and A_2 are constant matrices.

(a) Show that if $\hat{X}_1$ and $\hat{X}_2$ are unbiased (i.e., $E(\hat{X}_i) = \bar{X}$), then for $\hat{X}$ to be unbiased also it is sufficient that

$$A_2 = I - A_1 \tag{1.219}$$

(b) Define estimation errors as $\tilde{X}_1 = X - \hat{X}_1, \tilde{X}_2 = X - \hat{X}_2$, and $\tilde{X} = X - \hat{X}$. Show that, using Equation 1.219,

$$\tilde{X} = A_1\tilde{X}_1 + (I - A_1)\tilde{X}_2 \tag{1.220}$$

(c) Let the error covariance of $\hat{X}$ be P. Differentiate trace (P) to find that the optimal A_1 is

$$A_1 = P_2(P_1 + P_2)^{-1} \tag{1.221}$$

and hence

$$A_2 = P_1(P_1 + P_2)^{-1} \tag{1.222}$$

(d) Show that the overall error covariance P is given by

$$P^{-1} = P_1^{-1} + P_2^{-1} \tag{1.223}$$

(e) Show that

$$P^{-1}\hat{X} = P_1^{-1}\hat{X}_1 + P_2^{-1}\hat{X}_2 \tag{1.224}$$

(This is a generalized Millman's theorem. See Examples 1.8 and 1.9.)

17. Alternate Form of Estimate Update with Linear Measurements. Let unknown X be distributed as $(\bar{X}, P_X)$ and measurements be taken so that

$$Z = HX + V \tag{1.225}$$

with noise $V \sim (0, R)$ uncorrelated with X. Show that the linear mean-square estimate $\hat{X}$ is given by

$$P^{-1}\hat{X} = P_X^{-1}\bar{X} + H^T R^{-1} Z \tag{1.226}$$

with error covariance

$$P^{-1} = P_X^{-1} + H^T R^{-1} H \tag{1.227}$$

(If X and V are normal, then Equations 1.226 and 1.227 yield the optimal mean-square estimate.)

Section 1.4

18. Derive Equation 1.101.

19. Sequential Estimation with Block Diagonal R. Derive Equations 1.105, 1.106, and 1.108 for the case $X \in R^n$ and R block diagonal, so that $R =$ diag $\{S_i\}$, where S_i is $p_i \times p_i$ and nonsingular. Thus the components of Z are sequentially processed in groups of size p_i.

Section 1.5

20. An Alternative Design Method for Optimal Filters. A measurement $z(t)$ is taken of a signal $x(t)$ corrupted by additive noise $n(t)$. The signal and noise are uncorrelated, and

$$\Phi_X(\omega) = \frac{2\sigma^2}{1 + \omega^2} \tag{1.228}$$

$$\Phi_N(\omega) = \frac{r\omega^2}{1 + \omega^2} \tag{1.229}$$

An estimate $\hat{x}(t)$ is formed by passing $z(t)$ through the ideal low-pass filter

$$H(j\omega) = \begin{cases} 1, & |\omega| < \omega_c \\ 0, & \text{otherwise} \end{cases} \tag{1.230}$$

(a) Sketch $\Phi_X(\omega)$ and $\Phi_N(\omega)$. Show that $\Phi_X(\omega) = \Phi_N(\omega)$ at

$$\omega = \omega_0 = \sqrt{\frac{2\sigma^2}{r}} \tag{1.231}$$

(b) If $\tilde{x}(t) = x(t) - \hat{x}(t)$, show that

$$\tilde{X}(s) = [I - H(s)]X(s) - H(s)N(s) \tag{1.232}$$

and

$$\Phi_{\tilde{X}}(\omega) = |I - H(j\omega)|^2 \Phi_X(\omega) + |H(j\omega)|^2 \Phi_N(\omega) \tag{1.233}$$

Sketch $|H(j\omega)|^2$ and $|I - H(j\omega)|^2$.

(c) Find the mean-square error $P_{\tilde{X}} = \overline{\tilde{x}^2(t)}$. (Recall that $P_{\tilde{X}} = R_{\tilde{X}}(0)$.)

(d) Differentiate $P_{\tilde{X}}$ with respect to ω_c to find the optimal value of ω_c for the filter $H(\omega)$.

(e) For $r = 1$ and $\sigma^2 = 2$, evaluate the optimal ω_c and error covariance $P_{\tilde{X}}$.

21. Assume measurement $z(t)$ is derived from unknown $x(t)$ as in the previous problem.

(a) Find the Wiener filter $H(j\omega)$ to estimate $x(t)$. Sketch $|H(j\omega)|$. Note that its cutoff frequency is the same as the cutoff frequency in the previous problem.

(b) Find the error covariance $P_{\tilde{X}}$.

(c) For $r = 1$ and $\sigma^2 = 2$, find $H(j\omega)$ and $P_{\tilde{X}}$.

22. Wiener Filter with Prediction. If it is desired to estimate an unknown scalar signal $x(t)$ at time $t + d$ using data $z(t)$ up through time t, the Wiener–Hopf equation becomes

$$R_{XZ}(t + d) = \int_0^\infty h(\tau) R_Z(t - \tau) \mathrm{d}\tau; \quad 0 < t < \infty \tag{1.234}$$

Show that in this case the Wiener filter is given by

$$H(s) = [\phi_{XZ}(s) e^{sd} W(-s)]_+ W(s) \tag{1.235}$$

with $W(s)$, the stable factor, given by

$$\phi_Z^{-1}(s) = W(-s)W(s) \tag{1.236}$$

(Kailath 1981).

23. Wiener Filter for Second-Order Markov Process. The unknown $x(t)$ is a second-order Markov process satisfying

$$\dot{x} = x_2 + w \tag{1.237a}$$

$$\dot{x}_2 = -2x - 2x_2 \tag{1.237b}$$

with white process noise $w(t) \sim (0, 2)$. The data are generated using measurements given by

$$z(t) = x(t) + v(t) \tag{1.238}$$

with white measurement noise $v(t) \sim (0, 1)$ uncorrelated with $x(t)$.

(a) Find spectral densities $\phi_X(s), \phi_{XZ}(s)$, and $\phi_Z(s)$.

(b) Find optimal noncausal filter for reconstructing $x(t)$. Compute and sketch the impulse response.

ANSWER:

$$H(s) = \frac{2(-s^2 + 4)}{s^4 - 2s^2 + 12}$$

(c) Find a whitening filter for the data $z(t)$.

ANSWER:

$$W(s) = \frac{s^2 + 2s + 2}{s^2 + 2.988s + 3.4641}$$

(d) Find the Wiener filter. Compute and sketch impulse response.

2

Discrete-Time Kalman Filter

The object of this chapter is to estimate the state of a discrete linear system developing dynamically through time. We shall see that the optimal estimator, known as the Kalman filter, is itself a linear dynamical system and it is a natural extension of the estimators discussed in the previous chapter.

To introduce the basic ideas used in the discrete Kalman filter, we first review the deterministic state observer.

2.1 Deterministic State Observer

Suppose a dynamical plant is prescribed

$$x_{k+1} = Ax_k + Bu_k \tag{2.1a}$$

$$z_k = Hx_k \tag{2.1b}$$

where state $x_k \in R^n$, control input $u_k \in R^m$, output $z_k \in R^p$, and A, B, and H are known constant matrices of appropriate dimension. All variables are deterministic, so that if initial state x_0 is known then Equation 2.1 can be solved exactly for x_k, z_k for $k \geq 0$.

The *deterministic asymptotic estimation* problem is as follows. Design an estimator whose output $\hat{x}_k$ converges with k to the actual state x_k of Equation 2.1 when the initial state x_0 is unknown, but u_k and z_k are given exactly.

An estimator or *observer* that solves this problem has the form

$$\hat{x}_{k+1} = A\hat{x}_k + L(z_k - H\hat{x}_k) + Bu_k \tag{2.2}$$

which is shown in Figure 2.1. A, B, and H are given, so it is only necessary to find the constant observer matrix L. Note that the observer consists of two parts, a replica (A, B, H) of the dynamics of the plant whose state is to be estimated and an error-correcting portion described by $L(z_k - \hat{z}_k) = L\tilde{z}_k$, where the output estimate is $\hat{z}_k = H\hat{x}_k$ and $\tilde{z}_k = z_k - \hat{z}_k$ is the *residual*. The observer should be compared with the recursive estimators shown in Figures 1.4 and 1.5. Note, however, that here the gain L that multiplies the residual is a constant. Since we built the observer, its state $\hat{x}_k$ is available to us. (Plant state x_k is not available except through measurement z_k; therein lies the problem.)

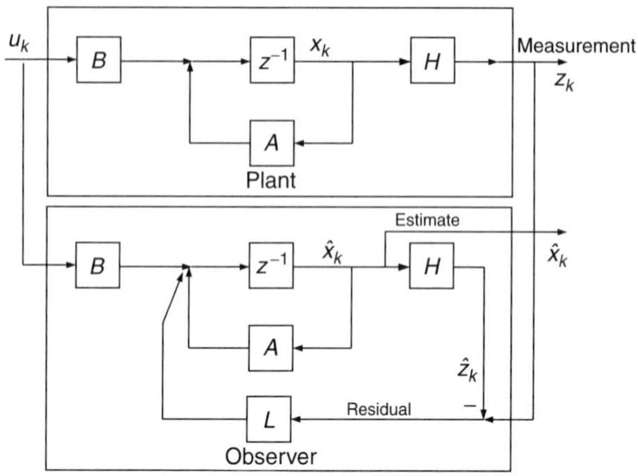

FIGURE 2.1
State observer.

To choose L so that the estimation error $\tilde{x}_k = x_k - \hat{x}_k$ goes to zero with k for all x_0, it is necessary to examine the dynamics of $\hat{x}_k$. Write

$$\begin{aligned}\tilde{x}_{k+1} &= x_{k+1} - \hat{x}_{k+1} \\ &= Ax_k + Bu_k - [A\hat{x}_k + L(z_k - H\hat{x}_k) + Bu_k] \\ &= A(x_k - \hat{x}_k) - L(Hx_k - H\hat{x}_k)\end{aligned}$$

so

$$\tilde{x}_{k+1} = (A - LH)\tilde{x}_k \tag{2.3}$$

describes the dynamics of the estimation error. System 2.3 is called the *error system*. It does not actually exist, but is a mathematical fiction that allows us to study the convergence properties of the observer. The initial condition of Equation 2.3 is $\tilde{x}_0 = x_0 - \hat{x}_0$, the unknown initial error.

It is now apparent that for $\tilde{x}_k$ to go to zero with k for any $\tilde{x}_0$, observer gain L must be selected so that $(A - LH)$ is stable. Thus the output injection problem is the key to deterministic observer design. L can be chosen so that $\tilde{x}_k \to 0$ if and only if (A, H) is detectable (i.e., the unstable poles can be arbitrarily assigned), for then $(A - LH)$ is stable for some L. If we want the error $\tilde{x}_k$ not simply to go to zero with k, but to die out as fast as we desire, then all the poles of $(A - LH)$ must be arbitrarily assignable, so that (A, H) must be observable.

There are many ways to compute L. In the single-output case, *Ackermann's formula*

$$L = \Delta^d(A)V_n^{-1}e_n \tag{2.4}$$

can be used, where e_n is the nth column of the $n \times n$ identity matrix, $\Delta^d(s)$ the desired error system characteristic polynomial, and

$$V_n = \begin{bmatrix} C \\ CA \\ \vdots \\ CA^{n-1} \end{bmatrix} \tag{2.5}$$

the observability matrix. Alternatively, we can use the next result, which applies for multioutput systems.

THEOREM 2.1

Let G and D be any matrices such that (A, G) is reachable and rank $([H, D]) =$ rank (D). Then (A, H) is detectable if and only if there exists a positive definite solution P_+ to the algebraic Riccati equation

$$P = APA^T - APH^T(HPH^T + DD^T)^{-1}HPA^T + GG^T \tag{2.6}$$

In this event

$$L = AP_+H^T(HPH^T + DD^T)^{-1} \tag{2.7}$$

stabilizes $(A - LH)$.

The proof of this result is fairly easy; it is part of a theorem we will prove in Section 2.5.

There is as yet no basis for selecting D and G; any matrices that satisfy the hypotheses will do. Different D and G will result in different locations for the error system poles. (One choice that satisfies the hypotheses is $D = I, G = I$.)

Example 2.1 Riccati Equation Design of Scalar Deterministic Observer

We can gain some insights on observer design using Equation 2.6 by considering the scalar case. Let the given plant be

$$x_{k+1} = ax_k + bu_k \tag{2.8}$$

with measurement

$$z_k = hx_k \tag{2.9}$$

where u_k, x_k, and z_k are scalars. If we let $q = g^2$ and $r = d^2$ then Equation 2.6 becomes

$$p = a^2p - \frac{a^2h^2p^2}{h^2p + r} + q \tag{2.10}$$

or

$$p = \frac{a^2rp}{h^2p + r} + q \tag{2.11}$$

which can be written as

$$h^2p^2 + [(1-a^2)r - h^2q]p - qr = 0 \tag{2.12}$$

This equation has two solutions given by

$$p = \frac{q}{2\lambda}\left[\pm\sqrt{(1-\lambda)^2 + \frac{4\lambda}{(1-a^2)}} - (1-\lambda)\right] \tag{2.13}$$

where

$$\lambda = \frac{h^2q}{(1-a^2)r} \tag{2.14}$$

(Compare λ to Equation 1.143 in Example 1.14.)

We must now consider two cases.

a. Original system stable

If $|a| < 1$, then $(1-a^2) > 0$ and $\lambda > 0$. In this case the positive (definite) solution to Equation 2.12 is

$$p_+ = \frac{q}{2\lambda}\left[\sqrt{(1-\lambda)^2 + \frac{4\lambda}{(1-a^2)}} - (1-\lambda)\right] \tag{2.15}$$

and the observer gain (Equation 2.7) is selected as

$$l = \frac{ahp_+}{h^2p_+ + r} \tag{2.16}$$

Then the error system matrix is

$$a^{\text{cl}} = a - lh = \frac{a}{1 + (h^2/r)p_+} \tag{2.17}$$

or

$$a^{\text{cl}} = \frac{a}{1 + \frac{1-a^2}{2}\left[\sqrt{(1-\lambda)^2 + \frac{4\lambda}{(1-a^2)}} - (1-\lambda)\right]} \tag{2.18}$$

(Superscript "cl" means "closed loop.")

Since the denominator is greater than or equal to 1, we have $|a^{\text{cl}}| \le |a| < 1$, so the observer error system is stable.

b. Original system unstable

If $|a| > 1$, then $(1-a^2) < 0$ and $\lambda < 0$. In this case the positive solution to Equation 2.12 is

$$p_+ = \frac{-q}{2\lambda}\left[\sqrt{(1-\lambda)^2 + \frac{4\lambda}{(1-a^2)}} + (1-\lambda)\right] \tag{2.19}$$

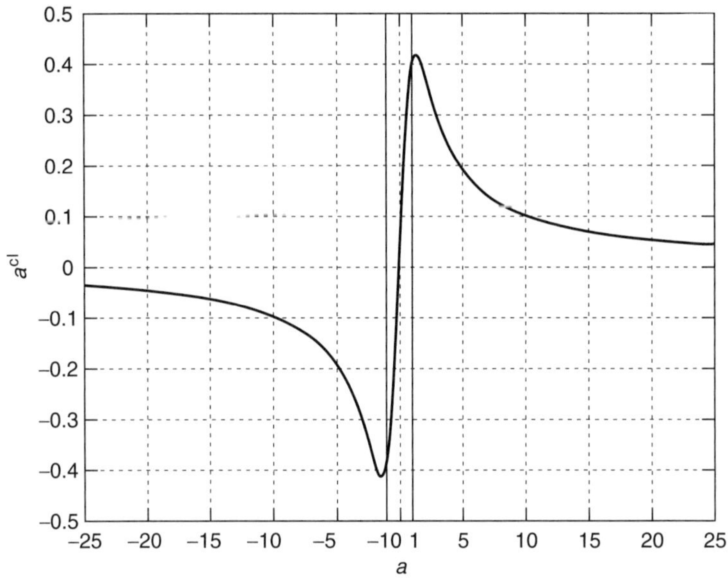

FIGURE 2.2
Closed-loop versus open-loop plant matrix.

Then the error system is given by

$$a^{\mathrm{cl}} = \frac{a}{1 - \frac{1-a^2}{2}\left[\sqrt{(1-\lambda)^2 + \frac{4\lambda}{(1-a^2)}} + (1-\lambda)\right]} \tag{2.20}$$

It is not quite so easy this time to show that $|a^{cl}| < 1$ (although it is if $\lambda \neq 0$). Note, however, that if $|a| \gg 1$ then $\lambda \approx 0$ and it is not difficult to show that

$$a^{\mathrm{cl}} \approx \frac{1}{a} \tag{2.21}$$

This is certainly stable if $|a| \gg 1$!

A plot of a^{cl} versus a for the case $h^2q/r = 1$ is shown in Figure 2.2. Note that a^{cl} is always stable.

Now consider the situation shown in Figure 2.3. *Process noise* w_k has been added to the plant and *measurement noise* v_k to the measurement. Under these more realistic circumstances, the observer designed deterministically as above cannot be expected to make $\tilde{x}_k \to 0$. It is somewhat surprising, however, that by using a structure virtually identical to Figure 2.1 the error can be forced to zero even in this noisy case. What is required in general is to find a time-varying observer gain L_k based on stochastic considerations such as those of the previous chapter. The resulting state estimator depends on a time-varying version of the Riccati Equation 2.6 and is known as the Kalman filter. $\square$

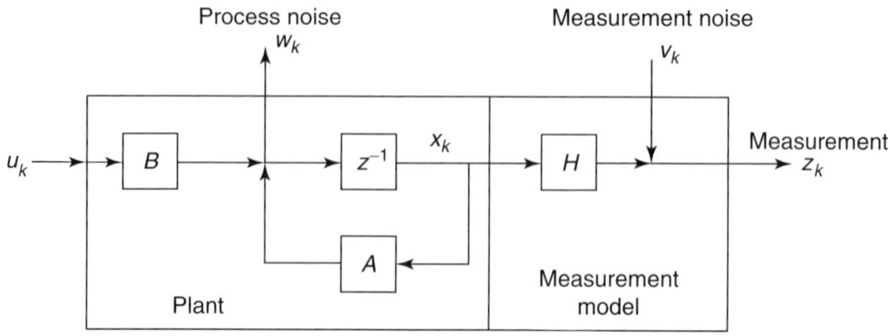

FIGURE 2.3
Plant and measurement with additive noise.

2.2 Linear Stochastic Systems

To describe a system with noise like the one shown in Figure 2.3, write the *discrete stochastic dynamical equation*

$$x_{k+1} = Ax_k + Bu_k + Gw_k \tag{2.22a}$$

$$z_k = Hx_k + v_k \tag{2.22b}$$

where $x_k \in R^n, u_k \in R^m, z_k \in R^p, w_k \in R^l, v_k \in R^p$. Let process noise w_k have mean $\bar{w}_k = 0$ and covariance $\overline{w_k w_k^T} = Q$, which is symbolized as $w_k \sim (0, Q)$. The overbar denotes expected value. Let *measurement noise* v_k be $(0, R)$. Both $\{w_k\}$ and $\{v_k\}$ are assumed to be stationary *white noise processes*, that is, their covariance functions satisfy

$$P_{w_{k_1} w_{k_2}} = \overline{w_{k_1} w_{k_2}^T} = Q\delta(k_1 - k_2) \tag{2.23}$$

where $\delta(k)$ is the Kronecker delta, and the corresponding version for v_k.

Suppose that initial state x_0 is unknown, but that there is available *a priori* knowledge that x_0 is $(\bar{x}_0, P_{x_0})$. We assume that x_0, w_k, and v_k are mutually uncorrelated, so that $\overline{w_k v_j^T} = 0$ for all j and k, and so on. Of course, P_{x_0}, Q, and R are symmetric and positive semidefinite. Signal u_k is a *deterministic* control input.

Under these circumstances, state x_k is a random variable with mean $\bar{x}_k$ and covariance P_{x_k}. Similarly, output z_k is random $(\bar{z}_k, P_{z_k})$. Notice that $\{x_k\}$ and $\{z_k\}$ are stochastic processes, in general nonstationary.

EXERCISE 2.1 Discrete Markov Process

A discrete first-order Markov process x_k can be represented as the state of the linear system (Equation 2.22a). Using this system, we can manufacture stationary or nonstationary Markov processes. Let x_k be a scalar for simplicity. Let $u_k = 0, g = 1$. (We use lowercase letters for scalar system parameters.)

a. Stationary Markov process

Let the initial condition at $k = -\infty$ equal zero and $w_k \sim (0, (1-a^2)\sigma_w^2)$. Show that x_k is a stationary Markov process with spectral density

$$\Phi_X(\omega) = \frac{(1-a^2)\sigma_w^2}{1 + a^2 - 2a\cos\omega} \tag{2.24}$$

and autocorrelation function

$$R_X(k) = \sigma_w^2 a^{|k|} \tag{2.25}$$

In this case the system must be stable.

b. Nonstationary Markov process

Let $w_k = 0$ and initial condition $x_0 \sim (0, \sigma_X^2)$. Show that x_k is a nonstationary Markov process with autocorrelation function

$$R_X(j,k) = \sigma_X^2 a^{(j+k)} \tag{2.26}$$

This result is valid for a stable or unstable system.

Note: If w_k is Gaussian, then so is x_k. In that case the statistics of x_k are completely described by the autocorrelation function.

2.2.1 Propagation of Means and Covariances

It is not difficult to find how the means and covariances of x_k and z_k propagate through time under the influence of the dynamics (Equation 2.22a).

For the mean of the state, simply write

$$\begin{aligned}\bar{x}_{k+1} &= \overline{Ax_k + Bu_k + Gw_k} \\ &= A\bar{x}_k + B\bar{u}_k + G\bar{w}_k\end{aligned}$$

or

$$\bar{x}_{k+1} = A\bar{x}_k + B\bar{u}_k \tag{2.27}$$

The initial condition of Equation 2.27 is $\bar{x}_0$, which is given. Therefore, the mean propagates exactly according to the deterministic dynamics (2.1a)!

To find how the state covariance propagates, write

$$\begin{aligned}P_{x_{k+1}} &= \overline{(x_{k+1} - \bar{x}_{k+1})(x_{k+1} - \bar{x}_{k+1})^T} \\ &= \overline{[A(x_k - \bar{x}_k) + Gw_k][A(x_k - \bar{x}_k) + Gw_k]^T} \\ &= A\overline{(x_k - \bar{x}_k)(x_k - \bar{x}_k)^T}A^T + G\overline{w_k(x_k - \bar{x}_k)^T}A^T + A\overline{(x_k - \bar{x}_k)w_k^T}G^T \\ &\quad + G\overline{w_k w_k^T}G^T\end{aligned}$$

or

$$P_{x_{k+1}} = AP_{x_k}A^T + GP_{w_k x_k}A^T + AP_{x_k w_k}G^T + GQG^T$$

Hence,

$$P_{x_{k+1}} = AP_{x_k}A^T + GQG^T \tag{2.28}$$

The last step follows from the uncorrelatedness of x_0 and w_k, and the whiteness of w_k through the following argument. State x_k depends on x_0 and the process noise w_j for $j \le k-1$. However w_k is uncorrelated with x_0 and all previous w_j, hence $P_{w_k x_k} = 0$. Equation 2.28 is a *Lyapunov equation* for P_{x_k}. The initial condition is P_{x_0}, which is given. (A Lyapunov equation is a linear matrix equation.)

Next, we want to find the mean and covariance of the output. Taking expectations in (Equation 2.22b) yields

$$\bar{z}_k = H\bar{x}_k \tag{2.29}$$

The cross covariance between state and output is

$$\begin{aligned} P_{x_k z_k} &= \overline{(x_k - \bar{x}_k)(z_k - \bar{z}_k)^T} \\ &= \overline{(x_k - \bar{x}_k)[H(x_k - \bar{x}_k) + v_k]^T} \end{aligned}$$

or

$$P_{x_k z_k} = P_{x_k}H^T \tag{2.30}$$

due to the uncorrelatedness of x_k and v_k (why are they uncorrelated?).

For the covariance of the output,

$$\begin{aligned} P_{z_k} &= \overline{(z_k - \bar{z}_k)(z_k - \bar{z}_k)^T} \\ &= \overline{[H(x_k - \bar{x}_k) + v_k][H(x_k - \bar{x}_k) + v_k]^T} \end{aligned}$$

so

$$P_{z_k} = HP_{x_k}H^T + R \tag{2.31}$$

where we again used the uncorrelatedness assumptions.

Now note, and this is the most important point of this section, that x_k and z_k are jointly distributed RV with moments given by

$$\begin{bmatrix} x_k \\ z_k \end{bmatrix} \sim \left(\begin{bmatrix} \bar{x}_k \\ \bar{z}_k \end{bmatrix}, \begin{bmatrix} P_{x_k} & P_{x_k z_k} \\ P_{z_k x_k} & P_{z_k} \end{bmatrix} \right) = \left(\begin{bmatrix} \bar{x}_k \\ H\bar{x}_k \end{bmatrix}, \begin{bmatrix} P_{x_k} & P_{x_k}H^T \\ HP_{x_k} & HP_{x_k}H^T + R \end{bmatrix} \right) \tag{2.32}$$

where $\bar{x}_k$ and P_{x_k} are determined recursively by Equations 2.27 and 2.28. If the RV of sequence z_k is measured and the related RV x_k is unknown, then we can apply the results of the previous chapter to find an estimate $\hat{x}_k$ for the state given the data z_k. (At this point, x_0, w_k, and v_k are assumed to have arbitrary statistics and only the first- and second-joint moments are given above, so the linear mean-square approach would be used.) In the next section, we pursue this to find the optimal linear mean-square estimate of x_k given the measurement sequence z_k.

The quantities $\bar{x}_k$ and P_{x_k} found using Equations 2.27 and 2.28 should actually be considered as the conditional mean and covariance given $\bar{x}_0, P_{x_0}$, and inputs $u_0, u_1, \ldots, u_{k-1}$.

All of these derivations have been for the case of general statistics. If it is assumed that x_0, w_k, and v_k are Gaussian, then since the system (Equation 2.22) is linear, x_k and z_k are also Gaussian for all k. Furthermore, x_k and z_k are jointly Gaussian, with moments given by Equation 2.32. In this case the first and second moments totally specify the joint PDF, so that complete information on the interdependence of state and output is available in Equation 2.32.

If all statistics are Gaussian, then x_k is described by the PDF

$$f_{x_k}(x,k) = \frac{1}{\sqrt{(2\pi)^n |P_{x_k}|}} \exp\left[-\frac{1}{2}(x-\bar{x}_k)^T P_{x_k}^{-1}(x-\bar{x}_k)\right] \tag{2.33}$$

Note that this is actually a conditional PDF, given $\bar{x}_0, P_{x_0}$, and inputs u_0, $u_1, \ldots, u_{k-1}$. Define the *hyperstate* as the conditional PDF of the state. Then in the Gaussian case, if no measurements are taken the hyperstate is completely described by the pair

$$(\bar{x}_k, P_{x_k}) \tag{2.34}$$

the conditional mean and covariance, and so Equations 2.27 and 2.28 provide an updating algorithm for the entire hyperstate in the absence of measurements.

Example 2.2 Propagation of Scalar Mean and Covariance

Let $x_0 \sim N(0,1)$ (i.e., normal) and x_k, z_k be given by Equation 2.22, where all quantities are scalars and $w_k \sim N(0,1), v_k \sim N(0, \frac{1}{2})$. Suppose that all of our assumptions hold. Let $u_k = u_{-1}(k)$ be the unit step. Take $b = 1, g = 1$ for simplicity.

Then the mean $\bar{x}_k$ is given by Equation 2.27, so

$$\bar{x}_{k+1} = a\bar{x}_k + u_k$$

which has the solution

$$\begin{aligned}
\bar{x}_k &= a^k \bar{x}_0 + \sum_{i=0}^{k-1} a^{k-1-i} \\
&= a^{k-1} \sum_{i=0}^{k-1} \left(\frac{1}{a}\right)^i \\
&= a^{k-1}\left[\frac{1-(1/a)^k}{1-1/a}\right] \\
\bar{x}_k &= \frac{a^k - 1}{a - 1}
\end{aligned} \tag{2.35}$$

This has a limiting value $\bar{x}_\infty$ if and only if $|a| < 1$, that is, if Equation 2.35 is stable. It is

$$\bar{x}_\infty = \frac{1}{1-a} \tag{2.36}$$

To find the covariance of the state use Equation 2.28,

$$P_{x_{k+1}} = a^2 P_{x_k} + 1, \qquad P_{x_0} = 1$$

Thus

$$\begin{aligned} P_{x_k} &= (a^2)^k + \sum_{i=0}^{k-1} (a^2)^{k-1-i} \\ &= a^{2k} + \frac{a^{2k}-1}{a^2-1} \\ &= \frac{a^{2(k+1)}-1}{a^2-1} \end{aligned} \tag{2.37}$$

There is a limiting solution P_{x_∞} if and only if Equation 2.35 is stable, and then

$$P_{x_\infty} = \frac{1}{1-a^2} \tag{2.38}$$

Since x_k, z_k are normal, the conditional PDF $f_{x_k}(x,k)$ is given by Equation 2.33 and we can sketch it as a function of k. This is done in Figure 2.4 for $a = 1/2$ and $a = 2$. It is evident that the effect of the process noise is to "flatten" the PDF for x_k as k increases, thus decreasing our confidence that x_k is near $\bar{x}_k$. There are two cases. If the system is stable, a limiting PDF $f_{x_k}(x)$ is reached and subsequent propagation does not inject further uncertainty. This limiting case is known as *statistical steady state*. In this case we can say that $\lim_{k\to\infty} x_k$ is near $\bar{x}_\infty$ to some tolerance described by P_{x_∞}. However, if Equation 2.35 is unstable, there is no limiting PDF and as k increases both the value of $\bar{x}_k$ and the probable spread of x_k about $\bar{x}_k$ go to infinity.

The analysis of the time-varying stochastic system

$$\begin{aligned} x_{k+1} &= A_k x_k + B_k u_k + G_k w_k \\ z_k &= H_k x_k + v_k \end{aligned}$$

where $x_0 \sim (\bar{x}_0, P_{x_0}), w_k \sim (0, Q_k), v_k \sim (0, R_k)$ is identical to the above. The results differ only in that subscripts k must be added to A, B, G, H, Q, and R. □

2.2.2 Statistical Steady-State and Spectral Densities

In this subsection, we are interested in examining the state and output of a discrete system when they are stationary random processes. We shall set the deterministic input u_k to zero.

The solution P_k to Equation 2.28 is in general time-varying. However, if A is asymptotically stable, then the Lyapunov equation has a steady-state

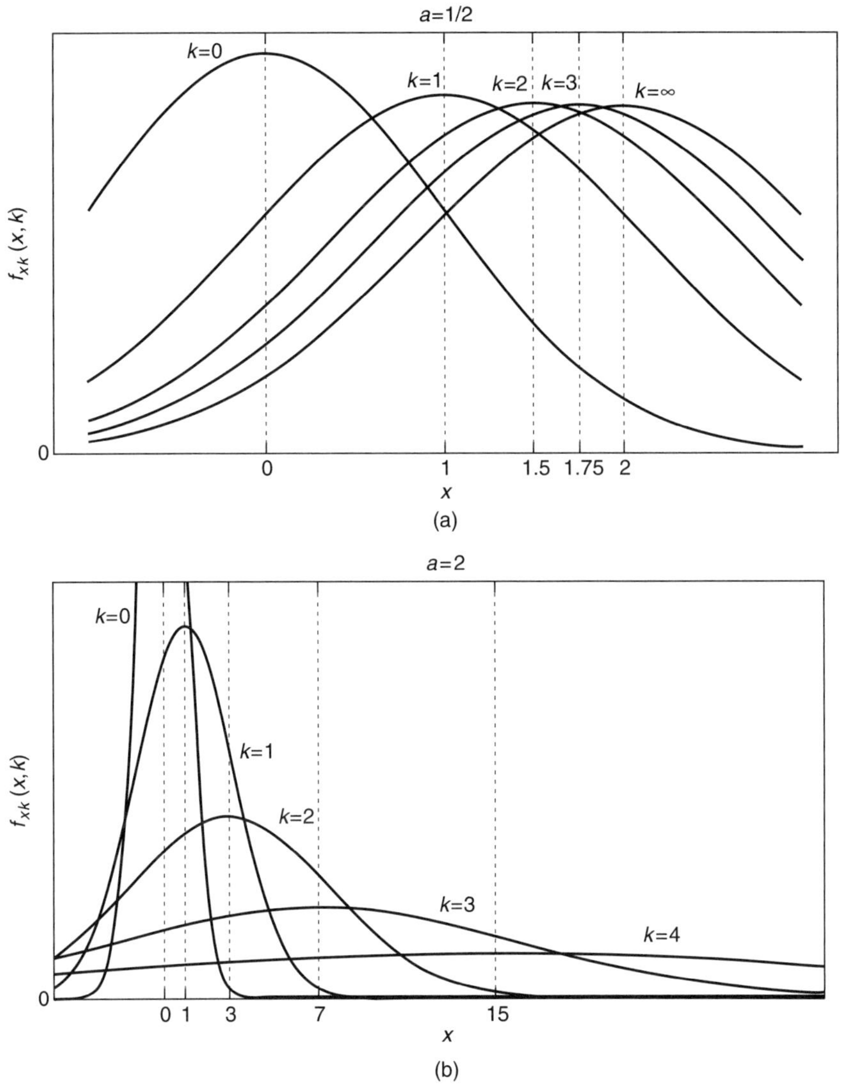

FIGURE 2.4
Propagation of the hyperstate. (a) Stable plant. (b) Unstable plant.

solution $P \triangleq P_{k+1} = P_k$ for large k. This steady-state covariance satisfies the *algebraic Lyapunov equation*

$$P = APA^T + GQG^T \tag{2.39}$$

Under these circumstances, as k increases the state x_k tends to a stationary Markov process with mean $\bar{x}_k = 0$ and covariance P. This limiting case is called *statistical steady state*. If A is unstable, then in general the covariance P_{x_k} increases without bound.

If the system has reached statistical steady state, then it is possible to discuss spectral densities. Using Equation 2.22a and the identities at the beginning of Section 1.5, the spectral density of the state x_k is given by

$$\Phi_X(z) = (zI - A)^{-1} GQG^T (z^{-1}I - A)^{-T} \tag{2.40}$$

where superscript $-T$ denotes the transpose of the inverse. (Note that $(zI - A)^{-1}G$ is the transfer function from w_k to x_k and $\Phi_w(z) = Q$.)

Likewise, the spectral density of the output z_k in statistical steady state is

$$\Phi_Z(z) = H(zI - A)^{-1} GQG^T (z^{-1}I - A)^{-T} H^T + R \tag{2.41}$$

(If we add two uncorrelated processes, their spectral densities add.) This can be written as

$$\Phi_Z(z) = \mathcal{H}(z)\mathcal{H}^T(z^{-1}) + R \tag{2.42}$$

where

$$\mathcal{H}(z) = H(zI - A)^{-1} G\sqrt{Q} \tag{2.43}$$

is the transfer function from w_k' to z_k, where $w_k = \sqrt{Q}w_k'$ and w_k' is white noise with unit spectral density. $\mathcal{H}(z)$ will be very useful to us later.

2.3 The Discrete-Time Kalman Filter

Consider the discrete-time stochastic dynamical system (Equation 2.22) with the attendant assumptions. All matrices describing the system, and $\bar{x}_0$ and all covariances are given, with $Q \geq 0, P_{x_0} \geq 0$. We shall assume that $R > 0$. The statistics are not necessarily Gaussian, but arbitrary.

According to our results from Section 1.1, in the absence of measurements the best mean-square estimate for the unknown state x_k is just the conditional mean $\bar{x}_k$ of the state given $\bar{x}_0$ and $u_0, u_1, \ldots, u_{k-1}$, which propagates according to the deterministic version of the system (Equation 2.27). Then the error covariance is just the (conditional) covariance of the state and it propagates according to the Lyapunov equation (Equation 2.28). To derive these results, we used only the state equation (Equation 2.22a).

However, according to Equations 1.51 and 1.53, and Example 1.3, given the linear measurement model (Equation 2.22b), the best linear estimator for the RV x_k, given measurement z_k and *a priori* statistics $\bar{x}_k, P_{x_k}$ on x_k is given by

$$P_{\tilde{x}_k} = \left(P_{x_k}^{-1} + H^T R^{-1} H\right)^{-1} \tag{2.44}$$

$$\hat{x}_k = \bar{x}_k + P_{\tilde{x}_k} H^T R^{-1}(z_k - H\bar{x}_k) \tag{2.45}$$

where the estimation error is $\tilde{x}_k = (x_k - \hat{x}_k)$. In the case of Gaussian statistics this is the conditional mean estimator, which is the *optimal* estimator.

In this section we must reconcile the *time update* (Equations 2.27 and 2.28), which describes the effects of the system dynamics (Equation 2.22a) with the *measurement update* (Equations 2.44 and 2.45) that describes the effects of measuring the output z_k according to Equation 2.22b. Since measurements z_k occur *at* each time k and the system dynamics describe the transition *between* times, we can do this as follows.

2.3.1 Kalman Filter Formulations

Define at each time k the *a priori* (before including measurement z_k) estimate $\hat{x}_k^-$ and error covariance P_k^-, and the *a posteriori* (after including measurement z_k) estimate $\hat{x}_k$ and error covariance P_k. The relation between P_k^- and P_k is illustrated in Figure 2.5. Then the optimal estimator for the state of Equation 2.22 with stated assumptions and given measurements z_k for all $k \geq 1$ is realized as follows, assuming that there is no measurement z_0 at time zero.

Discrete Kalman Filter Algorithm:

Set $\hat{x}_0 = \bar{x}_0, P_0 = P_{x_0}$. Given $\hat{x}_{k-1}$ and P_{k-1}, for each value of $k \geq 1$ apply time update (Equations 2.27 and 2.45) to obtain $\hat{x}_k^-, P_k^-$. Then apply measurement update (Equations 2.44 and 2.45) to obtain $\hat{x}_k, P_k$.

The resulting estimator is called the *discrete-time Kalman filter* and it is the best *linear* estimator if x_0, w_k, v_k have arbitrary statistics. If x_0, w_k, v_k are normal, then it is the optimal estimator. For convenient reference, the discrete-time Kalman filter is summarized in Table 2.1. The time-varying case, where all system matrices and noise covariances are functions of k, is presented for the sake of generality. Note that even for time-invariant systems and statistics, the Kalman filter itself is still time-varying since P_{k+1} depends on k!

The time update portion of the algorithm gives a prediction $\hat{x}_{k+1}^-$ of the state at time $k+1$, along with the associated error covariance P_{k+1}^-. The measurement update provides a *correction* based on the measurement z_{k+1} at

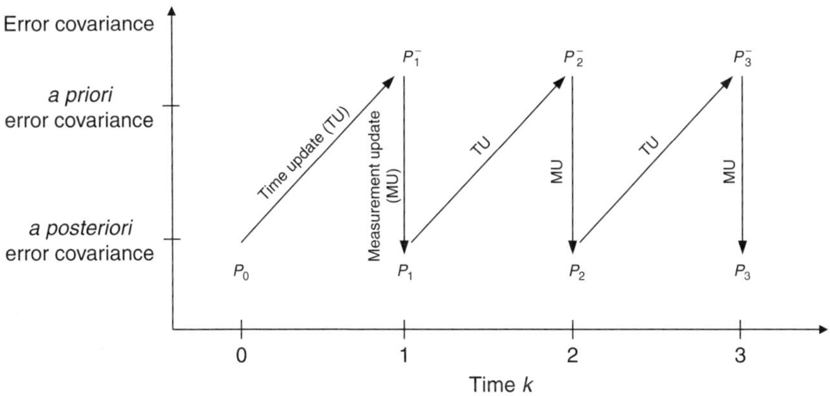

FIGURE 2.5
Error covariance update timing diagram.

TABLE 2.1
Discrete-Time Kalman Filter

System model and measurement model

$$x_{k+1} = A_k x_k + B_k u_k + G_k w_k \tag{2.46a}$$

$$z_k = H_k x_k + v_k \tag{2.46b}$$

$$x_0 \sim (\bar{x}_0, P_{x_0}), \qquad w_k \sim (0, Q_k), \qquad v_k \sim (0, R_k)$$

Assumptions

$\{w_k\}$ and $\{v_k\}$ are white noise processes uncorrelated with x_0 and with each other. $R_k > 0$.

Initialization

$$P_0 = P_{x_0}, \qquad \hat{x}_0 = \bar{x}_0$$

Time update (effect of system dynamics)

Error covariance $$P_{k+1}^- = A_k P_k A_k^T + G_k Q_k G_k^T \tag{2.47}$$

Estimate $$\hat{x}_{k+1}^- = A_k \hat{x}_k + B_k u_k \tag{2.48}$$

Measurement update (effect of measurement z_k)

Error covariance $$P_{k+1} = \left[(P_{k+1}^-)^{-1} + H_{k+1}^T R_{k+1}^{-1} H_{k+1}\right]^{-1} \tag{2.49}$$

Estimate $$\hat{x}_{k+1} = \hat{x}_{k+1}^- + P_{k+1} H_{k+1}^T R_{k+1}^{-1}(z_{k+1} - H_{k+1}\hat{x}_{k+1}^-) \tag{2.50}$$

---

time $k+1$ to yield the net *a posteriori* estimate $\hat{x}_{k+1}$ and its error covariance P_{k+1}. For this reason the equations in Table 2.1 are called the *predictor–corrector* formulation of the discrete Kalman filter.

The correction term (Equation 2.50) depends on the residual

$$\tilde{z}_k = z_k - H_k \hat{x}_k^- \tag{2.51}$$

The residual weighting coefficient is called the *Kalman gain*

$$K_k = P_k H_k^T R_k^{-1} \tag{2.52}$$

It is often useful to have an expression for K_k in terms of the *a priori* error covariance P_k^-. To derive this expression, write

$$\begin{aligned}
K_k &= P_k H_k^T R_k^{-1} \\
&= P_k H_k^T R_k^{-1}\left(R_k + H_k P_k^- H_k^T\right)\left(R_k + H_k P_k^- H_k^T\right)^{-1} \\
&= P_k\left(H_k^T + H_k^T R_k^{-1} H_k P_k^- H_k^T\right)\left(R_k + H_k P_k^- H_k^T\right)^{-1} \\
&= P_k\left[\left(P_k^-\right)^{-1} + H_k^T R_k^{-1} H_k\right] P_k^- H_k^T \left(R_k + H_k P_k^- H_k^T\right)^{-1} \\
&= P_k P_k^{-1} P_k^- H_k^T \left(R_k + H_k P_k^- H_k^T\right)^{-1}
\end{aligned}$$

so that

$$K_k = P_k^- H_k^T \left(H_k P_k^- H_k^T + R_k\right)^{-1} \tag{2.53}$$

There are many equivalent formulations for Equations 2.47 and 2.49. Using the matrix inversion lemma, the latter can be written as

$$P_{k+1} = P_{k+1}^- - P_{k+1}^- H_{k+1}^T \left(H_{k+1} P_{k+1}^- H_{k+1}^T + R_{k+1} \right)^{-1} H_{k+1} P_{k+1}^- \qquad (2.54)$$

The advantage of Equation 2.54 over Equation 2.49 is that inversion of only one $p \times p$ matrix is required. Equation 2.49 requires the inversion of two $n \times n$ matrices and the number of measurements p is usually less than n. If $|P_{k+1}^-| = 0$ then Equation 2.54 must be used in any event.

It is often convenient to replace the measurement update equations 2.49 and 2.50 by the alternative equations in Table 2.2.

TABLE 2.2
Alternative Measurement Update Equations

$K_{k+1} = P_{k+1}^- H_{k+1}^T \left(H_{k+1} P_{k+1}^- H_{k+1}^T + R_{k+1} \right)^{-1}$	(2.55)
$P_{k+1} = (I - K_{k+1} H_{k+1}) P_{k+1}^-$	(2.56)
$\hat{x}_{k+1} = \hat{x}_{k+1}^- + K_{k+1}(z_{k+1} - H_{k+1} \hat{x}_{k+1}^-)$	(2.57)

Equation 2.56 is equivalent (See the problems) to the Joseph Stabilized version.

$$P_{k+1} = (I - K_{k+1} H_{k+1}) P_{k+1}^- (I - K_{k+1} H_{k+1})^T + K_{k+1} R_{k+1} K_{k+1}^T \qquad (2.58)$$

This symmetric version guarantees the positive semidefiniteness of P_{k+1} in the presence of roundoff error and is often used for actual software implementation of the filter.

Note that the Kalman gain is the weighting which determines the influence of the residual in updating the estimate. Thus it plays the same role as the $1/(k+1)$ in the recursive sample mean equation in Example 1.10.

It is worth noting that although we demanded $|R_k| \neq 0$, this is not required since we only need $|H_k P_k^- H_k^T + R_k| \neq 0$.

An important point is that error covariance update equations 2.47 and 2.49 do not depend on the measurements. Thus, if A_k, G_k, H_k, Q_k, and R_k are known beforehand for all k, these equations can be used to find P_k off-line *before* the experiment is performed.

The time and measurement updates can be combined. The recursion for the *a posteriori* error covariance P_k involves a somewhat lengthy equation, so it is customary to use the recursion for the *a priori* error covariance P_k^-. To derive this, substitute Equation 2.54 into Equation 2.47. If we also use Equation 2.50 in Equation 2.48, there results the alternative filter formulation in Table 2.3, which manufactures the *a priori* estimate and error covariance. We have suppressed the time index on the plant matrices, as we shall usually do henceforth for convenience.

TABLE 2.3
Kalman Filter *A Priori* Recursive Formulation

Kalman Filter
$$\hat{x}_{k+1}^- = A(I - K_k H)\hat{x}_k^- + Bu_k + AK_k z_k \quad (2.59)$$
$$K_k = P_k^- H^T\left(HP_k^- H^T + R\right)^{-1} \quad (2.60)$$
Error covariance
$$P_{k+1}^- = A\left[P_k^- - P_k^- H^T\left(HP_k^- H^T + R\right)^{-1} HP_k^-\right]A^T + GQG^T \quad (2.61)$$

The *a priori* filter formulation is identical in structure to the deterministic observer in Section 2.1, and Figure 2.1 provides a block diagram if L is replaced by the time-varying gain AK_k. Even if the plant (Equation 2.46) is time-invariant, the Kalman filter is time-varying since the gain K_k depends on time. Some authors define AK_k as the Kalman gain. Equation 2.61 is a *Riccati equation*, which is a matrix quadratic equation. As we promised at the end of Section 2.1, it is just a time-varying version of Equation 2.6, which can be used to design the deterministic state observer!

In terms of the *a posteriori* estimate, we can write the recursion

$$\hat{x}_{k+1} = (I - K_{k+1}H)A\hat{x}_k + (I - K_{k+1}H)Bu_k + K_{k+1}z_{k+1} \quad (2.62)$$

It is worth remarking that if the unknown is constant, then (Equation 2.46a) becomes

$$x_{k+1} = x_k \quad (2.63)$$

so the time update is simply $\hat{x}_{k+1}^- = \hat{x}_k$, $P_{k+1}^- = P_k$. Then the error covariance measurement update in Table 2.1 is

$$P_{k+1}^{-1} = P_k^{-1} + H_{k+1}^T R_{k+1}^{-1} H_{k+1} \quad (2.64)$$

which has the simple solution for the information matrix

$$P_k^{-1} = P_0^{-1} + \sum_{i=1}^{k} H_i^T R_i^{-1} H_i \quad (2.65)$$

The Kalman filter is

$$\hat{x}_{k+1} = \hat{x}_k + K_{k+1}(z_{k+1} - H_{k+1}\hat{x}_k) \quad (2.66)$$

To get a maximum-likelihood estimate, we would set $P_0^{-1} = 0$, $\hat{x}_0 = 0$ to model a complete lack of *a priori* information. Use the matrix inversion lemma to compare Equations 2.64 and 2.66 to Equations 1.106 and 1.108.

The software for implementing the Kalman filter efficiently may be found in Bierman (1977). The number of computations required per iteration by the filter is on the order of n^3. As we shall see in subsequent examples, it is often possible in a particular application to do some *preliminary analysis* to simplify the equations that must actually be solved. Some short computer programs showing implementations of presimplified filters are included in the examples.

To gain some insight into the Kalman filter, let us consider a few examples.

Example 2.3 Ship Navigational Fixes

Suppose a ship is moving east at 10 mph. This velocity is assumed to be constant except for the effects of wind gusts and wave action. An estimate of the easterly position d and velocity $s = \dot{d}$ is required every hour. The navigator guesses at time zero that $d_0 = 0, s_0 = 10$, and his "guesses" can be modeled as having independent Gaussian distributions with variances $\sigma^2_{d_0} = 2$, $\sigma^2_{s_0} = 3$. Hence the initial estimates are $\hat{d}_0 = 0, \hat{s}_0 = 10$. The north-south position is of no interest. If d_k, s_k indicate position and velocity at hour k, we are required to find estimates $\hat{d}_k, \hat{s}_k$.

The first task is to model the system dynamics. During hour k the ship moves with velocity s_k mph, so its position will change according to

$$d_{k+1} = d_k + s_k \tag{2.67}$$

The quantity s_{k+1} should ideally be equal to s_k since velocity is constant at 10 mph; however, to model the unknown effects of wind and waves add a noise term w_k so that

$$s_{k+1} = s_k + w_k \tag{2.68}$$

To apply the Kalman filter, suppose w_k is white and independent of x_0 for all k. If, for example, there is a steady easterly current then w_k is not white. (We shall see later how to handle nonwhite noise. Basically, in this case a model of the noise process itself must be included in the system dynamics.)

Define the system state at time k as

$$x_k \triangleq \begin{bmatrix} d_k \\ s_k \end{bmatrix}$$

and write Equations 2.67 and 2.68 as one state equation

$$x_{k+1} = \begin{bmatrix} 1 & 1 \\ 0 & 1 \end{bmatrix} x_k + \begin{bmatrix} 0 \\ 1 \end{bmatrix} w_k \tag{2.69}$$

There is no deterministic input u_k. Suppose that previous observations of ships in the region have shown that w_k has on the average no effect, but that its effect on velocity can be statistically described by a Gaussian PDF with covariance 1. Then $w_k \sim N(0, Q) = N(0, 1)$. From the above, the initial condition is

$$x_0 \sim N(\bar{x}_0, P_0) = N\left(\begin{bmatrix} 0 \\ 10 \end{bmatrix}, \begin{bmatrix} 2 & 0 \\ 0 & 3 \end{bmatrix} \right)$$

This completes our modeling of the dynamics (Equation 2.46a).

Now, suppose that navigational fixes are taken at times $k = 1, 2, 3$. These fixes determine position to within an error covariance of 2, but yield directly no information about velocity. The fixes yield easterly positions of 9, 19.5, and 29 miles, respectively.

Before we can incorporate this new information into the estimates for d_k and s_k we need to model the measurement process. The situation described above corresponds to the output equation.

$$z_k = [1 \quad 0]x_k + v_k, \qquad v_k \sim N(0, 2) \tag{2.70}$$

where only the position is observed and it has been assumed that the fixes are corrupted by additive zero mean Gaussian noise with covariance of 2. Equation 2.70 corresponds to Equation 2.46b. We are now ready to estimate the state using the Kalman filter.

To incorporate the measurements $z_1 = 9, z_2 = 19.5, z_3 = 29$ into the estimate we use Table 2.1 to proceed as follows:

$k = 0$

Initial estimate:

$$\hat{x}_0 = \bar{x}_0 = \begin{bmatrix} 0 \\ 10 \end{bmatrix}$$

$$P_0 = \begin{bmatrix} 2 & 0 \\ 0 & 3 \end{bmatrix}$$

$k = 1$

Time update: propagate estimate to $k = 1$ using system dynamics

$$\hat{x}_1^- = A\hat{x}_0 + Bu_0 = \begin{bmatrix} 10 \\ 10 \end{bmatrix}, \qquad \textit{a prior} \text{ estimate at } k = 1$$

$$P_1^- = AP_0A^T + GQG^T = \begin{bmatrix} 5 & 3 \\ 3 & 4 \end{bmatrix}$$

Measurement update: include the effects of z_1

$$P_1 = \left[(P_1^-)^{-1} + H^TR^{-1}H\right]^{-1} = \begin{bmatrix} 1.429 & 0.857 \\ 0.857 & 2.714 \end{bmatrix}$$

$$\begin{aligned} \hat{x}_1 &= \hat{x}_1^- + P_1H^TR^{-1}(z_1 - H\hat{x}_1^-) \\ &= \begin{bmatrix} 9.286 \\ 9.571 \end{bmatrix}, \qquad \textit{a posteriori} \text{ estimate at } k = 1 \end{aligned}$$

$k = 2$

Time update: effects of system dynamics

$$\hat{x}_2^- = A\hat{x}_1 + Bu_1 = \begin{bmatrix} 18.857 \\ 9.571 \end{bmatrix}, \qquad \textit{a priori} \text{ estimate at time } k = 2$$

$$P_2^- = AP_1A^T + GQG^T = \begin{bmatrix} 5.857 & 3.571 \\ 3.571 & 3.714 \end{bmatrix}$$

Measurement update: effects of z_2

$$P_2 = \left[(P_2^-)^{-1} + H^T R^{-1} H\right]^{-1} = \begin{bmatrix} 1.491 & 0.909 \\ 0.909 & 2.091 \end{bmatrix}$$

$$\hat{x}_2 = \hat{x}_2^- + P_2 H^T R^{-1}(z_2 - H\hat{x}_2^-)$$

$$= \begin{bmatrix} 19.336 \\ 9.864 \end{bmatrix}, \qquad a\ posteriori \text{ estimate at } k = 2$$

$k = 3$

Time update: effects of system dynamics

$$\hat{x}_3^- = A\hat{x}_2 + Bu_2 = \begin{bmatrix} 29.2 \\ 9.864 \end{bmatrix}, \qquad a\ priori \text{ estimate}$$

$$P_3^- = AP_2A^T + GQG^T = \begin{bmatrix} 5.4 & 3 \\ 3 & 3.091 \end{bmatrix}$$

Measurement update: effects of z_3

$$P_3 = \left[(P_3^-)^{-1} + H^T R^{-1} H\right]^{-1} = \begin{bmatrix} 1.46 & 0.811 \\ 0.811 & 1.875 \end{bmatrix}$$

$$\hat{x}_3 = \hat{x}_3^- + P_3 H^T R^{-1}(z_3 - H\hat{x}_3^-)$$

$$= \begin{bmatrix} 29.054 \\ 9.783 \end{bmatrix}, \qquad a\ posteriori \text{ estimate}$$

□

Note that there are in fact two covariance sequences, the *a priori* sequence P_1^-, P_2^-, P_3^- and the *a posteriori* sequence P_0, P_1, P_2, P_3 as shown in Figure 2.5. As indicated in the figure, a time update generally increases the covariance due to the injection of uncertainty by the process noise w_k. A measurement update generally decreases the error covariance due to the injection of information by the measurement.

In this example, the error covariance sequences seem to be approaching a limit even after three iterations. In fact, limiting values to three decimal for P_k^- and P_k are reached after 10 iterations. They are

$$P_{10}^- = \begin{bmatrix} 4.783 & 2.604 \\ 2.604 & 2.836 \end{bmatrix}$$

$$P_{10} = \begin{bmatrix} 1.410 & 0.768 \\ 0.768 & 1.836 \end{bmatrix}$$

To give an indication of how the covariance develops, the (2, 2) components of P_k^- and P_k, that is, the *a priori* and *a posteriori* velocity error variances $\overline{(s_k - \hat{s}_k)(s_k - \hat{s}_k)^T}$, are plotted in Figure 2.6 using code written in MATLAB. It is thus apparent that even for an unstable system, taking measurements can make the error covariance converge to a finite value.

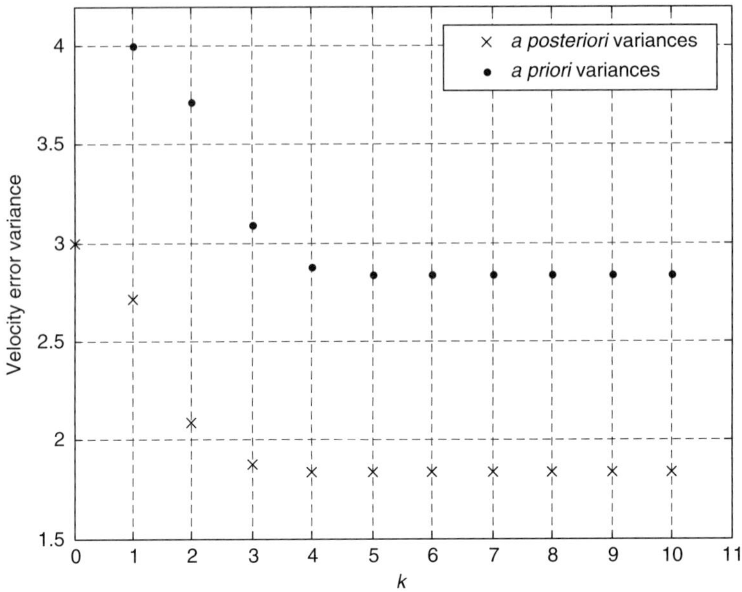


```
% Example 2.3
% Program to plot the apriori and aposteriori error covariance
x=[0;10]; p=[2 0 ; 0 3]; % Initialization
A=[1 1 ; 0 1]; G=[0;1]; Q=1; R=2; H=[1 0];
p22(1,1)= p(2,2);
for i=1:10
    pm= A*p*A'+ G*Q*G';
    pm22(i+1,:)= pm(2,2); % apriori error covariance p(2,2) for velocity s
    p= inv(inv(pm)+H'* inv(R)*H);
    p22(i+1,:)= p(2,2); % aposteriori error covariance p(2,2) for velocity s
end
plot([0:10],p22,'k.',[1:10],pm22([2:11]),'k*'); % Plotting
grid on; legend('aposteriori error covariance','apriori error covariance');
xlabel('k'); ylabel('Velocity error variances');
```

FIGURE 2.6
MATLAB code and plots of velocity error variances.

Example 2.4 Recursive Estimation of a Constant Scalar Unknown

To estimate the value of a constant scalar unknown x from measurements corrupted by additive white noise, the system and measurement models are

$$\begin{aligned} x_{k+1} &= x_k \\ z_k &= x_k + v_k \end{aligned} \tag{2.71}$$

where $x_k, z_k \in R$ and $v_k \sim (0, r)$ is white noise. Let $x_0 \sim (\bar{x}_0, p_0)$.

Before any measurements are taken, determine the error covariance off-line using Equations 2.47 and 2.49 (i.e., Equation 2.64). Thus (Gelb 1974),

$k = 0$

$$p_0 \quad \text{is given}$$

$k = 1$

Time update:

$$p_1^- = p_0$$

Measurement update:

$$p_1 = \left(\frac{1}{p_0} + \frac{1}{r}\right)^{-1} = \frac{p_0}{1 + p_0/r}$$

$k = 2$

Time update:

$$p_2^- = p_1 = \frac{p_0}{1 + p_0/r}$$

Measurement update:

$$\begin{aligned} p_2 &= \left(\frac{1}{p_1} + \frac{1}{r}\right)^{-1} \\ &= \frac{p_1}{1 + p_1/r} = \frac{p_0}{1 + 2p_0/r} \end{aligned}$$

$\vdots$

k

Time update:

$$p_k^- = \frac{p_0}{1 + (k-1)p_0/r}$$

Measurement update:

$$p_k = \frac{p_0}{1 + k(p_0/r)} \tag{2.72}$$

The error covariance is shown in Figure 2.7.

Now, to find the estimate $\hat{x}_k$ incorporating the *a priori* information on x_0 and all measurements through time k, use Equations 2.48 and 2.50 (or Equation 2.66), so that $\hat{x}_0 = \bar{x}_0$ and

$$\hat{x}_{k+1}^- = \hat{x}_k$$

$$\hat{x}_{k+1} = \hat{x}_k + \frac{p_{k+1}}{r}(z_{k+1} - \hat{x}_k) \tag{2.73}$$

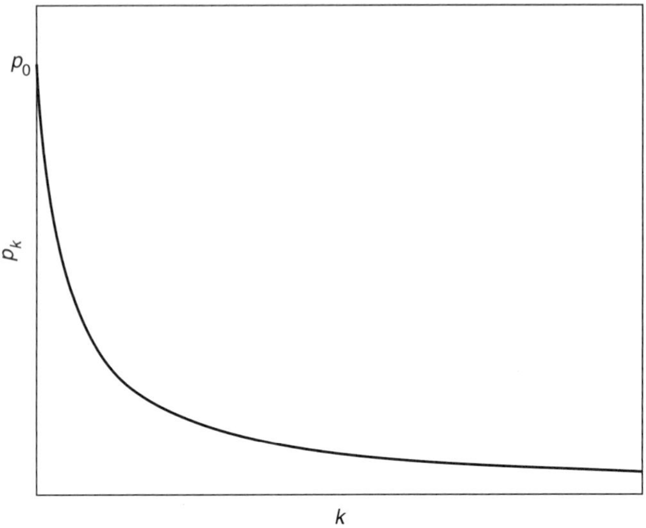

FIGURE 2.7
Error covariance for estimation of a constant.

Equations 2.72 and 2.73 constitute the Kalman filter for this example. Note that even in this case of time-invariant system and statistics, the filter is still varying since the Kalman gain (Equation 2.52) given by

$$K_k = \frac{p_k}{r} = \frac{p_0/r}{1 + k(p_0/r)} \tag{2.74}$$

is dependent on k. In fact, the weighting $K_{k+1} = p_{k+1}/r$ of the residual in Equation 2.73 goes to zero with k. This means that as more measurements are taken and our confidence in the estimate grows, subsequent measurements are increasingly discounted.

It is not difficult to find an explicit expression for $\hat{x}_k$. Putting Equation 2.74 into Equation 2.73 yields

$$\hat{x}_{k+1} = \hat{x}_k + \frac{p_0/r}{1 + (k+1)p_0/r}(z_{k+1} - \hat{x}_k) \tag{2.75}$$

Therefore

$$\begin{aligned}
\hat{x}_1 &= \hat{x}_0 + \frac{p_0/r}{1 + p_0/r}(z_1 - \hat{x}_0) = \frac{\hat{x}_0 + (p_0/r)z_1}{1 + p_0/r} \\
\hat{x}_2 &= \hat{x}_1 + \frac{p_0/r}{1 + 2p_0/r}(z_2 - \hat{x}_1) = \frac{(1 + p_0/r)\hat{x}_1 + (p_0/r)z_2}{1 + 2p_0/r} \\
&= \frac{\hat{x}_0 + (p_0/r)(z_1 + z_2)}{1 + 2p_0/r}
\end{aligned}$$

and in general

$$\hat{x}_k = \frac{1}{1 + kp_0/r}\left[\hat{x}_0 + (p_0/r)\sum_{i=1}^{k} z_i\right] \tag{2.76}$$

As $k \to \infty, \hat{x}_k$ becomes

$$\hat{x}_k \to \frac{1}{k}\sum_{i=1}^{k} z_i \tag{2.77}$$

the average of the data points. This is the *law of large numbers* (cf. Bryson and Ho 1975).

Now, let us consider two special cases. First let $r \to \infty$ so that the measurements are completely unreliable. Then $K_k \to 0$ so that the estimate (Equation 2.73) becomes $\hat{x}_{k+1} = \hat{x}_k = \bar{x}_0$. The filter is thus telling us to disregard all measurements and retain our initial estimate $\hat{x}_0 = \bar{x}_0$. The error covariance p_k is then constant p_0 for all k.

Suppose that $\bar{x}_0 = 0$ and $p_0 \to \infty$, corresponding to the lack of any *a priori* information on the unknown; the case of deterministic x. The filter Equations 2.72 and 2.73 now provides the maximum-likelihood estimate (if v_k is normal). Then $K_k \to 1/k$, so that the estimate (Equation 2.73) becomes

$$\hat{x}_0 = 0$$

$$\hat{x}_{k+1} = \hat{x}_k + \frac{1}{k+1}(z_{k+1} - \hat{x}_k) \tag{2.78}$$

This is exactly the recursive sample mean equation from Example 1.10! Note that the same result is obtained if $r \to 0$ (the case of perfect measurements).

If the statistics are time-varying so that $v_k \sim (0, r_k)$, then Equations 2.72 and 2.73 are replaced by

$$p_{k+1} = \frac{p_k}{1 + p_k/r_k}, \qquad \text{given } p_0 \tag{2.79}$$

$$\hat{x}_{k+1} = \hat{x}_k + \frac{p_{k+1}}{r_k}(z_{k+1} - \hat{x}_k), \qquad \hat{x}_0 = \bar{x}_0 \tag{2.80}$$

To summarize the insight gained from this example, the Kalman filter embodies many different estimators. If x_0, w_k, v_k have arbitrary statistics, it is the best linear estimator. If all statistics are normal, they yield the conditional mean estimate. If all statistics are normal and we set $\bar{x}_0 = 0, P_0 \to \infty$ then the filter yields the maximum-likelihood estimate.

If $P_0 \to \infty$, then Equation 2.49 should be used instead of Equation 2.54 for the measurement update, at least initially, since this uses the information matrix $(P_k^-)^{-1}$ which is bounded (i.e., $P_0^{-1} = 0$). □

Example 2.5 Filtering for a Scalar System

Let there be prescribed the scalar time-invariant system and measurement described by

$$x_{k+1} = ax_k + bu_k + w_k \tag{2.81}$$

$$z_k = hx_k + v_k \tag{2.82}$$

with $x_0 \sim (\bar{x}_0, p_0), w_k \sim (0, q), v_k \sim (0, r)$. Let $\{w_k\}$ and $\{v_k\}$ be white noise processes uncorrelated with each other and with x_0.

Using the formulation in Table 2.3, the Riccati equation becomes

$$p_{k+1}^- = a^2 p_k^- - \frac{a^2 h^2 (p_k^-)^2}{h^2 p_k^- + r} + q \tag{2.83}$$

or

$$p_{k+1}^- = \frac{a^2 r p_k^-}{h^2 p_k^- + r} + q \tag{2.84}$$

The Kalman gain becomes

$$K_{k+1} = \frac{h p_{k+1}^-}{h^2 p_{k+1}^- + r} \tag{2.85}$$

so that the (*a priori*) closed-loop formulation is

$$\hat{x}_{k+1}^- = \frac{ar}{h^2 p_k^- + r} \hat{x}_k^- + b u_k + \frac{a h p_k^-}{h^2 p_k^- + r} z_k \tag{2.86}$$

Even in this simple case, the closed-form solution to Equation 2.84 is difficult to deal with. Let us consider three special cases for which it can be solved without too much trouble.

a. Perfect measurements

Let $r \to 0$ so that our measurements become exact. Then Equation 2.84 becomes

$$p_{k+1}^- = q \tag{2.87}$$

and Equation 2.86 yields for the *a priori* estimate,

$$\hat{x}_{k+1}^- = \frac{a}{h} z_k + b u_k \tag{2.88}$$

We can understand Equation 2.88 by noting that in this case of no measurement noise, the best (*a posteriori*) estimate at time k of the state given z_k is just

$$\hat{x}_k = \frac{1}{h} z_k \tag{2.89}$$

The predicted value at time $k+1$ is therefore

$$\hat{x}_{k+1}^- = a \hat{x}_k + b u_k$$

or Equation 2.88. Thus, in the case of perfect measurements the estimate depends only on the latest measurement and the *a priori* error covariance is just the effect of the process noise over one time step. The initial information $(\bar{x}_0, p_0)$ is not used at all after $k = 0$.

b. Completely unreliable measurements

Let $r \to \infty$ so that the measurements are totally unreliable. Then Equation 2.84 becomes

$$p_{k+1}^- = a^2 p_k^- + q \tag{2.90}$$

The solution to this equation is

$$\begin{aligned} p_k^- &= a^{2k} p_0 + q \sum_{i=0}^{k-1} (a^2)^{k-i-1} \\ &= a^{2k} p_0 + q \left(\frac{1 - a^{2k}}{1 - a^2} \right) \end{aligned} \tag{2.91}$$

or

$$p_k^- = a^{2k} p_0 \left(1 + \frac{q}{p_0} \sum_{i=1}^{k} a^{-2i} \right); \qquad k \geq 1 \tag{2.92}$$

which converges to a limiting value if and only if a is stable. Equation 2.86 becomes

$$\hat{x}_{k+1}^- = a\hat{x}_k + bu_k, \qquad \text{with } \hat{x}_0^- = \bar{x}_0 \tag{2.93}$$

In this case, the estimate depends completely on the system dynamics. In fact, reexamine Example 2.2 to see that these two situations are identical!

c. No process noise

Let $q \to 0$. In this situation it is easier to propagate the information matrix than the error covariance. Combine Equations 2.47 and 2.49 to obtain

$$(p_{k+1}^-)^{-1} = \frac{p_k^{-1}}{a^2} = \frac{1}{a^2}(p_k^-)^{-1} + \frac{h^2}{ra^2} \tag{2.94}$$

which has solution

$$(p_k^-)^{-1} = \frac{1}{p_0 a^{2k}} + \sum_{i=1}^{k-1} \frac{h^2}{ra^2} \left(\frac{1}{a^2} \right)^{k-i-1}$$

for $k \geq 2$, with $(p_1^-)^{-1} = 1/a^2 p_0$. This is equivalent to

$$(p_k^-)^{-1} = \frac{1}{p_0 a^{2k}} + \frac{h^2}{r} \frac{(a^2 - a^{2k})}{(1 - a^2)a^{2k}} \tag{2.95}$$

which can in turn be manipulated to yield

$$p_k^- = \frac{a^{2k} p_0}{1 + \frac{h^2}{r} p_0 \left(\frac{a^2 - a^{2k}}{1 - a^2} \right)} \tag{2.96}$$

or

$$p_k^- = \frac{a^{2k} p_0}{1 + \frac{h^2}{r} p_0 \sum_{i=1}^{k-1} a^{2i}} \tag{2.97}$$

for $k \geq 2$. If $a = h = 1$ this is equivalent to Equation 2.72.

Now, compare Equation 2.96 to Equation 2.92. Equation 2.92 converges to a steady-state solution as k increases if and only if $|a| < 1$. However, Equation 2.96 can converge for unstable a. In fact, it converges for any value of a as long as h^2/r is nonzero! □

EXERCISE 2.2 Kalman Filter Derivation Using Orthogonality Principle

Assume that the system with measurement 2.46 is prescribed and let all statistics be Gaussian.

Suppose that $z_1, \ldots, z_{k-1}$ have been measured and that the optimal estimate $\hat{x}_k^-$ and error covariance P_k^- have been computed. Now another measurement z_k is made.

a. Use the orthogonality principle from Chapter 1 to derive the measurement update equations for the estimate and error covariance.
b. Prove that the residual $\tilde{z}_k$ is orthogonal to all the previous data $z_0, \ldots, z_{k-1}$. It therefore contains the new information in measurement z_k, which could not have been anticipated from the previous data. For this reason $\{\tilde{z}_k\}$ is also called the *innovations* process.
c. Now derive the time update equations by finding the best estimate $\hat{x}_{k+1}^-$ for the linear function Equation 2.46a of x_k.
d. Prove that the innovations sequence is a white noise process.

Notes: See Luenberger (1969) for an excellent treatment using these vector space methods.

If the statistics are not Gaussian, then the recursive filter derived in this exercise is the best *linear* estimator. It is also possible to derive the Kalman filter from the Wiener–Hopf equation. We shall do this in the continuous case in Chapter 3.

2.4 Discrete Measurements of Continuous-Time Systems

With the increasing sophistication of microprocessors, more and more control schemes are being implemented digitally. In a digital implementation, the state of the plant is required at discrete time instants. It therefore behooves us to discuss state estimation for continuous-time systems using discrete, or sampled, data.

2.4.1 Discretization of Continuous Stochastic Systems

Suppose the plant to be controlled is described by

$$\dot{x}(t) = Ax(t) + Bu(t) + Gw(t) \tag{2.98a}$$

with measurements

$$z(t) = Hx(t) + v(t) \tag{2.98b}$$

Let $x(0) \sim (\bar{x}_0, P_0), w(t) \sim (0, Q), v(t) \sim (0, R)$, where $\{w(t)\}$ and $\{v(t)\}$ are white noise processes uncorrelated with each other and with $x(0)$.

To control the plant using state feedback, we would first need to estimate its state $x(t)$. Suppose the microprocessor samples input $u(t)$ and measurement $z(t)$ every T s. We can apply the discrete Kalman filter to the continuous plant by first discretizing it. To do this, we begin with the solution to Equation 2.98a

$$x(t) = e^{A(t-t_0)}x(t_0) + \int_{t_0}^{t} e^{A(t-\tau)}Bu(\tau)\,\mathrm{d}\tau + \int_{t_0}^{t} e^{A(t-\tau)}Gw(\tau)\,\mathrm{d}\tau \tag{2.99}$$

To describe the state propagation between samples, let $t_0 = kT, t = (k+1)T$ for an integer k. Defining the sampled state function as $x_k \triangleq x(kT)$, we can write

$$\begin{aligned} x_{k+1} = e^{AT}x_k &+ \int_{kT}^{(k+1)T} e^{A[(k+1)T-\tau]}Bu(\tau)\,\mathrm{d}\tau \\ &+ \int_{kT}^{(k+1)T} e^{A[(k+1)T-\tau]}Gw(\tau)\,\mathrm{d}\tau \end{aligned} \tag{2.100}$$

Assuming that control input $u(t)$ is reconstructed from the discrete control sequence u_k by using a zero-order hold, $u(\tau)$ has a constant value of $u(kT) = u_k$ over the integration interval.

The third term is a smoothed (i.e., low-pass filtered) version of the continuous white process noise $w(t)$ weighted by the state transition matrix and the noise input matrix G. It is not difficult to show that this term describes a discrete white noise sequence. Hence, define

$$w_k = \int_{kT}^{(k+1)T} e^{A[(k+1)T-\tau]}Gw(\tau)\,\mathrm{d}\tau \tag{2.101}$$

Now Equation 2.100 becomes

$$x_{k+1} = e^{AT}x_k + \int_{kT}^{(k+1)T} e^{A[(k+1)T-\tau]}B\,\mathrm{d}\tau \cdot u_k + w_k$$

On changing variables twice ($\lambda = \tau - kT$ and then $\tau = T - \lambda$), we obtain

$$x_{k+1} = e^{AT}x_k + \int_{0}^{T} e^{A\tau}B\,\mathrm{d}\tau \cdot u_k + w_k \tag{2.102}$$

This is the sampled version of Equation 2.98a, which we can write as

$$x_{k+1} = A^s x_k + B^s u_k + w_k \tag{2.103}$$

with

$$A^s = e^{AT} \tag{2.104}$$

$$B^s = \int_0^T e^{A\tau} B \, d\tau \tag{2.105}$$

To find the covariance Q^s of the new noise sequence w_k in terms of Q, write

$$Q^s = \overline{w_k w_k^T}$$

$$\iint_{kT}^{(k+1)T} e^{A[(k+1)T-\tau]} \overline{Gw(\tau)w(\sigma)^T} G^T e^{A^T[(k+1)T-\sigma]} \, d\tau \, d\sigma$$

But

$$\overline{w(\tau)w(\sigma)^T} = Q\delta(\tau - \sigma)$$

so

$$Q^s = \int_{kT}^{(k+1)T} e^{A[(k+1)T-\tau]} GQG^T e^{A^T[(k+1)T-\tau]} \, d\tau$$

By changing variables twice (as above),

$$Q^s = \int_0^T e^{A\tau} GQG^T e^{A^T \tau} \, d\tau \tag{2.106}$$

It is worth noting that even if Q is diagonal, Q^s need not be. Sampling can destroy independence among the components of the process noise!

Discretizing the measurement equation is easy since it has no dynamics:

$$z_k = Hx_k + v_k \tag{2.107}$$

To find the covariance R^s of v_k in terms of the given R, we need to think a little. Define the *unit rectangle* as

$$\prod(t) = \begin{cases} 1, & -1/2 \le t \le 1/2 \\ 0, & \text{otherwise} \end{cases} \tag{2.108}$$

and note that $\lim_{T \to 0}(1/T)\Pi(t/T) = \delta(t)$, the Kronecker delta. We can write the covariance of v_k in terms of its spectral density R^s as

$$R_v(k) = R^s \delta(k) \tag{2.109}$$

Since $\delta(k)$ has a value of one at $k = 0$, in this discrete case the covariance is equal to R^s, a finite matrix. The covariance of $v(t)$, however, is given by

$$R\delta(t) \tag{2.110}$$

where R is its spectral density. For continuous white noise, the covariance is unbounded since $\delta(t)$ is unbounded at $t = 0$.

To make Equation 2.109 approach Equation 2.110 in the limit, we must write

$$R\delta(t) = \lim_{T \to 0} (R^s T) \frac{1}{T} \prod \left(\frac{t}{T} \right) \tag{2.111}$$

so that

$$R^s = \frac{R}{T} \tag{2.112}$$

By examining Equation 2.111 it is evident that Equation 2.112 is the relation required between R and R^s so that noise processes $v(t)$ and v_k have the same spectral densities. Note that if R is diagonal, then so is R^s.

Sampling does not affect the *a priori* information $\bar{x}_0, P_0$.

It is worthwhile to write down the first few terms of the infinite series expansions for the sampled matrices. They are

$$A^s = I + AT + \frac{A^2T^2}{2!} + \cdots \tag{2.113a}$$

$$B^s = BT + \frac{ABT^2}{2!} + \cdots \tag{2.113b}$$

$$Q^s = GQG^T T + \frac{(AGQG^T + GQG^T A^T)T^2}{2!} + \cdots \tag{2.113c}$$

If terms of order T^2 are disregarded, then the result is *Euler's approximation* to the sampled system. In Euler's approximation note that process noise covariance Q^s results from multiplication by T and measurement noise covariance R^s results from division by T.

Several examples are given to illustrate the application of the discrete Kalman filter to continuous systems. First we discuss the discretization of a continuous system which occurs quite often.

Example 2.6 Sampling Motion Governed by Newton's Laws

Consider the continuous system described by

$$\dot{x} = \begin{bmatrix} 0 & 1 \\ 0 & 0 \end{bmatrix} x + \begin{bmatrix} 0 \\ 1/m \end{bmatrix} u + \begin{bmatrix} 0 \\ 1 \end{bmatrix} w \tag{2.114}$$

If $u(t) = mg$ is a constant, then by an easy integration the components of $x = [d \quad s]^T$ satisfy

$$s(t) = s(0) + gt \tag{2.115}$$

$$d(t) = d(0) + s(0)t + \frac{1}{2}gt^2 \tag{2.116}$$

For an input force $u(t) = F = mg$, therefore, Equation 2.114 is just a formulation of Newton's law $F = ma$, with d the position and s the velocity.

Let $x(0) \sim (\bar{x}_0, p_0)$, $w \sim (0, Q)$, and measurements of position $d(t)$ be made so that

$$z = \begin{bmatrix} 1 & 0 \end{bmatrix} x + v \tag{2.117}$$

Measurement noise v is $(0, R)$ and all uncorrelatedness assumptions hold.

If measurements are made at intervals of T units, the discretized system can quickly be written down by noting that, since A is nilpotent (i.e., $A^2 = 0$), the sequences for the sampled plant matrices are all finite, so

$$A^s = e^{AT} = 1 + AT = \begin{bmatrix} 1 & T \\ 0 & 1 \end{bmatrix} \tag{2.118}$$

and

$$B^s = BT + \frac{ABT^2}{2} = \frac{1}{m} \begin{bmatrix} T^2/2 \\ T \end{bmatrix} \tag{2.119}$$

Letting

$$Q = \begin{bmatrix} q_1 & q_2 \\ q_2 & q_4 \end{bmatrix}$$

we get

$$Q^s = QT + \begin{bmatrix} q_2 & q_4/2 \\ q_4/2 & 0 \end{bmatrix} T^2 + \begin{bmatrix} q_4/3 & 0 \\ 0 & 0 \end{bmatrix} T^3 \tag{2.120}$$

Finally,

$$R^s = \frac{R}{T} \tag{2.121}$$

The sampled version of Equations 2.114 and 2.117 is thus

$$x_{k+1} = \begin{bmatrix} 1 & T \\ 0 & 1 \end{bmatrix} x_k + \frac{1}{m} \begin{bmatrix} T^2/2 \\ T \end{bmatrix} u_k + \begin{bmatrix} 0 \\ 1 \end{bmatrix} w_k \tag{2.122}$$

$$z_k = \begin{bmatrix} 1 & 0 \end{bmatrix} x_k + v_k \tag{2.123}$$

with noise covariance Q^s and R^s given above. Compare Equation 2.122 to Equation 2.69 of Example 2.3, where we used Euler's approximation.

Note that even if Q is diagonal, Q^s may not be. ▯

Example 2.7 Estimation of a Constant

If x is a time-invariant scalar random variable measured in additive white noise, then the system and measurement models are

$$\dot{x} = 0 \tag{2.124}$$

$$z = x + v \tag{2.125}$$

Let $x \sim (\tilde{x}_0, p_0)$ and $v(t) \sim (0, r^c)$ be uncorrelated. Compare this continuous formulation to the discrete formulation in Example 2.4.

To estimate x using measurements provided at sampling interval T, discretize Equations 2.124 and 2.125 to get

$$x_{k+1} = x_k \tag{2.126}$$

$$z_k = x_k + v_k \tag{2.127}$$

with $v_k \sim (0, r^c/T)$. These are identical to Equations 2.71 in Example 2.4, so from the results of those examples the error covariance is given by

$$p_k = \frac{p_0}{1 + kT(p_0/r^c)} \tag{2.128}$$

The Kalman gain is

$$K_k = \frac{p_0 T/r^c}{1 + kT(p_0/r^c)} \tag{2.129}$$

□

Example 2.8 Radar Tracking with Discrete Measurements

Consider the two-dimensional radar tracking problem illustrated in Figure 2.8. For simplicity, let velocity V be in the x direction.

a. State and measurement models

To derive the state model, write

$$\begin{aligned} r^2 &= x^2 + y^2 \\ 2r\dot{r} &= 2x\dot{x} + 2y\dot{y} \\ \dot{r} &= \frac{x}{r}\dot{x} + \frac{y}{r}\dot{y} \end{aligned}$$

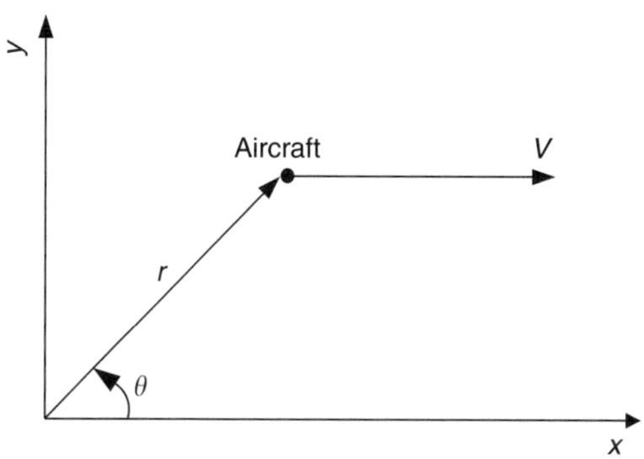

FIGURE 2.8
Radar tracking geometry.

but $x = r\cos\theta, \dot{x} = V$, and $\dot{y} = 0$, therefore

$$\dot{r} = V\cos\theta \tag{2.130}$$

Differentiating,

$$\ddot{r} = \dot{V}\cos\theta - V\dot{\theta}\sin\theta \tag{2.131}$$

For θ, write

$$\tan\theta = \frac{y}{x}$$

so

$$\dot{\theta}\sec^2\theta = \frac{x\dot{y} - \dot{x}y}{x^2}$$
$$\dot{\theta} = \frac{x\dot{y} - \dot{x}y}{x^2\sec^2\theta} = \frac{x\dot{y} - \dot{x}y}{r^2}$$

but $y = r\sin\theta, \dot{x} = V$, and $\dot{y} = 0$, therefore

$$\dot{\theta} = -\frac{V}{r}\sin\theta \tag{2.132}$$

Differentiating,

$$\ddot{\theta} = -\left(\frac{\dot{V}r - V\dot{r}}{r^2}\right)\sin\theta - \frac{V}{r}\dot{\theta}\cos\theta \tag{2.133}$$

State Equations 2.131 and 2.133 are nonlinear. We shall see in Chapter 5 how to apply the Kalman filter to nonlinear systems. In this example, we will apply the Kalman filter in Table 2.1 by taking advantage of its robustness and converting to a linear system as follows.

Assume that the aircraft is not accelerating, so that $\dot{V} = 0$. Then, if r and θ do not change much during each scan of the radar (about 1–10 s), we have $\ddot{r} \approx 0$ and $\ddot{\theta} \approx 0$. To account for the small changes in $\dot{r}$ and $\dot{\theta}$ during the scan time T, introduce random disturbances $w_1(t)$ and $w_2(t)$ so that

$$\frac{d}{dt}\dot{r} = w_1(t) \tag{2.134}$$
$$\frac{d}{dt}\dot{\theta} = w_2(t) \tag{2.135}$$

The robustness properties of the Kalman filter in the presence of process noise will now allow tracking of $\dot{r}$ and $\dot{\theta}$. Notice that process noise has been added to compensate for the *unmodeled dynamics* that we introduced by ignoring the nonlinear terms. We have more to say about this in Chapter 4. The state equations are now linear:

$$\frac{d}{dt}\begin{bmatrix} r \\ \dot{r} \\ \theta \\ \dot{\theta} \end{bmatrix} = \begin{bmatrix} 0 & 1 & 0 & 0 \\ 0 & 0 & 0 & 0 \\ 0 & 0 & 0 & 1 \\ 0 & 0 & 0 & 0 \end{bmatrix}\begin{bmatrix} r \\ \dot{r} \\ \theta \\ \dot{\theta} \end{bmatrix} + \begin{bmatrix} 0 \\ w_1(t) \\ 0 \\ w_2(t) \end{bmatrix} \tag{2.136}$$

Symbolize Equation 2.136 as

$$\dot{X}(t) = AX(t) + w(t) \tag{2.137}$$

(Do not confuse the state $X(t)$ with the x coordinate in Figure 2.8!) The process $w(t)$ is known as *maneuver noise.*

The radar measures range r and azimuth θ, so

$$z(t) = \begin{bmatrix} 1 & 0 & 0 & 0 \\ 0 & 0 & 1 & 0 \end{bmatrix} X(t) + v(t) \tag{2.138}$$

is the measurement model.

To find the maneuver noise covariance, suppose that the disturbance accelerations are independent and uniformly distributed between $\pm a$. Then (show this) their variances are $a^2/3$. Note that $w_2(t)$ is an angular acceleration, so $w_2(t)r$ is distributed according to the maneuver noise PDF and $w_2(t)$ has variance $a^2/3r^2$. However, $w_1(t)$ is a radial acceleration and its variance is $a^2/3$. Therefore, $w(t)$ has covariance

$$Q = \begin{bmatrix} 0 & 0 & 0 & 0 \\ 0 & a^2/3 & 0 & 0 \\ 0 & 0 & 0 & 0 \\ 0 & 0 & 0 & a^2/3r^2 \end{bmatrix} \tag{2.139}$$

Let the variance of the range and azimuth measurements be σ_r^2 and σ_θ^2 (for radar $\sigma_r^2 \ll \sigma_\theta^2$). Then the measurement noise has covariance

$$R = \begin{bmatrix} \sigma_r^2 & 0 \\ 0 & \sigma_\theta^2 \end{bmatrix} \tag{2.140}$$

b. Discretization

The measurements $z(t)$ are available at each scan. Let T be the scan time. Then, since $A^2 = 0$ it is easy to use the series representations for the sampled matrices to get

$$x_{k+1} = \begin{bmatrix} 1 & T & 0 & 0 \\ 0 & 1 & 0 & 0 \\ 0 & 0 & 1 & T \\ 0 & 0 & 0 & 1 \end{bmatrix} x_k + w_k \tag{2.141}$$

$$z_k = \begin{bmatrix} 1 & 0 & 0 & 0 \\ 0 & 0 & 1 & 0 \end{bmatrix} x_k + v_k \tag{2.142}$$

where

$$w_k \sim \left(0, \frac{a^2}{3} \begin{bmatrix} T^3/3 & T^2/2 & 0 & 0 \\ T^2/2 & T & 0 & 0 \\ 0 & 0 & T^3/3r^2 & T^2/2r^2 \\ 0 & 0 & T^2/2r^2 & T/r^2 \end{bmatrix} \right) \tag{2.143}$$

$$v_k \sim \left(0, \begin{bmatrix} \sigma_r^2/T & 0 \\ 0 & \sigma_\theta^2/T \end{bmatrix} \right) \tag{2.144}$$

Suppose the *a priori* information is $x_0 \sim (\bar{x}_0, p_0)$. To find $\bar{x}_0$ and P_0 we could take one initializing measurement $z_0 = (r_0 \quad \theta_0)^T$. Then

$$\bar{x}_0 = \begin{bmatrix} r_0 \\ 0 \\ \theta_0 \\ 0 \end{bmatrix}, \qquad P_0 = \frac{1}{T}\begin{bmatrix} \sigma_r^2 & 0 & 0 & 0 \\ 0 & \sigma_r^2 & 0 & 0 \\ 0 & 0 & \sigma_\theta^2 & 0 \\ 0 & 0 & 0 & \sigma_\theta^2 \end{bmatrix} \tag{2.145}$$

In fact it is often more convenient to select $\bar{x}_0 = 0, P_0 = I$, since (as we shall discover) the filter converges to the same stochastic steady state for any $\bar{x}_0$ and P_0, if it is properly designed.

c. Applying the discrete Kalman filter

Now all we need to do is enter these system and covariance matrices into our computer program, which implements the discrete Kalman filter. At each time kT we give the filter the radar readings $r(kT)$ and $\theta(kT)$, and it will give the best estimates of $r, \dot{r}, \theta$, and $\dot{\theta}$.

The next examples show how to perform some preliminary analysis to simplify the software implementation of the filter. □

Example 2.9 α–β Tracker for Radar

a. Parallel filtering

In the previous example we discussed the radar tracking problem with discrete measurements. By neglecting the nonlinear terms we obtained the discrete Equations 2.141 through 2.145 in that example. Note that these equations describe two completely decoupled systems, one for the range dynamics $\left[\begin{smallmatrix} r \\ \dot{r} \end{smallmatrix}\right]$ and one for the angle dynamics $\left[\begin{smallmatrix} \theta \\ \dot{\theta} \end{smallmatrix}\right]$. This means that we can considerably simplify things by running not one Kalman filter on a fourth-order system, but two parallel Kalman filters on two second-order systems. See Figure 2.9. The number of computations required per iteration by the Kalman filter is of the order of n^3, so one fourth-order filter requires about $4^3 = 64$ operations while two second-order filters require about $2 \times 2^3 = 16$ operations. Aside from this, using parallel processing a separate microprocessor can be assigned to each filter, resulting in a further saving of time.

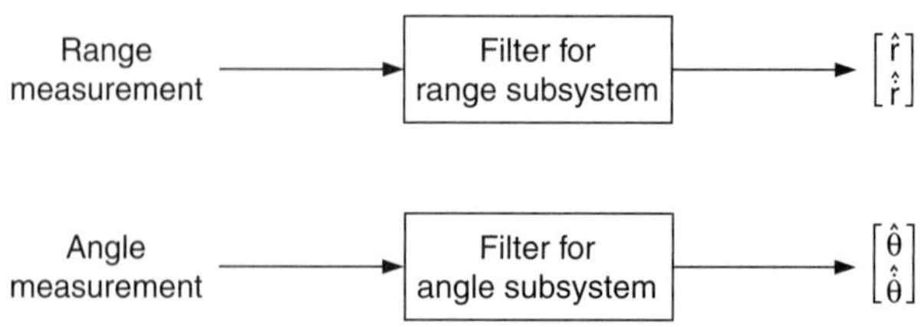

FIGURE 2.9
Parallel Kalman filtering.

b. α–β Tracker

An α–β tracker, or *second-order polynomial tracking filter*, is a Kalman filter to estimate the state of the system $\ddot{y}(t) = w$ with scalar measurements $z = y + v$ (Gelb 1974; Schooler 1975). This corresponds exactly to the equations for the range subsystem Equation 2.134 and angle subsystem Equation 2.135 in the previous example. The Kalman filters in Figure 2.9 are therefore both α–β trackers.

Let us consider the range subsystem in greater detail. We have the discretized system

$$x_{k+1} = \begin{bmatrix} 1 & T \\ 0 & 1 \end{bmatrix} x_k + w_k \tag{2.146}$$

$$z_k = \begin{bmatrix} 1 & 0 \end{bmatrix} x_k + v_k \tag{2.147}$$

with x_k now representing

$$\begin{bmatrix} r_k \\ \dot{r}_k \end{bmatrix}$$

The noises are described by

$$w_k \sim \left(0, \begin{bmatrix} \sigma_p^2 & \mu_{pv} \\ \mu_{pv} & \sigma_v^2 \end{bmatrix}\right) \tag{2.148}$$

$$v_k \sim \left(0, \quad \sigma_m^2\right) \tag{2.149}$$

The range (position) uncertainty is σ_p^2, the range velocity uncertainty is σ_v^2, and the measurement error variance is σ_m^2. We have shown a way to find quantities in the previous example. With this formulation, a Kalman filter program can now be run to provide filtered estimates of range and range rate. The Kalman filter for this problem has the form

$$\hat{x}_{k+1} = A\hat{x}_k + K_{k+1}(z_{k+1} - HA\hat{x}_k) \tag{2.150}$$

where the Kalman gain is traditionally represented in radar problems as

$$K_k = \begin{bmatrix} \alpha_k \\ \frac{\beta_k}{T} \end{bmatrix} \tag{2.151}$$

Hence the name "α–β tracker." We can write Equation 2.150 as

$$\hat{r}_{k+1} = \hat{r}_k + T\hat{\dot{r}}_k + \alpha_{k+1}(z_{k+1} - \hat{r}_k - T\hat{\dot{r}}_k) \tag{2.152a}$$

$$\hat{\dot{r}}_{k+1} = \hat{\dot{r}}_k + \frac{\beta_{k+1}}{T}(z_{k+1} - \hat{r}_k - T\hat{\dot{r}}_k) \tag{2.152b}$$

c. Zero process noise

Setting the process noise to zero in Equation 2.146 is not a good idea, since (as we shall see in Chapter 4) it can be very detrimental to the performance of the filter, especially if the target is maneuvering rapidly. We do it here to get a closed-form solution for K_k to obtain further insight.

Thus, suppose that the process noise covariance Q is zero and that there is no *a priori* state information so that $P_0 = \infty$. ($P_0 = \infty$ corresponds to maximum-likelihood estimation.) Then, since the plant matrix is nonsingular, we have for the information matrix

$$P_{k+1}^{-1} = A^{-T} P_k^{-1} A^{-1} + H^T R^{-1} H \tag{2.153}$$

To find a closed-form solution, let

$$P_k^{-1} = \begin{bmatrix} p_k^1 & p_k^2 \\ p_k^2 & p_k^4 \end{bmatrix} \tag{2.154}$$

Since

$$A^{-1} = \begin{bmatrix} 1 & -T \\ 0 & 1 \end{bmatrix} \tag{2.155}$$

we have

$$\begin{bmatrix} p_{k+1}^1 & p_{k+1}^2 \\ p_{k+1}^2 & p_{k+1}^4 \end{bmatrix} = \begin{bmatrix} 1 & 0 \\ -T & 1 \end{bmatrix} \begin{bmatrix} p_k^1 & p_k^2 \\ p_k^2 & p_k^4 \end{bmatrix} \begin{bmatrix} 1 & -T \\ 0 & 1 \end{bmatrix} + \begin{bmatrix} 1/\sigma_m^2 & 0 \\ 0 & 0 \end{bmatrix} \tag{2.156}$$

or

$$p_{k+1}^1 = p_k^1 + \frac{1}{\sigma_m^2} \tag{2.157}$$

$$p_{k+1}^2 = p_k^2 - T p_k^1 \tag{2.158}$$

$$p_{k+1}^4 = p_k^4 - 2T p_k^2 + T^2 p_k^1 \tag{2.159}$$

The solution to the first-order difference Equation 2.157 is ($P_0^{-1} = 0$)

$$p_k^1 = \sum_{i=0}^{k-1} \frac{1}{\sigma_m^2} = \frac{k}{\sigma_m^2} \tag{2.160}$$

Using this sequence as "input" to Equation 2.158, we can solve for p_k^2:

$$p_k^2 = -T \sum_{i=0}^{k-1} \left(\frac{i}{\sigma_m^2} \right) = \frac{-T}{2\sigma_m^2}(k-1)k \tag{2.161}$$

Finally, using Equations 2.160 and 2.161 in Equation 2.159,

$$\begin{aligned} p_{k+1}^4 &= p_k^4 + \frac{T^2}{\sigma_m^2}(k-1)k + \frac{T^2}{\sigma_m^2}k \\ &= p_k^4 + \frac{T^2}{\sigma_m^2}k^2 \end{aligned} \tag{2.162}$$

which has solution

$$p_k^4 = \frac{T^2}{\sigma_m^2} \sum_{i=0}^{k-1} i^2 = \frac{T^2}{\sigma_m^2} \frac{(k-1)k(2k-1)}{6} \tag{2.163}$$

Therefore

$$P_k^{-1} = \frac{k}{\sigma_m^2} \begin{bmatrix} 1 & (-T/2)(k-1) \\ (-T/2)(k-1) & (T^2/6)(k-1)(2k-1) \end{bmatrix} \tag{2.164}$$

Inverting this yields

$$P_k = \frac{2\sigma_m^2}{k(k+1)} \begin{bmatrix} 2k-1 & 3/T \\ 3/T & 6/[T^2(k-1)] \end{bmatrix} \tag{2.165}$$

The Kalman gain is given by

$$K_k = P_k H^T R^{-1} = \frac{2}{k(k+1)} \begin{bmatrix} 2k-1 \\ 3/T \end{bmatrix} \tag{2.166}$$

In the case of no process noise, the filter gains are thus given by the simple expressions

$$\alpha_k = \frac{2(2k-1)}{k(k+1)} \tag{2.167}$$

$$\beta_k = \frac{6}{k(k+1)} \tag{2.168}$$

Note that the steady-state gains with no process noise are zero, so that as time passes the measurements are increasingly discounted. This is because $Q = 0$ implies that the model $\ddot{r} = 0$ is exact, so that as time passes we rely to a greater extent on predictions based on this model. In a practical situation this would be undesirable, and Q should not be set to zero.

d. $\alpha - \beta - \gamma$ Tracker

If the aircraft is maneuvering rapidly so that $\ddot{r}$ and $\ddot{\theta}$ are nonzero, then we can obtain more robust performance by using an *α–β–γ tracker* in each channel in Figure 2.9. This is a Kalman filter to estimate the state of the system $y^{(3)}(t) = w(t)$ with measurements $z = y + v$. (The superscript means "third derivative".) □

Example 2.10 Filter Simplification by Preliminary Analysis

We can often considerably simplify the implementation of the Kalman filter by doing some *preliminary analysis.* Our object is basically to replace matrix equations by scalar equations. We demonstrate by implementing in MATLAB the α–β tracker in part b of the previous example. We only consider the computation of the error covariances and Kalman gain.

Suppose the *a posteriori* error covariance P_k is represented as

$$\begin{bmatrix} p_1 & p_2 \\ p_2 & p_4 \end{bmatrix} \tag{2.169}$$

For P_k^- we will add superscripts "–" to Equation 2.169. Let the Kalman gain components be represented as

$$\begin{bmatrix} K_1 \\ K_2 \end{bmatrix} \tag{2.170}$$

For the time update, time Equation 2.47 as

$$\begin{aligned} P_{k+1}^- &= AP_kA^T + Q \\ &= \begin{bmatrix} 1 & T \\ 0 & 1 \end{bmatrix} \begin{bmatrix} p_1 & p_2 \\ p_2 & p_4 \end{bmatrix} \begin{bmatrix} 1 & 0 \\ T & 1 \end{bmatrix} + \begin{bmatrix} \sigma_p^2 & 0 \\ 0 & \sigma_v^2 \end{bmatrix} \end{aligned} \tag{2.171}$$

which results in the scalar replacements

$$p_1^- = p_1 + 2Tp_2 + T^2p_4 + \sigma_p^2 \tag{2.172a}$$

$$p_2^- = p_2 + Tp_4 \tag{2.172b}$$

$$p_4^- = p_4 + \sigma_v^2 \tag{2.172c}$$

For the measurement update, use Table 2.2 to obtain

$$P_{k+1}^- H^T = \begin{bmatrix} p_1^- \\ p_2^- \end{bmatrix} \tag{2.173}$$

$$HP_{k+1}^- H^T + R = p_1^- + \sigma_m^2 \tag{2.174}$$

$$K_{k+1} = \frac{1}{p_1^- + \sigma_m^2} \begin{bmatrix} p_1^- \\ p_2^- \end{bmatrix} \tag{2.175}$$

or

$$K_1 = \frac{p_1^-}{p_1^- + \sigma_m^2} \tag{2.176a}$$

$$K_2 = \frac{p_2^-}{p_1^- + \sigma_m^2} \tag{2.176b}$$

and

$$\begin{aligned} P_{k+1} &= (I - K_{k+1}H)P_{k+1}^- \\ &= \begin{bmatrix} 1 - K_1 & 0 \\ -K_2 & 1 \end{bmatrix} \begin{bmatrix} p_1^- & p_2^- \\ p_2^- & p_4^- \end{bmatrix} \end{aligned} \tag{2.177}$$

which yields the scalar updates

$$p_1 = (1 - K_1)p_1^- \tag{2.178a}$$

$$p_2 = (1 - K_1)p_2^- \tag{2.178b}$$

```
% Example 2.10
% Define variables
T=1;
sigpsq=150; sigvsq=150; sigmsq=30;
A=[1 T ; 0 1];
Q=[sigpsq 0 ; 0 sigvsq];
H=[1 0];

% Initial values
p=[30 0 ; 0 30];
alpha(1)=0.5;
beta(1)=0;

% Calculate error covariance and filter gains
for i=1:10
    pm=A*p*A'+Q; % apriori error covariance
    k=pm*H'*inv(H*pm*H'+sigmsq); % kalman gain
    p=(eye(2)-k*H)*pm; % aposteriori error covariance
    alpha(i+1)=k(1);
    beta(i+1)=k(2);
end

% Plot the Filter Gains
plot([0:10],alpha,'k-','linewidth',2); hold on; plot([0:10],beta,'k--
','linewidth',2);
xlabel('k'); ylabel('Filter Gains');
legend('\alpha_k','\beta_k_/_T','location',[0.72 0.7 0.16 0.15])
```

FIGURE 2.10
MATLAB program for α–β tracker error covariance and gains.

or equivalently,

$$p_2 = p_2^- - K_2 p_1^- \tag{2.178c}$$

and

$$p_4 = p_4^- - K_2 p_2^- \tag{2.178d}$$

(Show that Equations 2.174b, and 2.174c are equivalent.)

Equations 2.172, 2.175, and 2.178 are MATLAB *replacements*, with quantities on the right representing old values and quantities on the left representing updated values. A MATLAB implementation of Equations 2.172, 2.175, and 2.178 appears in Figure 2.10. Note that we have been able to avoid using two sets of covariance elements, one for P_k and one for P_k^-, by careful attention to the order of the operations. Using this software and representative values for constants, the filter gains α_k and β_k/T were plotted. See Figure 2.11. □

Note that the steady-state values of α_k and β_k/T are reached in five scans. They are nonzero, in contrast with the situation in part c of the previous example, so that measurements are not discounted with increasing k. We shall see in Section 2.5 that this is a consequence of nonzero Q. In practice, it

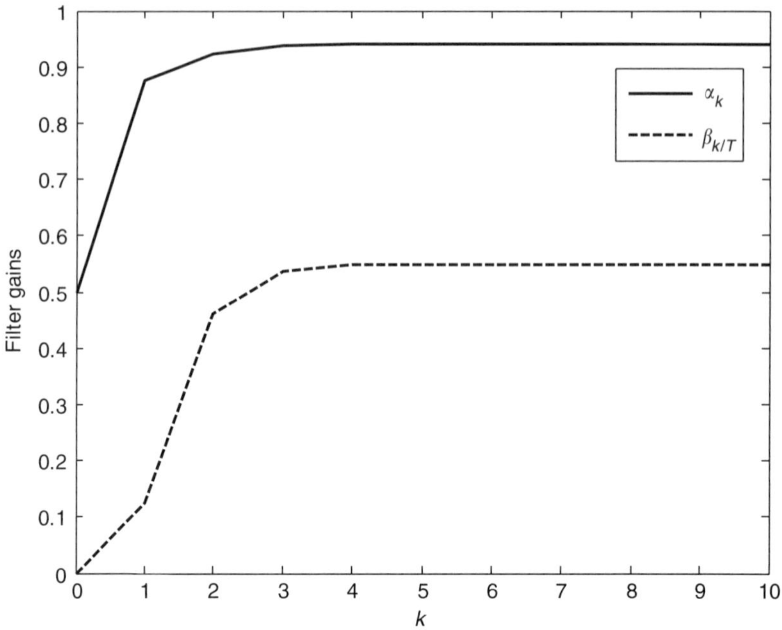

FIGURE 2.11
Filter gains for α–β tracker.

is wise to use a stabilized version of Equation 2.178 (Bierman 1977), since it can occur that Equations 2.178b and 2.178c do not give the same values due to round-off error. For greater numerical stability, the average of these two values can be taken as p_2.

2.4.2 Multiple Sampling Rates

We can considerably increase the usefulness of the method presented here by sampling the system at a higher rate than the data sampling rate.

Let the continuous system be discretized using a period of T in Equations 2.104 through 2.106, where T is much less than the data sampling period T_d. Let T be an integral divisor of T_d. Since the data arrive at the rate $1/T_d$, we should use the data sampling period T_d in Equation 2.112.

The Kalman filter would now be applied as follows. No data arrive between the times iT_d, where i is an integer, so during these intervals we should perform *only the time-update portion* of the Kalman filter equations. For this update, we would use the system dynamics discretized with period T.

When data are received, at times iT_d, we should perform the measurement update portion of the filter algorithm. For this update, we would use the measurement noise covariance $R^s = R/T_d$.

Two advantages of this technique immediately come to mind. First, an optimal estimate is available at every time iT, where i is an integer. Estimates

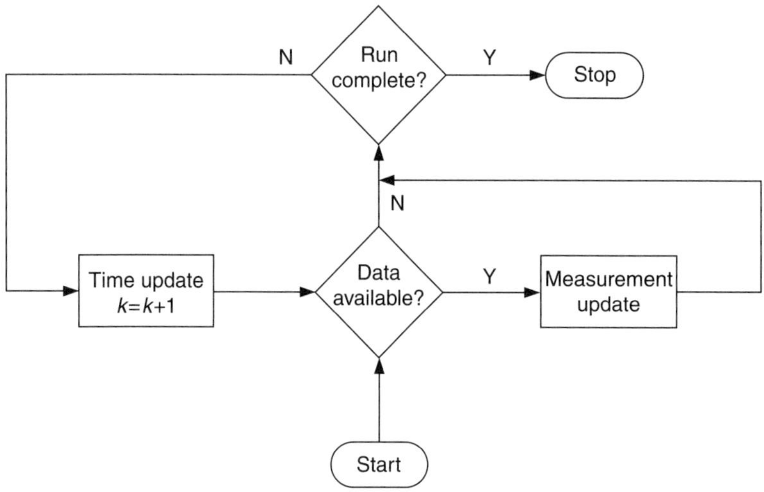

FIGURE 2.12
Flowchart for multiple-rate Kalman filter.

can thus be made available at points on as fine a division of the time line as we desire, independently of the sampling period T_d, which may not be up to us to choose. This means, for example, that in a radar tracking problem, range and azimuth estimates can be made available *between* scans. For a slowly scanning radar this can be quite useful!

The second advantage is that the data arrival times need not be uniform. For aperiodically sampled data we merely perform time updates until a data sample is received, then we perform a measurement update at the next integral multiple of T. If the discretization period T is chosen to be small, the data are incorporated virtually as soon as they arrive.

A flowchart/block diagram for the "multiple-rate Kalman filter" is given in Figure 2.12. The next example illustrates this technique.

Example 2.11 Multiple-Rate Kalman Filter for Radar Tracking

Suppose we discretize the range subsystem of the previous examples using a period of $T = 50$ ms. The data sampling period T_d is the radar period of revolution (which we called T in the previous examples). Suppose T_d is 1 s. Using the multiple-rate Kalman filter will then result in up-to-date optimal estimates every 50 ms although the data arrive only every 1 s.

If $T = 50$ ms is used in the error covariance time update Equation 2.170 in Example 2.10, and the measurement update (Equations 2.176 and 2.178) is performed only every $T_d = 1$ s, then position and velocity error variances like those in Figure 2.13 result. A simple modification of Figure 2.10 was used to obtain these curves. □

```
% Example 2.11
% Multi-rate Kalman Filter for radar tracking
% Time update every 50msec (T) and measurement update every 1 sec (Td)

% Define/Initialize variables
clear all; close all
i=1; T=0.05; Td=1;
sigpsq=150*T; sigvsq=150*T; sigmsq=30/Td; % Qs=Q*T, Rs=Q/Td
Q=[sigpsq 0 ; 0 sigvsq];
H=[1 0];
p=[30 0 ; 0 30]; p1(1)=p(1,1); p4(1)=p(2,2);

% Loop for calculating the error variances
for t=0:T:10-T
    A=[1 T ; 0 1]; i=i+1;
    p=A*p*A'+Q;
    k=p*H'*inv(H*p*H'+sigmsq);
    % Take Measurement at T=1,2...10 secs
    if (mod(t,1)==0 && t~=0)
        p=(eye(2)-k*H)*p;
    end
    p1(i)=p(1,1); p4(i)=p(2,2);
end

% Plotting
figure(1); plot([0:0.05:10],p1,'k'); hold on; plot([0:0.05:10],p4,'k-
','linewidth',2);
legend('p_1','p_4'); grid on;
xlabel('Time (Secs.)'); ylabel('Error Variances');
```

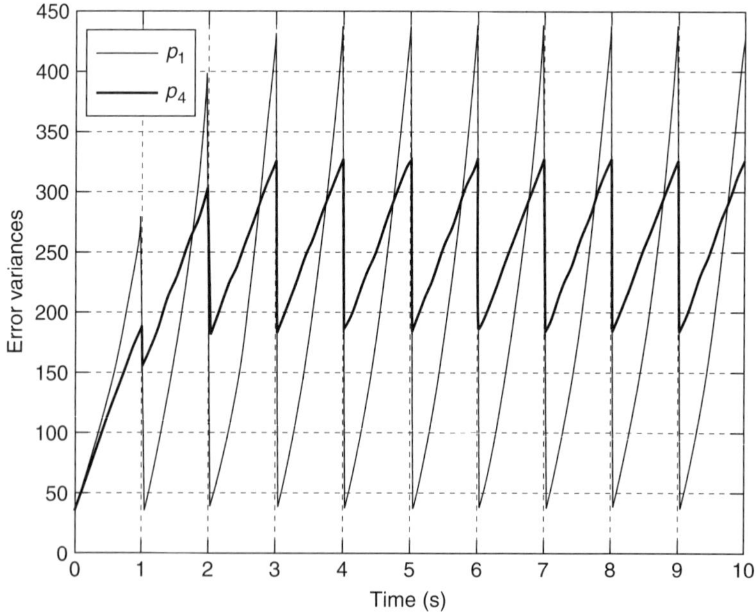

FIGURE 2.13
Range and range-rate error variances with time update every 0.05 s and measurement update every 1 s.

In Section 3.7 we present another approach to discrete filtering for continuous-time systems, which is known as the *continuous-discrete Kalman filter*. No preliminary discretization of the plant is required there. Instead, the estimate and error covariance are propagated between measurements *in continuous time* using equations we shall derive in Chapter 3.

2.4.3 Discretization of Time-Varying Systems

It is straightforward to find equations for discretizing time-varying continuous systems, as the next exercise indicates.

EXERCISE 2.3 Discretization of Time-Varying Systems

Suppose the prescribed plant is time-varying so that

$$\dot{x} = A(t)x + B(t)u + G(t)w \tag{2.179a}$$

$$z = H(t)x + v \tag{2.179b}$$

with $x(0) \sim (\bar{x}(0), P(0)), w(t) \sim (0, Q(t)), v(t) \sim (0, R(t))$, and the usual whiteness and uncorrelatedness assumptions.

Let $\phi(t, t_0)$ be the state-transition matrix of $A(t)$ defined by

$$\frac{d}{dt}\phi(t, t_0) = A(t)\phi(t, t_0), \qquad \phi(t_0, t_0) = I \tag{2.180}$$

Show that the discrete system and noise covariance matrices are time-varying matrices given by

$$A^s(k) = \phi((k+1)T, kT) \tag{2.181}$$

$$B^s(k) = \int_{kT}^{(k+1)T} \phi((k+1)T, \tau)B(\tau)\, d\tau \tag{2.182}$$

$$G^s(k) = I \tag{2.183}$$

$$Q^s(k) = \int_{kT}^{(k+1)T} \phi((k+1)T, \tau)G(\tau)Q(\tau)G^T(\tau)\phi^T((k+1)T, \tau)\, d\tau \tag{2.184}$$

$$H^s(k) = H(kT) \tag{2.185}$$

$$R^s(k) = \frac{R(kT)}{T} \tag{2.186}$$

2.5 Error Dynamics and Statistical Steady State

2.5.1 The Error System

To study the stability and convergence properties of the Kalman filter, we need to examine the *error system*. This will give us some of the insight we need to design Kalman filters for practical problems.

Define the *a priori estimation error* as

$$\tilde{x}_k^- = x_k - \hat{x}_k^- \tag{2.187}$$

the error at time k *before* the inclusion of data z_k. This is a one-step ahead prediction error. To find a recursion for $\tilde{x}_k^-$, we use Equations 2.46a, 2.46b, and 2.59 to write

$$\begin{aligned}\tilde{x}_{k+1}^- &= x_{k+1} - \hat{x}_{k+1}^- \\ &= Ax_k + Bu_k + Gw_k - A(I - K_kH)\hat{x}_k^- - Bu_k - AK_k(Hx_k + v_k)\end{aligned}$$

This simplifies to

$$\tilde{x}_{k+1}^- = A(I - K_kH)\tilde{x}_k^- + Gw_k - AK_kv_k \tag{2.188}$$

(We assume a time-invariant original plant for ease of notation.) Note that the error system is driven by both the process and measurement noises, but not by the deterministic input u_k.

It is easy to express the residual as a function of $\tilde{x}_k^-$; using Equations 2.46b and 2.51, we have

$$\tilde{z}_k = H\tilde{x}_k^- + v_k \tag{2.189}$$

We call Equations 2.188 and 2.189 the *a priori error system.* Note that the residual depends on $\tilde{x}_k^-$ exactly as the data z_k depends on x_k.

To find the covariance of $\tilde{x}_k^-$ note that

$$E(\tilde{x}_{k+1}^-) = A(I - K_kH)E(\tilde{x}_k^-) \tag{2.190}$$

so that $E(\tilde{x}_k^-) = 0$ if $E(\tilde{x}_0^-) = 0$. This, however, is guaranteed by the way we initialize the filter (Table 2.1). Hence the Kalman filter produces an unbiased estimate and so, from Equation 2.188,

$$\begin{aligned}P_{k+1}^- &= \overline{\tilde{x}_{k+1}^-(\tilde{x}_{k+1}^-)^T} \\ &= A\big[(I - K_kH)P_k^-(I - K_kH)^T + K_kRK_k^T\big]A^T + GQG^T\end{aligned} \tag{2.191}$$

By substituting for K_k from Equation 2.60, it is easy to see that we have just found another way to derive the Riccati equation (Equation 2.61). In fact, Equation 2.191 is called the Joseph stabilized formulation of the Riccati equation (cf. Equation 2.58).

Note that the Riccati equation is a symmetric equation (transpose both sides), so P_k^- is symmetric for all $k \geq 0$. P_k^- is also positive semidefinite for all k since it is a covariance matrix.

2.5.2 The Innovations Sequence

According to Equation 2.189, the residual has zero mean,

$$\bar{\tilde{z}}_k = 0 \tag{2.192}$$

Therefore, by the uncorrelatedness of $\tilde{x}_k^-$ and v_k the residual has covariance given by

$$P_{\tilde{z}_k} = HP_k^- H^T + R \tag{2.193}$$

where P_k^- satisfies the Riccati equation. If the measurements are not used to correct $\hat{x}(t)$, then Equation 2.193 still holds, but $P_k^- = P_k$ now satisfies the Lyapunov equation (Equation 2.28). (Why?)

Suppose that all statistics are Gaussian. Let

$$Z_k \triangleq \{z_1, z_2, \ldots, z_k\} \tag{2.194}$$

represent all the data up through time k. Then the Kalman filter manufactures the conditional mean and covariance

$$\hat{x}_k = \overline{x_k/Z_k} \tag{2.195a}$$

$$P_k = P_{x_k/Z_k} \tag{2.195b}$$

at each time k. (Recall from Section 1.1 that the error covariance P_k is equal to the conditional covariance P_{x_k/Z_k}.) In other words, in the Gaussian case the Kalman filter provides an update for the entire hyperstate. The conditional mean of the data is

$$\overline{z_k/Z_{k-1}} = H\hat{x}_k^- \tag{2.196}$$

so that

$$\tilde{z}_k = z_k - \overline{z_k/Z_{k-1}} \tag{2.197}$$

The residual is therefore the *new information* contained in z_k that could not have been anticipated by using the previous data Z_{k-1}. It is called the innovations process. See Exercise 2.2.

From the orthogonality principle, $\tilde{z}_k$ is orthogonal to (and hence uncorrelated with) Z_{k-1}. Since $\tilde{z}_j$ depends only on x_0, the input, and Z_j, it follows that $\tilde{z}_k$ is uncorrelated with $\tilde{z}_j$ for $j < k$. The innovation is therefore a *zero-mean white noise* process.

One good way to verify that the Kalman filter is performing as designed is to monitor the residual. If it is not white noise, there is something wrong with the design and the filter is not performing optimally.

2.5.3 The Algebraic Riccati Equation

Now consider the limiting behavior of the Kalman filter. As $k \to \infty$, the solution to the Riccati equation tends to a bounded steady-state value P if

$$\lim_{k\to\infty} P_k^- = P \tag{2.198}$$

is bounded. In this case, for large k, $P_{k+1}^{-} = P_k^{-} \triangleq P$ and Equation 2.61 tends to the *algebraic Riccati equation*

$$P = A[P - PH^T(HPH^T + R)^{-1}HP]A^T + GQG^T \tag{2.199}$$

It is clear that any limiting solution to Equation 2.61 is a solution to Equation 2.199. The algebraic Riccati equation can have more than one solution. Its solutions can be positive or negative, semidefinite or definite, indefinite, or even complex. There is quite a body of literature on conditions under which it has a *unique positive definite* solution. In the scalar case of course, a quadratic equation has only two solutions. See Example 2.1.

At this point, we can ask several questions:

1. When do bounded limiting solutions to Equation 2.61 exist for all choices of P_0? (That is, when is the *a priori* error covariance bounded for all initial conditions?)
2. Clearly, the limiting solution P depends on the initial condition P_0. When is P independent of our choice (possibly inaccurate!) of P_0?

And most important of all,

3. When is the error system (Equation 2.188) asymptotically stable? (That is, when does the estimate $\hat{x}_k$ *converge* to the true state x_k? This is the whole point of the Kalman filter design!)

We can answer these very important questions in terms of the dynamical properties of the original system (Equation 2.46). Recall that (A, H) is *observable* if the poles of $(A - LH)$ can be arbitrarily assigned by appropriate choice of the output injection matrix L. (A, H) is *detectable* if $(A - LH)$ can be made asymptotically stable by some matrix L. This is equivalent to the observability of the unstable modes of A.

Recall also that (A, B) is *reachable* if the poles of $(A-BK)$ can be arbitrarily assigned by appropriate choice of the feedback matrix K. (A, B) is *stabilizable* if $(A - BK)$ can be made asymptotically stable by some matrix K. This is equivalent to the reachability of all the unstable modes of A.

THEOREM 2.2

Let (A, H) be detectable. Then for every choice of P_0 there is a bounded limiting solution P to Equation 2.61. Furthermore, P is a positive semidefinite solution to the algebraic Riccati equation.

Proof:

Since (A, H) is detectable, there exists a constant output injection matrix L so that $(A - LH)$ is asymptotically stable. Define a suboptimal *a priori* estimate for x_k by

$$\hat{x}_{k+1}^L = (A - LH)\hat{x}_k^L + Bu_k + Lz_k \tag{2.200}$$

Then the associated *a priori* error system

$$\tilde{x}_{k+1}^L = (A - LH)\tilde{x}_k^L + Gw_k + Lv_k \tag{2.201}$$

is asymptotically stable.

Using a derivation like that of Equation 2.191, the error covariance S_k of $\hat{x}_k^L$ satisfies the Lyapunov equation

$$S_{k+1} = (A - LH)S_k(A - LH)^T + LRL^T + GQG^T \tag{2.202}$$

Since $(A - LH)$ is stable, the Lyapunov theory shows that S_k is bounded with a finite steady-state value.

However, the *optimal* estimate is given by the Kalman filter with time-varying gain K_k as shown in Table 2.3. Optimally implies that $P_k^- < S_k$ (i.e., $(S_k - P_k^-) > 0$) for all k, so that P_k^- is bounded above by a finite sequence.

It can be shown that the Riccati solution P_k^- is smooth, so that if it is bounded above by a finite sequence, then it converges to a finite limit P (see Casti 1977; or Kwakernaak and Sivan 1972).

Since the Riccati equation consists of a sum of positive semidefinite terms (see Equation 2.191) and $P_0 \geq 0$, we are guaranteed symmetry and positive semidefiniteness of P_k^-, and hence of P. Clearly, P is a solution to Equation 2.199. This result gives a neat test for the boundedness of the error covariance, especially if (A, H) satisfies the stronger condition of observability, which implies detectability. It is easy to test for observability by determining if the rank of the observability matrix (Equation 2.5) of (A, H) is equal to n.

Since the state estimation problem clearly depends on the observability properties of the original system, the theorem is not surprising. It reassures our intuition. The next result initially seems very strange. It was first derived by Kalman and Bucy (1961). It amounts to a restatement of Theorem 2.2, given a reachability hypothesis. This hypothesis makes the results much stronger. ▮

THEOREM 2.3

Let $\sqrt{Q}$ be a square root of the process noise covariance so that $Q = \sqrt{Q}\sqrt{Q}^T \geq 0$, and let the measurement noise have $R > 0$. Suppose $(A, G\sqrt{Q})$ is reachable. Then (A, H) is detectable if and only if

a. *There is a unique positive definite limiting solution P to Equation 2.61, which is independent of P_0. Furthermore, P is the* unique *positive* definite *solution to the algebraic Riccati equation.*

b. *The steady-state error system is defined as in Equation 2.188, with* steady-state Kalman gain

$$K = PH^T(HPH^T + R)^{-1} \tag{2.203}$$

asymptotically stable.

Proof:
Define D by $R = DD^T$ and note that $R > 0$ implies that $|D| \neq 0$. We first claim that if $(A, G\sqrt{Q})$ is reachable and $|D| \neq 0$, then for any L

$$((A - LH), [G\sqrt{Q} \quad LD]) \tag{2.204}$$

is also reachable. To show this, note that $|D| \neq 0$ implies that $H = DM$ for some M. The PBH test for reachability (Kailath 1980) implies that

$$\text{rank} \begin{bmatrix} zI - A & G\sqrt{Q} \end{bmatrix} = n \tag{2.205}$$

for all z and therefore

$$\begin{aligned} n &= \text{rank}\begin{bmatrix} zI - A & G\sqrt{Q} = LD \end{bmatrix} \\ &= \text{rank}\begin{bmatrix} zI - A & G\sqrt{Q} = LD \end{bmatrix} \begin{bmatrix} I & 0 & 0 \\ 0 & I & 0 \\ M & 0 & I \end{bmatrix} \\ &= \text{rank}\begin{bmatrix} zI - A + LH & G\sqrt{Q} = LD \end{bmatrix} \end{aligned} \tag{2.206}$$

This proves the claim. ∎

To prove necessity, let gain L define a suboptimal estimate $\hat{x}_k^L$ as an (1) of the proof of Theorem 2.2, with $(A - LH)$ asymptotically stable. By Theorem 2.2, (A, H) detectable implies $P_k^- \to P$ with P bounded and at least positive semidefinite. However, $E\|\tilde{x}_k^-\|^2 \leq E\|\tilde{x}_k^L\|^2$ for all L because of the optimality of the Kalman filter (note that $E[(\tilde{x}_k^-)^T\tilde{x}_k^-] = \text{trace } (P_k^-)$). Hence system $A(I-KH)$ is also asymptotically stable with $K = PH^T(HPH^T+R)^{-1}$.

Define $L' = AK$. Then we can write the algebraic Riccati equation as

$$\begin{aligned} P &= (A - L'H)P(A - L'H)^T + L'DD^T(L')^T + G\sqrt{Q}\sqrt{Q}^T G^T \\ &= (A - L'H)P(A - L'H)^T + \begin{bmatrix} G\sqrt{Q} & L'D \end{bmatrix}\begin{bmatrix} G\sqrt{Q} & L'D \end{bmatrix}^T \end{aligned} \tag{2.207}$$

We know that P is an at least positive semidefinite solution of Equation 2.207. However, Equation 2.207 is also a Lyapunov equation, for which we know that $(A-L'H)$ is stable and also that the pair $((A-L'H), [G\sqrt{Q} \quad L'D])$ is reachable. So the solution P is also unique and positive definite, and the gain $K = PH^T(HPH^T + R)^{-1}$ is uniquely defined. To show sufficiency, note that if $A(I - KH)$ is asymptotically stable, there is an $L = AK$ for which the system $(A - LH)$ is stable and hence the pair (A, H) is detectable.

This theorem is quite surprising. It says that if the state is *reachable by the process noise*, so that every mode is excited by w_k, then the Kalman filter is asymptotically stable if (A, H) is detectable. Thus, we can guarantee a stable filter by selecting the measurement matrix H correctly and ensuring that the process model is sufficiently corrupted by noise! This is the type of thing we need to know to develop sufficient intuition to apply the Kalman filter in practice.

To simplify the proof, we made the hypothesis of the theorem unnecessarily restrictive. In fact, it is only required that $(A, G\sqrt{Q})$ be stabilizable. In this

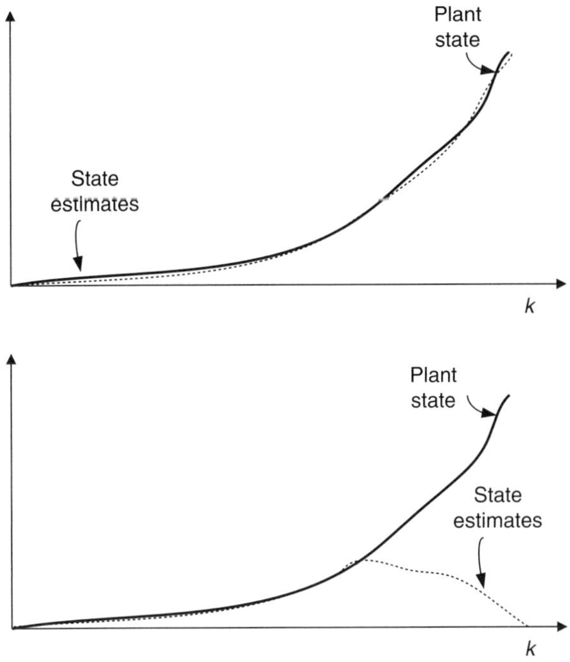

FIGURE 2.14
Relation between plant and filter stability. (a) Unstable plant, stable filter. (b) Unstable plant, unstable filter.

case, though, the unique limiting Riccati solution P can only be guaranteed to be positive semidefinite.

The difference between a stable original plant and a stable Kalman filter should be clearly realized. We are not concerned here with regulating the plant, but only with estimating its state. With a stable filter the estimate closely tracks the state of the plant, whether it is stable or not. Figure 2.14 shows the relation between an unstable plant and an unstable filter.

Example 2.12 Steady-State Behavior in the Scalar Case

Theorem 2.3 is illustrated by Example 2.1, where the algebraic Riccati equation was solved in the scalar case.

For scalars, detectability is equivalent to the condition $|a| < 1$ or $h \neq 0$. The reachability condition is equivalent to $q \neq 0$. In Example 2.1 part a, $|a| < 1$ so detectability is immediate regardless of the value of h. A solution $p > 0$ is given by Equation 2.15 when $q \neq 0$ and $h \neq 0$. If $q = 0$ and $h \neq 0$, then the solutions to Equation 2.12 are

$$p = 0, \qquad -1(1 - a^2)r/h^2 \tag{2.208}$$

so there is only a positive semidefinite solution (i.e., $p = 0$). If $h = 0$ but $q \neq 0$, then from Equation 2.12

$$p = q/(1 - a^2) \tag{2.209}$$

is greater than zero.

In Example 2.1 part b, $|a| > 1$ so for detectability we require $h \neq 0$ and for stabilizability $q \neq 0$. Indeed, Equation 2.209 is less than zero if a is unstable. If $h \neq 0$ and $q \neq 0$ then Riccati equation solution (Equation 2.19) is strictly positive. Furthermore, the closed-loop system (Equation 2.20) is stable if and only if $h \neq 0$ and $q \neq 0$. ▯

Example 2.13 Steady-State Filter for Damped Harmonic Oscillator

Many systems can be modeled, at least to a first approximation, by the equation

$$\ddot{y}(t) + 2\alpha\dot{y}(t) + \omega_n^2 y(t) = bu(t) \tag{2.210}$$

For instance, if $y(t)$ is aircraft pitch and $u(t)$ is elevator deflection, then Equation 2.210 corresponds to the short-period approximation to the longitudinal dynamics (Blakelock 1965). By selecting the state components as y and $\dot{y}$ the plant (Equation 2.210) can be put into the reachable canonical form

$$\dot{x}(t) = \begin{bmatrix} 0 & 1 \\ -\omega_n^2 & -2\alpha \end{bmatrix} x(t) + \begin{bmatrix} 0 \\ b \end{bmatrix} u(t) + w(t) \tag{2.211}$$

where we have added $(0, Q)$ process noise $w(t)$ to compensate for unmodeled system dynamics (as we did in Example 2.8). Let the measurement be $z = y + v(t)$, with noise $v(t) \sim (0, r)$. Let sampling period be $T \ll 1$ so that terms of order T^2 can be ignored. Then the sampled plant is approximately

$$x_{k+1} = \begin{bmatrix} 1 & T \\ -\omega_n^2 T & 1 - 2\alpha T \end{bmatrix} x_k + \begin{bmatrix} 0 \\ bT \end{bmatrix} u_k + w_k \tag{2.212}$$

$$z_k = [1 \quad 0]x_k + v_k \tag{2.213}$$

with $Q^S = QT, r^S = r/T$. (In practice we should use the exact discretized system, not Euler's approximation.)

To study the behavior of the Kalman filter error system, let $\omega_n = 0$, $\alpha = -0.1, r = 0.02, T = 20$ ms, and ignore u_k. Then the sampled plant is

$$x_{k+1} = \begin{bmatrix} 1 & 0.02 \\ 0 & 1.004 \end{bmatrix} x_k + w_k \triangleq A^s x_k + w_k \tag{2.214}$$

and $r^s = 1$. The poles are at

$$z = 1, 1.004 \tag{2.215}$$

so the plant is unstable.

Note that the plant is observable through the measurements. Let

$$Q^s = \begin{bmatrix} 0 & 0 \\ 0 & 0.02 \end{bmatrix} \tag{2.216}$$

so that (As, $\sqrt{Q^S}$) is reachable, allowing w_k to excite both modes ($\sqrt{Q^S} = [0 \quad \sqrt{0.02}]^T$).

A computer program was written to solve the Riccati equation (Equation 2.61). The solution sequence for the case $P_0^- = I$ converges within 4 s ($k = 200$) to a steady-state value of

$$P = \begin{bmatrix} 0.08247 & 0.16363 \\ 0.16363 & 0.61545 \end{bmatrix} \tag{2.217}$$

The same steady-state value is obtained if $P_0^- = 0$.

We can use the Kalman gain (Equation 2.60) to compute the error system matrix

$$A_k^{\text{cl}} \triangleq A^s(I - K_k H) \tag{2.218}$$

for each value of k. Clearly, the poles of the error system are functions of time k and we can plot a *root locus* for the error system as a function of k. The results are shown in Figure 2.15 for $P_0^- = I$. In this figure, the poles for $k = 0$ are marked by X's and the steady-state values of $0.9624 \pm j0.036117$, corresponding to a steady-state error system of

$$A^{\text{cl}} = \begin{bmatrix} 0.92079 & 0.02 \\ -1.15177 & 1.004 \end{bmatrix} \tag{2.219}$$

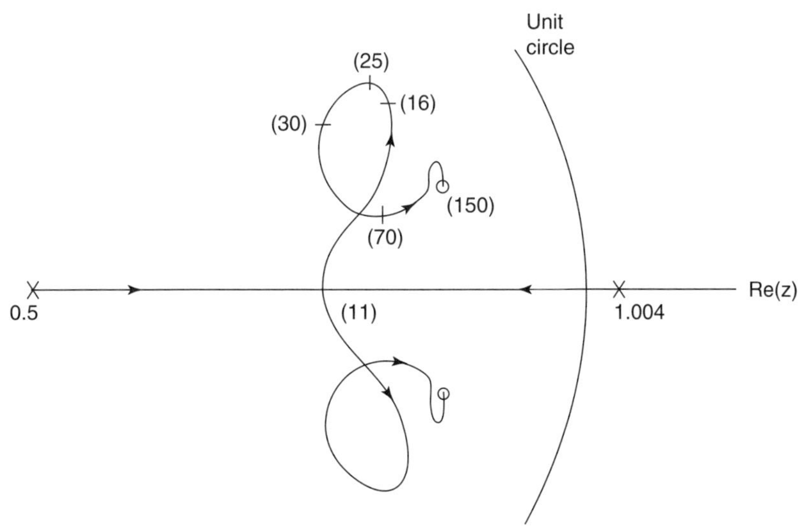

FIGURE 2.15
Locus of error system poles for detectable plant reachable by w_k.

This error system is stable and does not depend on P_0^-, so for any P_0^- the estimation error $\tilde{x}_k^-$ will be bounded with the steady-state variances of its components given by 0.08247 and 0.61545. □

EXERCISE 2.4 Weighted Minimum-Variance Estimate

Show that under the conditions of Theorem 2.3 the steady-state optimal gain minimizes

$$J_\infty = \lim_{k\to\infty} \overline{(\tilde{x}_k^-)^T W(\tilde{x}_k^-)} \tag{2.220}$$

for any $W \geq 0$ and that the minimum value is given by

$$J_\infty = \text{tr}(WP) \tag{2.221}$$

where P is the steady-state error covariance.

EXERCISE 2.5 A Posteriori Error System

Define the *a posteriori estimation error* as

$$\tilde{x}_k = x_k - \hat{x}_k \tag{2.222}$$

a. Show that its dynamics are described by

$$\tilde{x}_{k+1} = (I - K_{k+1}H)A\tilde{x}_k + (I - K_{k+1}H)Gw_k - K_{k+1}v_{k+1} \tag{2.223}$$

b. Show that the *a posteriori* error covariance satisfies

$$\begin{aligned} P_{k+1} = {} & (I - K_{k+1}H)AP_kA^T(I - K_{k+1}H)^T \\ & + (I - K_{k+1}H)GQG^T(I - K_{k+1}H)^T + K_{k+1}RK_{k+1}^T \end{aligned} \tag{2.224}$$

c. When does P_k tend to a steady-state value? When is Equation 2.223 asymptotically stable?

d. Show that $A(I - KH)$ and $(I - KH)A$ have the same eigenvalues.

2.5.4 Time-Varying Plant

If the original plant is time-varying, we need to redefine observability and reachability. Let the plant be given by

$$x_{k+1} = A_kx_k + B_ku_k + G_kw_k \tag{2.225}$$

$$z_k = H_kx_k + v_k \tag{2.226}$$

with $w_k \sim (0, Q_k)$ and $v_k \sim (0, R_k)$. Let $\phi(k, i)$ be the state transition matrix.

We say the plant is *uniformly completely observable*, or *stochastically observable*, if for every integer N the *observability gramian* satisfies

$$\alpha_0 I \leq \sum_{k=i}^{N-1} \phi^T(k,i)H_k^T R_k^{-1} H_k \phi(k,i) \leq \alpha_1 I \tag{2.227}$$

for some $i < N, \alpha_0 > 0$, and $\alpha_1 > 0$. (This guarantees the positive definiteness of both the gramian and its inverse.) We say the plant is *uniformly completely reachable* or *stochastically reachable*, if for every i the *reachability gramian* satisfies

$$\alpha_0 I \leq \sum_{k=i}^{N-1} \phi(N,k+1)G_k Q_k G_k^T \phi^T(N,k+1) \leq \alpha_1 I \tag{2.228}$$

for some $N > i, \alpha_0 > 0$, and $\alpha_1 > 0$.

If the plant is time-varying, then there is in general no constant steady-state solution to the Riccati equation. However, stochastic observability and stochastic reachability (and boundedness of A_k, Q_k, R_k) guarantee that for large k the behavior of P_k is unique independent of P_0. They also guarantee uniform asymptotic stability of the Kalman filter error system. See Kalman and Bucy (1961) and Kalman (1963).

It is quite illuminating to compute the Fisher information matrix for the system 2.225 with measurements 2.226 when $G_k = 0$ (i.e., no process noise) (Sorenson 1970b). Let us assume all statistics are Gaussian. If $u_k = 0$, then the solution to Equation 2.225 is $x_k = \phi(k,i)x_i$ for some initial time i, so that the measurements can be written in terms of the initial state as

$$z_k = H_k \phi(k,i)x_i + v_k \tag{2.229}$$

Suppose we take measurements from $k = i$ to $k = N - 1$. Let Z represent the entire collection of data, $Z = \{z_i, z_{i+1}, \ldots, z_{N-1}\}$. Then if v_k has the Gaussian PDF

$$f_{vk}(v_k) = c_v e^{v_k^T R_k^{-1} v_k/2} \tag{2.230}$$

with c_v a constant, the independence of the measurements means that

$$f_{Z/x_i}(Z/x_i) = c \exp\left\{-\frac{1}{2}\sum_{k=i}^{K-1}[z_k - H_k\phi(k,i)x_i]^T R_k^{-1}[z_k - H_k\phi(k,i)x_i]\right\} \tag{2.231}$$

with c a constant.

Now we can find the Fisher information matrix J_F (Section 1.3) by writing

$$\frac{\partial}{\partial x_i}\ln f_{Z/x_i}(Z/x_i) = \sum_{k=i}^{N-1} \phi^T(k,i)H_k^T R_k^{-1}[z_k - H_k\phi(k,i)x_i]$$

so

$$\begin{aligned} J_F &= -E\left[\frac{\partial^2 \ln f_{Z/x_i}(Z/x_i)}{\partial x_i^2}\right] \\ &= \sum_{k=i}^{N-1} \phi^T(k,i) H_k^T R_k^{-1} H_k \phi(k,i) \end{aligned} \tag{2.232}$$

This is exactly the observability gramian!

2.6 Frequency Domain Results

In the steady-state case, we can work in the frequency domain and derive two nice results; one gives an alternative "root locus" approach to the design of the optimal steady-state filter and the other shows the link between the Kalman filter and the Wiener filter.

2.6.1 A Spectral Factorization Result

The optimal steady-state filter for a time-variant plant is given by (see Table 2.3)

$$\hat{x}_{k+1}^- = A(I - KH)\hat{x}_k^- + Bu_k + AKz_k \tag{2.233}$$

where

$$K = PH^T(HPH^T + R)^{-1} \tag{2.234}$$

with P the positive semidefinite solution of

$$P = APA^T - APH^T(HPH^T + R)^{-1}HPA^T + GQG^T \tag{2.235}$$

It exists and is stable if (A, H) is detectable and (A, $G\sqrt{Q}$) is stabilizable. Note that (Appendix)

$$\begin{aligned} \Delta^{\text{cl}}(z) &\triangleq |zI - A(I - KH)| \\ &= |I + AKH(zI - A)^{-1}| \cdot |zI - A| \\ &= |I + H(zI - A)^{-1}AK| \cdot \Delta(z) \end{aligned} \tag{2.236}$$

where $\Delta(z)$, $\Delta^{cl}(z)$ are the open-loop (i.e., plant) and closed-loop (i.e., filter) characteristic polynomials, respectively.

To derive the result on which this section is based, note that

$$P - APA^T = (zI - A)P(z^{-1}I - A)^T + AP(z^{-1}I - A)^T + (zI - A)PA^T$$

Now use Equation 2.235 to write

$$(zI - A)P(z^{-1}I - A)^T + AP(z^{-1}I - A)^T + (zI - A)PA^T \\ + APH^T(HPH^T + R)^{-1}HPA^T = GQG^T$$

Premultiply this by $H(zI - A)^{-1}$ and postmultiply it by $(z^{-1}I - A)^{-T}\ H^T$ to get

$$HPH^T + H(zI - A)^{-1}APH^T + HPA^T(z^{-1}I - A)^{-T}H^T \\ + H(zI - A)^{-1}APH^T(HPA^T + R)^{-1}HPA^T(z^{-1}I - A)^{-T}H^T \\ = H(zI - A)^{-1}GQG^T(z^{-1}I - A)^{-T}H^T$$

Substituting from Equation 2.234, there results

$$HPH^T + H(zI - A)^{-1}AK(HPH^T + R) + (HPH^T + R)K^TA^T \\ \times (z^{-1}I - A)^{-T}H^T + H(zI - A)^{-1}AK(HPH^T + R)K^TA^T(z^{-1}I - A)^{-T}H^T \\ = H(zI - A)^{-1}GQG^T(z^{-1}I - A)^{-T}H^T$$

and by adding R to each side and factoring, we obtain (see Equation 2.41)

$$\Phi_Z(z) = H(zI - A)^{-1}GQG^T(z^{-1}I - A)^{-T}H^T + R \\ = [I + H(zI - A)^{-1}AK](HPH^T + R)[I + H(z^{-1}I - A)^{-1}AK]^T \tag{2.237}$$

Defining

$$W^{-1}(z) = [I + H(zI - A)^{-1}AK](HPH^T + R)^{1/2} \tag{2.238}$$

this can be written as

$$\Phi_Z(z) = W^{-1}(z)W^{-T}(z^{-1}) \tag{2.239}$$

To see why this is an important result, examine Equation 2.236. Clearly

$$|W^{-1}(z)| = |HPH^T + R|^{1/2} \cdot \frac{\Delta^{\text{cl}}(z)}{\Delta(z)} \tag{2.240}$$

and

$$|W(z)| = \frac{\Delta(z)}{\Delta^{\text{cl}}(z)|HPH^T + R|^{1/2}} \tag{2.241}$$

The optimal filter (Equation 2.233) is asymptotically stable, and this means that $|W^{-1}(z)|$ has all zeros inside $|z| = 1$. Thus $W(z)$ has all poles stable. If in addition the original plant is asymptotically stable, then $|W(z)|$ has all zeros inside $|z| = 1$, so that $W^{-1}(z)$ has all poles stable.

Under these circumstances, Equation 2.239 provides a minimum-phase factorization for the spectral density of z_k. Thus the Kalman filter can be viewed as an algorithm to compute a spectral factorization of the density $\Phi_Z(z)$ of the output process z_k of a stable plant. Solving the algebraic Riccati equation is equivalent to performing a spectral factorization!

The system with transfer function $W^{-1}(z)$ when driven by a white noise input v_k with unit spectral density yields a process with density $\Phi_Z(z)$, hence system $W(z)$ when driven by z_k results in the white noise process v_k. This means that $W(z)$ is a minimum-phase whitening filter for the data process z_k.

The Kalman filter can be viewed as a system that whitens the measurement sequence. The sequence v_k resulting when z_k is whitened by $W(z)$ is called the *innovations sequence* (cf. Section 1.5).

2.6.2 The Innovations Representation

It is not difficult to relate the innovations v_k to the residual $\tilde{z}_k$. Using Table 2.3 and Equation 2.51, we have

$$\hat{x}_{k+1}^- = A(I - K_k H)\hat{x}_k^- + Bu_k + AK_k z_k \tag{2.242}$$

$$\tilde{z}_k = -H\hat{x}_k^- + z_k \tag{2.243}$$

This formulation of the filter equations is called the *innovations representation.* It has as input the data z_k and as output the white residual $\tilde{z}_k$. This form makes it especially clear that the Kalman filter is a whitening filter.

Once the filter has reached steady state so that K_k is a constant K, the transfer function from z_k to $\tilde{z}_k$ can be written as

$$W'(z) = \{I - H[zI - A(I - KH)]^{-1}AK\} \tag{2.244}$$

We can use the matrix inversion lemma to see that

$$W'(z) = [I + H(zI - A)^{-1}AK]^{-1} \tag{2.245}$$

It is now clear (see Equation 2.238) that

$$\tilde{z}_k = (HPH^T + R)^{1/2}v_k \tag{2.246}$$

so that the residual is simply a scaled version of the innovations. The process $\tilde{z}_k$ has covariance

$$P_{\tilde{z}_k} = HPH^T + R \tag{2.247}$$

while v_k has covariance I.

Although technically the innovations and the residual are not the same, $\tilde{z}_k$ is commonly called the innovations.

2.6.3 Chang–Letov Design Procedure for the Kalman Filter

In this subsection, let us not assume that the original plant is stable. Then $W(z)$ is not necessarily minimum phase, but we can still derive a useful result from Equation 2.237. In fact, note that this equation along with Equation 2.236 yields the *Chang–Letov Equation* (Letov 1960)

$$\Delta^{\text{cl}}(z)\Delta^{\text{cl}}(z^{-1}) = |H(z)H^T(z^{-1}) + R\,| \cdot \Delta(z)\Delta(z^{-1}) \cdot |HPH^T + R|^{-1} \quad (2.248)$$

where

$$H(z) = H(zI - A)^{-1}G\sqrt{Q} \quad (2.249)$$

is the transfer function *in the original plant* from w'_k to z_k, where $w_k = \sqrt{Q}w'_k$ and w'_k is white noise with unit covariance. See Equation 2.43.

This is an extremely important result since it allows us an alternative method to steady-state Kalman filter design, which does not depend on solving an algebraic Riccati equation. Instead, we can compute the gain K by using tools such as Ackermann's formula (Equation 2.4). See Kailath (1980).

To see this, note that in the single-output case we can represent Equation 2.249 as

$$H(z) = \frac{N(z)}{\Delta(z)} \quad (2.250)$$

with the numerator polynomial $N(z) = H[adj(zI - A)]G\sqrt{Q}$, a row vector. Then Equation 2.248 becomes

$$\Delta^{\text{cl}}(z)\Delta^{\text{cl}}(z^{-1}) = \frac{N(z)N^T(z^{-1}) + r\Delta(z)\Delta(z^{-1})}{HPH^T + r} \quad (2.251)$$

Since $\Delta^{\text{cl}}(z)$ is asymptotically stable, to find the poles of the optimal steady-state observer we only need to form $N(z)N^T(z^{-1}) + r\Delta(z)\Delta(z^{-1})$ for the given covariances Q and r, and then take the stable roots of this polynomial. (The constant $HPH^T + r$ only plays a normalizing role and can be disregarded.) Then, Ackermann's formula or another design technique yields the value for the gain K. By this method, solution of the Riccati equation is completely avoided.

The zeros of Equation 2.251 are the same as the zeros of

$$1 + \frac{1}{r}H(z)H^T(z^{-1}) \quad (2.252)$$

Since this is in exactly the form required for a classical root locus analysis, it is evident that as $1/r$ goes from 0 to ∞, the optimal steady-state filter poles move from the stable poles of

$$G(z) \triangleq H(z)H^T(z)^{-1} \quad (2.253)$$

to its stable zeros.

The next example illustrates these notions.

Example 2.14 Chang–Letov Design of Steady-State Filter

Consider the discrete system

$$x_{k+1} = \begin{bmatrix} 1 & 0.02 \\ 0.1176 & 1 \end{bmatrix} x_k + \begin{bmatrix} 0 \\ 0.012/m \end{bmatrix} u_k + w_k \tag{2.254}$$

$$z_k = [5\ 0]x_k + v_k \tag{2.255}$$

where $x_k = [\theta_k\ \ \dot{\theta}_k]^T$ and we are measuring the angle θ_k. Let the process noise have covariance of

$$Q = \begin{bmatrix} 0.04 & 0 \\ 0 & 0.04 \end{bmatrix}$$

and the measurement noise have covariance of r. We want to find the best steady-state filter for estimating θ_k and $\dot{\theta}_k$ from z_k.

The plant is unstable with characteristic equation

$$\Delta(z) = (z - 0.9515)(z - 1.0485) \tag{2.256}$$

Since $(A,\ H)$ is observable and $(A,\ G\sqrt{Q})$ is reachable

$$\left(\text{note } G\sqrt{Q} = \begin{bmatrix} 0.2 & 0 \\ 0 & 0.2 \end{bmatrix}\right)$$

we know that there is an asymptotically stable steady-state Kalman filter. Let us use the Chang–Letov equation to find it.

The transfer function from the process noise to the measurement is

$$H(z) = H(zI - A)^{-1}G\sqrt{Q} = \frac{[z-1\ \ 0.02]}{\Delta(z)} \triangleq \frac{N(z)}{\Delta(z)} \tag{2.257}$$

so the Chang–Letov equation becomes

$$\begin{aligned}(HPH^T + r)\Delta^{\text{cl}}(z)\Delta^{\text{cl}}(z^{-1}) &= [z-1\ \ 0.02]\begin{bmatrix} z^{-1}-1 \\ 0.02 \end{bmatrix} \\ &\quad + r(z-0.9515)(z-1.0485)(z^{-1}-0.9515)(z^{-1}-1.0485)\end{aligned}$$

or

$$\begin{aligned}z^2(HPH^T + r)\Delta^{\text{cl}}(z)\Delta^{\text{cl}}(z^{-1}) &= (-z^3 + 2.0004z^2 - z) + r(0.99765z^4 \\ &\quad - 3.9953z^3 + 5.9953z^2 - 3.9953z + 0.99765) \\ &\triangleq n(z) + rd(z)\end{aligned} \tag{2.258}$$

The constant term $(HPH^T + r)$ only plays a normalizing role, and it can be disregarded. The "symmetric" form of the right-hand side of Equation 2.258 is worth noting. It means that if z is a root, then so is z^{-1}.

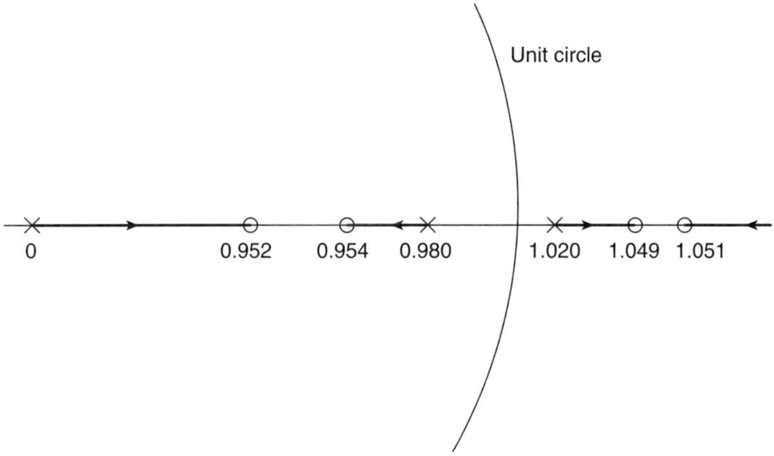

FIGURE 2.16
Chang–Letov root locus.

The roots of the right-hand side of Equation 2.258 depend on r. When $r = 0$, they are the roots of $n(z)$; these are (adding an ∞ for reasons we shall soon discover)

$$0, 0.9802, 1.0202, \infty \tag{2.259}$$

As the $r \to \infty$ the roots of the right-hand side of Equation 2.258 tend to the roots of $d(z)$; these are

$$0.9515, 0.95375, 1.0485, 1.051 \tag{2.260}$$

As r increases from 0 to ∞, the roots move from Equation 2.259 to Equation 2.260. The root locus is shown in Figure 2.16, where the roots (Equation 2.259) of $n(z) = N(z)N^T(z^{-1})$ are denoted by X's and the roots (Equation 2.260) of $d(z) = \Delta(z)\Delta(z^{-1})$ are denoted by 0's. We added ∞ to the roots of Equation 2.259 since there are four loci. Note also that both sets of roots Equations 2.259 and 2.260 are symmetric about $|z| = 1$, that is, if z is a root then so is z^{-1}. (This is why the factorization of Equation 2.251 into $\Delta^{\text{cl}}(z)$ and $\Delta^{\text{cl}}(z^{-1})$ works.)

According to the Chang–Letov procedure, we should select as the poles of the optimal steady-state filter the stable roots of Equation 2.258. Suppose $r = 1$, then the roots are

$$0.38083, 0.98011, 1.0203, 2.6258 \tag{2.261}$$

so we would select

$$\begin{aligned} \Delta^{\text{cl}}(z) &= (z - 0.38083)(z - 0.98011) \\ &= z^2 - 1.3609z + 0.37325 \end{aligned} \tag{2.262}$$

This is the desired characteristic equation of the closed-loop plant matrix $(A - LH)$. To find the required steady-state gain $L = AK$ for use in the filter equation in Table 2.3, use Ackermann's formula (Equation 2.4). Thus

$$V_2 = \begin{bmatrix} H \\ HA \end{bmatrix} = \begin{bmatrix} 5 & 0 \\ 5 & 0.1 \end{bmatrix} \tag{2.263}$$

and

$$\begin{aligned} L = AK &= \Delta^{\text{cl}}(A)V_2^{-1}\begin{bmatrix} 0 \\ 1 \end{bmatrix} \\ &= \begin{bmatrix} 0.12781 \\ 0.1467 \end{bmatrix} \end{aligned} \tag{2.264}$$

The same gain AK would be found by using the algebraic Riccati equation approach.

In the general multioutput case, it is necessary to use the *multivariable root locus* techniques to study the optimal steady-state filter poles as a function of R. In the case of $R = 0$, meaning perfect measurements, we can see from the Chang–Letov equation that the filter poles are the stable zeros of $|H(z)H^T(z^{-1})|$. In the limit as $R \to \infty$, the case of totally unreliable measurements, the filter poles are the original plant poles simply reflected inside the unit circle (i.e., unstable poles z are replaced by z^{-1}). □

2.6.4 Deriving the Discrete Wiener Filter

Consider the Wiener problem of estimating an unknown stationary stochastic process x_k from a stationary data process z_k that is known for all past time. From Section 1.5 we know that the Wiener filter that performs this task is given by

$$F(z) = \left[\Phi_{XZ}(z)W^T(z^{-1})\right]_+ W(z) \tag{2.265}$$

where

$$\Phi_Z^{-1}(z) = W^T(z^{-1})W(z) \tag{2.266}$$

By introducing a *shaping filter* for the unknown process, we can easily find $F(z)$ in terms of a solution to the algebraic Riccati equation. To wit, suppose x_k is generated by the shaping filter

$$x_{k+1} = Ax_k + Gw_k \tag{2.267a}$$

with $w_k \sim (0, Q)$ a white process noise. Suppose the data are related to x_k by

$$z_k = Hx_k + v_k \tag{2.267b}$$

with $v_k \sim (0, R)$ a white measurement noise. The object is now to estimate x_k in terms of z_k assuming Equation 2.267 has reached steady state. This requires of course that A be stable.

The first step in the solution of the problem is to find a whitening filter $W(z)$ to manufacture the innovations v_k from the data; but we have already solved this problem! We know that the whitening filter satisfying spectral factorization (Equation 2.266) is given by Equation 2.238, where K is the gain given in Equation 2.234.

From Example 1.3,

$$\Phi_{XZ}(z) = \Phi_X(z)H^T \tag{2.268}$$

and

$$\begin{aligned}\Phi_Z(z) &= H\Phi_X(z)H^T + R \\ &= H\Phi_{XZ}(z) + R\end{aligned} \tag{2.269}$$

therefore,

$$\Phi_Z(z) = H\Phi_{XZ}(z) + R = W^{-1}(z)W^{-T}(z^{-1})$$

or

$$H\Phi_{XZ}(z)W^T(z^{-1}) = -RW^T(z^{-1}) + W^{-1}(z) \tag{2.270}$$

To find the Wiener filter, we next require the realizable (i.e., stable) part of the left-hand side of this equation. Let us assume that

$$|\Phi_Z(z)| \neq 0 \quad \text{on} \quad |z| = 1 \tag{2.271}$$

so that $W(z)$ is minimum phase. Then, since the realizable part operator is linear

$$H[\Phi_{XZ}(z)W^T(z^{-1})]_+ = -[RW^T(z^{-1})]_+ + [W^{-1}(z)]_+$$

$W^{-1}(z)$ is realizable since $W(z)$ is minimum phase, and by using arguments like those in Example 1.16 it can be shown that the realizable (i.e., constant) part of $RW^T(z^{-1})$ is $(HPH^T + R)^{1/2}$. Therefore,

$$H[\Phi_{XZ}(z)W^T(z^{-1})]_+ = W^{-1}(z) - (HPH^T + R)^{1/2} \tag{2.272}$$

To complete the computation of $F(z)$, note that by the matrix inversion lemma

$$W(z) = (HPH^T + R)^{-1/2}[I - H(zI - A(I - KH))^{-1}AK] \tag{2.273}$$

so that by Equation 2.265

$$HF(z) = H[zI - A(I - KH)]^{-1}AK \tag{2.274}$$

This is the Wiener filter required to yield estimates of Hx_k from z_k. By construction it is stable and causal.

Now examine Equation 2.233 to see that this is exactly the transfer function from z_k to $H\hat{x}_k^-$ in the steady-state Kalman filter! The Wiener and the Kalman approach therefore give identical solutions for the optimal steady-state filter. If we examine our formulas, we will see that this is not the only interesting conclusion that follows from our line of thought. From Equations 2.265 and 2.272, we can write

$$HF(z) = I - (HPH^T + R)^{1/2}W(z) \tag{2.275}$$

This is important because it is possible to find $(HPH^T + R)$ and $W(z)$ without solving the algebraic Riccati equation for P. For, suppose we can factor $\Phi_Z^{-1}(z)$ as in Section 1.5 to obtain

$$\Phi_Z^{-1}(z) = [W'(z^{-1})]^T \Omega W'(z) \tag{2.276}$$

where $\Omega = \Omega^T > 0$ is a real matrix, $W'(z)$ is minimum phase, and $\lim_{z\to\infty} W'(z) = I$. Then the whitening filter is given by

$$W(z) = \Omega^{1/2}W'(z) \tag{2.277}$$

Now note that the term in square brackets in the factorization (Equation 2.237) satisfies the limiting condition which defines (Equation 2.276) [$\lim_{z\to\infty} W'(z) = I$ is equivalent to $\lim_{z\to\infty}[W'(z)]^{-1} = I$]. Therefore $\Omega = (HPH^T + R)^{-1}$. Hence in the case of linear measurements we have the nice expression for the Wiener filter

$$HF(z) = I - \Omega^{1/2}W(z) \tag{2.278}$$

where Ω and $W(z)$ are both found by factoring $\Phi_Z^{-1}(z)$. Alternatively,

$$HF(z) = I - W'(z) \tag{2.279}$$

Compare this to Equation 1.202, which was for the continuous case. Note that due to Equation 2.244, Equation 2.279 yields for the Wiener filter

$$HF(z) = H[zI - A(I - KH)]^{-1}AK \tag{2.280}$$

This is exactly Equation 2.274.

We have been discussing the Wiener filter for estimating $y_k = Hx_k$ from z_k. A natural question arises. If $\hat{y}_k = H\hat{x}_k^-$ is reconstructed by Equation 2.274, is $\hat{x}_k^-$ also provided by filtering the data by

$$F(z) = [zI - A(I - KH)]^{-1}AK \tag{2.281}$$

We should hope so, since this is the transfer function from z_k to $\hat{x}_k^-$ in the steady-state Kalman filter (Equation 2.233). Indeed, if (A, H) is observable, then this is the case. The demonstration is left to the problems.

The next example uses the simplified Equation 2.279 to find the Wiener filter and demonstrates that either the Wiener approach or the Kalman approach can be used to find the optimal steady-state filter.

Example 2.15 Kalman and Wiener Filters for Angle-of-Attack Measurements in the Presence of Wing Flutter

An aircraft wing experiences torsional flutter when it vibrates in a twisting mode. If y_k is the angle of attack, then this vibration might be modeled by the spectral density

$$\Phi_Y(\omega) = \frac{0.36(2 + 2\cos\omega)}{2.04 + 0.8\cos\omega + 2\cos 2\omega} \tag{2.282}$$

which corresponds to a resonant frequency ω_n of 1 rad/s (selected for illustration only!) and a Q factor $Q = \omega_n/2\alpha = 5$. See Figure 2.19.

Let measurements be made with an angle-of-attack indicator with additive white noise

$$z_k = y_k + v_k \tag{2.283}$$

$v_k \sim (0, 1)$.

We shall now find the optimal steady-state filter to estimate y_k using both the Wiener and the Kalman approaches.

a. Wiener filter

We will use Equation 2.278, which is based on a spectral factorization of $\Phi_Z^{-1}(z)$. Thus,

$$\Phi_Z(\omega) = \Phi_Y(\omega) + 1 = \frac{2.76 + 1.52\cos\omega + 2\cos 2\omega}{2.04 + 0.8\cos\omega + 2\cos 2\omega}$$

so

$$\begin{aligned}\Phi_Z^{-1}(z) &= \frac{(e^{-2j\omega} + 0.2e^{-j\omega} + 1)}{(e^{-2j\omega} + 0.239e^{-j\omega} + 0.46)}(0.4597)\frac{(e^{2j\omega} + 0.2e^{j\omega} + 1)}{(e^{2j\omega} + 0.239e^{j\omega} + 0.46)} \\ &\triangleq [W'(z^{-1})]^T \Omega W'(z) \end{aligned} \tag{2.284}$$

(Note that this factorization is not a trivial step! Much has been written about spectral factorization techniques and some techniques are based on solving Riccati equations.)

The whitening filter is therefore given by

$$W(z) = 0.678\frac{(z^2 + 0.2z + 1)}{(z^2 + 0.239z + 0.46)} \tag{2.285}$$

which is minimum phase and $\Omega = 0.4597$. According to Equation 2.278 (or Equation 2.279), the Wiener filter is given by

$$\begin{aligned} F(z) &= I - \Omega^{-1/2} W(z) \\ &= \frac{0.039z - 0.54}{z^2 + 0.239z + 0.46} \end{aligned} \tag{2.286}$$

This filter provides angle-of-attack estimates $\hat{y}_k$ if its input is the data z_k.

b. Steady-State Kalman filter

The Wiener approach uses a factorization of the spectral density of the data. The Kalman approach uses a state model of the data generation procedure.

From Figure 2.17, one state model corresponding to $\Phi_Y(\omega)$ given by Equation 2.282 is

$$\begin{aligned} x_{k+1} &= \begin{bmatrix} 0 & 1 \\ -1 & -0.2 \end{bmatrix} x_k + \begin{bmatrix} 1 \\ 0.8 \end{bmatrix} w_k \\ &\triangleq Ax_k + Gw_k \end{aligned} \tag{2.287}$$

$$y_k = \begin{bmatrix} 1 & 0 \end{bmatrix} x_k \triangleq Hx_k \tag{2.288}$$

with $w_k \sim (0, 0.36)$ a white noise process. This may be verified by finding $H(z) = H(zI - A)^{-1}\,G$ and using $\Phi_Y(z) = H(z)\Phi_w(z)H^T(z^{-1})$, with $\Phi_w(z) = 0.36$.

State model (Equations 2.287 and 2.288) contains the same stochastic information as the spectral density (Equation 2.282). These are equivalent models for the unknown stochastic process $\{y_k\}$. To incorporate the measurement noise into our state model, modify Equation 2.288 to read

$$z_k = Hx_k + v_k \tag{2.289}$$

with $v_k \sim (0, 1)$.

Since $(A,\ H)$ is observable and $(A, G\sqrt{Q})$ is reachable, there is a unique positive definite solution P to the algebraic Riccati equation, and the resulting error system $A(I - KH)$ is asymptotically stable. To find P, iteratively solve Equation 2.61 with any P_0 (we used $P_0 = I$) until a steady-state solution is reached. The result, achieved within 50 iterations, is

$$P = \begin{bmatrix} 1.1758 & 0.0855 \\ 0.0855 & 0.81915 \end{bmatrix} \tag{2.290}$$

Then, by Equation 2.234

$$K = \begin{bmatrix} 0.5404 \\ 0.03931 \end{bmatrix} \tag{2.291}$$

and the error system matrix is

$$A(I - KH) = \begin{bmatrix} -0.03931 & 1 \\ -0.45174 & -0.2 \end{bmatrix} \tag{2.292}$$

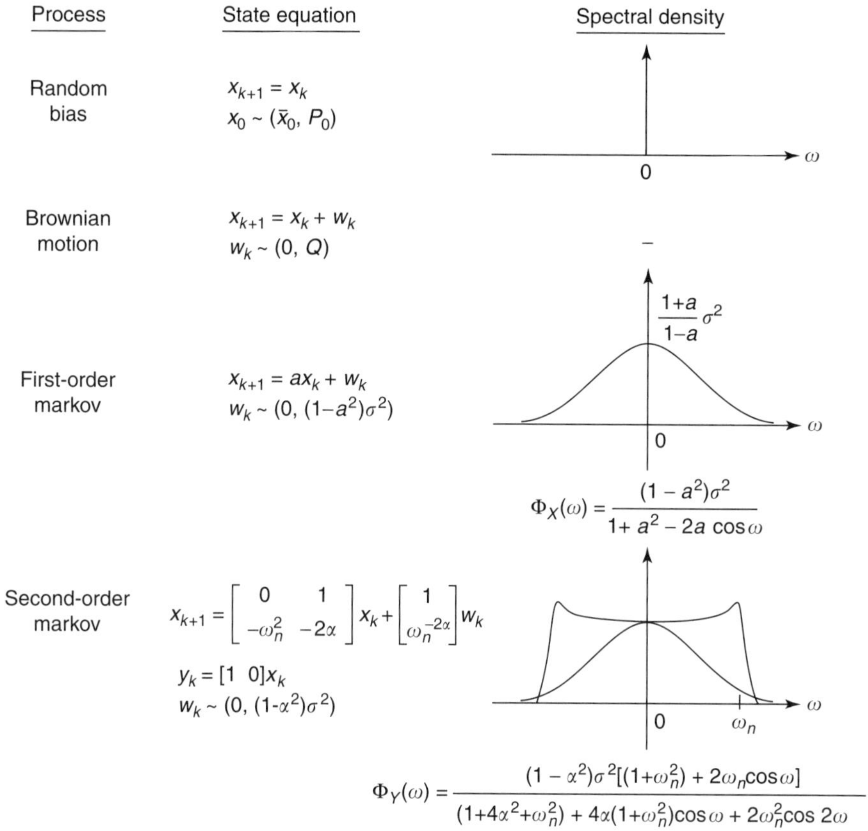

FIGURE 2.17
Some useful discrete spectrum-shaping filters.

The steady-state filter that reconstructs angle-of-attack y_k given z_k is therefore implemented by Equation 2.233, or

$$\hat{x}_{k+1}^- = \begin{bmatrix} -0.03931 & 1 \\ -0.45174 & -0.2 \end{bmatrix} \hat{x}_k^- + \begin{bmatrix} 0.03931 \\ -0.5483 \end{bmatrix} z_k$$
$$\hat{y}_k = \begin{bmatrix} 1 & 0 \end{bmatrix} \hat{x}_{k+1}^- \tag{2.293}$$

If we take the transfer function of the Kalman filter (Equation 2.293), we obtain exactly the Wiener filter (Equation 2.286). □

2.7 Correlated Noise and Shaping Filters

We have discussed the optimal filter for the ideal case of white uncorrelated noise sequences. Departure from this case can occur in three ways: the process noise can be nonwhite (i.e., correlated with itself), the measurement and process noises can be correlated with each other, or the measurement noise can be nonwhite.

2.7.1 Colored Process Noise

Suppose we are prescribed the plant

$$x_{k+1} = Ax_k + Bu_k + Gw_k \tag{2.294a}$$
$$z_k = Hx_k + v_k \tag{2.294b}$$

where $v_k \sim (0, R)$ is white; $x_0 \sim (\bar{x}_0, P_0)$; v_k, w_k, and x_0 are uncorrelated; but process noise w_k is not white.

A noise sequence w_k whose spectral density $\Phi_W(\omega)$ is not a constant is said to be *colored*. From the spectral factorization theorem, if $\phi_W(z)$ is rational and if $|\Phi_W(z)| \neq 0$ for almost every z, then there is a square, rational, asymptotically stable spectral factor $H(z)$ with zeros inside or on the unit circle such that

$$\phi_W(z) = H(z)H^T(z^{-1}) \tag{2.295}$$

If $|\Phi_W(z)| \neq 0$ on $|z| = 1$, then $H(z)$ is minimum phase (i.e., all zeros strictly inside $|z| = 1$). If the linear system $H(z)$ is driven by white noise w'_k with unit spectral density, then the output of the system has spectral density $\Phi_W(z)$. The system $H(z)$ that manufactures colored noise w_k with a given spectral density from white noise is called a spectrum shaping filter. See Figure 2.18.

In the continuous case, Equation 2.295 becomes

$$\Phi_W(s) = H(s)H^T(-s) \tag{2.296}$$

Given a spectral factorization for the density of w_k, we can find a state realization for $H(z)$,

$$H(z) = H'(zI - A')^{-1}G' + D' \tag{2.297}$$

(Kailath 1980). (Note that primes denote additional variables.) Then the noise w_k can be represented by

$$x'_{k+1} = A'x'_k + G'w'_k \tag{2.298a}$$
$$w_k = H'x'_k + D'w'_k \tag{2.298b}$$

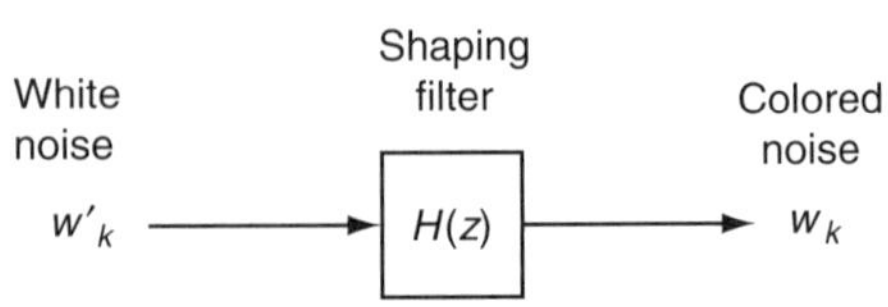

FIGURE 2.18
Spectral density shaping filter.

where $w'_k \sim (0, I)$ is white. *Augmenting* the state of Equation 2.294 by x'_k we get

$$\begin{bmatrix} x_{k+1} \\ x'_{k+1} \end{bmatrix} = \begin{bmatrix} A & GH' \\ 0 & A' \end{bmatrix} \begin{bmatrix} x_k \\ x'_k \end{bmatrix} + \begin{bmatrix} B \\ 0 \end{bmatrix} u_k + \begin{bmatrix} GD' \\ G' \end{bmatrix} w'_k \tag{2.299a}$$

$$z_k = \begin{bmatrix} H & 0 \end{bmatrix} \begin{bmatrix} x_k \\ x'_k \end{bmatrix} + v_k \tag{2.299b}$$

with w'_k and v_k white and uncorrelated. Now the Kalman filter in Table 2.1 can be run on this augmented plant. Note that it must estimate the states of the original plant and also of the noise-shaping filter.

In practice, to find $\Phi_W(z)$ in the first place one could assume $\{w_k\}$ is ergodic and use one sample function w_k to determine the autocorrelation function $R_w(k)$ by time averages. A Z transform yields the spectral density and then a curve fit yields $\Phi_W(z)$.

Example 2.16 Aircraft Longitudinal Dynamics with Gust Noise

The longitudinal dynamics of an aircraft can be represented in the short-period approximation by the harmonic oscillator

$$\dot{x} = \begin{bmatrix} 0 & 1 \\ -\omega_n^2 & -2\delta\omega_n \end{bmatrix} x + \begin{bmatrix} 0 \\ 1 \end{bmatrix} w \tag{2.300}$$

where $x = [\theta \quad \dot{\theta}]^T$ and θ is pitch angle. The process noise $w(t)$ represents wind gusts which change the angle of attack α and so influence pitch rate. The gust noise might have a spectral density that can be approximated by the low-frequency spectrum

$$\Phi_W(\omega) = \frac{2a\sigma^2}{\omega^2 + a^2}, \qquad a > 0 \tag{2.301}$$

This density is shown in Figure 2.19.

Performing a factorization on Equation 2.301, there results

$$\Phi_W(s) = \frac{\sqrt{2a}\sigma}{(s+a)} \cdot \frac{\sqrt{2a}\sigma}{(-s+a)} \tag{2.302}$$

so that the minimum-phase shaping filter is

$$H(s) = \frac{1}{s+a} \tag{2.303}$$

which must be driven by input noise $w'(t)$ with covariance $2a\sigma^2$. A state realization of Equation 2.303 is

$$\dot{x}' = -ax' + w' \tag{2.304a}$$

$$w = x' \tag{2.304b}$$

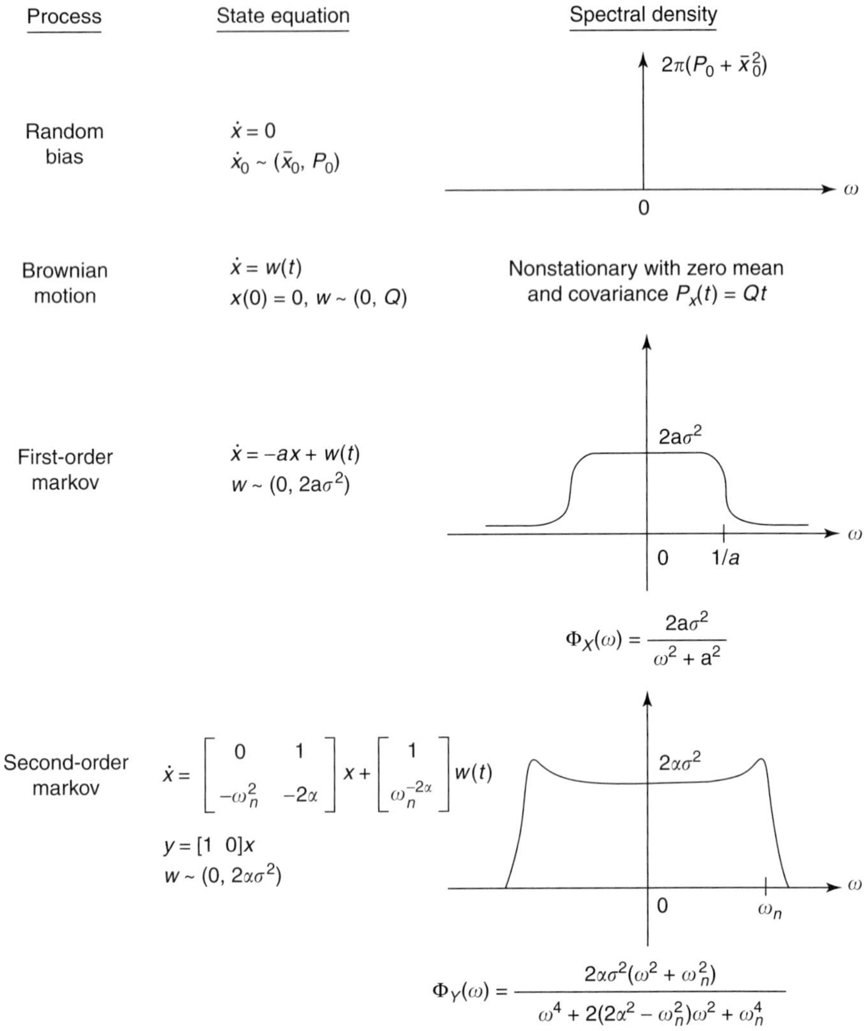

FIGURE 2.19
Some useful continuous spectrum-shaping filters.

Augmenting the plant (Equation 2.300) by the shaping filter (Equation 2.304) and redefining the state as $x \triangleq [\theta \quad \dot{\theta} \quad x']^T$ there results

$$\dot{x} = \begin{bmatrix} 0 & 1 & 0 \\ -\omega_n^2 & -2\delta\omega_n & 1 \\ 0 & 0 & -a \end{bmatrix} x + \begin{bmatrix} 0 \\ 0 \\ 1 \end{bmatrix} w' \tag{2.305}$$

If pitch θ is measured every T s, then

$$z = [1 \quad 0 \quad 0]x + v \tag{2.306}$$

suppose the measurement noise is white with $v \sim (0, r)$. If the sampling period T is small so that T^2 is negligible then the discretized plant is

$$x_{k+1} = \begin{bmatrix} 1 & T & 0 \\ -\omega_n^2 T & 1 - 2\delta\omega_n T & T \\ 0 & 0 & 1 - aT \end{bmatrix} x_k + \begin{bmatrix} 0 \\ 0 \\ 1 \end{bmatrix} w'_k \tag{2.307a}$$

$$z_k = [1 \quad 0 \quad 0] x_k + v_k \tag{2.307b}$$

with which noises $w'_k \sim (0, 2a\sigma^2 T)$, $v_k \sim (0, r/T)$.

Now the discrete Kalman filter can be run on Equation 2.307.

In practice the discretization would be performed using e^{AT}, not by the Euler's approximation $I + AT$, BT, QT, R/T. We use the latter method in the examples for simplicity only. The actual sampling process with zero-order-hold corresponds to discretization by e^{AT}, not $I + AT$.

Some useful continuous and discrete shaping filters are shown in Figures 2.17 and 2.19. In these figures, $w(t)$ and w_k are white noise processes.

The random bias could be used in a ship navigation problem where there is a constant unknown current. Brownian motion, or the Wiener process, is useful in describing the motion of particles under the influence of diffusion. It should also be used to model biases that are known to vary with time. First-order Markov processes are useful to model band-limited noises. The second-order Markov process has a periodic autocorrelation and is useful to describe oscillatory random processes such as fuel slosh or vibration (Maybeck 1979). □

2.7.2 Correlated Measurement and Process Noise

Now we consider the plant (Equation 2.294) with $w_k \sim (0, Q)$ and $v_k \sim (0, R)$ both white, but with

$$\overline{v_j w_k^T} = S\delta_{jk} \tag{2.308}$$

where δ_{jk} is the Kronecker delta. In this case we could write

$$\overline{\begin{bmatrix} w_k \\ v_k \end{bmatrix} \begin{bmatrix} w_k \\ v_k \end{bmatrix}^T} = \begin{bmatrix} Q & S^T \\ S & R \end{bmatrix} \tag{2.309}$$

To rederive the Kalman filter equations for this case it is easiest to begin with the measurement update. Since x_k and v_k in Equation 2.294b are still uncorrelated (x_k depends on w_{k-1} and hence on v_{k-1}), we obtain from Table 2.2 at time k

$$\hat{x}_k = \hat{x}_k^- + P_k^- H^T (H P_k^- H^T + R)^{-1} (z_k - H\hat{x}_k^-) \tag{2.310}$$

$$P_k = P_k^- - P_k^- H^T (H P_k^- H^T + R)^{-1} H P_k^- \tag{2.311}$$

The correlation (Equation 2.308) does not affect the measurement update. For the time update, we recall that the best estimate of a linear function is the same function of the estimate, that is,

$$\hat{f}(x) = f(\hat{x}) \tag{2.312}$$

if $f(\cdot)$ is linear. Hence from Equation 2.294a

$$\hat{x}_{k+1}^- = A\hat{x}_k + B\hat{u}_k + G\hat{w}_k \tag{2.313}$$

The effect of the correlatedness of w_k and v_k can be clearly understood at this point, for in contrast to our previous derivations, the estimated value $\hat{w}_k$ of w_k given z_k is no longer zero. To see this, use Example 1.2 to write

$$\hat{w}_k = \bar{w}_k + P_{WZ}P_Z^{-1}(z_k - \tilde{z}_k) \tag{2.314}$$

where

$$P_{WZ} = \overline{w_k(z_k - \tilde{z}_k)^T} = \overline{w_k(Hx_k + v_k - H\hat{x}_k^-)^T} = S^T \tag{2.315}$$

since w_k and x_k are uncorrelated. Therefore, we obtain as the estimate for the process noise given measurement z_k,

$$\hat{w}_k = S^T(HP_k^-H^T + R)^{-1}(z_k - H\hat{x}_k^-) \tag{2.316}$$

Now use this and Equation 2.310 in Equation 2.313 to get

$$\hat{x}_{k+1}^- = A\hat{x}_k^- + Bu_k + (AP_k^-H^T + GS^T)(HP_k^-H^T + R)^{-1}(z_k - H\hat{x}_k^-) \tag{2.317}$$

If we define a *modified Kalman gain* by

$$K_k = (AP_k^-H^T + GS^T)(HP_k^-H^T + R)^{-1} \tag{2.318}$$

then this becomes

$$\hat{x}_{k+1}^- = A\hat{x}_k^- + Bu_k + K_k(z_k - H\hat{x}_k^-) \tag{2.319}$$

Equations 2.319 and 2.59 are identical except for the gain terms; the only effect of measurement and process noise correlated as in (Equation 2.308) is to modify the Kalman gain.

To find the effect of the correlation (Equation 2.308) on the error covariance, write the new *a priori* error system as

$$\begin{aligned}
\tilde{x}_{k+1}^- &= x_{k+1} - \hat{x}_{k+1}^- \\
&= Ax_k + Bu_k + Gw_k \\
&= -A\hat{x}_k^- - Bu_k - K_k(z_k - H\hat{x}_k^-)
\end{aligned}$$

or

$$\tilde{x}_{k+1}^- = (A - K_kH)\tilde{x}_k^- + Gw_k - K_kv_k \tag{2.320}$$

with K_k as in Equation 2.318. Compare this to Equation 2.188. The only difference is the modified gain.

Now, since $\overline{\tilde{x}_{k+1}^-} = 0$, we have

$$\begin{aligned}P_{k+1}^- &= E(\tilde{x}_{k+1}^-(\tilde{x}_{k+1}^-)^T) \\ &= (A - K_k H)P_k^-(A - K_k H)^T + GQG^T \\ &\quad + K_k R K_k^T - GS^T K_k^T - K_k S G^T\end{aligned}$$

It takes only a few steps to show that this is equivalent to

$$P_{k+1}^- = AP_k^- A^T - K_k(HP_k^- H^T + R)K_k^T + GQG^T \tag{2.321}$$

Examine Equation 2.61 to see that it can be put into this same form, with a difference only in the Kalman gain definition. All of the results for this correlated noise case are, therefore, identical to our old ones, but the gain term AK_k with K_k as in Equation 2.60 must be replaced by the modified gain (Equation 2.318) and the formulation (Equation 2.321) of the Riccati equation must be used.

From Equation 2.321, it is clear that the error covariance is decreased due to the new information provided by the cross-correlation term S. Even if $S \neq 0$, we could still implement the filter of Table 2.1, but it would be suboptimal in this case.

EXERCISE 2.6 Measurement Noise Correlated with Delayed Process Noise

Suppose that instead of Equation 2.308, we have

$$\overline{v_j w_{k-1}^T} = S\delta_{jk} \tag{2.322}$$

This is sometimes useful if Equation 2.294 is a sampled continuous system, for it represents correlation between the process noise over a sample period and the measurement noise at the *end* of the period (see Equation 2.101).

Show that in this case:

a. The *time update* is still the same as in the uncorrelated noise case.
b. The measurement update is now given by

$$\hat{x}_k = \hat{x}_k^- + K_k(z_k - H\hat{x}_k^-) \tag{2.323}$$

with a modified gain defined by

$$K_k = (P_k^- H^T + GS^T)(HP_k^- H^T + R + HGS^T + SG^T H^T)^{-1} \tag{2.324}$$

Note that Equation 2.323 is identical to Equation 2.57.

c. The new error covariance is given by

$$P_{k+1}^1 = AP_k^- A^T - AK_k(HP_k^- H^T + R + HGS^T + SG^T H^T)K_k TA^T + GQG^T \tag{2.325}$$

2.7.3 Colored Measurement Noise

Suppose now that we are prescribed the plant (Equation 2.294) with $w_k \sim (0, Q)$ white and w_k and v_k uncorrelated; but that the measurement noise is

$$v_k = n_k + v'_k \tag{2.326}$$

where n_k is colored and v'_k is white.

If $\Phi_N(z)$ is the spectral density of n_k, then a spectral factorization yields

$$\Phi_N(z) = H(z)H^T(z^{-1}) \tag{2.327}$$

and we can find a state realization for the shaping filter so that

$$H(z) = H'(zI - A')^{-1}G' + D' \tag{2.328}$$

Then the measurement noise is modeled by

$$\begin{aligned} x'_{k+1} &= A'x'_k + G'w'_k \\ v_k &= H'x'_k + D'w'_k + v'_k \end{aligned} \tag{2.329}$$

where $w'_k \sim (0, I)$ is white. Let $v'_k \sim (0, R')$ with $R' > 0$. The new measurement noise in Equation 2.329 is $Dw'_k + v'_k$, and since w'_k and v'_k are uncorrelated,

$$R \triangleq \overline{(D'w'_k + v'_k)(D'w'_k + v'_k)^T} = R' + D'(D')^T \tag{2.330}$$

Augmenting the state of Equation 2.294 by Equation 2.329, there results the new plant

$$\begin{aligned} \begin{bmatrix} x_{k+1} \\ x'_{k+1} \end{bmatrix} &= \begin{bmatrix} A & 0 \\ 0 & A' \end{bmatrix} \begin{bmatrix} x_k \\ x'_k \end{bmatrix} + \begin{bmatrix} B \\ 0 \end{bmatrix} u_k + \begin{bmatrix} G & 0 \\ 0 & G' \end{bmatrix} \begin{bmatrix} w_k \\ w'_k \end{bmatrix} \\ z_k &= \begin{bmatrix} H & H' \end{bmatrix} \begin{bmatrix} x_k \\ x'_k \end{bmatrix} + (D'w'_k + v'_k) \end{aligned} \tag{2.331}$$

which describes the original dynamics and the measurement noise process.

The new process noise $[w_k^T \; w'^T_k]^T$ is white with

$$\begin{bmatrix} w_k \\ w'_k \end{bmatrix} \sim \left(0, \begin{bmatrix} Q & 0 \\ 0 & I \end{bmatrix}\right) \tag{2.332}$$

and the new measurement noise is white with

$$(D'w'_k + v'_k) \sim (0, R) \tag{2.333}$$

Now, however, the measurement and process noises are correlated with

$$E\left[(D'w'_k + v'_k) \begin{bmatrix} w_k \\ w'_k \end{bmatrix}^T\right] = \begin{bmatrix} 0 & D' \end{bmatrix} \tag{2.334}$$

so the modified gain of the previous subsection should be used for optimality. If the relative degree of $\Phi_N(z)$ is greater than zero, then $D' = 0$ and the new measurement and process noises are uncorrelated.

If $v'_k = 0$ in Equation 2.326, then $R' = 0$; however this approach can often still be used, since the discrete Kalman filter requires that $|[H \;\; H']P[H \;\; H']^T + R| \neq 0$.

In the case of colored measurement noise, there is an alternative to state augmentation. Suppose we are given plant (Equation 2.294) with $w_k \sim (0, Q)$ white, but with colored v_k generated by the shaping filter

$$\begin{aligned} x'_{k+1} &= A'x'_k + G'w'_k \\ v_k &= x'_k \end{aligned} \tag{2.335}$$

with $w'_k \sim (0, I)$ white and uncorrelated with w_k. (Presumably, this shaping filter was found by factoring the spectrum of v_k.) Define the *derived measurement* as

$$z'_{k+1} = z_{k+1} - A'z_k - HBu_k \tag{2.336}$$

Then

$$\begin{aligned} z'_{k+1} &= Hx_{k+1} + v_{k+1} - A'Hx_k - A'v_k - HBu_k \\ &= (HA - A'H)x_k + G'w'_k + HGw_k \end{aligned}$$

Defining a new measurement matrix as

$$H' = HA - A'H \tag{2.337}$$

we can write a new output equation in terms of the derived measurement as

$$z'_{k+1} = H'x_k + (G'w'_k + HGw_k) \tag{2.338}$$

In this equation, the measurement noise $(G'w'_k + HGw_k)$ is *white* with

$$(G'w'_k + HGw_k) \sim [0, (G'(G')^T + HGQG^TH^T)] \tag{2.339}$$

Therefore, Equations 2.294a and 2.338 define a new plant driven by white noises. In this plant, the measurement and process noises are correlated with

$$E\left[(G'w'_k + HGw_k)w_k^T\right] = HGQ \tag{2.340}$$

so the modified Kalman gain of the previous subsection should be used for optimality.

The advantage of the derived measurement approach is that the dimension of the state is not increased by adding a shaping filter. To see how to implement the Kalman filter with derived measurements, by Equations 2.319 and 2.336 we have

$$\begin{aligned} \hat{x}^-_{k+1} &= A\hat{x}^-_k + Bu_k + K_k(z'_k - H'\hat{x}^-_k) \\ &= (A - K_kH')\hat{x}^-_k + Bu_k - K_kA'z_{k-1} + K_k(z_k - HBu_{k-1}) \end{aligned} \tag{2.341}$$

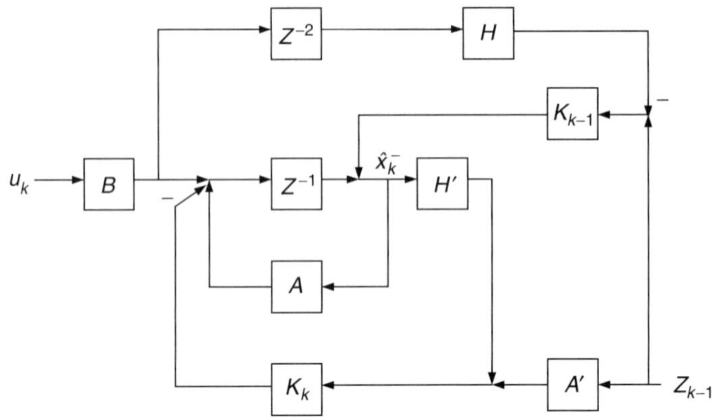

FIGURE 2.20
Implementing the Kalman filter with derived measurements.

We claim that an implementation of this recursion is given in Figure 2.20. To see this, note that the signal at the input to the delay z^{-1} is

$$(A - K_k H')\hat{x}_k^- + Bu_k - K_k A' z_{k-1} \tag{2.342}$$

so the output of the delay is this signal delayed by 1. Now it is easy to see from the figure that

$$\begin{aligned}\hat{x}_k^- &= (A - K_{k-1}H')\hat{x}_{k-1}^- + Bu_{k-1} - K_{k-1}A' z_{k-2} \\ &\quad + K_{k-1}(z_{k-1} - HBu_{k-2})\end{aligned} \tag{2.343}$$

which is a delayed version of Equation 2.341 (see Anderson and Moore 1979). Compare this figure to Figure 2.1.

2.8 Optimal Smoothing

This chapter has heretofore been devoted to the filtering problem. Given a linear stochastic system, the solution to the prediction problem is very straightforward and we deal with it in the problems. In this section we discuss optimal smoothing.

Suppose data are available over a time interval $(0, N]$. The determination of the state estimate $\hat{x}_k$ for $k \in (0, N)$ using *all* the data, past and future, is called the *smoothing problem*. It is a non-real-time operation where the available data are processed to obtain an estimate $\hat{x}_k$ for some past value of k. There are several types of smoothing, and here we shall consider *fixed interval smoothing*, where the final time N is fixed.

An application of smoothing is satellite orbit determination. The satellite's position is measured over a time interval and smoothed estimates over the interval of its radial and angular positions and velocities are computed. From these estimates the parameters of the orbit can be found. See Example 5.4.

Let the unknown x_k be described by the dynamics

$$x_{k+1} = Ax_k + Bu_k + Gw_k \tag{2.344a}$$

and let the data be

$$z_k = Hx_k + v_k \tag{2.344b}$$

White noises $w_k \sim (0, Q)$ and $v_k \sim (0, R)$ are uncorrelated with each other and with $x_0 \sim (\bar{x}_0, P_0)$. Let $Q > 0$, $R > 0$.

To find the optimal state estimate $\hat{x}_k$ at time k in the time interval $(0, N]$, we need to take account of the data z_j for $0 < j \leq k$ and also for $k < j \leq N$. We know now to incorporate the data for the subinterval prior to and including k. All we must do is use the Kalman filter. If we denote the estimate at time k based only on the data z_j for $0 < j \leq k$ by $\hat{x}_k^f$ (for "forward"), then we have $\hat{x}_k^f = \bar{x}_0$, $P_0^f = P_0$ and:

Time update

$$\hat{x}_{k+1}^{f-} = A\hat{x}_k^f + Bu_k \tag{2.345}$$

$$P_{k+1}^{f-} = AP_k^f A^T + GQG^T \tag{2.346}$$

Measurement update

$$\hat{x}_k^f = \hat{x}_k^{f-} + K_k^f\left(z_k - H\hat{x}_k^{f-}\right) \tag{2.347}$$

$$P_k^f = P_k^{f-} - P_k^{f-}H^T\left(HP_k^{f-}H^T + R\right)^{-1}HP_k^{f-} \tag{2.348a}$$

$$= \left[(P_k^{f-})^{-1} + H^T R^{-1} H\right]^{-1} \tag{2.348b}$$

$$K_k^f = P_k^{f-}H^T\left(HP_k^{f-}H^T + R\right)^{-1} \tag{2.349a}$$

$$= P_k^f H^T R^{-1} \tag{2.349b}$$

What we now propose to do is the following. First we will construct a filter that runs *backward* in time beginning at N to make an estimate $\hat{x}_k^b$ incorporating the data z_j for $k < j \leq N$. This is the *information formulation* of the Kalman filter. Then we will discuss how to combine $\hat{x}_k^f$ and $\hat{x}_k^b$ into the overall smoothed estimate $\hat{x}_k$ that incorporates all the data over the entire interval.

The smoother we will develop provides the best linear estimate of x_k given the data in $(0, N]$. If all statistics are normal, it provides the optimal estimate. It also extends directly to the time-varying case.

2.8.1 The Information Filter

In the problems for Section 2.3, we discuss a filter based on the information matrix P_k^{-1}. This filter uses A^{-1}. Here we shall see that the natural direction

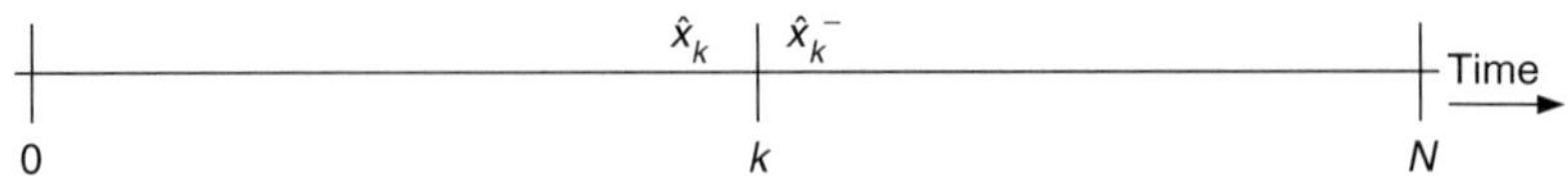

FIGURE 2.21
Backward filter update diagram.

for time flow in the information formulation is backward. If we reverse time, then A appears instead of A^{-1}. The key to solving the smoothing problem by our approach is to derive a convenient backward form of the optimal filter, and this form is provided by the information formulation.

We now derive a filter that provides the best estimate of x_k given the data on the subinterval $k < j \leq N$. All $\hat{x}$, K, and P in this subsection of the book should be considered as having a superscript "b" (for "backward"), which we suppress for notational brevity. In the derivation we assume $|A| \neq 0$, though the final form of the information filter does not require this.

Write the system Equation 2.344a as

$$x_k = A^{-1}x_{k+1} - A^{-1}Bu_k - A^{-1}Gw_k \tag{2.350}$$

This is a recursion for decreasing k beginning at $k = N$. At time k, before the data z_k are included, we have the *a priori* estimate $\hat{x}_k^-$. After z_k is included we have the *a posteriori* estimate $\hat{x}_k$. Since we are proceeding backward, the timing diagram is as depicted in Figure 2.21. Note that $\hat{x}_k^-$ occurs immediately to the right of k, while $\hat{x}_k$ occurs to its left.

The error covariance measurement update is

$$P_k^{-1} = (P_k^-)^{-1} + H^T R^{-1} H \tag{2.351}$$

The time update for stochastic system (Equation 2.350) is easily seen to be

$$P_k^- = A^{-1}(P_{k+1} + GQG^T)A^{-T} \tag{2.352}$$

where $A^{-T} \triangleq (A^{-1})^T$. By using the matrix inversion lemma on Equation 2.352 and defining information matrices

$$S_k \triangleq P_k^{-1} \tag{2.353a}$$

$$S_k^- \triangleq (P_k^{-1})^{-1} \tag{2.353b}$$

we can write these updates as

$$S_k = S_k^- + H^T R^{-1} H \tag{2.354}$$

$$S_k^- = A^T[S_{k+1} - S_{k+1}G(G^T S_{k+1} G + Q^{-1})^{-1} G^T S_{k+1}]A \tag{2.355}$$

If a gain matrix is defined as

$$K_k = S_{k+1}G(G^T S_{k+1} G + Q^{-1})^{-1} \tag{2.356}$$

then the time update becomes

$$S_k^- = A^T(I - K_kG^T)S_{k+1}A \tag{2.357}$$

A Joseph stabilized version is

$$S_k^- = A^T(I - K_kG^T)S_{k+1}(I - K_kG^T)^TA + A^TK_kQ^{-1}K_k^TA \tag{2.358}$$

Note that these equations do not involve A^{-1}.

For the estimate updates, it is convenient to define intermediate variables as

$$\hat{y}_k = P_k^{-1}\hat{x}_k = S_k\hat{x}_k \tag{2.359a}$$

$$\hat{y}_k^- = \left(P_k^-\right)^{-1}\hat{x}_k^- = S_k^-\hat{x}_k^- \tag{2.359b}$$

According to the problems for Section 1.2, the measurement update of the Kalman filter can be written as

$$P_k^{-1}\hat{x}_k = (P_k^-)^{-1}\hat{x}_k^- + H^TR^{-1}z_k \tag{2.360}$$

or

$$\hat{y}_k = \hat{y}_k^- + H^TR^{-1}z_k \tag{2.361}$$

The backward time update for the estimate for stochastic system (Equation 2.350) is

$$\hat{x}_k^- = A^{-1}\hat{x}_{k+1} - A^{-1}Bu_k \tag{2.362}$$

$$\hat{y}_k^- = S_k^-\hat{x}_k^- = S_k^{-1}A^{-1}(S_{k+1}^-\hat{y}_{k+1} - Bu_k) \tag{2.363}$$

Taking into account Equation 2.357 yields

$$\hat{y}_k^- = A^T(I - K_kG^T)(\hat{y}_{k+1} - S_{k+1}Bu_k) \tag{2.364}$$

Equations 2.354 through 2.356, Equation 2.361, and Equation 2.364 comprise the information filter. Note the duality between this version and the Kalman filter. In the latter, time updates are simple while in the information formulation measurement updates are simple.

It is sometimes convenient to have another expression for the gain K_k. To find it, write

$$K_k = (AP_k^-A^T)^{-1}(AP_k^-A^T)S_{k+1}G(G^TS_{k+1}G + Q^{-1})^{-1}$$

But using Equation 2.352

$$\begin{aligned}(AP_k^-A^T)S_{k+1}G &= (P_{k+1} + GQG^T)P_{k+1}^{-1}G\\ &= GQ(Q^{-1} + G^TP_{k+1}^{-1}G)\end{aligned}$$

so that

$$K_k = (AP_k^-A^T)^{-1}GQ \tag{2.365}$$

Compare this to Equation 2.349b.

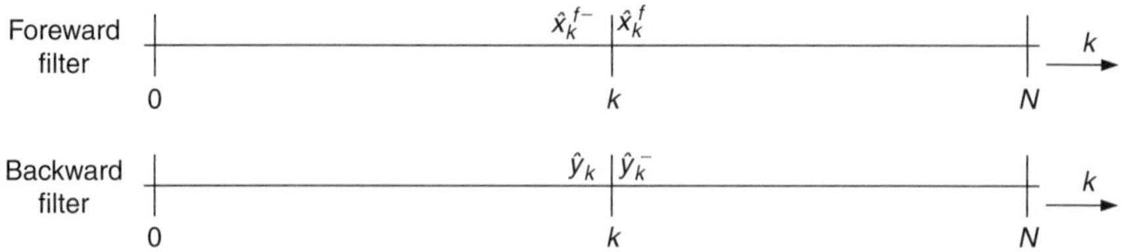

FIGURE 2.22
Smoother timing diagram.

2.8.2 Optimal Smoothed Estimate

Let us now suppose that a Kalman filter has been run on the interval $[0, N]$ to compute for each k an estimate $\hat{x}_k^f$ based on data z_j for $0 < j \le k$. Let us also suppose that a separate information filter has been run backward on the interval to find for each k an estimate $\hat{x}_k^{b-}$ based on data z_j for $k < j \le N$. It is now necessary to combine $\hat{x}_k^f$ and $\hat{x}_k^{b-}$ and their respective error covariances to determine an optimal overall estimate $\hat{x}_k$ and error covariance that incorporate all the available data. See Figure 2.22.

This is very easy to do. Since w_k and v_k are white, estimates $\hat{x}_k^f$ and $\hat{x}_k^{b-}$ are independent and we can use the "Millman's theorem" from the problems in Section 1.2. (For simplicity, we suppress the superscript "b" on $\hat{y}$ and S.)

The *a posteriori* smoothed error covariance is given by

$$P_k = \left[\left(P_k^f\right)^{-1} + \left(P_k^{b-}\right)^{-1}\right]^{-1} \tag{2.366a}$$

$$P_k = \left[\left(P_k^f\right)^{-1} + S_k^-\right]^{-1} \tag{2.366b}$$

The matrix inversion lemma can be employed to write this in several alternative ways, two of which are

$$P_k = P_k^f - P_k^f\left(P_k^f + P_k^{b-}\right)^{-1}P_k^f \tag{2.367}$$

and

$$P_k = P_k^f - P_k^f S_k^-\left(I + P_k^f S_k^-\right)^{-1}P_k^f \tag{2.368}$$

Note that the smoothed covariance P_k is smaller than or equal to the Kalman filter covariance P_k^f, so that using future data adds information. See Figure 2.23.

The *a posteriori* smoothed estimate is (by "Millman's theorem")

$$\hat{x}_k = P_k\left[\left(P_k^f\right)^{-1}\hat{x}_k^f + \left(P_k^{b-}\right)^{-1}\hat{x}_k^{b-}\right] \tag{2.369}$$

Use Equation 2.366b to write

$$\hat{x}_k = \left(I + P_k^f S_k^-\right)^{-1}\hat{x}_k^f + P_k\hat{y}_k^- \tag{2.370}$$

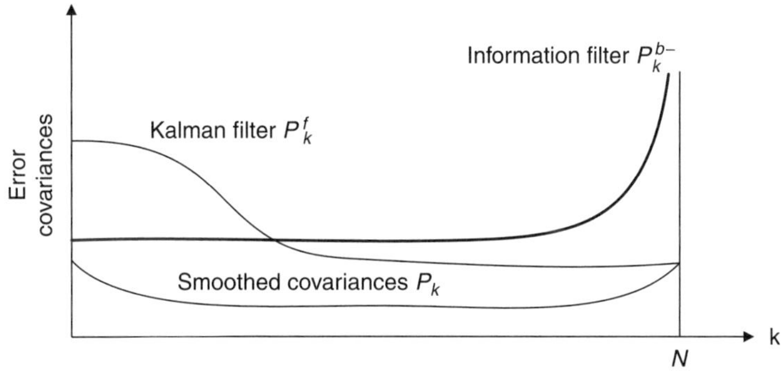

FIGURE 2.23
Relation of forward, backward, and smoothed error covariances.

so that by the matrix inversion lemma (or we can use Equation 2.368 in Equation 2.369)

$$\begin{aligned}\hat{x}_k &= (I - K_k)\hat{x}_k^f + P_k\hat{y}_k^- \\ &= \hat{x}_k^f + \left(P_k\hat{y}_k^- - K_k\hat{x}_k^f\right)\end{aligned} \tag{2.371}$$

where the "smoother gain" is

$$K_k = P_k^f S_k^- \left(I + P_k^f S_k^-\right)^{-1} \tag{2.372}$$

In terms of this gain, smoother update (Equation 2.368) becomes

$$P_k = (I - K_k)P_k^f \tag{2.373}$$

The entire smoothing scheme, which includes a Kalman filter, an information filter, and a smoother is shown in Table 2.4. A corresponding scheme can be derived for the *a priori* smoothed estimate $\hat{x}_k^-$.

A word on the boundary conditions is in order. The initial conditions on the forward filter are $\hat{x}_0^f = \bar{x}_0, P_0^f = P_0$. At time N we have $P_N = P_N^f$, since the filtered and smoothed estimates for x_N are the same. According to Equation 2.366b then, $S_N^- = 0$. The boundary condition $\hat{x}_N^{b-}$ is not known, but according to Equation 2.359b, $\hat{y}_N^- = 0$. These are the boundary conditions needed to start up the backward filter. (Now it is evident why we propagate $\hat{y}_k^-$ instead of $\hat{x}_k^{b-}$ in the information filter; we do not know the boundary condition $\hat{x}_N^{b-}$.)

It is possible in some cases that $P_k = P_k^f$, therefore the smoothed estimate is no better than the filtered estimate. It is also possible for only some of the directions of the state $\hat{x}_k$ to be improved by smoothing. These are called *smoothable* states. It can be shown that only those states which are $(A, G\sqrt{Q})$ reachable are smoothable. In particular, when $Q = 0$, smoothing does not improve the estimate. For further discussion, see Gelb (1974) and Fraser (1967). Additional references on smoothing are Anderson and Moore (1979), Bryson and Ho (1975), and Maybeck (1979).

TABLE 2.4
Optimal Smoothing Scheme

System and measurement model

$$\begin{aligned} x_{k+1} &= Ax_k + Bu_k + Gw_k \\ z_k &= Hx_k + v_k, \\ x &\sim (\bar{x}_0, P_0), w_k \sim (0, Q), v_k \sim (0, R). \\ & x_0, w_k, v_k, \text{ uncorrelated.} \end{aligned}$$

Forward filter
Initialization

$$\hat{x}_0^f = \bar{x}_0, P_0^f = P_0$$

Time update

$$\begin{aligned} P_{k+1}^{f-} &= AP_k^f A^T + GQG^T \\ \hat{x}_{k+1}^{f-} &= A\hat{x}_k^f + Bu_k \end{aligned}$$

Measurement update

$$\begin{aligned} K_k^f &= P_k^{f-} H^T (HP_k^{f-} H^T + R)^{-1} \\ P_k^f &= (I - K_k^f H) P_k^{f-} \\ \hat{x}_k^f &= \hat{x}_k^{f-} + K_k^f (z_k - H\hat{x}_k^{f-}) \end{aligned}$$

Backward filter
Initialization

$$\hat{y}_N^- = 0, S_N^- = 0$$

Measurement update

$$\begin{aligned} S_k &= S_k^- + H^T R^{-1} H \\ \hat{y}_k &= \hat{y}_k^- + H^T R^{-1} z_k \end{aligned}$$

Time update

$$\begin{aligned} K_k^b &= S_{k+1} G (G^T S_{k+1} G + Q^{-1})^{-1} \\ S_k^- &= A^T (I - K_k^b G^T) S_{k+1} A \\ \hat{y}_k^- &= A^T (I - K_k^b G^T)(\hat{y}_{k+1} - S_{k+1} Bu_k) \end{aligned}$$

Smoother

$$\begin{aligned} K_k &= P_k^f S_k^- (I + P_k^f S_k^-)^{-1} \\ P_k &= (I - K_k) P_k^f \\ \hat{x}_k &= (I - K_k)\hat{x}_k^f + P_k \hat{y}_k^- \end{aligned}$$

2.8.3 Rauch–Tung–Striebel Smoother

The smoothing scheme in Table 2.4 requires three components: a forward filter, a separate backward filter, and a separate smoother. It is possible to devise a scheme with only two components: a forward filter and a backward smoother algorithm that incorporates both the backward filter and the smoother. This

Rauch–Tung–Striebel formulation is more convenient than the scheme we have just presented. See Rauch, Tung, and Striebel (1965).

To derive a backward recursion for the smoothed error covariance P_k that depends only on the forward filter covariances, we can eliminate $S_k^- = (P_k^{b-})^{-1}$ and $S_k = (P_k^b)^{-1}$ from the smoother equations as follows.

First note that by using Equations 2.351, 2.348b, and 2.366a we have

$$\begin{aligned} P_{k+1}^b &= \left[\left(P_{k+1}^b\right)^{-1} + H^T R^{-1} H\right]^{-1} \\ &= \left[\left(P_{k+1}^{b-}\right)^{-1} + \left(P_{k+1}^f\right)^{-1} - \left(P_{k+1}^{f-}\right)^{-1}\right]^{-1} \\ &= \left[\left(P_{k+1}^{-1}\right) - \left(P_{k+1}^{f-}\right)^{-1}\right]^{-1} \end{aligned} \tag{2.374}$$

Therefore, by Equations 2.352, 2.346, and 2.374

$$\begin{aligned} \left\{P_k^f + P_k^{b-}\right\}^{-1} &= \left\{P_k^f + A^{-1}\left(P_{k+1}^b + GQG^T\right)A^{-T}\right\}^{-1} \\ &= A^T\left\{AP_{k+1}^f A^T + GQG^T + P_{k+1}^b\right\}^{-1} A \\ &= A^T\left\{P_{k+1}^{f-} + P_{k+1}^b\right\}^{-1} A \\ &= A^T\left\{P_{k+1}^{f-} + \left[P_{k+1}^{-1} - \left(P_{k+1}^{f-}\right)^{-1}\right]^{-1}\right\}^{-1} A \\ &= A^T\left(P_{k+1}^{f-}\right)^{-1}\left\{\left(P_{k+1}^{f-}\right)^{-1}\right. \\ &\quad \left. + \left(P_{k+1}^{f-}\right)^{-1}\left[P_{k+1}^{-1} - \left(P_{k+1}^{f-}\right)^{-1}\right]^{-1}\left(P_{k+1}^{f-}\right)^{-1}\right\}^{-1}\left(P_{k+1}^{f-}\right)^{-1} A \end{aligned}$$

According to the matrix inversion lemma, there results

$$\left\{P_k^f + P_k^{b-}\right\}^{-1} = A^T\left(P_{k+1}^{f-}\right)^{-1}\left\{P_{k+1}^{f-} - P_{k+1}\right\}\left(P_{k+1}^{f-}\right)^{-1} A$$

Using this in Equation 2.367 yields

$$P_k = P_k^f - P_k^f A^T\left(P_{k+1}^{f-}\right)^{-1}\left(P_{k+1}^{f-} - P_{k+1}\right)\left(P_{k+1}^{f-}\right)^{-1} A P_k^f \tag{2.375}$$

By defining the gain matrix

$$F_k = P_k^f A^T\left(P_{k+1}^{f-}\right)^{-1} \tag{2.376}$$

this becomes

$$P_k = P_k^f - F_k\left(P_{k+1}^{f-} - P_{k+1}\right)F_k^T \tag{2.377}$$

Equations 2.376 and 2.377 are the desired backward recursion for the smoothed error covariance P_k. They depend only on $\{P_k^f\}$ and $\{P_k^{f-}\}$, so that only the forward filter must be run before using them.

By the same sort of manipulations, the recursive smoother measurement update can be determined to be

$$\hat{x}_k = \hat{x}_k^f + F_k\left(\hat{x}_{k+1} - \hat{x}_{k+1}^{f-}\right) \tag{2.378}$$

The Rauch–Tung–Striebel smoother algorithm is thus as follows. First run the usual Kalman filter on the data. Then run the recursive smoother defined by Equations 2.376 through 2.378 backward for $k = N-1, N-2, \ldots$. The initial conditions for the smoother are

$$P_N = P_N^f, \qquad \hat{x}_N = \hat{x}_N^f \tag{2.379}$$

It is worth remarking that the backward recursive smoother depends neither on the data nor on the deterministic input.

Problems

Section 2.1

1. Stabilization via Riccati Equation. The plant is

$$x_{k+1} = \begin{bmatrix} 1 & 1 \\ 0 & 0 \end{bmatrix} x_k + \begin{bmatrix} 0 \\ 1 \end{bmatrix} u_k \tag{2.380}$$

$$z_k = \begin{bmatrix} 1 & 0 \end{bmatrix} x_k$$

and we want to design a stable observer.

(a) Solve the algebraic Riccati equation with $DD^T = 1$, $G^T = [0 \;\; \sqrt{q}]$. Hint: Substitute

$$P = \begin{bmatrix} p_1 & p_2 \\ p_2 & p_3 \end{bmatrix}$$

and solve for the p_i.

(b) Find the observer gain L.

(c) Plot the observer poles as q goes from 0 to ∞.

Section 2.2

2. Writing the Lyapunov Equation as a Vector Equation. Let $\otimes$ denote Kronecker product (Appendix) and $s(\cdot)$ denote the stacking operator. Show that the Lyapunov (matrix) equation $P_{k+1} = AP_kA^T + GG^T$ is equivalent to the *vector* equation

$$s(P_{k+1}) = (A \otimes A)s(P_k) + s(GG^T)$$

3. Solutions to Algebraic Lyapunov Equations. Use the results of the previous problem to show that $P = APA^T + GG^T$ has a unique solution if and only if $\lambda_i\lambda_j \neq 1$ for all i and j, where λ_i is an eigenvalue of A. Show that the stability of A is sufficient to guarantee this.

4. (a) Find *all* possible solutions to $P = APA^T + GG^T$ if

$$A = \begin{bmatrix} \frac{1}{2} & 1 \\ 0 & -1 \end{bmatrix}, \qquad G = \begin{bmatrix} 2 \\ 0 \end{bmatrix}$$

Interpret this in the light of the previous problem. Classify the solutions as positive definite, negative definite, and so on.

(b) Repeat if $G = [2 \quad 1]^T$.

(c) Repeat if

$$A = \begin{bmatrix} \frac{1}{2} & 1 \\ 0 & -\frac{1}{2} \end{bmatrix}, \qquad G = \begin{bmatrix} 2 \\ 1 \end{bmatrix}$$

5. Analytic Solution to the Lyapunov Equation. Show that the solution to $P_{k+1} = AP_kA^T + GQG^T$ for $k \geq 0$ is

$$P_k = A^k P_0 (A^T)^k + \sum_{i=0}^{k-1} A^i GQG^T (A^T)^i$$

6. Stochastic System

$$x_{k+1} = \begin{bmatrix} \frac{1}{2} & 1 \\ 0 & -\frac{1}{2} \end{bmatrix} x_k + \begin{bmatrix} 2 \\ 1 \end{bmatrix} w_k \tag{2.381}$$

$$z_k = \begin{bmatrix} 1 & 0 \end{bmatrix} x_k + v_k$$

where $w_k \sim (0, 1)$, $v_k \sim (0, 2)$ are white and uncorrelated.

(a) Find analytic expressions for $\bar{x}_k$ and P_{x_k} if $\bar{x}_0 = [1 \quad 1]^T$, $P_{x_0} = I$.

(b) Find steady-state values $\bar{x}_\infty$, P_{x_∞}.

(c) Find spectral densities of x_k and z_k in steady state.

Section 2.3

7. Recursion for *A Posteriori* Error Covariance

(a) Show that a recursion for P_k is given by

$$\begin{aligned} P_{k+1} = (AP_kA^T + GQG^T) - (AP_kA^T + GQG^T)H^T \\ \times [H(AP_kA^T + GQG^T)H^T + R]^{-1} H(AP_kA^T + GQG^T) \end{aligned} \tag{2.382}$$

or, if $Q = 0$,

$$P_{k+1} = A(P_k^{-1} + A^T H^T R^{-1} HA)^{-1} A^T \tag{2.383}$$

(b) Show that if $|A| \neq 0$ (which is always true if A is obtained by sampling a continuous system) then a recursion for the *a posteriori* information matrix when $Q = 0$ is given by

$$P_{k+1}^{-1} = A^{-T} P_k^{-1} A^{-1} + H^T R^{-1} H \tag{2.384}$$

8. Riccati Equation as Lyapunov Equation for Closed-Loop System. Show that Equations 2.53 and 2.58 are equivalent to Equation 2.54.

9. Information Matrix Formulation. The Kalman filter measurement update in terms of information matrices is

$$P_k^{-1} = (P_k^-)^{-1} + H^T R^{-1} H \tag{2.385}$$

Show that the information matrix time update is

$$(P_{k+1}^-)^{-1} = S_k - S_k G(G^T S_k G + Q^{-1})^{-1} G^T S_k \tag{2.386}$$

where

$$S_k \triangleq A^{-T} P_k^{-1} A^{-1} \tag{2.387}$$

Note that the measurement update is easier to perform in terms of measurement matrices, while the time update is easier in terms of covariance matrices.

10. Square Root Riccati Formulation. Define square roots S_k, S_k^- by $P_k = S_k S_k^T$, $P_k^- = S_k^- (S_k^-)^T$. The following results show how to propagate S_k, S_k^- instead of P_k, P_k^-, resulting in stable algorithms and a reduced numerical range. See Schmidt (1967, 1970), Businger and Golub (1965), Dyer and McReynolds (1969), Bierman (1977), and Morf and Kailath (1975).

(a) Show that time update 2.47 is equivalent to

$$[AS_k \quad C]T_1 = [S_{k+1}^- \quad 0] \tag{2.388}$$

where $CC^T = GQG^T$ and T_1 is any orthogonal transformation (i.e., $T_1 T_1^T = I$) selected so that S_{k+1}^- has n columns.

(b) Show that measurement update Equation 2.54 is equivalent to

$$\begin{bmatrix} D & HS_k^- \\ 0 & S_k^- \end{bmatrix} T_2 = \begin{bmatrix} (HP_k^- H^T + R)^{1/2} & 0 \\ \bar{K}_k & S_k \end{bmatrix} \tag{2.389}$$

where $R = DD^T$ and T_2 is any orthogonal transformation so that S_k has n columns. (You will find an expression for $\bar{K}_k$ during the demonstration.)

(c) Show that both updates can be expressed together (i.e., as in Equation 2.61) in square root form as

$$\begin{bmatrix} D & HS_k^- & 0 \\ 0 & AS_k^- & C \end{bmatrix} T_3 = \begin{bmatrix} (HP_k^- H^T + R)^{1/2} & 0 & 0 \\ A\bar{K}_k & S_{k+1}^- & 0 \end{bmatrix} \tag{2.390}$$

where T_3 is orthogonal.

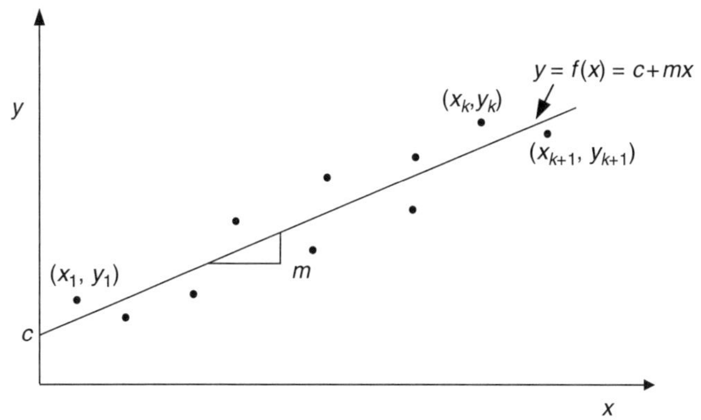

FIGURE 2.24
Recursive line of best fit.

11. Recursive Line of Best Fit. There is prescribed a set of data points (x_k, y_k) as in Figure 2.24 to which it is desired to fit a line $y = mx+c$. We want to use the Kalman filter to determine recursively the best estimates $\hat{m}_k, \hat{c}_k$ for the slope and intercept given the data through tie k. Define the state as

$$X_k = \begin{bmatrix} c \\ m \end{bmatrix} \tag{2.391}$$

Then $X_{k+1} = X_k$.

Suppose the measurements x_k are exact and y_k are corrupted by $(0, r)$ white noise v_k so that

$$z_k \triangleq y_k = H_k X_k + v_k \tag{2.392}$$

with $H_k = [1 \quad x_k]$.

(a) Write the filter equations.

(b) To obtain the maximum-likelihood estimate, set $P_0^{-1} = 0, \bar{X}_0 = 0$. Then the information matrix P_k^{-1} is given by Equation 2.65. Compute it.

(c) For the special case of uniformly spaced abscissas, we may have $x_k = k$. Find the error covariance in this case and show that the Kalman gain is

$$K_k = \frac{2}{k(k+1)} \begin{bmatrix} -(k+1) \\ 3 \end{bmatrix}$$

Write down the Kalman filter.

12. Prediction and Intermittent Measurements. Let

$$x_{k+1} = ax_k + w_k \tag{2.393}$$

$$z_k = x_k + v_k \tag{2.394}$$

with $w_k \sim (0, q)$, $v_k \sim (0, r_k)$ white and uncorrelated. Let $x_0 \sim (\bar{x}_0, p_0)$.

(a) Let $a = 0.9$, $q = 1$, $r_k = \frac{1}{2}$, and $x_0 \sim (1, 2)$. The data are $z_1 = 0.9$, $z_2 = 0.8$, $z_3 = 0.7$, $z_4 = 0.7$, $z_5 = 0.6$. Find filtered estimates $\hat{x}_k$ and error covariances p_k for $k = 0$ through 8. (Note, no data are available beyond $k = 5$.)

(b) Find the filtered maximum-likelihood estimate $\hat{x}_k$, $k = 0, 1, 2, 3, 4, 5$, for (a) if all statistics are Gaussian.

(c) Repeat (a) if the available data are $z_1 = 0.9$, $z_3 = 0.7$, $z_4 = 0.7$, $z_6 = 0.53$, $z_8 = 0.43047$.

(d) Repeat (a) if $r_k = 0.9^k/2$, that is, the data get better with time.

(e) Now let $a = 1$, and $q = 0$, so there is no process noise. Show that after two measurements

$$p_2 = \left(\frac{1}{p_0} + \frac{1}{r_1} + \frac{1}{r_2}\right)^{-1} \tag{2.395}$$

$$\hat{x}_2 = p_2 \left(\frac{\bar{x}_0}{p_0} + \frac{z_1}{r_1} + \frac{z_2}{r_2}\right) \tag{2.396}$$

Compare to Millman's theorem.

13. The plant is

$$x_{k+1} = \begin{bmatrix} 0.9 & -1 \\ 0 & 0.8 \end{bmatrix} x_k + \begin{bmatrix} 0 \\ 1 \end{bmatrix} u_k + \begin{bmatrix} 0 \\ 1 \end{bmatrix} w_k$$

$$z_k = \begin{bmatrix} 1 & 0 \end{bmatrix} x_k + v_k$$

Let $w_k \sim (0, 2)$ and $v_k \sim (0, 1)$ be white, and $x_0 \sim (0, 1)$, with all three uncorrelated.

(a) Find $\bar{x}_k$ if $\hat{x}_0 = 0$, $u_k = u_{-1}(k)$ the unit step for $k = 0$ through 5.

(b) Find $\hat{x}_k$ and P_k for $k = 0$ through 5 if $u_k = u_{-1}(k)$ and the data are $z_1 = 0.1$, $z_2 = -1$, $z_3 = -2$, $z_4 = -5$, $z_5 = -7.2$.

(c) If all statistics are normal, find the filtered maximum-likelihood estimate $\hat{x}_k$, $k = 0, 1, 2, 3$, for (b).

14. Time-Varying System. Consider the time-varying system

$$x_{k+1} = 2^k x_k + (-1)^k u_k + w_k$$

$$z_k = x_k + v_k$$

with $x_0 \sim (1, 2)$, $w_k \sim (0, 1.1^k)$, $v_k \sim (0, 0.9^k)$. (That is, as time passes, the process noise increases and the measurement noise decreases.)

(a) Find error covariances for $k = 0, 1, 2, 3$.

(b) Suppose $u_k = 1$ for $k = 0, 1, 2, \ldots$ and the measurements are given by $z_1 = -5$, $z_2 = -10$, $z_3 = -30$. Find the state estimates $\hat{x}_k$ for $k = 0, 1, 2, 3$.

15. Scalar Algebraic Riccati Equation. Consider the scalar system in Examples 2.5 and 2.12.

(a) Let $h = 0$ so there are no measurements. Sketch the steady-state error variance p as a function of the plant matrix a. Interpret this.

(b) Let $h = 1$, $r = 1$, $a = \frac{1}{2}$. Write the algebraic Riccati equation. Plot the root locus of this equation as q varies from zero to infinity. For what values of q does there exist a positive root to the algebraic Riccati equation?

Section 2.4

16. There is prescribed the system

$$\dot{x} = \begin{bmatrix} 0 & 1 \\ -2 & -3 \end{bmatrix} x + \begin{bmatrix} 0 \\ 2 \end{bmatrix} u + w$$

$$z = \begin{bmatrix} 1 & 1 \end{bmatrix} x + v$$

with $w(t)$, $v(t)$, and $x(0)$ having means of zero and covariances of

$$Q = \begin{bmatrix} 1 & -2 \\ -2 & 4 \end{bmatrix}, \qquad R = 0.4, \qquad P_0 \begin{bmatrix} 2 & 1 \\ 1 & 1 \end{bmatrix}$$

respectively. The usual uncorrelatedness assumptions apply.

(a) Measurements z_k are taken every $T = 0.1$ s. Find the discretized system (not Euler's approximation).

(b) For the continuous system, find the mean $\bar{z}(t)$ if $u(t) = u_{-1}(t)$ the unit step.

(c) For the discretized system, find $\bar{z}_k$ for $k = 0, 1, 2, 3, 4$ if $u_k = u_{-1}(k)$ the unit step. Sketch $\bar{z}_k$ and $\bar{z}(t)$ on the same graph. Do they agree?

(d) Find error covariances P_k and Kalman gains K_k for $k = 0, 1, 2$ for the discretized system assuming no measurement at time $k = 0$.

(e) If $u_k = u_{-1}(k)$ and the data are $z_1 = 0.181$, $z_2 = 0.330$ find the estimates $\hat{x}_k$ for $k = 0, 1, 2$.

(f) Now, after completing (d) you discover that there is a change in plant and a measurement z_0 is taken. It is equal to 0.05. The remaining data and the input are given as in (e). Find the new optimal linear estimates $\hat{x}_k$ for $k = 0, 1, 2$.

17. Vibration Isolation System. A cargo vibration isolation system on an aircraft is shown in Figure 2.25a. The cargo mass is M, k is a spring constant, B is a damper, $y(t)$ is vertical position, and $f(t)$ is an input control force. An electrical analog is shown in Figure 2.25b, where $F(t)$ is the integral of $f(t)$.

(a) Write the state equation if the state is defined as $x = [y \quad \dot{y}]^T$. To compensate for unmodeled dynamics, add a white process noise $w(t) \sim (0, \frac{1}{2})$ to $\ddot{y}(t)$.

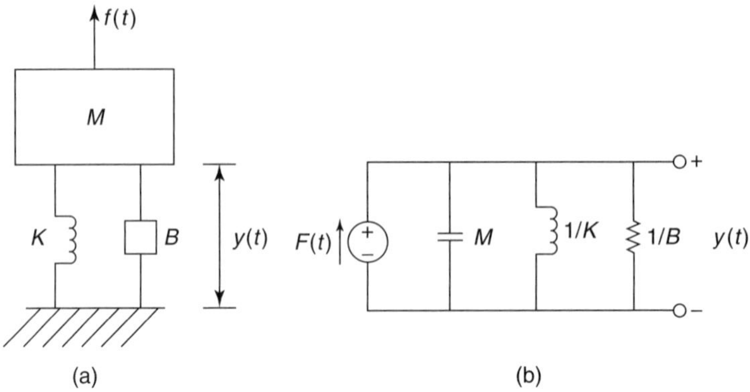

FIGURE 2.25
(a) Vibration isolation system. (b) Electrical analog.

(b) We are interested in the vertical position $y(t)$ of the cargo. Unfortunately, we can only measure the velocity $\dot{y}(t)$, which is done by using a device with a gain of 1 that adds white measurement noise $v(t) \sim (0, \frac{1}{2})$. Write the measurement equation.

(c) The onboard flight computer samples the velocity measurement every $T = 50$ ms. Find the discretized system (not Euler's approximation) which we can use to run a Kalman filter on. Let $M = 2$, $K = 4$, $B = 6$. Note that the process noise in the discretized system excites *both* states, although $w(t)$ excites only the velocity in the continuous system.

(d) The next step is to find the Kalman gains to store *a priori* in the computer so that we can save time during the online computation of the estimates. Assume no measurement at time $k = 0$. Find P_k and K_k for $k = 0, 1, 2, 3$.

(e) Let

$$x(0) \sim \left(\begin{bmatrix} 0 \\ 1 \end{bmatrix}, \begin{bmatrix} 0.5 & 0 \\ 0 & 0.5 \end{bmatrix} \right)$$

and suppose the data samples are $z_1 = 0.429$, $z_2 = 0.366$, $z_3 = 0.311$. Find estimates for the position $y(kT)$ for $k = 0, 1, 2, 3$ if $f(t) = 0$.

18. (a) Find the differential equation relating $x(t)$ to $u(t)$ in Figure 2.26.

(b) We take measurements $z(t)$ of $x(t)$ every 50 ms and get the results

Time t (s)	$u(t)$	$z(t)$
0.05	0.95	0.05
0.10	1.00	0.20
0.15	1.10	0.24
0.20	1.01	0.30

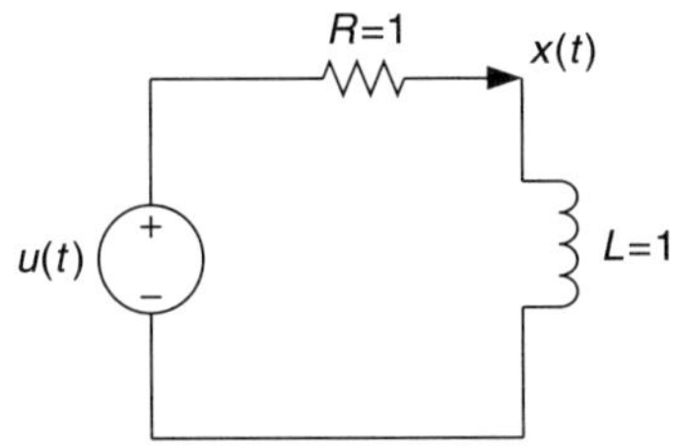

FIGURE 2.26
An electric circuit.

The measurements are corrupted by additive zero-mean white noise with a variance of 0.1. We know that $x(0) \sim (0, 0.2)$. Let $u(0) = 1.05$.

Assume that the plant has process noise which we can model as being zero-mean white noise with variance of 2 added to the derivative of $x(t)$. The usual uncorrelatedness assumptions hold.

Find the best estimates for $x(t)$ at $t = 0, 0.05, 0.10, 0.15$. (Use three decimal places.)

Section 2.5

19. Limiting Behavior. Let

$$x_{k+1} = \begin{bmatrix} 4 & 0 \\ 0 & 1 \end{bmatrix} x_k + w_k, \qquad w_k \sim \left(0, \begin{bmatrix} 0 & 0 \\ 0 & 0.5 \end{bmatrix}\right), \qquad x_0 \sim (0, P_0)$$

$$z_k = Hx_k + v_k, \qquad v_k \sim (0, 1)$$

(a) Let $H = [0 \quad 3]$. Find P_k for $k = 0, 1, 2, 3$ if

i. $P_0 = 0$.

ii. $P_0 = I$.

(b) Repeat (a) if $H = [3 \quad 3]$.

(c) Discuss. What is going on here?

Section 2.6

20. Wiener Filter for Optimal Estimation of the State. Prove the statements accompanying Equation 2.281. (Assume A and H are known. Then suppose that the optimal filter for estimating $y_k = Hx_k$ is given by Equation 2.274 using a gain of K_1. Suppose further that $\hat{x}_k$ is provided by Equation 2.281 with a gain K_2. Show that observability implies that $K_1 = K_2$. See Anderson and Moore (1979).)

21. Kalman and Wiener Filters. In this problem we consider a rather artificial discrete system which is simple enough to obtain analytic results.

Let

$$x_{k+1} = \begin{bmatrix} 0 & 1 \\ 0 & 0 \end{bmatrix} x_k + \begin{bmatrix} q \\ 1 \end{bmatrix} w_k$$

$$z_k = \begin{bmatrix} 1 & 0 \end{bmatrix} x_k + v_k$$

with $w_k \sim (0,1)$, $v_k \sim (0,1)$ white and uncorrelated. Let $q = 1$.

(a) Solve the ARE to determine the optimal steady-state Kalman filter. Find the filter poles and the transfer function from data z_k to estimate $\hat{x}_k^-$.

(b) Find spectral density $\Phi_Z(z)$. Factor to obtain the whitening filter $W(z)$. Hence, find the Wiener filter. Compare with (a).

(c) Use the Chang–Letov equation to find the steady-state Kalman filter.

(d) Find the steady-state Kalman filter if $q = 0$ so that the process noise does not drive state component one.

Section 2.7

22. Colored Measurement Noise due to Tachometer Vibration. A DC servo motor is described by the transfer function

$$\frac{\theta}{e} = \frac{1}{s(1 + 0.2s)} \tag{2.397}$$

where $e(t)$ is applied voltage and θ is shaft angle. Angular velocity $\dot{\theta}(t)$ is measured using a tachometer which yields a voltage of $y(t) = \dot{\theta}(t)$. Tachometer vibration results in an additive measurement noise $v(t)$ with spectrum

$$\Phi_v(\omega) = \frac{0.2(\omega^2 + 4\pi^2)}{\omega^4 + 4(0.01 - 2\pi^2)\omega^2 + 16\pi^4} \tag{2.398}$$

The sampling period for the measurements is $T = 20$ ms. Assume that terms of the order of T_2 can be ignored.

(a) Set up the continuous state and measurement equations.

(b) Find a measurement noise shaping filter in state variable form.

(c) Use state augmentation to convert to an equivalent problem with white measurement noise. Sample this system and set up the discrete Kalman filter equations. (Use Euler's approximation to simplify this.)

(d) Sample the plant and shaping filter in (a) and (b). Now use the derived measurement approach to convert to an equivalent system driven by white noises. Find all covariances and set up the discrete Kalman filtering problem.

23. Riccati Equation as Lyapunov Equation for Closed-Loop System. Define K_k by Equation 2.318. Show that if $Q = I$ and $R = SS^T$ then Equation 2.321 can be written as

$$P_{k+1}^- = (A - K_k H)P_k^-(A - K_k H)^T + (G - K_k S)(G - K_k S)^T$$

This is the Joseph stabilized formulation. Note that this is a *Lyapunov* equation for the closed-loop system formed by applying the time-varying output injection K_k.

24. Correlated Measurement and Process Noise. In Problem 13, let $\overline{v_k w_k} = -0.5$. Repeat (b) of the problem.

25. Colored Process Noise. The plant

$$x_{k+1} = 0.9x_k + w_k$$
$$z_k = x_k + v_k$$

has $v_k \sim (0, 1)$ white, but the process noise has spectral density of

$$\phi_w(\omega) = \frac{2 + 2\cos\omega}{\frac{4}{3} - \cos\omega}$$

The two noises v_k and w_k are uncorrelated.

(a) Find a shaping filter for w_k.

(b) Write the augmented state equations and set up the Kalman filter equations.

Section 2.8

26. *A Priori* Optimal Smoother. Derive a smoothing scheme like the one in Table 2.4 for the *a priori* smoothed estimate $\hat{x}_k^-$.

27. Find optimal *smoothed* estimates for Problem 12(a) using:

(a) The scheme of Table 2.4.

(b) The Rauch–Tung–Striebel smoother.

28. Repeat Problem 27 for the system of Problem 13.

3

Continuous-Time Kalman Filter

If discrete measurements are taken, whether they come from a discrete or a continuous system, the discrete Kalman filter can be used. The continuous Kalman filter is used when the measurements are continuous functions of time. Discrete measurements arise when a system is sampled, perhaps as part of a digital control scheme. Because of today's advanced microprocessor technology and the fact that microprocessors can provide greater accuracy and computing power than analog computers, digital control is being used more frequently instead of the classical analog control methods. This means that for modern control applications, the discrete Kalman filter is usually used.

However, a thorough study of optimal estimation must include the continuous Kalman filter. Its relation to the Wiener filter provides an essential link between classical and modern techniques, and it yields some intuition which is helpful in a discussion of nonlinear estimation.

3.1 Derivation from Discrete Kalman Filter

There are several ways to derive the continuous-time Kalman filter. One of the most satisfying is the derivation from the Wiener–Hopf equation presented in Section 3.3. In this section we present derivation that is based on "unsampling" the discrete-time Kalman filter. This approach proves an understanding of the relation between the discrete and continuous filters. It also provides insight into the behavior of the discrete Kalman gain as the sampling period goes to zero.

Suppose there is prescribed the continuous time-invariant plant

$$\dot{x}(t) = Ax(t) + Bu(t) + Gw(t) \tag{3.1a}$$

$$z(t) = Hx(t) + v(t) \tag{3.1b}$$

with $w(t) \sim (0, Q)$ and $v(t) \sim (0, R)$ white; $x(0) \sim (\bar{x}_0, P_0)$; and $w(t), v(t)$ and $x(0)$ mutually uncorrelated. Then if sampling period T is small, we can use Euler's approximation to write the discretized version of Equation 3.1 as

$$x_{k+1} = (I + AT)x_k + BTu_k + Gw_k \tag{3.2a}$$

$$z_k = Hx_k + v_k \tag{3.2b}$$

with $w_k \sim (0, QT)$ and $v_k \sim (0, R/T)$ white; $x(0) \sim (\bar{x}_0, P_0)$; and $x(0)$ mutually uncorrelated.

By using Tables 2.1 and 2.2 we can write the covariance update equations for Equation 3.2 as

$$P_{k+1}^- = (I + AT)P_k(I + AT)^T + GQG^T T \tag{3.3}$$

$$K_{k+1} = P_{k+1}^- H^T \left(HP_{k+1}^- H^T + \frac{R}{T} \right)^{-1} \tag{3.4}$$

$$P_{k+1} = (I - K_{k+1}H)P_{k+1}^- \tag{3.5}$$

We shall manipulate these equations and then allow T to go to zero to find the continuous covariance update and Kalman gain.

Let us first examine the behavior of the discrete Kalman gain K_k as T tends to zero. By Equation 3.4 we have

$$\frac{1}{T}K_k = P_k^- H^T (HP_k^- H^T T + R)^{-1} \tag{3.6}$$

so that

$$\lim_{T \to 0} \frac{1}{T} K_k = P_k^- H^T R^{-1} \tag{3.7}$$

This implies that

$$\lim_{T \to 0} K_k = 0 \tag{3.8}$$

the discrete Kalman gain tends to zero as the sampling period becomes small. This result is worth remembering when designing discrete Kalman filters for continuous-time systems. For our purposes now, it means that the continuous Kalman gain $K(t)$ should not be defined by $K(kT) = K_k$ in the limit as $T \to 0$.

Turning to Equation 3.3, we have

$$P_{k+1}^- = P_k + (AP_k + P_k A^T + GQG^T)T + 0(T^2) \tag{3.9}$$

where $O(T^2)$ represents terms of order T^2. Substituting Equation 3.5 into this equation there results

$$P_{k+1}^- = (I - K_k H)P_k^- + [A(I - K_k H)P_k^- + (I - K_k H)P_k^- A^T + GQG^T]T + 0(T^2)$$

or, on dividing by T,

$$\begin{aligned} \frac{1}{T}(P_{k+1}^- - P_k^-) = (&AP_k^- + P_k^- A^T + GQG^T - AK_k HP_k^- \\ &- K_k HP_k^- A^T) - \frac{1}{T} K_k HP_k^- + 0(T) \end{aligned} \tag{3.10}$$

In the limit as $T \to 0$ the continuous-error covariance $P(t)$ satisfies

$$P(kT) = P_k^- \tag{3.11}$$

so letting T tend to zero in Equation 3.10 there results (use Equations 3.7 and 3.8)

$$\dot{P}(t) = AP(t) + P(t)A^T + GQG^T - P(t)H^T R^{-1} HP(t) \tag{3.12}$$

This is the *continuous-time Riccati equation* for propagation of the error covariance. It is the continuous-time counterpart of Equation 2.61.

The discrete estimate update for Equation 3.2 is given by Equation 2.62, or

$$\hat{x}_{k+1} = (I + AT)\hat{x}_k + BTu_k + K_{k+1}[z_{k+1} - H(1 + AT)\hat{x}_k - HBTu_k] \tag{3.13}$$

which on dividing by T can be written as

$$\frac{\hat{x}_{k+1} - \hat{x}_k}{T} = A\hat{x}_k + Bu_k + \frac{K_{k+1}}{T}[z_{k+1} - H\hat{x}_k - H(A\hat{x}_k + Bu_k)T] \tag{3.14}$$

Since, in the limit as $T \to 0, \hat{x}(t)$ satisfies

$$\hat{x}(kT) = \hat{x}_k \tag{3.15}$$

we obtain in the limit

$$\dot{\hat{x}}(t) = A\hat{x}(t) + Bu(t) + P(t)H^T R^{-1}[z(t) - H\hat{x}(t)] \tag{3.16}$$

This is the estimate update equation. It is a differential equation for the estimate $\hat{x}(t)$ with initial condition $\hat{x}(0) = \bar{x}_0$. If we define the continuous Kalman gain by

$$K(kT) = \frac{1}{T} K_k \tag{3.17}$$

in the limit as $T \to 0$, then

$$K(t) = P(t)H^T R^{-1} \tag{3.18}$$

and

$$\dot{\hat{x}} = A\hat{x} + Bu + K(z - H\hat{x}) \tag{3.19}$$

Equations 3.12, 3.18, and 3.19 are the continuous-time Kalman filter for Equation 3.1. They are summarized in Table 3.1. If matrices A, B, G, Q, and R are time-varying, these equations still hold.

TABLE 3.1
Continuous-Time Kalman Filter

System model and measure model

$$\dot{x} = Ax + Bu + Gw \tag{3.20a}$$

$$z = Hx + v \tag{3.20b}$$

$$x(0) \sim (\bar{x}_0, P_0), w \sim (0, Q), v \sim (0, R)$$

Assumptions

$\{w(t)\}$ and $\{v(t)\}$ are white noise processes uncorrelated with $x(0)$ and with each other. $R > 0$.

Initialization

$$P(0) = P_0, \hat{x}(0) = \bar{x}_0$$

Error covariance update

$$\dot{P} = AP + PA^T + GQG^T - PH^T R^{-1} HP \tag{3.21}$$

Kalman gain

$$K = PH^T R^{-1} \tag{3.22}$$

Estimate update

$$\dot{\hat{x}} = A\hat{x} + Bu + K(z - H\hat{x}) \tag{3.23}$$

It is often convenient to write $P(t)$ in terms of $K(t)$ as

$$\dot{P} = AP + PA^T + GQG^T - KRK^T \tag{3.24}$$

It should be clearly understood that while, in the limit as $T \to 0$, the discrete-error covariance sequence P_k^- is a sampled version of the continuous-error covariance $P(t)$, the discrete Kalman gain is *not* a sampled version of the continuous Kalman gain. Instead, K_k represents the samples of $TK(t)$ in the limit as $T \to 0$.

Figure 3.1 shows the relation between $P(t)$ and the discrete covariances P_k and P_k^- for the case when there is a measurement z_0. As $T \to 0, P_k$ and P_{k+1}^- tend to the same value since $(I + AT) \to I$ and $QT \to 0$ (see Equation 2.47). They both approach $P(kT)$.

A diagram of Equation 3.23 is shown in Figure 3.2. Note that it is a linear system with two parts: a model of the system dynamics (A, B, H) and an error correcting portion $K(z - H\hat{x})$. The Kalman filter is time-varying even when the original system (Equation 3.1) is time invariant. It has exactly the same form as the deterministic state observer of Section 2.1. It is worth remarking that according to Figure 3.2 the Kalman filter is a low-pass filter with time-varying feedback.

If all statistics are Gaussian, then the continuous Kalman filter provides the optimal estimate $\hat{x}(t)$. In general, for arbitrary statistics it provides the best *linear* estimate.

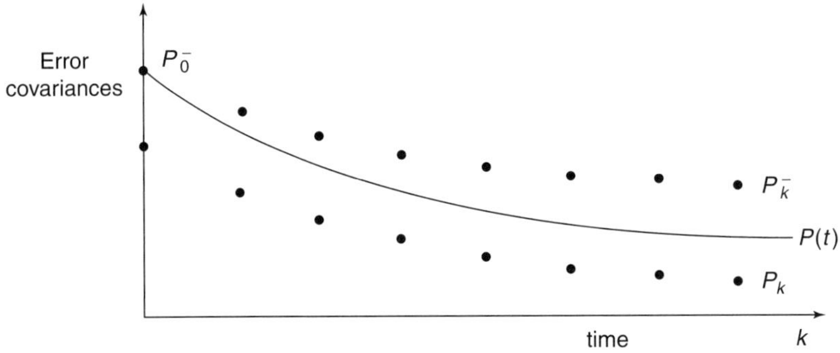

FIGURE 3.1
Continuous and discrete error covariances.

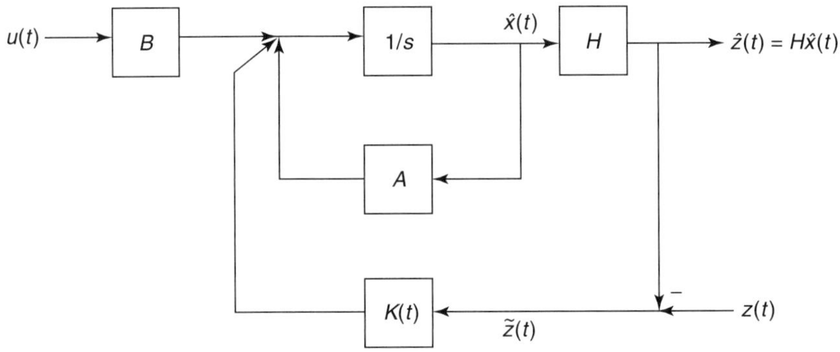

FIGURE 3.2
Continuous-time Kalman filter.

If all system matrices and noise covariances are known *a priori*, then $P(t)$ and $K(t)$ can be found and stored *before* any measurements are taken. This allows us to evaluate the filter design before we build it, and also saves computation time during implementation.

The continuous-time *residual* is defined as

$$\tilde{z}(t) = z(t) - H\hat{x}(t) \tag{3.25}$$

since $\hat{z}(t) = H\hat{x}(t)$ is an estimate for the data. We shall subsequently examine the properties of $\tilde{z}(t)$.

It should be noted that Q and R are not covariance matrices in the continuous time case; they are spectral density matrices. The covariance of $v(t)$, for example, is given by $R(t)\delta(t)$. (The continuous Kronecker delta has units of s^{-1}.)

In the discrete Kalman filter it is not strictly required that R be nonsingular; all we required was $|HP_k^- H^T + R| \neq 0$ for all k. In the continuous case, however, it is necessary that $|R| \neq 0$. If R is singular, we must use the *Deyst filter* (Example 3.8).

The continuous Kalman filter cannot be split up into separate time and measurement updates; there is no "predictor-corrector" formulation in the continuous-time case. (Note that as T tends to zero, P_{k+1}^- tends to P_k; so that in the limit the *a priori* and *a posteriori* error covariance sequences become the same sequence.) We can say, however, that the term $-PH^TR^{-1}HP$ in the error covariance update represents the decrease in $P(t)$ due to the measurements. If this term is deleted, we recover the analog of Section 2.2 for the continuous case. The following example illustrates.

Example 3.1 Linear Stochastic System

If $H = 0$ so that there are no measurements, then

$$\dot{P} = AP + PA^T + GQG^T \tag{3.26}$$

represents the propagation of the error covariance for the linear stochastic system

$$\dot{x} = Ax + Bu + Gw \tag{3.27}$$

It is a *continuous-time Lyapunov equation* which has similar behavior to the discrete version (Equation 2.28).

When $H = 0$ so that there are no measurements, Equation 3.19 becomes

$$\dot{\hat{x}} = A\hat{x} + Bu \tag{3.28}$$

so that the estimate propagates according to the deterministic version of the system. This is equivalent to Equation 2.27.

In Part II of the book, we shall need to know how the mean-square value of the state

$$X(t) = \overline{x(t)x^T(t)} \tag{3.29}$$

propagates. Let us derive a differential equation satisfied by $X(t)$ when the input $u(t)$ is equal to zero and there are no measurements.

Since there are no measurements, the optimal estimate is just the mean value of the unknown so that $\bar{x} = \hat{x}$. Then

$$P = \overline{(x - \bar{x})(x - \bar{x})^T} = X - \overline{x}\overline{x}^T$$

so that by Equation 3.28

$$\dot{P} = \dot{X} - \dot{\overline{x}}\overline{x}^T - \overline{x}\dot{\overline{x}}^T = \dot{X} - A\overline{x}\overline{x}^T - \overline{x}\overline{x}^TA^T$$

Equating this to Equation 3.26 yields

$$\dot{X} = AX + XA^T + GQG^T \tag{3.30}$$

The initial condition for this is

$$X(0) = P_0 + \bar{x}_0\bar{x}_0^T \tag{3.31}$$

Thus, in the absence of measurements and a deterministic input, $X(t)$ and $P(t)$ satisfy the same Lyapunov equation.

Let us now assume the scalar case $x \in R$ and repeat Example 2.2 for continuous time. Suppose $x_0 \sim (0, p_0)$ and $u(t) = u_{-1}(t)$, the unit step. Let the control weighting $b = 1$, and suppose $g = 1$ and $w(t) \sim (0, 1)$. Then, by using the usual state equation solution there results from Equation 3.28

$$\hat{x}(s) = \frac{1}{s-a} \cdot \frac{1}{s} = \frac{1/a}{s-a} + \frac{-1/a}{s}$$

so that

$$\hat{x}(t) = \frac{1}{a}(e^{at} - 1)u_{-1}(t) \tag{3.32}$$

This reaches a finite steady-state value only if $a < 0$; that is, if the system is asymptotically stable. In this case $\hat{x}(t)$ behaves as in Figure 3.3a.

Equation 3.26 becomes

$$\dot{p} = 2ap + 1 \tag{3.33}$$

with $p(0) = p_0$. This may be solved by *separation of variables*. Thus,

$$\int_{p0}^{p(t)} \frac{\mathrm{d}p}{2ap + 1} = \int_0^t \mathrm{d}t$$

or

$$p(t) = \left(p_0 + \frac{1}{2a}\right) e^{2at} - \frac{1}{2a} \tag{3.34}$$

There is a bounded limiting solution if and only if $a < 0$. In this case $p(t)$ behaves as in Figure 3.3b. ▯

3.2 Some Examples

The best way to gain some insight into the continuous Kalman filter is to look at some examples.

Example 3.2 Estimation of a Constant Scalar Unknown

This is the third example in the natural progression, which includes Examples 2.4 and 2.7. If x is a constant scalar unknown that is measured in the presence of additive noise, then

$$\begin{aligned} \dot{x} &= 0 \\ z &= x + v \end{aligned} \tag{3.35}$$

$v \sim (0, r^c)$. Let $x(0) \sim (x_0, p_0)$ be uncorrelated with $v(t)$.

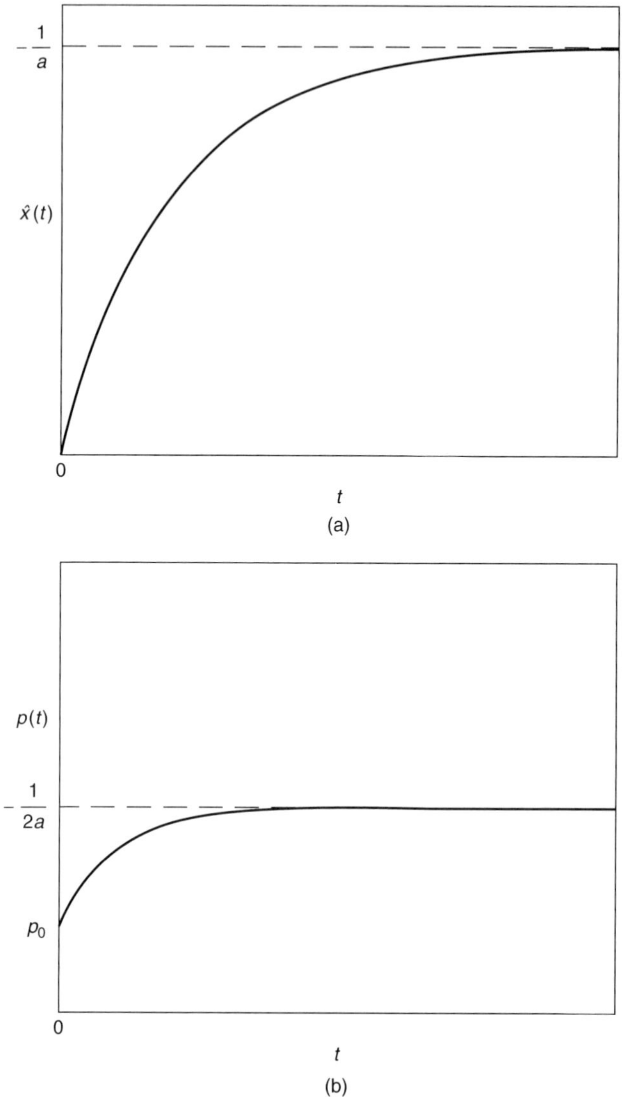

FIGURE 3.3
Propagation of (a) estimate and (b) error covariance.

Equation (3.21) becomes

$$\dot{p} = \frac{-p^2}{r^c} \tag{3.36}$$

which can be solved by separation of variables:

$$\int_{p0}^{p(t)} \frac{1}{p^2} \mathrm{d}p = -\frac{1}{r^c} \int_0^t \mathrm{d}t \tag{3.37}$$

so that

$$p(t) = \frac{p_0}{1 + (p_0/r^c)t} \tag{3.38}$$

Comparing this to Example 2.7, Equation 2.128

$$p(kT) = p_k \tag{3.39}$$

so that the discrete covariance sequence p_k simply corresponds to the samples of the continuous covariance $p(t)$.

The continuous Kalman gain is

$$K(t) = \frac{p_0/r^c}{1 + (p_0/r^c)t} \tag{3.40}$$

which is shown in Figure 3.4. Note that $K(t)$ vanishes as $t \to \infty$, meaning that for large t the measurements are disregarded as we become more and more certain of the estimate. As is seen from Example 2.7,

$$K(kT) = \frac{K_k}{T} \tag{3.41}$$

In this example, p_k and K_k correspond exactly to the samples of $p(t)$ and $TK(t)$ because the plant matrix and process noise in Equation 3.35 are equal to zero (so that the discretized plant matrix is equal to zero). By Equation 2.47 this means that $p_k = p_{k+1}^-$. □

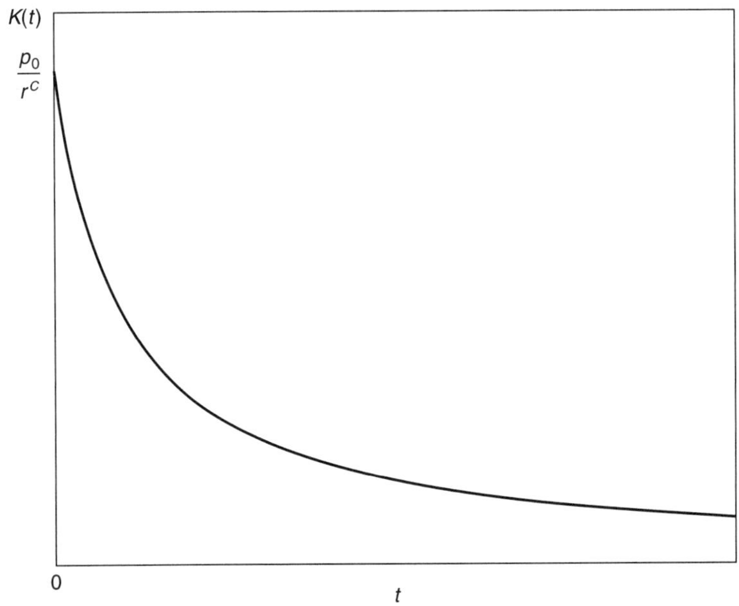

FIGURE 3.4
Kalman gain for estimation of a constant.

Example 3.3 Continuous Filtering for a Scalar System

a. Solution of Riccati equation

In the case of $x(t) \in R$, it is not difficult to solve the continuous Riccati equation analytically. We have, with $g = 1, p(0) = p_0$ (lowercase letters emphasize these are scalars)

$$\dot{p} = -\frac{h^2}{r}p^2 + 2ap + q \tag{3.42}$$

so that by separation of variables

$$\int_{p0}^{p(t)} \frac{dp}{(h^2/r)p^2 - 2ap - q} = -\int_0^t \mathrm{d}t \tag{3.43}$$

Integrating,

$$\left[\frac{(h^2/r)p - a - \beta}{(h^2/r)p - a + \beta}\right]\left[\frac{(h^2/r)p_0 - a + \beta}{(h^2/r)p_0 - a - \beta}\right] = e^{-2\beta t} \tag{3.44}$$

where

$$\beta = \sqrt{a^2 + h^2 q/r} \tag{3.45}$$

By defining (Bryson and Ho 1975)

$$p_1 = \frac{r}{h^2}(\beta - a), \quad p_2 = \frac{r}{h^2}(\beta + a) \tag{3.46}$$

this yields

$$p(t) = p_2 + \frac{p_1 + p_2}{[(p_0 + p_1)/(p_0 - p_2)]e^{2\beta t} - 1} \tag{3.47}$$

A sketch is provided in Figure 3.5.

The steady-state value of $p(t)$ is given by p_2, or, if $a > 0$,

$$p_\infty = \frac{2q}{\lambda}\left(1 + \sqrt{1 + \frac{\lambda}{2a}}\right) \tag{3.48}$$

where

$$\lambda = \frac{2h^2 q}{ar} \tag{3.49}$$

is a signal-to-noise ratio. A similar expression holds if $a < 0$. (See Example 2.1.)

The steady-state value p_2 is bounded if $h \neq 0$, corresponding to observability. It is strictly positive if $q \neq 0$, which corresponds to reachability, and $h \neq 0$. The conditions for existence and uniqueness of a steady-state solution for the continuous Riccati equation are the same as those found in Section 2.5 for the discrete equation, as we shall discuss in Section 3.4.

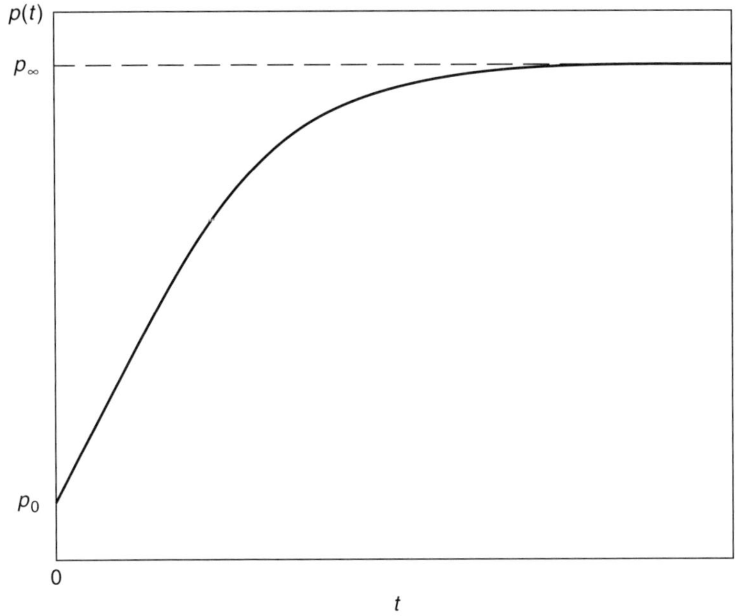

FIGURE 3.5
Error covariance for first-order system.

The entire Kalman filter, including the Riccati equation, for the first-order system is shown in Figure 3.6. Note that the Riccati equation manufacturing $p(t)$ is a sort of "doubled" or "squared" replica of the filter manufacturing $\hat{x}(t)$. This figure helps make the point that in the continuous Kalman filter, the data $z(t)$ are a continuous-time signal and all quantities vary continuously with time. To implement Figure 3.6, we would use either an analog computer, or a digital computer with an integration routine such as Runge–Kutta, as found in MATLAB.

We now consider some special cases. The next two parts of this example should be compared to Example 2.5.

b. Completely unreliable measurements

Let $r \to \infty$ so that the measurements are total unreliable. Then the Riccati equation is

$$\dot{p} = 2ap + q \tag{3.50}$$

This solution is

$$p(t) = e^{2at}p_0 + \int_0^t e^{2a(t-\tau)}q\mathrm{d}\tau \tag{3.51}$$

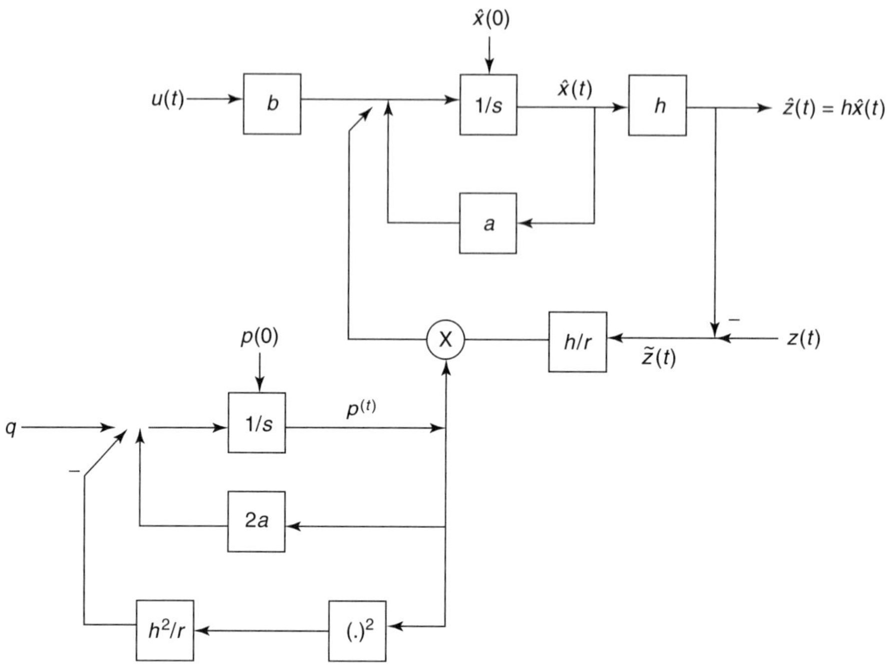

FIGURE 3.6
Kalman filter for first-order system.

or

$$p(t) = \left(p_0 + \frac{q}{2a}\right)e^{2at} - \frac{q}{2a} \tag{3.52}$$

There is a steady-state solution of

$$p_\infty = -\frac{q}{2a} \tag{3.53}$$

if and only if the original plant is stable, $a < 0$.

In this case the Kalman gain $K(t)$ is equal to zero, so the estimate update is

$$\dot{\hat{x}} = a\hat{x} + bu \tag{3.54}$$

with $\hat{x}(0) = \overline{x(0)}$. The estimate is a prediction based purely on the plant dynamics. This is as in Example 3.1.

c. No process noise

Let $q = 0$ so that the plant dynamics are purely deterministic. Then the solution of a applies with $\beta = |a|$.

If $a > 0$, the plant is unstable, then

$$\beta = a, \quad p_1 = 0, \quad p_2 = \frac{2ar}{h^2} \tag{3.55}$$

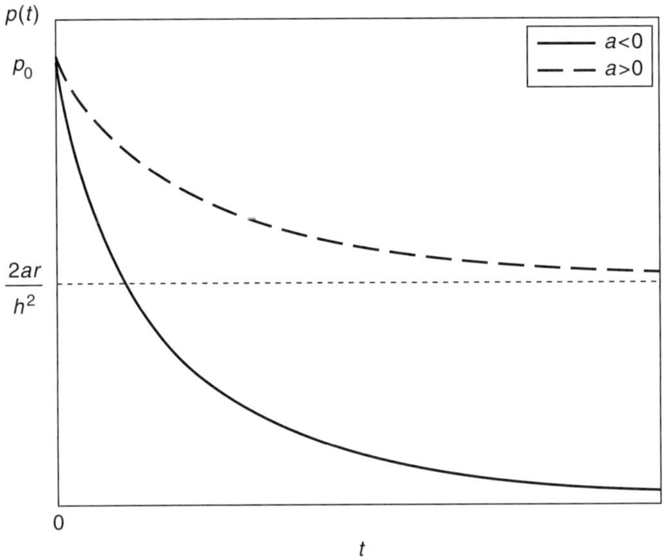

FIGURE 3.7
Error covariance with zero process noise for unstable plant and stable plant.

and $p(t)$ is given by Equation 3.42, which can be written as

$$p(t) = \frac{p_0}{p_0 h^2/2ar + (1 - p_0 h^2/2ar)e^{-2at}} \tag{3.56}$$

The steady-state solution is p_2, or

$$p_\infty = \frac{2ar}{h^2} \tag{3.57}$$

which is nonzero, and is bounded if $h \neq 0$.

If $a < 0$, the plant is stable, then

$$\beta = -a, \quad p_1 = \frac{2ar}{-h^2}, \quad p_2 = 0 \tag{3.58}$$

□

In this case $p(t)$ is also given by Equation 3.56, however, now

$$p_\infty = 0 \tag{3.59}$$

Figure 3.7 shows $p(t)$ for both unstable and stable plant. In either case, $p(t)$ is bounded if $h \neq 0$. If a is stable, then $K(t) \to 0$ and the measurements are discounted for large t. If a is unstable, then $K(t)$ approaches a nonzero steady-state value of

$$K_\infty = \frac{2a}{h} \tag{3.60}$$

so that the measurements are not discounted for large t.

d. Estimating a scalar Wiener process

Let $a = 0$ and $h = 1$ so that the plant is

$$\dot{x} = w \tag{3.61a}$$

$$z = x + v \tag{3.61b}$$

$w \sim (0, q), v \sim (0, r)$. Then

$$\beta = \sqrt{\frac{q}{r}}, p_1 = p_2 = \sqrt{qr}$$

The error covariance is

$$p(t) = \sqrt{qr} + \frac{2\sqrt{qr}}{[(p_0 + \sqrt{qr})/(p_0 - \sqrt{qr})]e^{2\sqrt{q/rt}} - 1} \tag{3.62}$$

Since $q \neq 0$ the process noise excites $x(t)$ and the error covariance reaches a nonzero limiting value. The steady-state error covariance and Kalman gain are

$$p_\infty = \sqrt{qr} \tag{3.63}$$

$$K_\infty = \sqrt{\frac{q}{r}} \tag{3.64}$$

□

Example 3.4 Continuous Filtering for Damped Harmonic Oscillator

In Example 2.13 we applied the discrete Kalman filter to the discretized harmonic oscillator. Let us discuss here the continuous Kalman filter. Suppose the plant is given by

$$\dot{x} = \begin{bmatrix} 0 & 1 \\ -\omega_n^2 & -2\alpha \end{bmatrix} x + \begin{bmatrix} 0 \\ 1 \end{bmatrix} w \tag{3.65}$$

$$z = [0 \quad 1]x + v \tag{3.66}$$

$w(t) \sim (0, q), v(t) \sim (0, r), x(0) \sim (\overline{x(0)}, P_0)$. This is a harmonic oscillator driven by white noise with a velocity measurement.

a. Riccati equation solution

To convert the Riccati equation into scalar equations, suppose that (P is symmetric)

$$P = \begin{bmatrix} p_1 & p_2 \\ p_2 & p_4 \end{bmatrix} \tag{3.67}$$

Then, Equation 3.21 yields the system of three coupled nonlinear differential equations

$$\dot{p}_1 = 2p_2 - \frac{1}{r}p_2^2 \tag{3.68a}$$

$$\dot{p}_2 = p_4 - \omega_n^2 p_1 - 2\alpha p_2 - \frac{1}{r}p_2 p_4 \tag{3.68b}$$

$$\dot{p}_4 = -2\omega_n^2 p_2 - 4\alpha p_4 - \frac{1}{r}p_4^2 + q \tag{3.68c}$$

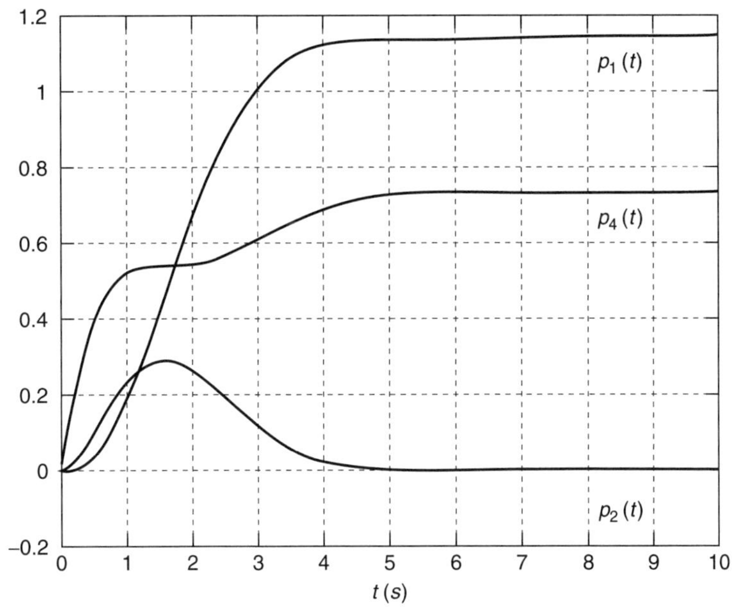

FIGURE 3.8
Error covariance terms for a harmonic oscillator.

These equations cannot easily be solved in closed form. The MATLAB code that uses a Runge–Kutta integrator is shown in Figure 3.9 and error covariance are plotted in Figure 3.8.

The Kalman gain is

$$K(t) = \frac{1}{r}\begin{bmatrix} p_2(t) \\ p_4(t) \end{bmatrix} \tag{3.69}$$

According to Figure 3.8, the first component of $K(t)$ tends to zero, while its second component tends to a nonzero steady-state value. The filter is

$$\dot{\hat{x}} = A\hat{x} + \frac{1}{r}\begin{bmatrix} p_2 \\ p_4 \end{bmatrix}(z - \hat{x}_2)$$

where x_2 is the second component of x. During the transient phase where $p_2(t) \neq 0$, the velocity measurement has a direct effect on the position estimate $\hat{x}_1$. At steady-state $p_2 = 0$ and the measurement has a direct effect only on the velocity estimate $\hat{x}^2$. □

b. Steady-state solution

It is easy to find the steady-state error covariance. At steady state, $\dot{p}_1 = \dot{p}_2 = \dot{p}_3 = 0$, so that Equation 3.68 becomes an algebraic equation. The steady-state

```
% Example 3.4
function f3p2d6()
    % Runge Kutta Integrator for Continuous time Kalman Filter
    options = odeset('RelTol',1e-4,'AbsTol',[1e-8 1e-8 1e-8]);
    % Runge-Kutta Integrator
    [t p] = ode45(@errcov,[0 10],[0 0 0],options);
    % Plot the error covariances
    plot(t,p(:,1),t,p(:,2),t,p(:,3));
    legend('p_1','p_2','p_3'); grid on; xlabel('t (secs.)');
return

function dp=errcov(t,p)
    dp = zeros(3,1);
    q=1; r=1; wns=0.64; a=0.16; % Constants
    dp(1)=2*p(2)-(1/r)*p(2)^2; % Differential Equations
    dp(2)=p(3)-wns*p(1)-2*a*p(2)-p(2)*p(3)/r;
    dp(3)=-2*wns*p(2)-4*a*p(3)-(1/r)*p(3).^2+q;
return
```

FIGURE 3.9
Harmonic oscillator error covariance MATLAB code using ode45.

solutions given by

$$p_1(\infty) = \frac{2\alpha r}{\omega_n^2}\left(\sqrt{1+\frac{q}{4\alpha^2 r}}-1\right) \tag{3.70a}$$

$$p_2(\infty) = 0 \tag{3.70b}$$

$$p_4(\infty) = 2\alpha r\left(\sqrt{1+\frac{q}{4\alpha^2 r}}-1\right) \tag{3.70c}$$

where we have selected positive roots (assuming $\alpha > 0$) since $p(\infty) \geq 0$. Note that

$$\lambda = \frac{q}{\alpha r} \tag{3.71}$$

can be interpreted as a signal-to-noise ratio. See Bryson and Ho (1975) and Parkus (1971).

This is a good example of simplifying a filter by *preliminary analysis* prior to implementation. The scalar equations (Equations 3.68a,b,c) are easier to program and require fewer operations to solve than the matrix Riccati equation. Even fairly complicated systems can be implemented by these techniques if their order is not too high.

3.3 Derivation from Wiener–Hope Equation

This section provides the complete connection between the Wiener filter and the Kalman filter. In section 1.5 we saw that the optimal linear filter $h(t,\tau)$ for estimating an unknown process $x(t)$ from a data process $z(t)$ given on

(t_0, t) is the solution to the Weiner–Hopf equation

$$R_{XZ}(t, u) = \int_{t_0}^{t} h(t, \tau) R_Z(\tau, u) \mathrm{d}\tau; \quad t_0 < u < t \tag{3.72}$$

We have selected the final time for which data are available as t to focus on the filtering problem. In terms of the time-varying filter $h(t,\ \tau)$, the optimal estimate for $x(t)$ is given by

$$\hat{x}(t) = \int_{t_0}^{t} h(t, \tau) z(\tau) \mathrm{d}\tau \tag{3.73}$$

The associated estimation error is defined as

$$\tilde{x}(t) = x(t) - \hat{x}(t) \tag{3.74}$$

with covariance given by

$$P(t) = R_X(t, t) - \int_{t_0}^{t} h(t, \tau) R_{ZX}(\tau, t) \mathrm{d}\tau \tag{3.75}$$

Wiener solved Equation 3.72 for the steady-state case with $t_0 = -\infty$, and $\{z\,(t)\}$ and $\{x(t)\}$ stationary processes. We derived the solution for the Wiener filter in Section 1.5. In the steady-state case the optimal filter is time invariant, and Wiener's solution was in the frequency domain. In 1961 Kalman and Bucy solved the Weiner–Hopf equation in the general nonstationary case. We follow here the solution procedure in Van Trees (1968).

The solution is somewhat roundabout, and the result is not an expression for $h(t, \tau)$. Instead, by working in the time domain a differential equation is found for $P(t)$. This turns out to be the Riccati equation. From $P(t), h(t, t)$ can be found, which leads to a method of finding $\hat{x}(t)$. The full impulse response $h(t, \tau)$ is not explicitly constructed. The solution is laid out in several subsections. They are

1. Introduction of a shaping filter
2. A differential equation for the optimal impulse response
3. A differential equation for the estimate
4. A differential equation for the error covariance
5. Discussion

3.3.1 Introduction of a Shaping Filter

The approach of Kalman and Bucy depends on assuming that the unknown $x(t)$ is generated by the *shaping filter*

$$\dot{x}(t) = Ax(t) + Gw(t) \tag{3.76a}$$

with $x(0) \sim (x_0, P_0)$. It is also assumed that the data $z(t)$ are related to $x(t)$ by the *linear measurement model*

$$z(t) = Hx(t) + v(t) \tag{3.76b}$$

The processes $w(t)$ and $v(t)$ are white, with $w \sim (0, Q), v \sim (0, R)$, and $Q \geq 0, R > 0$. Suppose $w(t), v(t)$, and $x(0)$ are uncorrelated. Under these circumstances, the required autocorrelations are given by

$$R_{XZ}(t,u) = \overline{x(t)z^T(u)} = \overline{x(t)[Hx(u) + v(u)]^T}$$

or

$$R_{XZ}(t,u) = R_X(t,u)H^T \tag{3.77}$$

and

$$\begin{aligned} R_Z(t,u) &= \overline{z(t)z^T(u)} \\ &= \overline{[Hx(t) + v(t)][Hx(u) + v(u)]^T} \end{aligned}$$

or

$$R_Z(t,u) = HR_X(t,u)H^T + R\delta(t-u) \tag{3.78}$$

The frequency domain solution of Wiener required a knowledge of $\Phi_{XZ}(\omega)$ and $\Phi_Z(\omega)$. By assuming Equation 3.76, we have substituted for this requirement the alternative requirement of knowledge of the matrices A, G, H, P_0, Q, and R. By using the expression for $R_{XZ}(t,u)$ in the Wiener–Hopf equation, we obtain

$$R_X(t,u)H^T = \int_{t_0}^{t} h(t,\tau)R_Z(\tau,u)\mathrm{d}\tau; \qquad t_0 < u < t \tag{3.79}$$

3.3.2 A Differential Equation for the Optimal Impulse Response

The solution does not provide an explicit expression for $h(t,\tau)$, but it depends on the fact that a differential equation can be found for the optimal impulse response. Differentiate the Wiener–Hopf Equation 3.79 using Leibniz's rule (Bartle 1976) to get

$$\frac{\partial R_X(t,u)}{\partial t}H^T = h(t,t)R_Z(t,u) + \int_{t_0}^{t} \frac{\partial h(t,\tau)}{\partial t}R_Z(\tau,u)\mathrm{d}\tau; \qquad t_0 < u < t \tag{3.80}$$

Now use Equation 3.78 to write for $u < t$

$$h(t,t)R_Z(t,u) = h(t,t)HR_X(t,u)H^T$$

By using Equation 3.79, this can be written as

$$h(t,t)R_Z(t,u) = \int_{t_0}^{t} h(t,t)Hh(t,\tau)R_Z(\tau,u)\mathrm{d}\tau; \qquad t_0 < u < t \tag{3.81}$$

Now, use Equation 3.76a to say that

$$\begin{aligned}\frac{\partial R_X(t,u)}{\partial t} &= E\left[\frac{\mathrm{d}x(t)}{\mathrm{d}t}x^T(u)\right] \\ &= \overline{[Ax(t)+Gw(t)]x^T(u)} \\ &= AR_X(t,u); \quad u<t\end{aligned} \tag{3.82}$$

(We have $R_{WX}(t,u)=0$ for $u<t$ since $w(t)$ is white.) Therefore, by Equation 3.79

$$\frac{\partial R_X(t,u)}{\partial t}H^T = \int_{t_0}^{t} Ah(t,\tau)R_Z(\tau,u)\mathrm{d}\tau; \qquad t_0<u<t \tag{3.83}$$

Substituting into Equation 3.80 using Equations 3.81 and 3.83 yields

$$0=\int_{t_0}^{t}\left[-Ah(t,\tau)+h(t,t)Hh(t,\tau)+\frac{\partial h(t,\tau)}{\partial t}\right]R_Z(\tau,u)\mathrm{d}r; \qquad t_0<u<t \tag{3.84}$$

By Equation 3.78

$$\begin{aligned}0=\int_{t_0}^{t}&\left[-Ah(t,\tau)+h(t,t)Hh(t,\tau)+\frac{\partial h(t,\tau)}{\partial t}\right]HR_X(\tau,u)H^T\mathrm{d}\tau \\ &+\left[-Ah(t,u)+h(t,t)Hh(t,u)+\frac{\partial h(t,u)}{\partial t}\right]R; \qquad t_0<u<t\end{aligned}$$

Since $|R|\neq 0$, this is equivalent to

$$\frac{\partial h(t,\tau)}{\partial t}=Ah(t,\tau)-h(t,t)Hh(t,\tau); \qquad t_0<\tau<t \tag{3.85}$$

This is a differential equation satisfied by the optimal impulse response $h(t,\tau)$; it is the result we were seeking in this subsection. We will not find $h(t,\tau)$ explicitly, but will use Equation 3.85 in the next subsection to find a differential equation for $\hat{x}(t)$.

3.3.3 A Differential Equation for the Estimate

To find a differential equation for the optimal estimate $\hat{x}(t)$, differentiate Equation 3.73 using Leibniz's rule to get

$$\frac{\mathrm{d}\hat{x}(t)}{\mathrm{d}t}=h(t,\,t)z(t)+\int_{t_0}^{t}\frac{\partial h(t,\tau)}{\partial t}z(t)\mathrm{d}\tau \tag{3.86}$$

By using Equation 3.85, this becomes

$$\dot{\hat{x}}(t)=h(t,t)z(t)+\int_{t_0}^{t}[Ah(t,\tau)-h(t,t)Hh(t,\tau)]z(\tau)\mathrm{d}\tau$$

Now substitute from Equation 3.73 to obtain

$$\dot{\hat{x}}(t) = A\hat{x}(t) + h(t,t)[z(t) - H\hat{x}(t)] \tag{3.87}$$

This is a differential equation for the estimate; now things are beginning to look familiar!

All that remains is to find an expression for $h(t,t)$, for then Equation 3.87 provides a realization of the optimal filter $h(t,\tau)$. Unfortunately, we must tackle this is a roundabout manner. In the remainder of this subsection we shall note only that we can express $h(t,t)$ in terms of the error covariance $P(t)$.

Use Equation 3.78 in the Wiener–Hopf Equation 3.79 to obtain

$$R_X(t,u)H^T = \int_{t_0}^{t} h(t,\tau)[HR_X(\tau,u)H^T + R\delta(\tau - u)]\mathrm{d}\tau$$

or with $u = t$,

$$R_X(t,t)H^T - \int_{t_0}^{t} h(t,\tau)HR_X(\tau,t)H^T\mathrm{d}\tau = h(t,t)R$$

According to Equation 3.75 for the error covariance, and Equation 3.77, this becomes

$$P(t)H^T = h(t,t)R$$

so that

$$h(t,t) = P(t)H^T R^{-1} \tag{3.88}$$

At this point it is clear that $h(t,t)$ is nothing but the Kalman gain

$$K(t) = h(t,t) = P(t)H^T R^{-1} \tag{3.89}$$

By using $K(t)$ in Equation 3.87 we can manufacture the estimate $\hat{x}(t)$ from the data $z(t)$. To compute $K(t)$ we need to find the error covariance $P(t)$, which we do in the next subsection.

3.3.4 A Differential Equation for the Error Covariance

To find a differential equation for the error covariance $P(t)$, we first need to find the *error system*. From Equations 3.74, 3.76, and 3.87 we get

$$\begin{aligned}\dot{\tilde{x}}(t) &= \dot{x}(t) - \dot{\hat{x}}(t)\\ &= Ax(t) + Gw(t) - A\hat{x}(t) - K(t)[Hx(t) + v(t) - H\hat{x}(t)]\end{aligned}$$

or

$$\dot{\tilde{x}}(t) = [A - K(t)H]\tilde{x}(t) - K(t)v(t) + Gw(t) \tag{3.90}$$

This is the error system.

Let $\phi(t, t_0)$ be the state-transition matrix for $[A - K(t)H]$. Then the solution to Equation 3.90 is

$$\tilde{x}(t) = \phi(t, t_0)\tilde{x}(t_0) - \int_{t_0}^{t} \phi(t,\tau)K(\tau)v(\tau)\mathrm{d}\tau + \int_{t_0}^{t} \phi(t,\tau)Gw(\tau)\mathrm{d}\tau \tag{3.91}$$

We shall soon need the cross-correlations $R_{V\tilde{X}}(t,t)$ and $R_{W\tilde{X}}(t,t)$. To find them use Equation 3.91 and the uncorrelatedness assumptions. Thus

$$\begin{aligned} R_{V\tilde{X}}(t,t) &= \overline{v(t)\tilde{x}^T(t)} \\ &= -\int_{t_0}^{t} \overline{v(t)v^T(\tau)}K^T(\tau)\phi^T(t,\tau)\mathrm{d}\tau \end{aligned} \tag{3.92}$$

Note that

$$R_V(t,\tau) = R\delta(t-\tau) \tag{3.93}$$

but the integral in Equation 3.92 has an upper limit of t. Recall that the unit impulse can be expressed as

$$\delta(t) = \lim_{T \to 0} \frac{1}{T}\Pi\left(\frac{t}{T}\right) \tag{3.94}$$

where the rectangle function is

$$\frac{1}{T}\Pi\left(\frac{t}{T}\right) = \begin{cases} \frac{1}{T}, & |t| < T/2 \\ 0, & \text{otherwise} \end{cases} \tag{3.95}$$

Therefore, only half of the area of $\delta(t)$ should be considered as being to the left of $t = 0$. Hence, Equation 3.92 is $[\phi(t,t) = I]$

$$R_{V\tilde{X}}(t,t) = -\frac{1}{2}RK^T(t) \tag{3.96}$$

Similarly,

$$\begin{aligned} R_{W\tilde{X}}(t,t) &= \overline{w(t)\tilde{x}^T(t)} \\ &= \int_{t_0}^{t} \overline{w(t)w^T(\tau)}G^T\phi^T(t,\tau)\mathrm{d}\tau \end{aligned}$$

or

$$R_{W\tilde{X}}(t,t) = \frac{1}{2}QG^T \tag{3.97}$$

To find a differential equation for $P(t)$, write

$$\dot{P}(t) = E\left[\frac{\mathrm{d}\tilde{x}(t)}{\mathrm{d}t}\tilde{x}^T(t)\right] + E\left[\tilde{x}(t)\frac{\mathrm{d}\tilde{x}^T(t)}{\mathrm{d}t}\right] \tag{3.98}$$

According to the error dynamics in Equation 3.90, the first term is

$$E\left[\frac{\mathrm{d}\tilde{x}(t)}{\mathrm{d}t}\tilde{x}^T(t)\right] = [A - K(t)H]P(t) + \frac{1}{2}K(t)RK^T(t) + \frac{1}{2}GQG^T \tag{3.99}$$

where we have used Equations 3.96 and 3.97.

To Equation 3.99 add its transpose to get

$$\dot{P} = (A - KH)P + P(A - KH)^T + KRK^T + GQG^T \tag{3.100}$$

Finally, using Equation 3.89 yields

$$\dot{P} = AP + PA^T - PH^TR^{-1}HP + GQG^T \tag{3.101}$$

This is just the Riccati equation for the error covariance!

3.3.5 Discussion

The solution for $\hat{x}(t)$ provided by Equations 3.101, 3.89, and 3.87 is the Kalman filter in Table 3.1. If the shaping filter (Equation 3.76) and covariances are time-varying, these equations still hold. If Equation 3.76a has a deterministic input, it is easy to carry it through the derivation.

The Kalman filter provides the optimal estimate in the general nonstationary case. It is time-varying even for a time-invariant shaping filter (Equation 3.76) since the Kalman gain $K(t) = h(t, t)$ in Equation 3.87 is dependent on t. This accounts for the fact that the optimal filter cannot be found in the nonsteady-state case using frequency domain methods.

The Kalman filter does not explicitly provide an expression for $h(t, \tau)$, however a realization for the optimal filter is given by Equation 3.87.

In the limit as $t \to \infty$, the error covariance tends to a unique positive definite solution P_∞ under appropriate reachability and observability conditions (Section 3.4). Then the steady-state gain

$$K_\infty = P_\infty H^T R^{-1} \tag{3.102}$$

is a constant. The steady-state Kalman filter is

$$\begin{aligned}\dot{\hat{x}} &= (A - K_\infty H)\hat{x} + K_\infty z\\ \hat{y} &= H\hat{x}\end{aligned} \tag{3.103}$$

where $y = Hx$. Under these circumstances the stationarity assumptions hold if A is stable, and the transfer function

$$HF(s) = H[sI - (A - K_\infty H)]^{-1}K_\infty \tag{3.104}$$

is exactly the Wiener filter found by frequency domain methods in Section 1.5.

An analogous derivation of the discrete Kalman filter exists. In Section 2.6 we took a somewhat different (frequency domain) approach to show that the steady-state Kalman filter is the Wiener filter.

Example 3.5 Comparison of Kalman Filter and Wiener Filter for Scalar System

Suppose the scalar unknown $x(t)$ has spectral density

$$\Phi_X(\omega) = \frac{2a\sigma^2}{\omega^2 + a^2} \tag{3.105}$$

and that data $z(t)$ are generated by

$$z(t) = x(t) + v(t) \tag{3.106}$$

with white noise $v(t) \sim (0, r)$ uncorrelated with $x(t)$.

a. Wiener filter

In Example 1.15 we found that the Wiener filter is

$$H(s) = \frac{2\gamma}{1+\sqrt{1+\lambda}} \cdot \frac{1}{s + a\sqrt{1+\lambda}} \tag{3.107}$$

with signal-to-noise ratios

$$\gamma = \frac{\sigma^2}{r} \tag{3.108}$$

and

$$\lambda = \frac{2\gamma}{a} = \frac{2\sigma^2}{ar} \tag{3.109}$$

This can also be written as

$$H(s) = \frac{a(\sqrt{1+\lambda} - 1)}{s + a\sqrt{1+\lambda}} \tag{3.110}$$

(It is easy to verify that $2\gamma/[a(\sqrt{1+\lambda}+1)(\sqrt{1+\lambda}-1)] = 1$.) The Wiener filter is a low-pass filter with a cutoff frequency of

$$\omega_c = a\sqrt{1+\lambda} \tag{3.111}$$

Recall that the Wiener filter is found by first factoring $\Phi_Z^{-1}(s)$ to find a whitening filter $W(s)$:

$$\Phi_Z^{-1}(\omega) = W(-j\omega)W(j\omega) \tag{3.112}$$

b. Kalman filter

To find a shaping filter for the unknown $x(t)$, we can perform a spectral factorization on $\Phi_X(\omega)$ to obtain

$$\begin{aligned}\Phi_X(\omega) &= \frac{1}{j\omega + a}(2a\sigma^2)\frac{1}{-j\omega + a} \\ &= G(j\omega)(2a\sigma^2)G(-j\omega)\end{aligned} \tag{3.113}$$

The shaping filter therefore has transfer function of

$$G(s) = \frac{1}{s+a} \tag{3.114}$$

for which a state realization is

$$\dot{x} = -ax + w \tag{3.115}$$

with $w \sim (0, 2a\sigma^2)$. Let $x(0) \sim (x_0, p_0)$. The plant on which the Kalman filter should be run is Equations 3.115 and 3.106.

For this plant, the Riccati equation is

$$\dot{p} = -\frac{1}{r}p^2 - 2ap + 2a\sigma^2 \tag{3.116}$$

The solution to this is given in Example 3.3 (with a replaced by $-a$). The Kalman gain is

$$K(t) = \frac{1}{r}p(t) \tag{3.117}$$

and the Kalman filter is

$$\dot{\hat{x}} = -[a + K(t)]\hat{x} + K(t)z \tag{3.118}$$

In the limit as $t \to \infty$ the steady-state filter is reached. It is

$$p_\infty = p_2 = ar(\sqrt{1+\lambda} - 1) \tag{3.119}$$

$$= \frac{2\sigma^2}{1 + \sqrt{1+\lambda}} \tag{3.120}$$

$$K_\infty = a(\sqrt{1+\lambda} - 1) \tag{3.121}$$

and

$$\dot{\hat{x}} = -a\sqrt{1+\lambda}\hat{x} + a(\sqrt{1+\lambda} - 1)z \tag{3.122}$$

The transfer function of the steady-state Kalman filter (Equation 3.122) is

$$H(s) = \frac{a(\sqrt{1+\lambda} - 1)}{s + a\sqrt{1+\lambda}} \tag{3.123}$$

which is exactly the Wiener filter (Equation 3.110). Moreover, the steady-state error covariance (Equation 3.120) agrees with the Wiener filter error covariance in Example 1.15.

Note that, since the measurement matrix $h = 1$, the numerator of the Wiener filter is the steady-state Kalman gain. □

Example 3.6 Estimating Motion Governed by Newton's Laws

Newton's law

$$\ddot{y}(t) = \frac{u(t)}{m} \tag{3.124}$$

can be expressed, by defining the state as $x = \begin{bmatrix} y & \dot{y} \end{bmatrix}^T$, in the reachable canonical form

$$\dot{x} = \begin{bmatrix} 0 & 1 \\ 0 & 0 \end{bmatrix} x + \begin{bmatrix} 0 \\ 1/m \end{bmatrix} u + \begin{bmatrix} 0 \\ 1 \end{bmatrix} w \tag{3.125a}$$

where white process noise $w(t) \sim (0, \sigma^2)$ has been added to compensate for modeling errors. If position measurements are taken, then

$$z = \begin{bmatrix} 1 & 0 \end{bmatrix} x + v \tag{3.125b}$$

Let white noise $v(t) \sim (0, \rho^2)$ be uncorrelated with $w(t)$ and with the initial position $y(0)$ and velocity $\dot{y}(0)$. If we present the error covariance as

$$P = \begin{bmatrix} p_1 & p_2 \\ p_2 & p_4 \end{bmatrix} \tag{3.126}$$

then the Riccati equation yields, after simplification,

$$\dot{p}_1 = 2p_2 - \frac{p_1^2}{\rho^2} \tag{3.127}$$

$$\dot{p}_2 = p_4 - \frac{p_1 p_2}{\rho^2} \tag{3.128}$$

$$\dot{p}_4 = -\frac{p_2^2}{\rho^2} + \sigma^2 \tag{3.129}$$

A simulation could be run to plot $p_1(t), p_2(t), p_4(t)$ exactly as in Example 3.4, but we are concerned here with some rather interesting behavior at steady state that relates to some classical control concepts.

After the Kalman filter has been running for some time, a steady-state value of error covariance P_∞ is reached for this system. (We investigate under what conditions this occurs in Section 3.4.) At steady state $\dot{P} = 0$; so let us set $\dot{p}_1$, $\dot{p}_2$, and $\dot{p}_4$ to zero. Then Equations 3.127 through 3.129 become an algebraic set of equations which is easily solved to yield

$$p_2 = \sigma\rho \tag{3.130}$$

$$p_1 = \rho^2 \sqrt{\frac{2\sigma}{\rho}} \tag{3.131}$$

$$p_4 = \sigma\rho \sqrt{\frac{2\sigma}{\rho}} \tag{3.132}$$

Defining signal-to-noise ratio

$$\gamma = \frac{\sigma}{\rho} \tag{3.133}$$

the steady-state Kalman gain can be written as

$$\begin{aligned} K_\infty &= P_\infty H^T R^{-1} = \frac{1}{\rho^2} \begin{bmatrix} p_1 \\ p_2 \end{bmatrix} \\ &= \begin{bmatrix} \sqrt{2\gamma} \\ \gamma \end{bmatrix} \end{aligned} \tag{3.134}$$

The steady-state Kalman filter is therefore

$$\dot{\hat{x}} = (A - K_\infty H)\hat{x} + Bu + K_\infty z \tag{3.135}$$

or

$$\dot{\hat{x}} = \begin{bmatrix} -\sqrt{2\gamma} & 1 \\ -\gamma & 0 \end{bmatrix}\hat{x} + \begin{bmatrix} 0 \\ 1/m \end{bmatrix}u + \begin{bmatrix} \sqrt{2\gamma} \\ \gamma \end{bmatrix}z \tag{3.136}$$

These equations describe in state-variable form the Wiener filter for reconstructing state $x(t)$ from the data.

The steady-state filter (Equation 3.136) is time invariant, so the transfer function from data $z(t)$ to estimate $\hat{x}(t)$ can be determined. It is

$$H(s) = (sI - A + K_\infty H)^{-1}K_\infty \tag{3.137}$$

$$= \frac{1}{s^2 + \sqrt{2\gamma}s + \gamma}\begin{bmatrix} \sqrt{2\gamma}s + \gamma \\ \gamma s \end{bmatrix} \tag{3.138}$$

This is the conventional transfer-function form of the Wiener filter. Now, notice something quite interesting. The characteristic polynomial of a second-order system is of the form

$$s^2 + 2\delta\omega_n s + \omega_n^2 \tag{3.139}$$

where δ is the damping ratio and ω_n the natural frequency. By comparing Equations 3.138 and 3.139, we see that

$$\omega_n = \sqrt{\gamma} \tag{3.140}$$

$$\delta = \frac{1}{\sqrt{2}} \tag{3.141}$$

so that the natural frequency of the observer system is dependent on the ratio of process to measurement noise variances. Furthermore, the damping ratio of the optimal steady-state filter is always $1/\sqrt{2}$ independent of the noise! This certainly provides some justification for the classical design criterion of placing poles so that $\delta = 1/\sqrt{2}$.

The roots of the characteristic equation are

$$s = -\sqrt{\frac{\gamma}{2}} \pm j\sqrt{\frac{\gamma}{2}} \tag{3.142}$$

and a root-locus plot for the steady-state filter poles as γ increases from zero is shown in Figure 3.10. Note that when the process noise is zero, the poles of the system (Equation 3.125a) and the Wiener filter (Equation 3.136) coincide. ▯

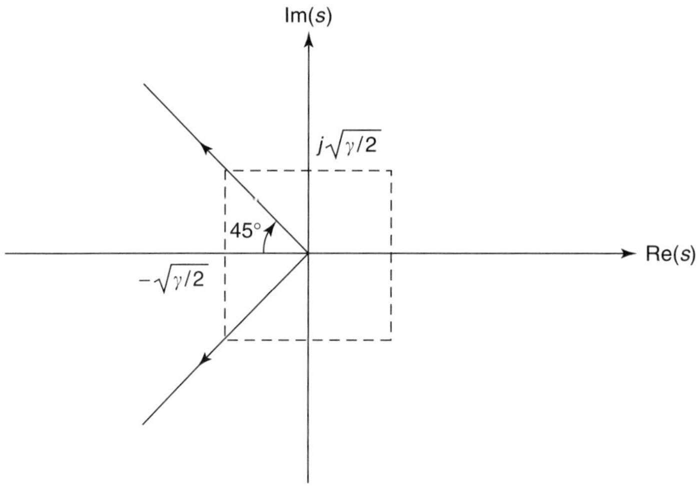

FIGURE 3.10
Root locus for Wiener as signal-to-noise ratio increases.

3.4 Error Dynamics and Statistical Steady State

All of our notions of the steady-state behavior of the discrete Kalman filter also hold for the continuous case. Some of the details are different. This section provides a summary of results; the derivations and proofs are very similar to those in the discrete development and are for the most part omitted. A time-invariant system is assumed here.

3.4.1 The Error System

The convergence and stability properties of the Kalman filter are defined in terms of the error system. The estimation error is

$$\tilde{x}(t) = x(t) - \hat{x}(t) \tag{3.143}$$

An expression for its dynamical behavior is given by Equation 3.90;

$$\dot{\tilde{x}} = (A - KH)\tilde{x} + Gw - Kv \tag{3.144}$$

This equation is driven by the process and the measurement noise. If $[A - K(t)H]$ tends to an asymptotically stable matrix as $t \to \infty$, then in the limit the estimate $\hat{x}(t)$ converges to the expected true plant state $x(t)$.

The residual is given by

$$\tilde{z} = H\tilde{x} + v \tag{3.145}$$

The continuous error system is given by Equations 3.144 and 3.145. The error covariance is given by (Equation 3.100):

$$\dot{P}(t) = (A - KH)P + P(A - KH)^T + KRK^T + GQG^T \tag{3.146}$$

Compare this to the discrete version (Equation 2.191). If we use $K(t) = P(t)H^TR^{-1}$ then Equation 3.146 reduces to Equation 3.21. Note that $\overline{\tilde{x}(t)} = 0$ so that the estimate is unbiased.

If all statistics are Gaussian, then the Kalman filter manufactures the conditional mean $\hat{x}(t)$ and covariance $P(t)$ of $x(t)$ given all previous data. In this case it provides a scheme for updating the entire hyperstate (i.e., the conditional PDF of $x(t)$).

3.4.2 The Innovations Sequence

By Equation 3.145, the residual has zero mean

$$\overline{\tilde{z}(t)} = 0 \tag{3.147}$$

The residual $\tilde{z}(t)$ is the new information in $z(t)$ which cannot be anticipated by using the data $z(\tau)$ for $\tau < t$. It is a *zero-mean white noise* process. It is also called the *innovations*.

The covariance of $\tilde{z}(t)$ has a form which makes the continuous Kalman filter simpler than the discrete filter. To find it, note that $\tilde{z}(t)$ is the limit as $T \to 0$ of the discrete innovations $\tilde{z}_k$. By Equation 2.193, if R is the covariance of the *continuous* noise $v(t)$, then

$$\begin{aligned} P_{\tilde{z}_k} &= HP_k^-H^T + \frac{R}{T} \\ &= \frac{HP_k^-H^TT + R}{T} \end{aligned} \tag{3.148}$$

To make a discrete noise sequence equivalent in the limit to a continuous noise process, we must multiply its covariance by T (R is a spectral density matrix, but R/T is a covariance matrix in the limit). Therefore, letting $T \to 0$, we obtain

$$P_{\tilde{z}}(t) = R \tag{3.149}$$

This accounts for the fact that in the continuous filter the factor R^{-1} appears, but in the discrete filter the corresponding factor is $(HP_k^-H^T + R)^{-1}$. (An alternative derivation of Equation 3.149 is discussed in the problems.)

One way to test for the proper performance of an optimal filter is to monitor $\tilde{z}(t)$. If the innovations is not white noise, then the filter is performing suboptimality.

Technically, the innovations $v(t)$ are defined by

$$v(t) = R^{-1/2}\tilde{z}(t) \tag{3.150}$$

so that

$$P_v(t) = I \tag{3.151}$$

The residual is, however, usually considered as the innovations; particularly because the measurement can always be prewhitened (Sections 1.4 and 4.4) to make $R = I$.

For an excellent discussion of these ideas see Kailath (1981).

3.4.3 The Algebraic Riccati Equation

The solution of the Riccati equation tends to a bounded steady-state value P if

$$\lim_{t \to \infty} P(t) = P \tag{3.152}$$

is bounded. In this case, for large t, $\dot{P}(t) = 0$ so that Equation 3.21 tends to the continuous *algebraic Riccati equation (ARE)*

$$0 = AP + PA^T - PH^T R^{-1} HP + GQG^T \tag{3.153}$$

Any limiting solution to Equation 3.21 is a solution to Equation 3.153, although the converse is not true. The solutions to Equation 3.153 are explored further in the problems.

The conditions under which:

1. Bounded limiting solutions to Equation 3.21 exist for all choices of $P(0)$
2. The limiting solution to Equation 3.21 is unique, independent of the choice of $P(0)$
3. The error system Equation 3.144 is asymptotically stable in the limit

are the same as for the discrete filter. Now, however, detectability and stabilizability are defined with "stable" poles meaning poles in the open-left-half plane. Thus we have the next results.

THEOREM 3.1
Let (A, H) be detectable. Then for every choice $P(0)$ there is a bounded limiting solution P to Equation 3.21. Furthermore, P is a positive semidefinite solution to the algebraic Riccati equation 3.153.

By imposing a reachability condition, we can make this result much stronger.

THEOREM 3.2
Let $Q = \sqrt{Q}\sqrt{Q}^T \geq 0$, and $R > 0$. Suppose $(A, G\sqrt{Q})$ is reachable. Then (A, H) is detectable if and only if:

a. *There is a unique positive definite limiting solution P to Equation 3.21, which is independent of $P(0)$. Furthermore, P is the* unique *positive* definite *solution to the algebraic Riccati equation.*
b. *The steady-state error system defined by Equation 3.144 with the* steady-state *Kalman gain*

$$K = PH^T R^{-1} \tag{3.154}$$

is asymptotically stable.

The proofs of these results are similar to the discrete-time case in Section 2.5. These theorems of course explain the limiting behavior noted in the examples of Section 3.2.

If we relax the requirement in Theorem 3.2 to stabilizability of $(A, G\sqrt{Q})$, then the theorem still holds but the unique solution P can then only be guaranteed to be positive semidefinite.

3.4.4 Time-Varying Plant

If the plant is time-varying then we must redefine observability and reachability. Suppose the plant is

$$\dot{x} = A(t)x + B(t)u + G(t)w \tag{3.155a}$$

$$z = H(t)x + v \tag{3.155b}$$

with $w \sim [0, Q(t)]$ and $v \sim [0, R(t)]$. Let $\phi(t, t_0)$ be the state transition matrix of A.

We say the plant is *uniformly completely observable* or *stochastically observable*, if for every final time T the *observability gramian* satisfies

$$\alpha_0 I \leq \int_{t_0}^{T} \phi^T(\tau, t_0) H^T(\tau) R^{-1} H(\tau) \phi(\tau, t_0) \mathrm{d}\tau \leq \alpha_1 I \tag{3.156}$$

for some $t_0 < T, \alpha_0 > 0$, and $\alpha_1 > 0$. We say the plant is *uniformly completely reachable*, or *stochastically reachable*, if for every initial time t_0 the *reachability gramian* satisfies

$$\alpha_0 I \leq \int_{t_0}^{T} \phi(t, \tau) G(\tau) G^T(\tau) \phi^T(t, \tau) \mathrm{d}\tau \leq \alpha_1 I \tag{3.157}$$

for some $T > t_0, \alpha_0 > 0$, and $\alpha_1 > 0$.

Stochastic observability and reachability (and boundedness of $A(t), Q(t), R(t)$) guarantee that for large t the behavior of $P(t)$ is unique, independent of $P(0)$. They also guarantee the uniform asymptotic stability of the error system, which means that the estimate $\hat{x}(t)$ converges to the expected value of $x(t)$. See Kalman and Bucy (1961).

3.5 Frequency Domain Results

The presentation in this section parallels that in Section 2.6, and the proofs follow the same philosophy as in the discrete case. Because the continuous innovations process has a covariance equal simply to R, the measurement noise covariance, the results here are somewhat simpler than in Section 2.6.

We assume here a time-invariant plant, and begin with some frequency domain results for continuous linear stochastic systems corresponding to those at the end of Section 2.2. Then we derive a spectral factorization result and discuss further the relation between the Wiener and Kalman filters. Next we present the Chang–Letov design procedure for continuous systems, which gives a connection with the classical root-locus design method.

3.5.1 Spectral Densities for Linear Stochastic Systems

Consider the linear stochastic system

$$\dot{x} = Ax + Bu + Gw \tag{3.158a}$$

$$z = Hx + v \tag{3.158b}$$

with $w(t) \sim (0, Q)$ and $v(t) \sim (0, R)$ white; $x(0) \sim (x_0, P_0)$; $R > 0$; and $w(t)$, $v(t)$, and $x(0)$ uncorrelated.

If A is asymptotically stable, then there is a positive semidefinite solution to the algebraic Lyapunov equation

$$0 = AP + PA^T + GQG^T \tag{3.159}$$

In this event, as $t \to \infty$ the effects of the initial condition $x(0)$ die away and if $w(t)$ is stationary (let us suppose for simplicity that $u(t) = 0$) then $x(t)$ also tends to a stationary process with covariance P.

Under these conditions of *stochastic steady state*, we can discuss spectral densities. According to Equation 3.158, the spectral density of $x(t)$ is

$$\Phi_X(s) = (sI - A)^{-1}GQG^T(-sI - A)^{-T} \tag{3.160}$$

(See Section 1.5.) Likewise, the spectral density of the output $z(t)$ is

$$\Phi_Z(s) = H(sI - A)^{-1}GQG^T(-sI - A)^{-T}H^T + R \tag{3.161}$$

Let the white noise $w'(t)$ with unit spectral density be defined by $w(t) = \sqrt{Q}w'(t)$, and define the transfer function from $w'(t)$ to $z(t)$ as

$$H(s) = H(sI - A)^{-1}G\sqrt{Q} \tag{3.162}$$

Then we can say

$$\Phi_Z(s) = H(s)H^T(-s) + R \tag{3.163}$$

3.5.2 A Spectral Factorization Result

As in the discrete case, the Kalman filter manufactures a minimum-phase factorization for the spectral density of the data $z(t)$. Therefore, solution of the

Riccati equation in the time domain is equivalent to a spectral factorization in the frequency domain. To show this, we proceed as follows:

The steady-state Kalman filter is given by

$$\dot{\hat{x}} = (A - KH)\hat{x} + Bu + Kz \tag{3.164}$$

where the steady-state Kalman gain satisfies $K = PH^TR^{-1}$ with P the positive semidefinite solution to the algebraic Riccati equation 3.153, which exists if (A, H) is detectable and $(A, G\sqrt{Q})$ is stabilizable.

Using a derivation like the one for Equation 2.236 we have

$$\Delta^{\mathrm{cl}}(s) = |I + H(sI - A)^{-1}K|\Delta(s) \tag{3.165}$$

where the plant and filter characteristic polynomials are

$$\Delta(s) = |sI - A| \tag{3.166}$$

and

$$\Delta^{\mathrm{cl}}(s) = |sI - (A - KH)| \tag{3.167}$$

We can also show

$$\begin{aligned}\Phi_Z(s) &= H(s)H^T(-s) + R \\ &= [I + H(sI - A)^{-1}K]R[I + H(-sI - A)^{-1}K]^T\end{aligned} \tag{3.168}$$

with plant transfer function $H(s)$ as in Equation 3.162. By defining

$$W^{-1}(s) = [I + H(sI - A)^{-1}K]R^{1/2} \tag{3.169}$$

this can be written as

$$\Phi_Z(s) = W^{-1}(s)W^{-T}(-s) \tag{3.170}$$

which is a spectral factorization of $\Phi_Z(s)$.

The Kalman filter is guaranteed to be stable under the detectability and stability conditions. If in addition the plant is stable, then by Equation 3.165, $W^{-1}(s)$ is a minimum-phase spectral factor of $\Phi_Z(s)$.

The *innovations representation* is given by

$$\dot{\hat{x}} = [A - K(t)H]\hat{x} + Bu + K(t)z \tag{3.171a}$$

$$\tilde{z} = -H\hat{x} + z \tag{3.171b}$$

which at steady state has a transfer function from z to $\tilde{z}$ of

$$W'(s) = I - H[sI - (A - KH)]^{-1}K \tag{3.172a}$$

By the matrix inversion lemma

$$W'(s) = [I + H(sI - A)^{-1}K]^{-1} \tag{3.172b}$$

so that

$$W(s) = R^{-1/2}W'(s) \tag{3.173}$$

The residual $\tilde{z}(t)$ is white noise with covariance

$$P_{\tilde{z}} = R \tag{3.174}$$

so that Equation 3.171 is a *whitening filter* that turns the data $z(t)$ into the white noise process $\tilde{z}(t)$. It is clear then that $W(s)$ is a whitening filter that turns the data $z(t)$ into the innovations (Equation 3.150).

In terms of $W'(s)$ we can write

$$\Phi_Z(s) = [W'(s)]^{-1}R[W'(-s)]^{-T} \tag{3.175}$$

Note that

$$\lim_{s\to\infty} W'(s) = I \tag{3.176}$$

since $H(sI - A)^{-1}K$ has a relative degree of at least 1. Hence, Equation 3.175 is the spectral factorization defined by Equation 1.119.

These results show that the Kalman filter is a whitening filter that converts the data $z(t)$ to the residual $\tilde{z}(t)$, simultaneously providing the optimal estimate $\hat{x}(t)$. It can be viewed alternatively as a time-domain algorithm for computing a spectral factorization for the density of $z(t)$. For further discussion of these ideas see Shaked (1976).

We showed in the previous section that the Wiener filter and steady-state Kalman filter are identical, since they both provide the solution to the stationary Wiener–Hopf equation. Another demonstration of this follows.

In the case of linear measurements Equation 3.158b, the Wiener filter that estimates $y(t) = Hx(t)$ is given by (see Section 1.5) the simple expression

$$HF(s) = I - R^{1/2}W(s) \tag{3.177}$$

where

$$\Phi_Z^{-1}(s) = W^T(-s)W(s) \tag{3.178}$$

In terms of $W'(s)$, therefore, the Wiener filter is

$$HF(s) = I - W'(s) \tag{3.179}$$

According to Equation 3.172, this is just

$$HF(s) = H[sI - (A - KH)]^{-1}K \tag{3.180}$$

which is the transfer function of the Kalman filter (Equation 3.164) from $z(t)$ to $\hat{z}(t) = H\hat{x}(t)$.

Recall from Section 2.6 that if Equation 3.180 provides the optimal estimate of $y = Hx$ and (A, H) is observable, then the optimal estimate $\hat{x}$ of the state is given by the filter

$$F(s) = [sI - (A - KH)]^{-1}K \tag{3.181}$$

3.5.3 Chang–Letov Design Procedure

We do not assume here that the plant matrix A is stable. By manipulation of our equations, we can drive a design procedure that provides a link with classical root-locus techniques.

By using Equations 3.168 and 3.165, we obtain the continuous *Chang–Letov* equation

$$\Delta^{\text{cl}}(s)\Delta^{\text{cl}}(-s) = |H(s)H^T(-s) + R| \cdot \Delta(s)\Delta(-s) \cdot |R|^{-1} \tag{3.182}$$

where

$$H(s) = H(sI - A)^{-1}G\sqrt{Q} \tag{3.183}$$

All matrices on the right-hand side are known so that this represents a method of determining the closed-loop poles $\Delta^{\text{cl}}(s) = 0$ of the optimal filter (Equation 3.164). We know that the steady-state filter is stable if (A, H) is observable and $(A, G\sqrt{Q})$ is reachable. Thus, it is only necessary to compute the right-hand side of the Chang–Letov equation and find the roots of the resulting polynomial. The stable roots are the roots of $\Delta^{\text{cl}}(s)$, and the unstable roots are the roots of $\Delta^{\text{cl}}(-s)$. Once the desired closed-loop poles have been determined in this fashion, a technique such as Ackermann's formula can be used to find the required output injection K.

This procedure allows determination of the optimal filter by a method that does not involve either the solution of a Riccati equation or a spectral factorization.

In the single-output case we can represent $H(s)$ as

$$H(s) = \frac{N(s)}{\Delta(s)} \tag{3.184}$$

where $N(s) = H\ \text{adj}(sI - A)G\sqrt{Q}$ is a row vector. Then the Chang–Letov equation is

$$\Delta^{\text{cl}}(s)\Delta^{\text{cl}}(-s) = \frac{1}{r}N(s)N^T(-s) + \Delta(s)\Delta(-s) \tag{3.185}$$

The right-hand side is in exactly the canonical form for a root-locus design: as r increases from 0 (perfect measurements) to ∞ (totally unreliable measurements), the roots of $\Delta^{\text{cl}}(s)\Delta^{\text{cl}}(-s)$ move from the roots of $N(s)N^T(-s)$ to the roots of $\Delta(s)\Delta(-s)$. This means the following:

- If $r = 0$, then we examine the original plant *zeros* considering $w(t)$ as the input. If any of them are in the right-half of the s plane, we reflect them into the left-half plane by making them negative. The resulting n stable complex numbers are the required poles of $(A - KH)$.
- If $r \to \infty$, then we examine the original plant *poles*. Unstable plant poles are reflected into the left-half plane to form a set of n stable poles for $(A - KH)$.

- If $0 < r < \infty$, then the roots of $\Delta^{\text{cl}}(s)$ are on loci between the above two sets of stable numbers.

The following variation of the above procedure shows that the Chang–Letov method does not apply only to the state-variable formulation.

Example 3.7 Chang–Letov Design for Lateral Aircraft Dynamics

a. The plant

The transfer function from rudder displacement $\delta_r(t)$ to sideslip angle $\beta(t)$ in a typical jet transport flying straight and level at 300 mph at sea level might be given by (Blakelock 1965)

$$H(s) = \frac{\beta}{\delta_r} = \frac{0.0364s^2 + 1.3737s - 0.0137}{s^3 + 0.376s^2 + 1.8115s - 0.00725} \tag{3.186}$$

$$= \frac{0.0364(s - 0.01)(s + 37.75)}{[(s + 0.19)^2 + 1.333^2](s - 0.004)} \tag{3.187}$$

(Sideslip is the angle in the horizontal plane between the relative wind and the aircraft longitudinal axis. It should be zero in straight and level flight to conserve fuel.) The pole at $s = 0.004$ corresponds to the unstable spiral mode. The complex-conjugate pole pair corresponds to the Dutch roll mode, which causes the nose of the aircraft to move in a damped figure-eight pattern with a period of $0.75 = 1/1.333\,\text{s}$. The magnitude portion of the Bode plot for Equation 3.186 is given in Figure 3.11.

Wind gusts acting on the rudder are a source of perturbing noise. This noise is not white, but is low-frequency noise, so we should add a noise-shaping filter to our model (Section 3.6). Rather than complicate things in this matter, let us make the following approximation. The plant (Equation 3.186) is itself a low-pass filter, so let us assume the wind-gust noise is white. According to Figure 3.11, only the low-frequency components of this white noise will have

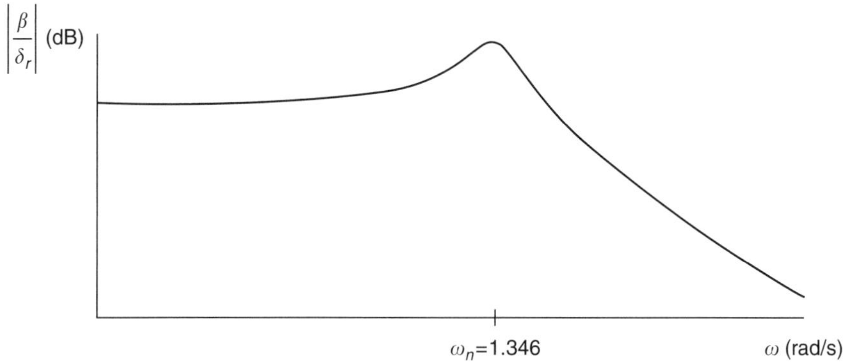

FIGURE 3.11
Bode magnitude plot for lateral dynamics.

an appreciable effect on $\beta(t)$. If the wind-gust noise contains frequencies up to at least about $10\omega_n = 13.46\,\text{rad/s}$, this is a satisfactory approximation. Let the noise $\delta_r(t)$ have covariance of q.

Suppose sideslip angle $\beta(t)$ is measured in additive white noise so that

$$z(t) = \beta(t) + v(t) \tag{3.188}$$

with $v(t) \sim (0, r)$.

b. Chang–Letov design of optimal steady-state filter

To design the Wiener filter to estimate $\beta(t)$ from the data $z(t)$, let us use the Chang–Letov approach.

According to Equation 3.185 we have

$$\begin{aligned}\Delta^{\text{cl}}(s)\Delta^{\text{cl}}(-s) &= 0.0364^2 \frac{q}{r}(s-0.01)(s+37.75)(s+0.01)(s-37.75) \\ &\quad + [(s+0.19)^2 + 1.333^2](s-0.004)[(-s+0.19)^2 \\ &\quad + 1.333^2](-s-0.004)\end{aligned} \tag{3.189}$$

where we have included in the numerator of $H(s)$ the factor $\sqrt{q}$ required in Equation 3.183. Symbolize Equation 3.189 as

$$\Delta^{\text{cl}}(s)\Delta^{\text{cl}}(-s) = cn(s) + d(s) \tag{3.190}$$

where gain c is $0.0364^2 q/r$.

The root-locus plot for Equation 3.190 as c varies from 0 to ∞ can be found by the conventional rules. It is shown in Figure 3.12, where 0's represent the roots of $n(s)$ and x's represent the roots of $d(s)$. Note that the roots of Equation 3.190 move from the poles of $n(s)/d(s)$ to the zeros of $n(s)/d(s)$. [Observe from Equation 3.189 that the last factor in $d(s)$ is negative, so that we must actually compute the conventional root locus with negative gain values $-c$.]

Now suppose that $q = 10$, $r = 0.1$. In this case, $c = 0.1325$ so that Equation 3.189 yields

$$\Delta^{\text{cl}}(s)\Delta^{\text{cl}}(-s) = -s^6 - 3.3491s^4 - 192.11s^2 + 0.018935 \tag{3.191}$$

Note the even structure of this polynomial which has roots symmetric about the $j\omega$ axis. The roots of Equation 3.191 are

$$s = 0.0099, -0.0099, 2.468 \pm j2.787, -2.468 \pm j2.787 \tag{3.192}$$

and assigning the stable roots to the optimal filter yields

$$\begin{aligned}\Delta^{\text{cl}}(s) &= (s+0.0099)[(s+2.468)^2 + 2.787^2] \\ &= s^3 + 4.946s^2 + 13.907s + 0.137\end{aligned} \tag{3.193}$$

This is the closed-loop characteristic polynomial of the desired Wiener filter.

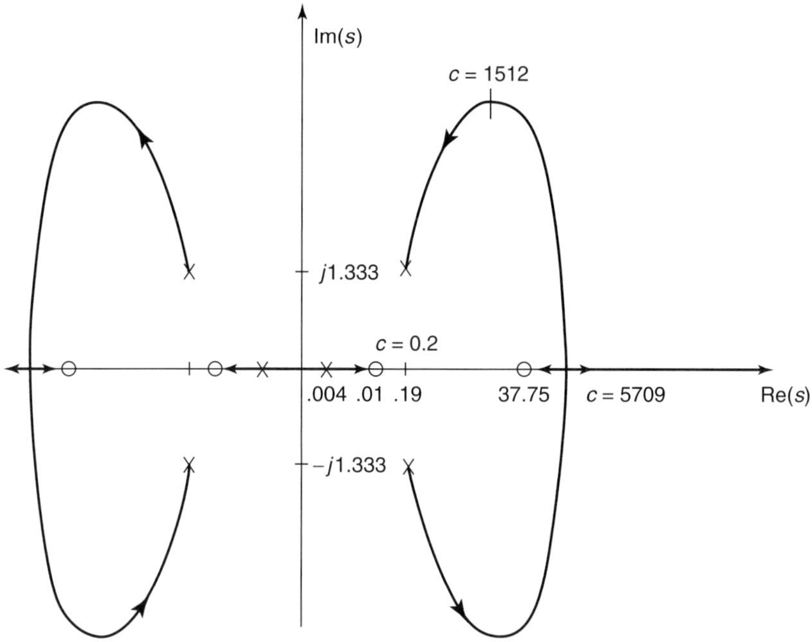

FIGURE 3.12
Chang–Letov root locus.

c. Computation of optimal filter

The denominator of the filter transfer function is given by Equation 3.193.

To find the numerator of the filter transfer function, realize Equation 3.186 in observable canonical form (Kailath 1980). By inspection,

$$\dot{x} = \begin{bmatrix} -0.376 & 1 & 0 \\ -1.8115 & 0 & 1 \\ 0.00725 & 0 & 0 \end{bmatrix} x + \begin{bmatrix} 0.0364 \\ 1.3737 \\ -0.0137 \end{bmatrix} \delta_r$$

$$\beta = \begin{bmatrix} 1 & 0 & 0 \end{bmatrix} x \tag{3.194}$$

With Kalman gain $K = [K_1 \;\; K_2 \;\; K_3]^T$, we have

$$A - KH = \begin{bmatrix} -0.376 - K_1 & 1 & 0 \\ -1.8115 - K_2 & 0 & 1 \\ 0.00725 - K_3 & 0 & 0 \end{bmatrix} \tag{3.195}$$

By comparison with the desired coefficients (Equation 3.193), therefore,

$$K = \begin{bmatrix} -0.376 \\ -1.8115 \\ 0.00725 \end{bmatrix} + \begin{bmatrix} 4.946 \\ 13.907 \\ 0.137 \end{bmatrix} = \begin{bmatrix} 4.57 \\ 12.0955 \\ 0.14425 \end{bmatrix} \tag{3.196}$$

Now, by Equation 3.180 the optimal filter for estimating the sideslip angle $\beta(t)$ from the measurements is

$$\frac{\hat{\beta}}{z} = \frac{4.57s^2 + 12.0955s + 0.14425}{s^3 + 4.946s^2 + 13.907s + 0.137}$$

(Since we are using observable canonical form, the numerator coefficients are just the coefficients of the Kalman gain.) □

3.6 Correlated Noise and Shaping Filters

The philosophy in this section is the same as for the discrete situation. Where the details are different they are explicitly given.

3.6.1 Colored Process Noise

Let the plant be

$$\dot{x} = Ax + Bu + Gw \tag{3.197a}$$

$$z = Hx + v \tag{3.197b}$$

with $v(t) \sim (0, R)$ white and uncorrelated with $x(0) \sim (x_0, P_0)$. If process noise $w(t)$ is uncorrelated with $x(0)$ and $v(t)$ but not white, then we should proceed as follows:

Factor the spectral density of $w(t)$ as

$$\Phi_W(s) = H(s)H^T(-s) \tag{3.198}$$

If $|\Phi_W(s)| \neq 0$ for $\text{Re}(s) = 0$, then $H(s)$ is minimum phase. Find a state representation for $H(s)$:

$$H(s) = H'(sI - A')^{-1}G' + D' \tag{3.199}$$

where primes denote additional variables. Then a *shaping filter* for $w(t)$ is given by

$$\dot{x}' = A'x' + G'w' \tag{3.200a}$$

$$w = H'x' + D'w' \tag{3.200b}$$

In this shaping filter, $w'(t) \sim (0, I)$ is white noise. To describe the dynamics of both the unknown $x(t)$ and the process noise $w(t)$, write the *augmented system*

$$\begin{bmatrix} \dot{x} \\ \dot{x}' \end{bmatrix} = \begin{bmatrix} A & GH' \\ 0 & A' \end{bmatrix} \begin{bmatrix} x \\ x' \end{bmatrix} + \begin{bmatrix} B \\ 0 \end{bmatrix} u + \begin{bmatrix} GD' \\ G' \end{bmatrix} w' \tag{3.201a}$$

$$z = \begin{bmatrix} H & 0 \end{bmatrix} \begin{bmatrix} x \\ x' \end{bmatrix} + v \tag{3.201b}$$

The continuous Kalman filter is now applied to this augmented system.

Example 2.16 demonstrates the procedure, since the continuous Kalman filter can be applied to Equations 2.305 and 2.306. See Figure 2.18 for some useful shaping filters.

3.6.2 Correlated Measurement and Process Noise

Now let the plant be described by Equation 3.197 with $w(t) \sim (0, Q)$ and $v(t) \sim (0, R)$ both white, but with $v(t)$ and $w(t)$ correlated according to

$$\overline{v(t)w^T(\tau)} = S\delta(t-\tau) \tag{3.202}$$

Then

$$\overline{\begin{bmatrix} w(t) \\ v(t) \end{bmatrix} \begin{bmatrix} w(\tau) \\ v(\tau) \end{bmatrix}^t} = \begin{bmatrix} Q & S^T \\ S & R \end{bmatrix} \delta(t-\tau) \tag{3.203}$$

Rather than proceeding as in Section 2.7, let us use an interesting alternative (Gelb 1974).

We can convert this problem into an equivalent problem with uncorrelated process and measurement noises by adding zero to the right-hand side of Equation 3.197a to get

$$\dot{x} = Ax + Bu + Gw + D(z - Hx - v) \tag{3.204}$$

This results in the modified plant

$$\dot{x} = (A - DH)x + Dz + Bu + (Gw - Dv) \tag{3.205a}$$

$$z = Hx + v \tag{3.205b}$$

which has process noise of $(Gw - Dv)$. This noise is white since it is zero mean and its covariance is given by

$$\begin{aligned} &\overline{[Gw(t) - Dv(t)][Gw(\tau) - Dv(\tau)]^T} \\ &\quad = (GQG^T + DRD^T - DSG^T - GS^TD^T)\delta(t-\tau) \end{aligned} \tag{3.206}$$

The motivation behind these manipulations is that D can be selected to make $(Gw - Dv)$ and v uncorrelated. To wit,

$$\overline{[Gw(t) - Dv(t)]v^T(\tau)} = (GS^T - DR)\delta(t-\tau) \tag{3.207}$$

which is zero if we select

$$D = GS^TR^{-1} \tag{3.208}$$

With this choice, the modified plant is

$$\dot{x} = (A - GS^TR^{-1}H)x + GS^TR^{-1}z + Bu + (Gw - GS^TR^{-1}v) \tag{3.209a}$$

$$z = Hx + v \tag{3.209b}$$

where $GS^TR^{-1}z(t)$ is an additional deterministic input. The new process noise autocorrelation (Equation 3.206) is

$$\overline{[Gw(t) - Dv(t)][Gw(\tau) - Dv(\tau)]^T} = G(Q - S^TR^{-1}S)G^T\delta(t-\tau) \quad (3.210)$$

which is less than the GQG^T term arising from the original process noise $w(t)$.

Applying the continuous Kalman filter to Equation 3.209, there results the estimate update

$$\dot{\hat{x}} = (A - GS^TR^{-1}H)\hat{x} + Bu + GS^TR^{-1}z + PH^TR^{-1}(z - H\hat{x})$$

or

$$\dot{\hat{x}} = A\hat{x} + Bu + (PH^T + GS^T)R^{-1}(z - H\hat{x}) \quad (3.211)$$

Defining a *modified Kalman gain* by

$$K = (PH^T + GS^T)R^{-1} \quad (3.212)$$

this becomes

$$\dot{\hat{x}} = A\hat{x} + Bu + K(z - H\hat{x}) \quad (3.213)$$

which is identical in form to the estimate update equation in Table 3.1.

The error covariance update equation for Equation 3.209 with process noise autocorrelation (Equation 3.210) is

$$\begin{aligned}\dot{P} = &(A - GS^TR^{-1}H)P + P(A - GS^TR^{-1}H)^T \\ &+ G(Q - S^TR^{-1}S)G^T - PH^TR^{-1}HP\end{aligned}$$

This can be written in terms of the modified gain as

$$\dot{P} = AP + PA^T + GQG^T - KRK^T \quad (3.214)$$

which is identical to Equation 3.24. Nonzero S results in a smaller error covariance than the uncorrelated noise case due to the additional information provided by the cross-correlation term S.

We have found that if the measurement and process noises are correlated, then it is only necessary to define a modified Kalman gain and use the formulation Equation 3.214 for the Riccati equation; all other equations are the same as for the uncorrelated case. This is the same result as for the discrete Kalman filter.

3.6.3 Colored Measurement Noise

If the plant (Equation 3.197) has $w(t) \sim (0, Q)$ white with $w(t)$ and $v(t)$ uncorrelated, but nonwhite $v(t)$ of the form Equation 2.326 then a continuous analog to Equation 2.331 can be written. However, if the term v' in Equation 2.326

is zero then this approach cannot be used unless $|R| = |D'(D')^T| \neq 0$. An alternative to state augmentation is to define a *derived measurement* that has white measurement noise.

Suppose that, by spectral factorization or some other means, a shaping filter for $v(t)$ has been found to be

$$\dot{x}' = A'x' + G'w' \tag{3.215a}$$

$$v = x' \tag{3.215b}$$

with $w'(t) \sim (0, I)$ white and uncorrelated with plant process noise $w(t)$. Define the derived measurement by

$$z' = \dot{z} - A'z - HBu \tag{3.216}$$

Then, if H is time-invariant

$$\begin{aligned} z' &= H(Ax + Bu + Gw) + \dot{v} - A'(Hx + v) - HBu \\ &= (HA - A'H)x + G'w' + HGw \end{aligned}$$

Defining a new measurement matrix by

$$H' = HA - A'H \tag{3.217}$$

we can write

$$z' = H'x + (G'w' + HGw) \tag{3.218}$$

This is a new output equation in terms of the derived measurement. The reason for these manipulations is that the new measurement noise is white with

$$[G'w'(t) + HGw(t)] \sim \{0, [G'(G')^T + HGQG^TH^T]\} \tag{3.219}$$

so that Equations 3.197a and 3.218 define a new plant driven by white noises. The order of this plant is still n, which is a reason for preferring the derived measurement approach over the augmented state approach.

In this new plant, the process and measurement noises are correlated with

$$E\{[G'w'(t) + HGw(t)]w^T(\tau)\} = HGQ \tag{3.220}$$

so that for optimality the modified Kalman gain of the previous subsection should be used. The optimal filter is therefore given by

$$\dot{\hat{x}} = A\hat{x} + Bu + K(\dot{z} - A'z - HBu - H'\hat{x}) \tag{3.221}$$

$$\dot{P} = AP + PA^T + GQG^T - K[G'(G')^T + HGQG^TH^T]K^T \tag{3.222}$$

$$K = [P(H')^T + GQG^TH^T][G'(G')^T + HGQG^TH^T]^{-1} \tag{3.223}$$

A disadvantage of Equation 3.221 is that the data must be differentiated. To avoid this (Gelb 1974), note that

$$\frac{\mathrm{d}}{\mathrm{d}t}(Kz) = \dot{K}z + K\dot{z} \tag{3.224}$$

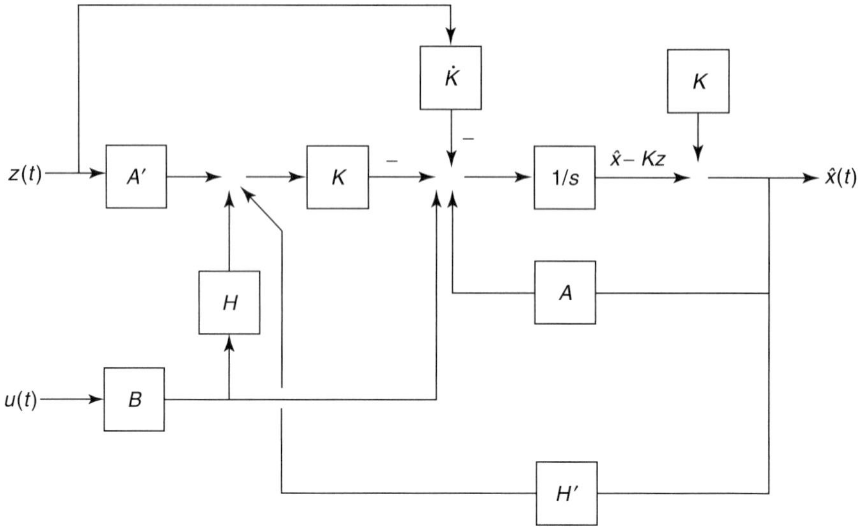

FIGURE 3.13
Continuous Kalman filtering with derived measurements.

so we can write Equation 3.221 in the alternative form

$$\frac{\mathrm{d}}{\mathrm{d}t}(\hat{x} - Kz) = A\hat{x} + Bu - \dot{K}z - K(A'z + HBu + H'\hat{x}) \tag{3.225}$$

Figure 3.13 gives a block diagram of this version of the optimal filter.

To initialize the derived filter, we must use

$$\hat{x}(0^+) = x_0 + P_0H^T(HP_0H^T + R)^{-1}[z(0) - Hx_0] \tag{3.226}$$

and

$$P(0)^+ = P_0 - P_0H^T(HP_0H^T + R)^{-1}HP_0 \tag{3.227}$$

Since the instant after data are available ($t = 0^+$), discontinuities in $\hat{x}(t)$ and $P(t)$ occur. These equations amount to a discrete Kalman filter measurement update. For details see Bucy (1967), Bryson and Johansen (1965), and Stear and Stubberud (1968).

Example 3.8 Continuous Filtering with No Measurement Noise

In the continuous Kalman filter, it is required that the measurement noise covariance R be nonsingular. Reexamining the last few pages, it becomes apparent that the derived measurement approach allows us to derive a filter that does not require $|R| \neq 0$.

Let the plant be

$$\dot{x} = Ax + Bu + Gw \tag{3.228a}$$

$$z = Hx \tag{3.228b}$$

$w(t) \sim (0, Q)$ with the usual assumptions. Then in the "measurement-noise shaping filter" (Equation 3.215) we evidently have $A' = G' = 0$, and the derived measurement is given by Equation 3.216, or

$$z' = \dot{z} - HBu \tag{3.229}$$

According to Equations 3.221 through 3.223, the optimal filter is now

$$\dot{\hat{x}} = A\hat{x} + Bu + K(\dot{z} - HBu - HA\hat{x}) \tag{3.230a}$$

$$= (I - KH)A\hat{x} + (I - KH)Bu + K\dot{z} \tag{3.230b}$$

$$\dot{P} = AP + PA^T + GQG^T - K(HGQG^TH^T)K^T \tag{3.231}$$

$$K = (PA^T + GQG^T)H^T(HGQG^TH^T)^{-1} \tag{3.232}$$

If $|HGQG^TH^T| = 0$, then we must differentiate the data once more to define a new derived measurement and obtain the resulting filter.

This is basically the *Deyst* filter. See Deyst (1969) and Maybeck (1979). □

3.7 Discrete Measurements of Continuous-Time Systems

In this section, we discuss an alternative to the method of Section 2.4 for discrete filtering of continuous-time systems. In that section we proceeded by first discretizing the continuous system. Here, we show how to avoid system discretization by propagating the estimate and error covariance between measurements in *continuous time* using an integration routine such as Runge–Kutta. This approach yields the optimal estimate continuously at all times, including times *between* the data arrival instants.

Let the plant be given by

$$\dot{x} = Ax + Bu + Gw \tag{3.233a}$$

with $w(t) \sim (0, Q)$. Suppose measurements are given at discrete times t_k according to

$$z_k = Hx_k + v_k \tag{3.233b}$$

where we represent $x(t_k)$ by x_k. Let white noises $w(t) \sim (0, Q)$ and $v_k \sim (0, R)$ be uncorrelated with each other and with $x(0) \sim (\bar{x}_0, P_0)$. Note that R is a covariance matrix while Q is a spectral density matrix.

The approach we are about to describe applies to time-varying systems, though for ease of notation all plant and noise matrices are written as constants here.

Applying the Kalman filter directly to Equation 3.233 with no preliminary discretization is very easy since the estimate at each time is the conditional mean given all previous data. The time update between measurements for the continuous linear stochastic system (Equation 3.233a) is found from

TABLE 3.2
Continuous-Discrete Kalman Filter

System and measurement models

$$\dot{x} = Ax + Bu + Gw$$
$$z_k = Hx_k + v_k$$
$$x(0) \sim (\bar{x}_0, P_0), w(t) \sim (0, Q), v_k \sim (0, R)$$

Assumptions

$\{w(t)\}$ and $\{v_k\}$ are white noise processes uncorrelated with each other and with $x(0)$.

Initialization

$$P(0) = P_0, \hat{x}(0) = \bar{x}_0$$

Time update between measurements

$$\dot{P} = AP + PA^T + GQG^T$$
$$\dot{\hat{x}} = A\hat{x} + Bu$$

Measurement update at times t_k

$$K_k = P^-(t_k)H^T[HP^-(t_k)H^T + R]^{-1}$$
$$P(t_k) = (I - K_kH)P^-(t_k)$$
$$\hat{x}_k = \hat{x}_k^- + K_k(z_k - H\hat{x}_k^-)$$

Table 3.1 by setting $H = 0$;

$$\dot{\hat{x}} = A\hat{x} + Bu \tag{3.234}$$

$$\dot{P} = AP + PA^T + GQG^T \tag{3.235}$$

See Example 3.1. The measurement update to incorporate a discrete data sample z_k is given in Table 2.2.

The optimal filter for the plant (Equation 3.233) is given by combining these two updates. The resulting *continuous-discrete Kalman filter* is summarized in Table 3.2. In the Gaussian case, this is the optimal filter. In the case of arbitrary statistics, it is the best *linear* filter.

Note that we do not require $|R| > 0$ as is the case for continuous measurements. All we require is $R \geq 0$ and $|HP^-(t_k)H^T + R| \neq 0$ for each k.

One advantage of the continuous-discrete filter over the alternative approach using system discretization is that in the former, it is not necessary for the sample times t_k to be equally spaced. This means that the cases of irregular and intermittent measurements are easy to handle. In the absence of data, the optimal *prediction* is given by performing only the time update portion of the algorithm.

A block diagram/flowchart of a continuous-discrete Kalman filter using a Runge–Kutta integrator with step size Δt is shown in Figure 3.14. The measurement update routine can be based on one of the measurement update

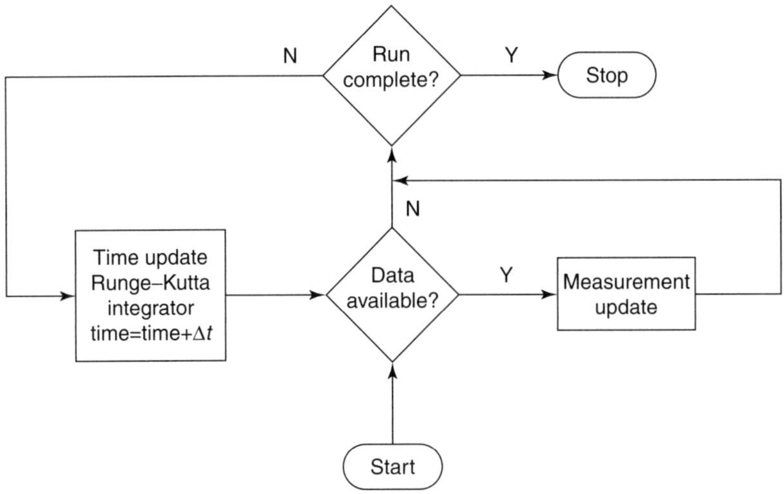

FIGURE 3.14
Software implementation of the continuous-discrete Kalman filter.

routines (stabilized Kalman, UDU^T) in Bierman (1977). Figure 3.14 should be compared to Figure 2.12.

As an illustration of the continuous-discrete Kalman filter, let us supply it to one of the examples we did by discretization in Section 2.4.

Example 3.9 Continuous-Discrete α-β Tracker

Consider the range subsystem in the radar tracking problem of Example 2.8:

$$\dot{x} = \begin{bmatrix} 0 & 1 \\ 0 & 0 \end{bmatrix} x + \begin{bmatrix} 0 \\ 1 \end{bmatrix} w \tag{3.236}$$

where $x \triangleq [r \;\; \dot{r}]^T$. Let $w \sim (0, q)$ and range r be measured every T s so that $t_k = kT$ and

$$z_k = \begin{bmatrix} 1 & 0 \end{bmatrix} x_k + v_k \tag{3.237}$$

$v_k \sim (0, R)$. We use the notation $x_k \triangleq x(kT)$.

In Examples 2.9 and 2.10 we derived simple scalar updates implementing the optimal filter by first discretizing Equation 2.146, respectively. Here, we intend to avoid this discretization by using the continuous-discrete optimal filter formulation. In this simple example, discretization is quite easy since $e^{AT} = I + AT$. In general, however, it can be quite complicated to find e^{AT} and if a good integration routine is available this is not required. An additional advantage we will have in the solution we are about to present is that up-to-date estimates for range and range rate will be available at all times, including times *between* the appearance of the target on the radar screen. With slowly scanning radars this can be quite a valuable benefit!

a. Time update

Between measurements, the optimal estimate propagates according to

$$\dot{\hat{x}} = A\hat{x} \tag{3.238}$$

or

$$\frac{d\hat{r}}{dt} = \hat{\dot{r}} \tag{3.239a}$$

$$\frac{d}{dt}\hat{\dot{r}} = 0 \tag{3.239b}$$

The error covariance obeys the Lyapunov equation $\dot{P} = AP + PA^T + GQG^T$. Let

$$P \triangleq \begin{bmatrix} p_1 & p_2 \\ p_2 & p_4 \end{bmatrix} \tag{3.240}$$

and substitute to obtain the scalar equations

$$\dot{p}_1 = 2p_2 \tag{3.241a}$$

$$\dot{p}_2 = p_4 \tag{3.241b}$$

$$\dot{p}_4 = q \tag{3.241c}$$

We could easily solve these analytically, but shall deliberately refrain from doing so to make the point that this is not required to implement the optimal filter.

b. Measurement update

At a data point the Kalman gain is found as follows. Let

$$\delta \triangleq HP^{-1}(t_k)H^T + R \tag{3.242}$$

Write

$$P^-(t_k) \triangleq \begin{bmatrix} p_1^- & p_2^- \\ p_2^- & p_4^- \end{bmatrix} \tag{3.243}$$

and, therefore

$$\delta = p_1^- + R \tag{3.244}$$

Then

$$K \triangleq \begin{bmatrix} k_1 \\ k_2 \end{bmatrix} = \begin{bmatrix} \frac{p_1^-}{\delta} \\ \frac{p_2^-}{\delta} \end{bmatrix} \tag{3.245}$$

The error covariance update is

$$P(t_k) = \begin{bmatrix} 1 - k_1 & 0 \\ -k_2 & 1 \end{bmatrix} P^-(t_k) \tag{3.246}$$

or

$$p_1 = (1 - k_1)p_1^- = \frac{Rp_1^-}{\delta} \tag{3.247a}$$

$$p_2 = (1 - k_1)p_2^- = \frac{Rp_2^-}{\delta} \tag{3.247b}$$

$$p_4 = p_4^- - k_2 p_2^- \tag{3.247c}$$

The estimate update is (omitting the time subscript k)

$$\hat{r} = \hat{r}^- + k_1(z - \hat{r}^-) \tag{3.248a}$$

$$\hat{\dot{r}} = \hat{\dot{r}}^- + k_2(z - \hat{r}^-) \tag{3.248b}$$

c. Filter implementation

Our preliminary analysis has yielded scalar updates that are easy to program and faster to run than matrix updates. MATLAB implementations of the time update between measurements and the measurement update are given in Figure 3.15. The argument X is a vector containing both the estimate and the error covariance entries. We have been able to avoid using two sets of covariance entries, one for $P(t_k)$ and one for $P^-(t_k)$, by paying careful attention to the order of the operations.

The error variances for range and range rate shown in Figure 3.16 were obtained. Measurements were taken every second, and each time a measurement update was made, the error variances decreased. Steady state has nearly attained after 6 s. The Runge–Kutta integration step size was 10 ms. This figure is similar in concept to Figure 2.13, though different parameters were used.

□

3.8 Optimal Smoothing

We have discussed the *filtering* problem at some length. Given a stochastic system, the solution to the *prediction* problem is straightforward. All we must do to find the optimal prediction is use only the time update portion of the Kalman filter, which simply involves setting $H = 0$ in Table 3.1. Here we discuss optimal *smoothing*. This is the continuous version of Section 2.8. We shall basically follow the presentation in Gelb (1974).

Suppose data are available over a time interval $[0, T]$, and we desire optimal estimate $\hat{x}(t)$ for every t within the interval. All smoothed estimates are based somehow on filtered estimates, so the first step is to run the usual Kalman filter on the data to generate a filtered estimate $\hat{x}_f(t)$. We proceed as follows:

Let the unknown $x(t)$ be described by

$$\dot{x} = Ax + Bu + Gw \tag{3.249a}$$

```
% Example 3.9
% Multi-rate Continuous Discrete KF implementation for alpha-beta tracker
% A=2x2, G=2x1, X0=2x1, P0=2x2, Q=1x1, ttu=ix1, H=1x2, z=mx1, R=1x1,
% tmu=mx1
% Example: cdkf([0 1;0 0],[0;1],[0;0],[10 10 10 10],1,[0:0.1:10],[1 0],[5 ; 10
; 15 ; 20 ; 25 ; 30 ; 35 ; 40 ; 45 ; 50]+randn(10,1),1,[1:10])
function cdkf(A,G,X0,P0,Q,ttu,H,Z,R,tmu)
     global P1;
    X=X0'; P=P0; j=1;
    for i=1:length(ttu)-1
        % Time update for states
        X(i+1,:)=(A*X(i,:)')';
        % Time update for error covariance
        P1=vectortomatrix(P(i,:)); P1=A*P1+P1*A'+G*Q*G'; P1=matrixtovector(P1);
        [t1 P1] = ode45(@tu_ecov,[ttu(i) ttu(i+1)],P(i,:));
        P(i+1,:)=P1(end,:);
        % Measurement Update
        if (ttu(i+1)>=tmu(j)) % if true then do measurement update

[X(i+1,:),P(i+1,:)]=measurement_update(X(i+1,:)',P(i+1,:),Z(j,1),H,R);
            j=j+1;
        end
    end
    % Plotting
    plot(ttu,P(:,1),'k-',ttu,P(:,4),'k--');
    legend('p_1','p_4'); grid on;
return

% Function used by ode45() for finding P-
function Pd=tu_ecov(t,P)
    global P1;
    Pd=P1';
return

% Measurement Update
function [X,P]=measurement_update(X,P,Z,H,R)
    P=vectortomatrix(P);
    K=P*H'*inv(H*P*H'+R);
    P=(eye(length(P))-K*H)*P;
    X=X+K*(Z-H*X);
    P=matrixtovector(P);
    X=X';
return

function P1=matrixtovector(P)
    P1=[];
    for i=1:length(P)
        P1=[P1 P(i,:)];
    end
return

function P1=vectortomatrix(P)
    lng=sqrt(length(P)); P1=[];
    for k=1:lng:length(P)
        P1=[P1 ; P(k:k+lng-1)];
    end
return
```

FIGURE 3.15
Continuous-discrete-tracker time and measurement update subroutines.

with measurements

$$z = Hx + v \tag{3.249b}$$

White noises $w(t) \sim (0, Q)$ and $v(t) \sim (0, R)$ are uncorrelated with each other and with $x(0) \sim (\bar{x}_0, P_0)$. Let $Q \geq 0$ and $R > 0$. (In contrast to the discrete case, we will not require $|Q| \neq 0$.)

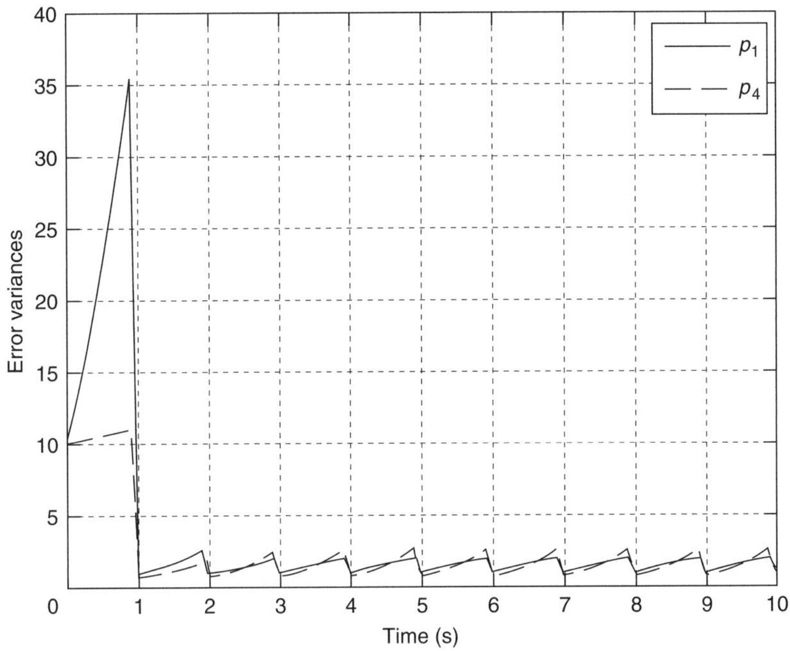

FIGURE 3.16
Continuous-discrete-tracker range and range-rate error variances.

To find the optimal smoothed estimate $\hat{x}(t)$ for $t \in [0, T]$ we must take into account the data $z(\tau)$ for $0 \leq \tau \leq t$ and for $t < \tau \leq T$. Suppose that the data for the subinterval preceding time t have been used to construct via the Kalman filter an estimate $\hat{x}_f(t)$. This estimate is given by

$$\dot{P}_f = AP_f + P_f A^T + GQG^T - P_f H^T R^{-1} HP_f \tag{3.250}$$

$$K_f = P_f H^T R^{-1} \tag{3.251}$$

$$\dot{\hat{x}}_f = A\hat{x}_f + Bu + K_f(z - H\hat{x}_f) \tag{3.252}$$

Then we can use the filtered estimate $\hat{x}_f(t)$ to manufacture the smoothed estimate $\hat{x}(t)$, which includes the effects of future data.

To do this, we first construct a filter that runs *backward* in time to generate an estimate $\hat{x}_b(t)$ that includes the data $z(\tau)$ for $t < \tau \leq T$. This is known as the *information filter*. Then we use a smoother algorithm to combine $\hat{x}_f(t)$ and $\hat{x}_b(t)$ into the smoothed estimate $\hat{x}(t)$.

The smoother scheme we will present gives the best linear estimate if the statistics are arbitrary, and the optimal estimate if all statistics are normal. It extends directly to the time-varying case.

3.8.1 The Information Filter

All $\hat{x}$ and P in this subsection should be considered as having a subscript "b" (i.e., "backward"), which we suppress for convenience.

To reverse time, define $\tau = T - t$. Then

$$\frac{\mathrm{d}}{\mathrm{d}\tau}x = -\frac{\mathrm{d}}{\mathrm{d}t}x \tag{3.253a}$$

so that

$$\frac{\mathrm{d}}{\mathrm{d}\tau}x = -Ax - Bu - Gw \tag{3.253b}$$

replaces plant equation 3.249a. The measurement equation remains the same. Evidently, then, the filter equations are simply

$$\frac{\mathrm{d}}{\mathrm{d}\tau}P = -AP - PA^T + GQG^T - PH^TR^{-1}HP \tag{3.254}$$

$$\frac{\mathrm{d}}{\mathrm{d}\tau}\hat{x} = -A\hat{x} - Bu + PH^TR^{-1}(z - H\hat{x}) \tag{3.255}$$

It is more convenient to express these equations in terms of the information matrix

$$S(\tau) = P^{-1}(\tau) \tag{3.256}$$

and to define an intermediate variable

$$\hat{y}(\tau) = P^{-1}(\tau)\hat{x}(\tau) = S(\tau)\hat{x}(\tau) \tag{3.257}$$

By using

$$\frac{\mathrm{d}}{\mathrm{d}\tau}P^{-1} = -P^{-1}\left(\frac{\mathrm{d}}{\mathrm{d}\tau}P\right)P^{-1} \tag{3.258}$$

Equation 3.254 yields

$$\frac{\mathrm{d}}{\mathrm{d}\tau}P^{-1} = P^{-1}A + A^TP^{-1} - P^{-1}GQG^TP^{-1} + H^TR^{-1}H \tag{3.259}$$

or

$$\frac{\mathrm{d}}{\mathrm{d}\tau}S = SA + A^TS - SGQG^TS + H^TR^{-1}H \tag{3.260}$$

Differentiating Equation 3.257 with respect to τ and using Equations 3.255 and 3.260 yields

$$\frac{\mathrm{d}}{\mathrm{d}\tau}\hat{y} = (A^T - SGQG^T)\hat{y} - SBu + H^TR^{-1}z \tag{3.261}$$

By defining the *backward filter gain*

$$K_b = SGQ \tag{3.262}$$

we can write

$$\frac{\mathrm{d}}{\mathrm{d}\tau}\hat{y} = (A^T - K_bG^T)\hat{y} - SBu + H^TR^{-1}z \tag{3.263}$$

Equations 3.260 and 3.261 are the information formulation of the continuous Kalman filter. The information filter runs backward in time beginning at time $t = T$, or $\tau = 0$.

3.8.2 Optimal Smoothed Estimate

Suppose now that a Kalman filter has been run forward on the interval $[0, T]$ to compute for each t in $[0, T]$, a filtered estimated $\hat{x}_f(t)$ based on data prior to t. Suppose also that a separate information filter has been run backward on $[0, T]$ to find for each t an estimate $\hat{x}_b(t)$ based on data subsequent to t. To combine $\hat{x}_f(t)$ and $\hat{x}_b(t)$ into an overall smoothed estimate $\hat{x}(t)$ which takes into account all the available data is very simple.

Since $w(t)$ and $v(t)$ are white, the estimate $\hat{x}_f(t)$ and $\hat{x}_b(t)$ are independent. We can therefore use the "Millman's theorem" result from the problems of Chapter 1 to write for the smoothed error covariance

$$P = \left(P_f^{-1} + P_b^{-1}\right)^{-1} \tag{3.264}$$

or

$$P = \left(P_f^{-1} + S\right)^{-1} \tag{3.265}$$

(We shall suppress subscripts "b" on $S(t)$ and $\hat{y}(t)$.) By the matrix inversion lemma

$$P = P_f - P_f S(I + P_f S)^{-1} P_f \tag{3.266}$$

or

$$P = (I - K)P_f \tag{3.267}$$

where the *smoother gain* is

$$K = P_f S(I + P_f S)^{-1} \tag{3.268}$$

The smoothed estimate is

$$\hat{x} = P\left(P_f^{-1}\hat{x}_f + P_b^{-1}\hat{x}_b\right) \tag{3.269}$$

or, using Equations 3.267 and 3.257,

$$\hat{x} = (I - K)\hat{x}_f + P\hat{y} \tag{3.270}$$

$$\hat{x} = \hat{x}_f + (P\hat{y} - K\hat{x}_f) \tag{3.271}$$

The smoothing scheme contains a Kalman filter, an information filter, and a smoother; it is shown in its entirely in Table 3.3.

Since for time T the filtered and smoothed estimates are the same, we have the boundary condition $S(\tau = 0) = 0$. According to Equation 3.257 then, $\hat{y}(\tau = 0) = 0$. (The boundary condition $\hat{x}_b(T)$ is unknown, and this is the reason for using the intermediate variable $\hat{y}(\tau)$ in the backward filter instead of $\hat{x}_b(\tau)$.)

TABLE 3.3
Optimal Smoothing Scheme

System and measurement models

$$\dot{x} = Ax + Bu + Gw$$
$$z = Hx + v,$$

$$x(0) \sim (\bar{x}_0, P_0), w(t) \sim (0, Q), v(t) \sim (0, R).$$

$x(0)$, $w(t)$, $v(t)$ uncorrelated.

Forward filter:

Initialization

$$\hat{x}_f(0) = \bar{x}_0, P_f(0) = P_0$$

Update

$$\begin{aligned} \dot{P}_f &= AP_f + P_f A^T + GQG^T - P_f H^T R^{-1} H P_f \\ K_f &= P_f H^T R^{-1} \\ \dot{\hat{x}}_f &= A\hat{x}_f + Bu + K_f(z - H\hat{x}_f) \end{aligned}$$

Backward filter ($\tau = T - t$):

Initialization

$$\hat{y}(0) = 0, S(0) = 0$$

Update

$$\begin{aligned} \frac{\mathrm{d}}{\mathrm{d}\tau} S(T-\tau) &= S(T-\tau)A + A^T S(T-\tau) \\ &\quad - S(T-\tau)GQG^T S(T-\tau) + H^T R^{-1} H \\ K_b(T-\tau) &= S(T-\tau)GQ \\ \frac{d}{d\tau}\hat{y}(T-\tau) &= [A^T - K_b(T-\tau)G^T]\hat{y}(T-\tau) \\ &\quad - S(T-\tau)Bu(T-\tau) + H^T R^{-1} z(T-\tau) \end{aligned}$$

Smoother

$$\begin{aligned} K(t) &= P_f(t)S(t)[I + P_f(t)S(t)]^{-1} \\ P(t) &= (I - K(t)]P_f(t) \\ \hat{x}(t) &= [I - K(t)\hat{x}_f(t) + P(t)\hat{y}(t)] \end{aligned}$$

In some cases P may be equal to P_f, so that the smoothed and filtered estimates are the same. In this case, smoothing does not offer an improvement over filtering alone. States whose estimates are improved by smoothing are said to be *smoothable*. It can be shown that if the plant is $(A, G\sqrt{Q})$ reachable so that all states are distributed by the process noise, then all states are smoothable. See Gelb (1974) and Fraser (1967). Figure 2.23 shows the relation between forward, backward, and smoothed error covariances.

There are other types of smoothers ("fixed-point," "fixed-lag"). The type we are discussing in which the final time T is fixed is known as the *fixed-interval* smoother.

3.8.3 Rauch–Ting–Striebel Smoother

The smoothing scheme in Table 3.3 requires a forward filter, a backward filter, and a smoother. We can quickly derive a more convenient scheme which combines the backward filter and the smoother into a single backward smoother recursion (Rauch, Tung, and Striebel 1965).

To this end, it is necessary to eliminate S and $\hat{y}$ from the smoother equations. To accomplish this, use Equation 3.258 to obtain from Equation 3.264

$$\frac{\mathrm{d}}{\mathrm{d}t}P^{-1} = -P_f^{-1}\dot{P}_f P_f^{-1} + \frac{\mathrm{d}}{\mathrm{d}t}(P_b^{-1}) \tag{3.272}$$

As in Equation 3.253a, we have

$$\frac{\mathrm{d}}{\mathrm{d}\tau}P^{-1} = -P^{-1}\dot{P}P^{-1} = P_f^{-1}\dot{P}_f P_f^{-1} + \frac{d}{\mathrm{d}\tau}(P_b^{-1}) \tag{3.273}$$

so that Equations 3.250, 3.259, and 3.264 imply

$$\begin{aligned}
P^{-1}\dot{P}P^{-1} &= P_f^{-1}A + A^T P_f^{-1} + P_f^{-1}GQG^T P_f^{-1} - H^T R^{-1} H \\
&\quad + P_b^{-1}A + A^T P_b^{-1} - P_b^{-1}GQG^T P_b^{-1} + H^T R^{-1} H \\
&= P^{-1}A + A^T P^{-1} + P_f^{-1}GQG^T P_f^{-1} \\
&\quad - \left(P^{-1} - P_f^{-1}\right) GQG^T \left(P^{-1} - P_f^{-1}\right)
\end{aligned}$$

or

$$\dot{P} = AP + PA^T + PP_f^{-1}GQG^T + GQG^T P_f^{-1}P - GQG^T$$

Finally,

$$\dot{P} = (A + GQG^T P_f^{-1})P + P(A + GQG^T P_f^{-1})^T - GQG^T \tag{3.274}$$

Defining a smoother gain as

$$F = GQG^T P_f^{-1} \tag{3.275}$$

this becomes

$$\dot{P} = (A + F)P + P(A + F)^T - GQG^T \tag{3.276}$$

In a similar fashion, we obtain the equation for the smoothed estimate

$$\dot{\hat{x}} = A\hat{x} + F(\hat{x} - \hat{x}_f) \tag{3.277}$$

The Rauch–Tung–Striebel smoother algorithm is then as follows. First run the usual Kalman filter on the given data. Then run the recursive smoother defined by Equations 3.275 through 3.277 backward for $t \leq T$ to generate the smoothed estimate $\hat{x}(t)$ and error covariance $P(t)$. Since at $t = T$, the filtered and smoothed estimates are identical, the initial conditions for the backward recursion are

$$P(T) = P_f(T), \qquad \hat{x}(T) = \hat{x}_f(T)$$

Problems

Section 3.1

1. The Lyapunov Equation Written as a Vector Equation. Let $\otimes$ denote the Kronecker product (Appendix) and $s(\cdot)$ the stacking operator. Show that the continuous matrix Lyapunov equation $\dot{P} = AP + PA^T + CC^T$ is equivalent to the *vector* equation

$$\frac{\mathrm{d}}{\mathrm{d}t}s(P) = [(A \otimes I) + (I \otimes A)]s(P) + s(CC^T)$$

2. Solutions to Algebraic Lyapunov Equations. Use the results of the previous problem to show that $0 = AP + PA^T + CC^T$ has a unique solution if and only if $\lambda_i + \lambda_j \neq 0$ for all i and j, where λ_i is an eigenvalue of A. Show that the stability of A is sufficient to guarantee this.

(a) Find all solutions to $0 = AP + PA^T + CC^T$ if

$$A = \begin{bmatrix} 0 & 0 \\ 1 & -1 \end{bmatrix}, \qquad C = \begin{bmatrix} 0 \\ 1 \end{bmatrix}$$

Interpret this in light of the previous problem.

(b) Find all symmetric solutions.

(c) Find the positive definite solution(s).

(d) Which eigenvalues of A are (A, C) reachable?

3. For the A and C of the previous problem, solve $\dot{P} = AP + PA^T + CC^T$ for $t \geq 0$ if $P(0) = 0$ and if $P(0) = I$. Now let t go to infinity. To which of the solutions found in the previous problem do the solutions for $P(0) = 0$ and $P(0) = I$ converge?

4. System Described by Newton's Laws. A system is described by $\ddot{x}(t) = w(t)$, with white noise $w(t) \sim (0, q)$ independent of the initial state

$$\begin{bmatrix} x(0) \\ \dot{x}(0) \end{bmatrix} \sim \left(\begin{bmatrix} \bar{x}_0 \\ \bar{v}_0 \end{bmatrix}, \begin{bmatrix} p_x(0) & 0 \\ 0 & p_{\dot{x}}(0) \end{bmatrix} \right)$$

(v_0 means initial velocity.)

(a) Find and sketch the means $\bar{x}(t), \bar{\dot{x}}(t)$.

(b) Find and sketch the variances $p_x(t)$, $p_{\dot{x}}(t)$, and $p_{x\dot{x}}(t)$.

5. Computer Simulation of Stochastic Systems. Let

$$\dot{x} = \begin{bmatrix} 0 & 1 \\ -\omega_n^2 & -2\alpha \end{bmatrix} x + \begin{bmatrix} 0 \\ 1 \end{bmatrix} w$$

$$z = \begin{bmatrix} 1 & 0 \end{bmatrix} x + v$$

with $w(t) \sim (0,1)$, $v(t) \sim (0,1/2)$ independent white noise processes with *uniform* distribution.

Write the MATLAB code to simulate the system.

6. Stochastic Harmonic Oscillator. Solve analytically for the mean and covariance of $x(t)$ in Problem 5 if $x(0) \sim ([1\ \ 1]^T, I)$. $\omega_n = 1$, $\alpha = 0.1$.

7. Periodic Solution of Riccati Equations

(a) Find the general solution to $\dot{p} = ap^2 + bp + c$ for the case $q = 4ca - b^2 > 0$ if the initial condition is $p(0) = p_0$.

(b) Show that $p(t)$ exhibits periodic behavior and never converges to a steady-state value. Sketch $p(t)$.

8. (Friedland 1969). Let $\bar{P}$ be any solution to Equation 3.21. Then show that any other solution can be expressed as

$$P = \bar{P} + VMV^T \tag{3.278}$$

where

$$\dot{V} = (A - \bar{P}H^T R^{-1} H)V \tag{3.279}$$

and

$$\dot{M} = -MV^T H^T R^{-1} HVM \tag{3.280}$$

Section 3.2

9. Kalman Filter Simulation. Consider the plant in Example 3.4. Write the MATLAB code to simulate the entire Kalman filter. You must integrate the Riccati equation and the actual filter dynamics (Equation 3.23). To run your program, use $\omega_n^2 = 2$, $\alpha = 1.5$, $q = 0.1$, $r = 0.1$, $\bar{x}(0) = [1\ \ 1]^T$, $P_0 = I$. For data, use $z(t) = -29e^{-1.1t} + 41e^{-1.9t}$; $t \geq 0$.

10. Harmonic Oscillator with Position Measurements. In Example 3.4, suppose measurements of position instead of velocity are taken so that

$$z = [1\ \ 0]x + v$$

Let $\alpha = 1$, $\omega_n = 2$, $q = 1$, $r = 1$.

(a) Write Riccati equation as three coupled scalar equations.

(b) Find steady-state error covariance.

(c) Find steady-state Kalman filter (Equation 3.23).

(d) Write MATLAB code to simulate the entire time-varying filter.

11. Filtering with No Process Noise. Show that if there is no process noise so that $Q = 0$, the Riccati equation can be written in terms of the information matrix $P^{-1}(t)$ as the Lyapunov equation

$$\frac{\mathrm{d}}{\mathrm{d}t}(P^{-1}) = -P^{-1}A - A^T P^{-1} + H^T R^{-1} H$$

(Hint: $\mathrm{d}/\mathrm{d}t(P^{-1}) = -P^{-1}\dot{P}P^{-1}$.)

12. Filtering with No Process Noise. Consider Newton's system

$$\ddot{x} = u$$
$$z = x + v$$

with $v(t) \sim (0, r)$ white and $x(0) \sim (\overline{x(0)}, P_0)$.

(a) Find the error covariance $P(t)$ for the maximum-likelihood estimator. Note that no *a priori* information about $x(t)$ means $\bar{x}(0) = 0$, $P(0) = \infty$, which can more conveniently be expressed as $P^{-1}(0) = 0$.

(b) Find the Kalman gain and write down the maximum-likelihood Kalman filter.

A filter to estimate the state of this system is called a second-order polynomial tracking filter.

13. Repeat Problem 12 for the harmonic oscillator in Example 3.4 with no process noise.

14. Discrete Polynomial Tracking Filter. Repeat Problem 12, first discretizing the plant using a sampling period of T s.

Section 3.3

15. Steady-State Filter for Harmonic Oscillator. Consider the harmonic oscillator in Example 3.4, where velocity measurements were made.

(a) Find steady-state Kalman filter. Show it is always stable (consider the two cases $\alpha > 0$ and $\alpha < 0$ separately).

(b) Find transfer function of Wiener filter.

(c) Plot filter root locus as q/r goes from 0 to ∞. Show that the damping ratio of the filter is $\sqrt{\alpha^2 + q/4r}/\omega_n$ and that the natural frequency is the same as that of the original plant.

16. Harmonic Oscillator with Position Measurements. Redo Problem 15 if position measurements are taken instead of velocity measurements. Note that now the filter natural frequency changes with q/r, while the real part of the poles stays constant.

17. Newton's System with Velocity Measurements. Repeat Problem 15 for the system in Example 3.6 if position and velocity measurements are taken so that

$$z = [1 \;\; 1]x + v$$

Section 3.4

18. Algebraic Riccati Equation Analytic Solution

(a) Show that a solution to the equation

$$0 = AP^T + PA^T + PR^{-1}P^T + Q \tag{3.281}$$

for $R > 0$ is given by

$$P = \sqrt{ARA^T - Q}L\sqrt{R}^T - AR \tag{3.282}$$

where L is any orthogonal matrix, $LL^T = I$. Show that this reduces to the well-known quadratic equation solution in the scalar case.

(b) Show that if we want $P = P^T$ in Equation 3.282, then L must be selected as the solution to a Lyapunov equation.

19. Covariance of Residual. Use Equation 3.91 (with $\tilde{x}(t_0) = 0$) to show that $P_{\tilde{Z}}(t, u) = \overline{\tilde{z}(t)\tilde{z}^T(u)} = R\delta(t - u)$.

(Hint: let the error system have state-transition matrix $\phi(t, t_0)$. Then show $\overline{\tilde{x}(t)\tilde{x}^T(u)} = \phi(t, u)P(u)$ for $t \geq u$ (Bryson and Ho 1975).)

20. Estimation of a Voltage Using Continuous and Discrete Filters. Consider the circuit of Figure 3.17 where $u(t)$ is an input voltage and $w(t) \sim (0, q)$ is a noise source. The initial state is $x(0) \sim (0, p_0)$, uncorrelated with $w(t)$.

(a) Write state equation.

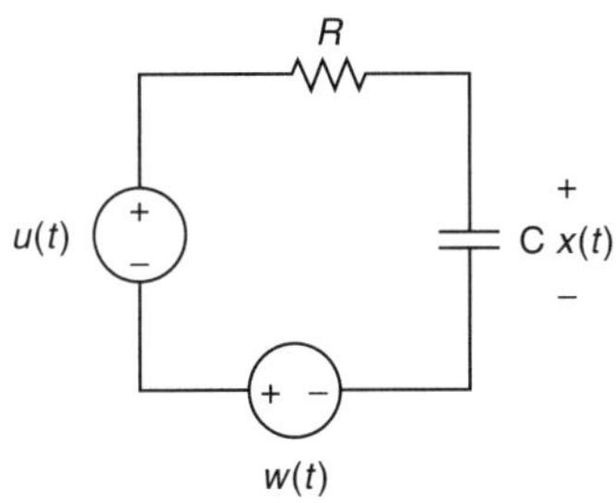

FIGURE 3.17
An electric circuit.

(b) Find and sketch $x(t)$ if $u(t) = u_{-1}(t)$ the unit step, for the deterministic case $p_0 = 0$ and $w(t) = 0$. Now let $RC = 1$ and measurements be taken according to

$$z(t) = x(t) + v(t) \tag{3.283}$$

with noise $v(t) \sim (0, r)$ uncorrelated with $w(t)$ and $x(0)$.

(c) Solve the algebraic Riccati equation for p_∞, the steady-state error variance. When is p_∞ greater than zero? Find a differential equation for the estimate $\hat{x}(t)$ in the steady-state filter. Find the transfer function of the Wiener filter.

(d) Let $q = 3$, $r = 1$. Using the steady-state filter, find $\hat{x}(t)$ if $u(t) = u_{-1}(t)$ and the data are $z(t) = 0.75(1 - e^{-1.2t})u^{-1}(t)$.

(e) Find the differential equation $\hat{x}(t)$ in the steady-state filter if r tends to infinity. Comment. What happens as r goes to zero so that the measurements are perfect?

(f) Let $q = 3$, $r = 1$. Use separation of variables to solve the Riccati equation for $p(t)$ if $p_0 = 2$. Sketch $p(t)$. Verify that the limiting behavior agrees with p_∞ in part c.

(g) Now, convert to the discretized system with $T = 0.1$ s. Let $q = 3$, $r = 1$, $p_0 = 2$.

(h) For the discrete system, find p_k for $k = 0$–5. Do these values agree with $p(kT)$ in (f)?

(i) For the discrete system, use data $z_k = z(kT)$ with $z(t)$ as in (d).

(Disregard z_0.) Let $u_k = u_{-1}(k)$. Find optimal estimate $\hat{x}_k$ for $k = 0, 1, 2, 3$. Compare to the (suboptimal) estimates $\hat{x}(kT)$ in (d).

21. An Alternative Design Method for Steady-State Filters. An observer for the plant 3.76 is given by

$$\dot{\hat{x}} = A\hat{x} + K(z - H\hat{x})$$

The observer gain K can be chosen, for example, to place the observer poles so that $\hat{x}(t)$ converges to $x(t)$ as quickly as desired. The error system using this observer is given by Equation 3.90.

Write an equation that may be solved for P_∞, the steady-state error covariance. Now select K to minimize P_∞. One way to do this is to differentiate the error covariance equation with respect to K, using $\partial P_\infty / \partial K = 0$ for a minimum (Schultz and Melsa 1967).

22. ARE Solutions

(a) Find all symmetric solutions to Equation 3.153 if

$$A = \begin{bmatrix} 0 & 8 \\ 1 & -2 \end{bmatrix}, \quad G = \begin{bmatrix} \sqrt{2} & 0 \\ 0 & 1 \end{bmatrix}, \quad H = \begin{bmatrix} 0 & 1 \end{bmatrix}, \quad Q = I, \quad R = I$$

(b) Check the stability of the closed-loop system $(A - KH)$, where $K = PH^TR^{-1}$, for all your solutions.

Section 3.5

23. Prove Equation 3.168.

24. Steady-State Filters. Design an optimal steady-state filter to separate a signal $x(t)$ with spectral density

$$\Phi_X(\omega) = \frac{\omega^2 + 5}{\omega^4 + 6\omega^2 + 25}$$

from additive white noise $v(t) \sim (0, r)$:

(a) Using Kalman theory.

(b) Using Wiener theory.

(c) Sketch filter poles as a function of r.

(d) Sketch $\Phi_X(\omega)$, noise spectral density, and filter frequency response for $r = 1$.

25. Filter Design Using Hamiltonian Matrix. The Hamiltonian matrix

$$M = \begin{bmatrix} A & -GQG^T \\ -H^TR^{-1}H & -A^T \end{bmatrix}$$

provides an alternative design procedure for the optimal steady-state filter.

(a) Perform a state-space transformation

$$T = \begin{bmatrix} I & P \\ 0 & I \end{bmatrix}$$

Show that if P satisfies the ARE, then

$$TMT^{-1} = \begin{bmatrix} A - KH & 0 \\ -H^TR^{-1}H & -(A - KH)^T \end{bmatrix}$$

where $K = PH^TR^{-1}$.

(b) Hence, show that

$$|sI - M| = \Delta^{\text{cl}}(s)\Delta^{\text{cl}}(-s)$$

where $\Delta^{\text{cl}}(s) = |sI - (A - KH)|$ is the characteristic polynomial of the optimal steady-state filter. Therefore, the filter poles are the stable eigenvalues of M. Knowing these, a technique like Ackermann's formula can be used to determine the optimal filter gain K without solving the ARE for P (Kailath 1980).

26. Filtering with No Process Noise. Use Problem 25 to show that if $GQG^T = 0$, then the optimal steady-state filter poles are the "stabilized" poles of the plant (i.e., unstable plant poles s_i are replaced by $-s_i$).

27. Steady-State Filter Design. In Example 3.4, let $\omega_n^2 = 5, \alpha = 1$, $q = 1$, $r = 1$. Design the steady-state filter using:

(a) Chang–Letov.

(b) The results of Problem 25.

(c) Wiener theory.

28. Redo Problem 24 using:

(a) The Hamilton matrix.

(b) The Chang–Letov procedure.

29. Steady-State Filter Poles. Let

$$\dot{x}_1 = -x_1 + 0.5x_2$$
$$\dot{x}_2 = 8.5x_1 - 2x_2 + w$$
$$z = x_1 + v$$

where $w(t) \sim (0, q), v \sim (0, r)$ are uncorrelated white noises. Sketch the poles of the optimal steady-state filter as q/r goes from 0 to ∞. Find the optimal filter if $q/r = 2$.

Section 3.6

30. Derived Measurement for Time-Varying Measurement Model. Redo the derived measurement analysis for the case of time-varying measurement matrix $H(t)$.

31. Colored Measurement Noise. Design an optimal time-invariant filter to estimate a signal with spectral density

$$\Phi_Y(\omega) = \frac{3\omega^2}{\omega^4 + 4}$$

which is measured in the presence of additive noise with spectral density

$$\Phi_V(\omega) = \frac{2}{\omega^2 + 4}$$

(a) Using Kalman filter theory with state augmentation.

(b) Using Wiener filter theory.

(c) Using the Kalman filter with derived measurements.

(d) Sketch the signal and noise spectral densities.

32. Aircraft Longitudinal Dynamics with Gust Noise. Consider the situation in Example 2.16.

(a) Sketch $\Phi_w(\omega)$ and find and sketch the autocorrelation function $R_w(t)$.

(b) Find a shaping filter for $w(t)$. Augment the state Equation 2.300.

(c) If $\omega_n = 3$ rad/s, $\delta = 0.1, a = 1, \sigma^2 = 2$, use Chang–Letov to plot the poles of the optimal steady-state filter as r goes from ∞ to 0.

(d) Find the steady-state filter if $r = 2$.

(e) Find the optimal estimate for pitch $\theta(t)$ if the data are $z(t) = 0.1 \cos 4t u_{-1}(t)$. Use the steady-state filter.

(f) Find steady-state error covariance.

33. Correlated Process and Measurement Noise. Let

$$\dot{x} = x + u + w$$
$$z = x + v$$

where $w(t) \sim (0, 1), v(t) \sim (0, r)$, and $\overline{vw} = \mu$.

(a) Find optimal filter gain $K(t)$ and error covariance $P(t)$.

(b) Find optimal steady-state filter and error covariance.

(c) Sketch poles of steady-state filter as μ/r goes from 0 to ∞.

(d) Let $\mu = 1, r = 2$. Find optimal estimate for $x(t)$ if the data are $z(t) = e^{0.9t} u_{-1}(t)$ and the input $u(t) = 0$. Use steady-state filter.

Section 3.7

34. Newton's System. In Example 3.9, let $x(0) \sim ([1\ \ 1]^T, I), q = 1$, $r = 0.5, T = 0.5$ s. Solve analytically for and sketch the components of $P(t)$ for $t = 0$ to 2 s if the continuous-discrete filter is used.

35. Harmonic Oscillator

(a) Repeat Problem 34 for the harmonic oscillator in Example 3.4 if $x(0) \sim ([1\ \ 1]^T, I), q = 1, r = 0.5, T = 0.5$ s, $\omega_n = \sqrt{2}$ rad/s, $\alpha = 1$.

(b) Write MATLAB code for this example to implement the continuous-discrete Kalman filter.

36. Comparison of Continuous, Discrete, and Continuous-Discrete Filters. Let

$$\dot{x} = 0.2x + w$$
$$z = x + v$$

with $w(t) \sim (0, 1)$ and $v(t) \sim (0, 0.5)$ uncorrelated white noises. Let $x(0) \sim (1, 1)$.

(a) Write the continuous filter equations. Solve for the error covariance $p(t)$.

(b) Measurements are now taken every second. Discretize and write the discrete Kalman filter equations. Solve for error covariance p_k for $t = 0$ to 5 s.

(c) Write the continuous-discrete filter equations. Solve for $p(t)$ for $t = 0$ to 5 s.

(d) Plot all three error covariance on the same graph.

37. Convergence for Continuous-Discrete Filter. Show that if $(A, G\sqrt{Q})$ is reachable and (A, H) is detectable, then in the continuous-discrete filter, $P(t)$ displays a bounded steady-state behavior independent of $P(0)$. Furthermore, the Kalman filter is stable. (What does stability mean in this case?)

Section 3.8

38. Smoothing for Scalar System

$$\dot{x} = ax + bu + w$$
$$z = x + v$$

with $w(t) \sim (0, q), v(t) \sim (0, r)$.

(a) Formulate the two-filter smoothing in Table 3.3. Now, if $a = 0, q = 1$, $r = 0.5$, final time $T = 5$ s, $x(0) \sim (1, 1)$ solve analytically for smoothed error covariance $p(t)$. Sketch $p_f(t), s(t)$, and $p(t)$.

(b) Repeat using the Rauch–Tung–Striebel formulation.

39. Smoothing for Newton's System. Let

$$\ddot{x} = w$$
$$z = x + v$$

$w(t) \sim (0, q), v(t) \sim (0, r)$. Repeat Problem 38. For the numerical work, let $x(0) \sim (1, I), q = 0, r = 1$.

4

Kalman Filter Design and Implementation

In this chapter we discuss some practical aspects of Kalman filter design and implementation. There are two general topics. In Section 4.1, we analyze the effects of assuming a model that does not accurately describe the physical system whose state is to be estimated. This topic is also discussed in Part II (Robust Estimation) of the book. In the remaining sections we present some techniques for designing filters of simplified structure.

4.1 Modeling Errors, Divergence, and Exponential Data Weighting

We have discussed Kalman filter design for the deal situation when all plant dynamics and noise statistics are exactly known. If this is not the case serious divergence problems can occur where the theoretical behavior of the filter and its actual behavior do not agree. The Kalman filter has no mechanism built into its theory for dealing with inaccurate dynamics A, B, G, and H, or statistics P_0, Q, and R. If the plant is inaccurately known we should, to be theoretically rigorous, apply not Kalman filter theory but adaptive control theory, which can result in more complicated designs.

Fortunately, by understanding the properties of the Kalman filter we can often use its robustness in the face of process noise to compensate for inaccurately known plant dynamics. Alternatively, we can modify the filter to get better performance. These topics are the subject of this section.

4.1.1 Modeling Errors

For this discussion we need to define two concepts. They must be clearly distinguished from each other. The *plant* is the actual physical system we are interested in. The *model* is the mathematical description of the physical system, which we use for the filter design stage. (The plant is also called the "truth model.")

Let the model be given by

$$x_{k+1} = Ax_k + Bu_k + Gw_k \tag{4.1a}$$

$$z_k = Hx_k + v_k \tag{4.1b}$$

with w_k and v_k white and uncorrelated with each other and with x_0. Let $w_k \sim (0, Q)$, $v_k \sim (0, R)$, and $x_0 \sim [E(x_0), P_0]$. The Kalman filter designated on the basis of this model is given by

Time update:

$$P_{k+1}^- = AP_kA^T + GQG^T \tag{4.2}$$

$$\hat{x}_{k+1}^- = A\hat{x}_k + Bu_k \tag{4.3}$$

Measurement update:

$$K_k = P_k^- H^T \left(HP_k^- H^T + R\right)^{-1} \tag{4.4}$$

$$P_k = (I - K_kH)P_k^-(I - K_kH)^T + K_kRK_k^T \tag{4.5}$$

$$\hat{x}_k = \hat{x}_k^- + K_k(z_k - H\hat{x}_k^-) \tag{4.6}$$

The *a priori* error covariance is defined as

$$P_k^- = E[(x_k - \hat{x}_k^-)(x_k - \hat{x}_k^-)^T] \tag{4.7}$$

Now suppose that in reality the plant is described not by Equation 4.1 but by the state x'_k with dynamics

$$x'_{k+1} = \bar{A}x'_k + \bar{B}u'_k + \bar{G}w'_k \tag{4.8a}$$

$$z'_k = \bar{H}x'_k + v'_k \tag{4.8b}$$

Assume w'_k and v'_k are white and uncorrelated with each other and with x'_0, and let $w'_k \sim (0, \bar{Q})$, $v'_k \sim (0, \bar{R})$, and $x'_0 \sim [E(x'_0), \bar{P}_0]$. We assume here that the model and the plant have the same order n. Matrices $\bar{A}, \bar{B}, \bar{G}, \bar{H}, \bar{P}_0, \bar{Q}$, and $\bar{R}$ are in general all different from the values we assumed in our model (Equation 4.1), which we used to design the "optimal" filter. (In this section, overbars denote not expected values, but plant quantities.)

The word "optimal" is in inverted commas because, although it is indeed optimal for the assumed model, it is not necessarily optimal for the actual plant. In fact, if the filter (Equations 4.2 through 4.7) is applied to the plant (Equation 4.8) some very serious problems may result.

Now we need to develop a theoretical background for analyzing *mismatch* between the model and the plant.

Define the *a priori* and *a posteriori* errors between the estimate generated by the filter (Equations 4.2 through 4.7) and the state of the *actual plant* (Equation 4.8) as

$$\tilde{x}_k^- = x'_k - \hat{x}_k^- \tag{4.9}$$

and

$$\tilde{x}_k = x'_k - \hat{x}_k \tag{4.10}$$

These are *not* the same as the *predicted* error $(x_k - \hat{x}_k^-)$ between the estimate and the *model* state whose covariance P_k^- in Equation 4.7 is minimized by the filter (Equations 4.2 through 4.7)! They are, however, the errors we are interested in. If the filter is designed and implemented correctly then the *actual* errors $\tilde{x}_k^-$ and $\tilde{x}_k$ should be acceptably small during the identification run.

To find expressions for the *actual error covariances* $E[\tilde{x}_k^-(\tilde{x}_k^-)^T]$ and $E(\tilde{x}_k\tilde{x}_k^T)$ we proceed as follows. First consider the time update. We have

$$\begin{aligned}\tilde{x}_{k+1}^- &= x'_{k+1} - \hat{x}_{k+1}^- \\ &= \bar{A}x'_k + \bar{B}u'_k + \bar{G}w'_k - A\hat{x}_k - Bu_k\end{aligned}$$

By defining the *model mismatch system matrix*

$$\Delta A = \bar{A} - A \tag{4.11}$$

This can be written as

$$\tilde{x}_{k+1}^- = A\tilde{x}_k + \Delta A x'_k + \bar{B}u'_k - Bu_k + \bar{G}w'_k \tag{4.12}$$

Now append the error onto the plant state to define an augmented state, and write Equations 4.8a and 4.12 in one equation (Gelb 1974),

$$\begin{bmatrix} x'_{k+1} \\ \tilde{x}_{k+1}^- \end{bmatrix} = \begin{bmatrix} \bar{A} & 0 \\ \Delta A & A \end{bmatrix} \begin{bmatrix} x'_k \\ \tilde{x}_k \end{bmatrix} + \begin{bmatrix} \bar{B} & 0 \\ \bar{B} & -B \end{bmatrix} \begin{bmatrix} u'_k \\ u_k \end{bmatrix} + \begin{bmatrix} \bar{G} \\ \bar{G} \end{bmatrix} w'_k \tag{4.13}$$

Define correlation matrices by

$$\begin{bmatrix} U_k & (V_k^-)^T \\ V_k^- & S_k^- \end{bmatrix} = E \begin{bmatrix} x'_k \\ \tilde{x}_k^- \end{bmatrix} \begin{bmatrix} x'_k \\ \tilde{x}_k^- \end{bmatrix}^T \tag{4.14}$$

and

$$\begin{bmatrix} U_k & V_k^T \\ V_k & S_k \end{bmatrix} = E \begin{bmatrix} x'_k \\ \tilde{x}_k \end{bmatrix} \begin{bmatrix} x'_k \\ \tilde{x}_k \end{bmatrix}^T \tag{4.15}$$

Then from Equation 4.13

$$\begin{bmatrix} U_{k+1} & (V_{k+1}^-)^T \\ V_{k+1}^- & S_{k+1}^- \end{bmatrix} = \begin{bmatrix} \bar{A} & 0 \\ \Delta A & A \end{bmatrix} \begin{bmatrix} U_k & V_k^T \\ V_k & S_k \end{bmatrix} \begin{bmatrix} \bar{A}^T & \Delta A^T \\ 0 & A^T \end{bmatrix} + \begin{bmatrix} \bar{G} \\ \bar{G} \end{bmatrix} \bar{Q} [\bar{G}^T \quad \bar{G}^T]$$

where we have set $u'_k = 0$ and $u_k = 0$ for simplicity. Multiplying this out there results the time-update equations

$$U_{k+1} = \bar{A}U_k\bar{A}^T + \bar{G}\bar{Q}\bar{G}^T \tag{4.16}$$

$$V_{k+1}^- = \Delta A U_k \bar{A}^T + AV_k\bar{A}^T + \bar{G}\bar{Q}\bar{G}^T \tag{4.17}$$

$$S_{k+1}^- = AS_kA^T + \bar{G}\bar{Q}\bar{G}^T + \Delta A U_k \Delta A^T + \left(AV_k\Delta A^T + \Delta A V_k^T A^T\right) \tag{4.18}$$

We should make a digression to ensure we have a feel for these equations before proceeding.

Note that U_k is the plant state correlation matrix $E[x'_k(x'_k)^T]$. Also note that if $\Delta A = 0$, $G = \bar{G}$, $Q = \bar{Q}$ so that there is no model mismatch, then Equation 4.18 is identical to Equation 4.2. The extra terms in Equation 4.18 represent the increase in the error covariance due to modeling inaccuracies. We are not directly interested in the plant state correlation U_k or the cross-correlation matrix V_k^-; they are required only to update the quantity of interest S_k^-, which is the correlation matrix of the actual error $\tilde{x}_k^- = x'_k - \hat{x}_k^-$. In the presence of modeling errors, the actual error correlation depends on the plant state correlation.

If S_k^- is small then our estimate $\hat{x}_k^-$ matches well with the actual plant state x'_k, which is what we really want.

Now let us derive the measurement update equations. We have

$$\tilde{x}_k = x'_k - \hat{x}_k = x'_k - \hat{x}_k^- - K_k(z'_k - H\hat{x}_k^-) \tag{4.19}$$

A comment needs to be made at this point. What we are doing is running the designed filter (Equations 4.2 through 4.7) on the plant (Equation 4.8). The measurements are therefore the actual plant output. For this reason, z'_k and not the model output z_k is used in Equation 4.19. The model is used only to design the filter.

Proceeding, Equation 4.19 becomes

$$\tilde{x}_k = \tilde{x}_k^- - K_k\bar{H}x'_k - K_k v'_k + K_k H\hat{x}_k^-$$

If we define the *model mismatch output matrix*

$$\Delta H = \bar{H} - H \tag{4.20}$$

this can be written as

$$\tilde{x}_k = (I - K_k H)\tilde{x}_k^- - K_k\Delta H x'_k - K_k v'_k \tag{4.21}$$

By defining an augmented state vector we can write

$$\begin{bmatrix} x'_k \\ \tilde{x}_k \end{bmatrix} = \begin{bmatrix} I & 0 \\ -K_k\Delta H & I - K_k H \end{bmatrix} \begin{bmatrix} x'_k \\ \tilde{x}_k^- \end{bmatrix} + \begin{bmatrix} 0 \\ -K_k \end{bmatrix} v'_k \tag{4.22}$$

Now take correlations of both sides to obtain

$$\begin{bmatrix} U_k & V_k^T \\ V_k & S_k \end{bmatrix} = \begin{bmatrix} I & 0 \\ -K_k\Delta H & I - K_k H \end{bmatrix} \begin{bmatrix} U_k & (V_k^-)^T \\ V_k^- & S_k^- \end{bmatrix} \times \begin{bmatrix} I & -\Delta H^T K_k^T \\ 0 & (I - K_k H)^T \end{bmatrix} + \begin{bmatrix} 0 \\ -K_k \end{bmatrix} \bar{R} [0 \quad -K_k^T] \tag{4.23}$$

which yields the measurement update

$$U_k = U_k \tag{4.24}$$

$$V_k = (I - K_k H)V_k^- - K_k\Delta H U_k \tag{4.25}$$

$$
\begin{aligned}
S_k = (I - K_k H)S_k^-(I - K_k H)^T + K_k\bar{R}K_k^T + K_k\Delta H U_k \Delta H^T K_k^T \\
- \left[(I - K_k H)V_k^- \Delta H^T K_k^T + K_k \Delta H (V_k^-)^T (I - K_k H)^T\right]
\end{aligned}
\tag{4.26}
$$

Note that if $\Delta H = 0$ and $R = \bar{R}$ so that there are no modeling errors, then updating Equation 4.26 for the actual error correlation is equal to update Equation 4.5 for the *predicted covariance.* Again, we are not interested in U_k and V_k, but we are forced to carry them as "excess baggage" to find the quantity of interest S_k.

Since initially the uncertainty in the estimate is the same as the uncertainty in the plant state, we would initialize the recursion by

$$
S_0 = -V_0 = U_0 = \bar{P}_0 \tag{4.27}
$$

By combining Equations 4.12 and 4.21 we obtain the *error system* for the actual error

$$
\begin{aligned}
\tilde{x}_{k+1}^- = A(I - K_k H)\tilde{x}_k^- + (\Delta A - AK_k\Delta H)x_k' + \bar{B}u_k' - Bu_k \\
+ \bar{G}w_k' - AK_k v_k'
\end{aligned}
\tag{4.28}
$$

Several things are worth mentioning about this equation. First, it should be compared to Equation 2.188, to which it reduces if the model is an exact description of the plant. Second, it is driven by the *plant* noise processes. Third, if the model mismatch matrices ΔA and ΔH are not both zero, the error system is driven by the plant state x_k'. Unless x_k' is bounded, the error $\tilde{x}_k^-$ will grow. We are thus faced with the realization that in the presence of modeling inaccuracies, the plant must in most cases be stable to avoid causing an unbounded estimation error.

The effect of an inaccurately known deterministic input is also clearly seen in Equation 4.28, for the error system is driven by the term $(\bar{B}u_k' - Bu_k)$. If this is not zero, unbounded errors can again occur in $\tilde{x}_k^-$.

It is evident from the brief discussion that even if $A(I - K_k H)$ is stable for large k, which is guaranteed by the steady-state properties of the Kalman filter, modeling inaccuracies can lead to nonconvergence and unbounded actual estimation errors.

A final point is worth making. The expected error is given by

$$
E(\tilde{x}_{k+1}^-) = A(I - K_k H)E(\tilde{x}_k^-) + (\Delta A - AK_k\Delta H)E(x_k') + \bar{B}u_k' - Bu_k \tag{4.29}
$$

Even if $E(\tilde{x}_0^-) = 0$, it is not guaranteed that $E(\tilde{x}_k^-) = 0$ in the presence of modeling inaccuracies. Therefore U_k, V_k, and S_k should not be considered as covariances, but as correlation matrices. Modeling inaccuracies can result in biased estimates.

We shall call P_k the *predicted* error covariance given by Equations 4.2 through 4.7, and S_k the *actual* error correlation given by Equations 4.16 through 4.18 and Equations 4.24 through 4.26.

Example 4.1 Inaccurate Modeling of Wiener Process as Constant Unknown

To illustrate the use of Equations 4.16 through 4.18 and Equations 4.24 through 4.26 in the analysis of model mismatch, suppose the model is assumed to be the scalar system

$$x_{k+1} = x_k \tag{4.30}$$

$$z_k = x_k + v_k \tag{4.31}$$

$v_k \sim (0, r), x_0 \sim (m_0, p_0)$.

Then according to Example 2.4, using Equations 4.2 through 4.7 to design a filter for this model results in an error covariance given by

$$p_k = \frac{p_0}{1 + k(p_0/r)} \tag{4.32}$$

and in Kalman gain given by

$$K_k = \frac{p_0/r}{1 + k(p_0/r)} \tag{4.33}$$

The predicted error covariance (Equation 4.32) is sketched in Figure 4.1 for $p_0 = 1$, $r = 10$; it shows the anticipated decrease to zero of the error as more and more measurements are made. The Kalman gain also goes to zero with k; this means that as more measurements are made, the estimate theoretically

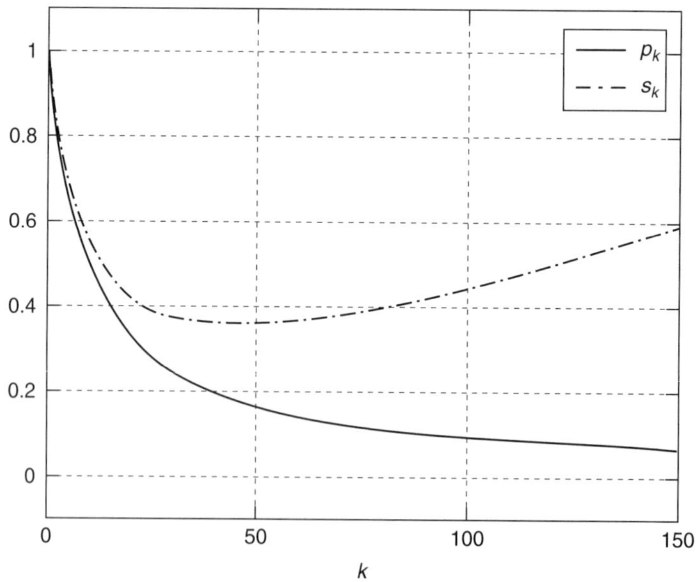

FIGURE 4.1
Behavior of error covariances with model mismatch. Predicted error covariance p_k and actual error covariance s_k are shown.

becomes more reliable so that each subsequent measurement has less and less effect.

Now let the actual plant, which is supposedly modeled by Equations 4.30 and 4.31, be given by

$$x'_{k+1} = x'_k + w'_k \tag{4.34}$$

$$z'_k = x'_k + v'_k \tag{4.35}$$

$w'_k \sim (0, q)$, $v'_k \sim (0, r)$, $x'_0 \sim (m_0, p_0)$. Thus, the actual unknown is a Wiener process, which we have incorrectly modeled as a random constant.

Applying the Kalman filter (Equation 4.33) to the actual plant results in

$$\hat{x}_{k+1} = \hat{x}_k + \frac{p_0/r}{1 + (k+1)(p_0/r)}(z'_{k+1} - \hat{x}_k) \tag{4.36}$$

To find the resulting *actual* error correlation for $\tilde{x}_k^- = x'_k - \hat{x}_k^-$, we need to use Equations 4.16 through 4.18 and Equations 4.24 through 4.26. Since $\Delta A = 0$ and $\Delta H = 0$, we do not need to compute U_k, V_k, and V_k^-, but can simply use the recursion for the actual error correlation s_k:

$$s_0 = p_0 \tag{4.37}$$

Time update:

$$s_{k+1}^- = s_k + q \tag{4.38}$$

Measurement update:

$$s_k = (1 - K_k)^2 s_k^- + K_k^2 r \tag{4.39}$$

$$= \left(\frac{1 + (k-1)p_0/r}{1 + kp_0/r}\right)^2 s_k^- + \frac{p_0^2/r}{(1 + kp_0/r)^2} \tag{4.40}$$

We get

$k = 0$

$$s_0 = p_0 \tag{4.41}$$

$k = 1$

Time update:

$$s_1^- = p_0 + q \tag{4.42}$$

Measurement update:

$$s_1 = \frac{p_0 + q + p_0^2/r}{(1 + p_0/r)^2} = \frac{p_0(1 + p_0/r) + q}{(1 + p_0/r)^2} \tag{4.43}$$

$k = 2$

Time update:

$$s_2^- = \frac{p_0(1 + p_0/r) + q\left[1 + (1 + p_0/r)^2\right]}{(1 + p_0/r)^2} \tag{4.44}$$

Measurement update:

$$
\begin{aligned}
s_2 &= \frac{p_0(1+p_0/r) + q\big[1+(1+p_0/r)^2\big] + p_0^2/r}{(1+2p_0/r)^2} \\
&= \frac{p_0(1+2p_0/r) + q\big[1+(1+p_0/r)^2\big]}{(1+2p_0/r)^2}
\end{aligned}
\tag{4.45}
$$

And, in general

$$
\begin{aligned}
s_k &= \frac{p_0(1+kp_0/r) + q\sum_{i=0}^{k-1}(1+ip_0/r)^2}{(1+kp_0/r)^2} \\
&= p_k + \frac{q\sum_{i=0}^{k-1}(1+ip_0/r)^2}{(1+kp_0/r)^2}
\end{aligned}
\tag{4.46}
$$

This actual error correlation is also shown in Figure 4.1 for $p_0=1$, $r=10$, and $q=0.01$. The behavior of the actual correlation s_k is quite different from the behavior of the predicted covariance p_k!

To find the limiting behavior of s_k, use the identities

$$
\sum_{i=0}^{k-1} i = \frac{(k-1)k}{2} \tag{4.47}
$$

$$
\sum_{i=0}^{k-1} i^2 = \frac{(k-1)k(2k-1)}{6} \tag{4.48}
$$

to write

$$
\begin{aligned}
s_k &\to \frac{q\sum_{i=0}^{k-1}\left[1+(2ip_0/r)+i^2(p_0^2/r^2)\right]}{1+2k(p_0/r)+k^2(p_0^2/r^2)} \\
&\to \frac{q\big[k^2(p_0/r)+(k^3/3-k^2/2)(p_0^2/r^2)\big]}{k^2p_0^2/r^2} \\
&= q\left[\left(\frac{r}{p_0}-\frac{1}{2}\right)+\frac{k}{3}\right]
\end{aligned}
\tag{4.49}
$$

While we think that all is well for large k because of (4.32), the actual error correlation s_k is increasing linearly with k!

The program used to make Figure 4.1 is shown in Figure 4.2. It implements the predicted error covariance recursion

$$
p_{k+1} = p_k - \frac{p_k^2}{p_k+r} \tag{4.50}
$$

and the actual error covariance recursion (Equations 4.38 and 4.39). Note that carrying out a *performance analysis* often requires only simple programs. The analysis we have done subsequent to Equation 4.39 is not required in a practical performance study! □

```
% Example 4.1
% Program for Kalman Filter performance study
clear all; close all;
k=[0:150];
p0=1; r=10; q=0.01; s=p0;
p=p0./(1+k*(p0/r)); % Predicted error Covariance p
K=p./r;% Kalman Gain K
for i=2:length(k)
    s(i)=s(i-1)+q;
    s(i)=(1-K(i)).^2*s(i)+K(i).^2*r; % Actual Error covariances
end
plot(k,p,k,s); grid on; legend('p_k','s_k'); xlabel('k');% Plot p
```

FIGURE 4.2
MATLAB program for Kalman filter performance study.

EXERCISE 4.1 Model Mismatch in Reduced-Order Filters

Let the actual plant be given by Equation 4.8 and the assumed model by Equation 4.1. It can happen that the order n_m of the model is less than the order n_p of the plant. This can occur in two ways. Either we can fail to account for all the states through ignorance of the plant, or we can deliberately select a reduced-order model of describing the plant by omitting some of the plant states. If done properly, this reduced-order modeling can result in a (nonoptimal!) reduced-order filter that has a simplified structure and yet gives us the essential information we require about the plant state (Gelb 1974, D'Appolito 1971, Nash *et al.* 1972). The design of reduced-order filters is discussed further later on in this chapter. Here we discuss the actual error correlation equations.

Let $T \in R^{n_m \times n_p}$ be the state-space transformation that takes the actual plant state x'_k to the model state x_k (i.e., $x_k = Tx'_k$). Let T have full row rank n_m. Define modified model mismatch matrices by

$$\Delta A = \bar{A} - T^+ AT \tag{4.51}$$

$$\Delta H = \bar{H} - HT \tag{4.52}$$

where T^+ is the Moore–Penrose matrix inverse of T. Note that $TT^+ = I$ since T has full row rank, so T^+ is a right inverse here. [Since T has full row rank, its right inverse is given by $T^+ = T^T(TT^T)^{-1}$.] Then the actual errors are defined as the n_p vectors

$$\tilde{x}_k^- = x'_k - T^+\hat{x}_k^- \tag{4.53}$$

$$\tilde{x}_k = x'_k - T^+\hat{x}_k \tag{4.54}$$

a. Show that the actual error correlation S_k^- for a reduced-order filter that does not estimate the full plant state is given by Equations 4.16 through 4.18 with A replaced everywhere by T^+AT, and Equations 4.24 through 4.26 with H and K_k replaced everywhere by HT and T^+K_k, respectively. Let $u'_k = 0$, $u_k = 0$ for simplicity.

b. Show that the actual error for the reduced-order filter develops according to Equation 4.28 with the above replacements, and with B also replaced by T^+B.

EXERCISE 4.2 Actual Error Correlation for Continuous Kalman Filter

In the continuous-time case, define the actual error as

$$\tilde{x}(t) = x'(t) - \hat{x}(t) \tag{4.55}$$

where $x'(t)$ is the actual plant state and $\hat{x}(t)$ the estimate produced by the filter designed based on the model. Define correlation matrices by

$$\begin{bmatrix} U & V^T \\ V & S \end{bmatrix} = E\begin{bmatrix} x' \\ \tilde{x} \end{bmatrix}\begin{bmatrix} x' \\ \tilde{x} \end{bmatrix}^T \tag{4.56}$$

a. Show that the actual error correlation $S(t)$ is given by

$$\begin{aligned} \dot{S} &= (A - KH)S + S(A - KH)^T + \bar{G}\bar{Q}\bar{G}^T \\ &\quad + K\bar{R}K^T + (\Delta A - K\Delta H)V^T + V(\Delta A - K\Delta H)^T \end{aligned} \tag{4.57a}$$

$$\dot{V} = V\bar{A}^T + (A - KH)V + (\Delta A - K\Delta H)U + \bar{G}\bar{Q}\bar{G}^T \tag{4.57b}$$

$$\dot{U} = \bar{A}U + U\bar{A}^T + \bar{G}\bar{Q}\bar{G}^T \tag{4.57c}$$

where overbars denote actual plant quantities, and ΔA and ΔH are the same as for the discrete case. The initial conditions are given by Equation 4.27.

b. Show that the actual error system is

$$\dot{\tilde{x}} = (A - KH)\tilde{x} + (\Delta A - K\Delta H)x' + \bar{B}u' - Bu + \bar{G}w' - Kv' \tag{4.58}$$

EXERCISE 4.3 Model Mismatch in Continuous Reduced-Order Filters

For the continuous Kalman filter the results of Exercise 4.1 are also valid.

a. Defining the actual error as

$$\tilde{x} = x' - T^+\hat{x} \tag{4.59}$$

show that the actual error correlation $S(t)$ for a filter that does not estimate the entire plant state is given for the special case $T^+ = T^T$ by

$$\begin{aligned} \dot{S} &= T^T(A - KH)TS + ST^T(A - KH)^T T + \bar{G}\bar{Q}\bar{G}^T + T^T K\bar{R}K^T T \\ &\quad + (\Delta A - T^T K\Delta H)V^T + V(\Delta A - T^T K\Delta H)^T \end{aligned} \tag{4.60}$$

$$\dot{V} = V\bar{A}^T + T^T(A - KH)TV + (\Delta A - T^T K \Delta H)U + \bar{G}\bar{Q}\bar{G}^T \quad (4.61)$$

$$\dot{U} = \bar{A}U + U\bar{A}^T + \bar{G}\bar{Q}\bar{G}^T \quad (4.62)$$

The model mismatch matrices ΔA and ΔH are as defined in Exercise 4.1.

b. Show that the actual error evolves according to

$$\begin{aligned}\dot{\tilde{x}} &= T^T(A - KH)T\tilde{x} + (\Delta A - T^T K \Delta H)x' \\ &\quad + \bar{B}u' - T^T Bu + \bar{G}w' - T^T Kv' \end{aligned} \quad (4.63)$$

4.1.2 Kalman Filter Divergence

Filter divergence occurs when the actual error correlation S_k is not bounded as expected from the predicted error covariance P_k (Gelb 1974, Anderson and Moore 1979). In *apparent divergence*, S_k remains bounded, but it approaches a larger bound than does P_k (see Figure 4.3). This occurs when there are modeling errors that cause the implemented filter to be suboptimal. In *true divergence*, P_k remains bounded while S_k tends to infinity (see Figure 4.1). This occurs when the filter is not stable (Section 2.5), or when there are unmodeled unstable plant states. An illustration of an unstable filter is provided by Example 4.1 ($(A, G\sqrt{Q})$ is not reachable in that example). An illustration of unmodeled unstable states is provided in the next example.

Example 4.2 Unmodeled Unstable States

Suppose the assumed model is the scalar random bias

$$x_{k+1} = x_k \quad (4.64)$$

$$z_k = x_k + v_k \quad (4.65)$$

$v_k \sim (0, r)$, $x_0 \sim (\tilde{x}_0, p_0)$. Then by the results of Example 2.4, the predicted error covariance (for $(x_k - \hat{x}_k)$) is

$$p_k = \frac{p_0}{1 + k(p_0/r)} \quad (4.66)$$

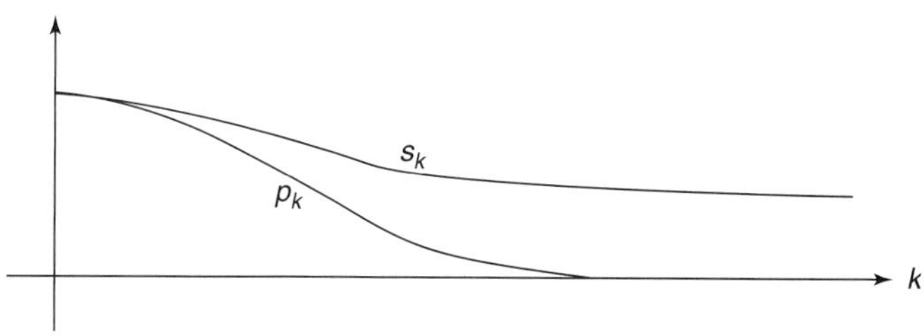

FIGURE 4.3
Apparent filter divergence.

and the Kalman gain is

$$K_k = \frac{p_0/r}{1 + k(p_0/r)} \tag{4.67}$$

As k increases P_k and K_k go to zero.

Now suppose the actual plant is given by

$$x'_{k+1} = \begin{bmatrix} 1 & \tau \\ 0 & 1 \end{bmatrix} x'_k \tag{4.68}$$

$$z'_k = \begin{bmatrix} 1 & 0 \end{bmatrix} x'_k + v'_k \tag{4.69}$$

$v'_k \sim (0, r)$, $x'_0 \sim (\bar{x}'_0, \bar{P}_0)$, where τ is the sampling period. Thus, the plant has an unmodeled random ramp behavior. The zero-input response of Equation 4.68 is shown in Figure 4.4, where $x'_k \triangleq [x'_1(k) \;\; x'_2(k)]^T$.

Applying the Kalman filter designed based on model (Equations 4.64 and 4.65) to the plant (Equations 4.68 and 4.69) there results

$$\hat{x}_{k+1} = \hat{x}_k + \frac{p_0/r}{1 + (k+1)(p_0/r)}(z'_{k+1} - \hat{x}_k) \tag{4.70}$$

as the estimate for $x'_1(k)$.

We can use Exercise 4.1 to investigate the behavior of the actual error in this situation. What we have inadvertently done is to assume a reduced-order model related to the plant by $x_k = Tx'_k$ with

$$T = \begin{bmatrix} 1 & 0 \end{bmatrix} \tag{4.71}$$

Hence $T^+ = [1 \; 0]^T$ is a right inverse for T and

$$T^+AT = \begin{bmatrix} 1 & 0 \\ 0 & 0 \end{bmatrix} \tag{4.72}$$

$$HT = \begin{bmatrix} 1 & 0 \end{bmatrix} \tag{4.73}$$

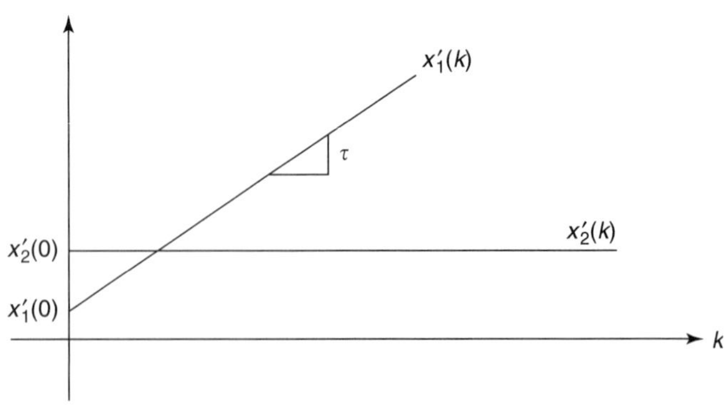

FIGURE 4.4
Zero-input response of plant with random ramp.

The model-mismatch matrices are

$$\Delta A = \bar{A} - T^{+}AT = \begin{bmatrix} 0 & \tau \\ 0 & 1 \end{bmatrix} \tag{4.74}$$

and

$$\Delta H = \bar{H} - HT = 0 \tag{4.75}$$

and

$$T^{+}K_k = \begin{bmatrix} \dfrac{p_0/r}{1 + k(p_0/r)} \\ 0 \end{bmatrix} \tag{4.76}$$

The actual error system therefore becomes (see Equation 4.28 and Exercise 4.1)

$$\tilde{x}_{k+1}^{-} + \begin{bmatrix} \dfrac{1 + (k-1)(p_0/r)}{1 + k(p_0/r)} & 0 \\ 0 & 0 \end{bmatrix} \tilde{x}_k^{-} + \begin{bmatrix} 0 & \tau \\ 0 & 1 \end{bmatrix} x'_k + \begin{bmatrix} \dfrac{p_0/r}{1 + k(p_0/r)} \\ 0 \end{bmatrix} v'_k \tag{4.77}$$

If the plant were indeed described by Equations 4.64 and 4.65, then the error system would be only the scalar system obtained by keeping the (1, 1) entries of the coefficient matrices. In this example, however, we are not so fortunate. Error system (Equation 4.77) is driven by the plant state x'_k. According to Figure 4.4, $x'_k(k)$ is constant at $x'_2(0)$, so this term tends to make the first component of $\tilde{x}_k^{-}$ increase linearly with k at the rate τ. In the limit of large k, the Kalman gain tends to zero and Equation 4.77 becomes

$$\tilde{x}_{k+1}^{-} = \begin{bmatrix} 1 & 0 \\ 0 & 0 \end{bmatrix} \tilde{x}_k^{-} + \begin{bmatrix} 0 & \tau \\ 0 & 1 \end{bmatrix} x'_k \tag{4.78}$$

so that the error increases without bound because of the unmodeled unstable plant dynamics!

It is also easy to see this behavior in terms of an unbounded actual error correlation matrix S_k by using Equations 4.16 through 4.18 and Equations 4.24 through 4.26. Let us suppose $\bar{P}_0 = I$. Then $S_0 = -V_0 = U_0 = I$.

According to Equations 4.16 and 4.24, the plant state correlation matrix develops by the recursion

$$U_{k+1} = \bar{A}U_k\bar{A}^T \tag{4.79}$$

which has solution

$$U_k = \bar{A}^k U_0 (\bar{A}^T)^k = \begin{bmatrix} 1 & k\tau \\ 0 & 1 \end{bmatrix} \begin{bmatrix} 1 & 0 \\ k\tau & 1 \end{bmatrix} = \begin{bmatrix} 1 + k^2\tau^2 & k\tau \\ k\tau & 1 \end{bmatrix} \tag{4.80}$$

The correlation of $x'_1(k)$ increases quadratically with k. Define the components of V_k and S_k by

$$V_k \triangleq \begin{bmatrix} V_1(k) & V_2(k) \\ V_3(k) & V_4(k) \end{bmatrix} \tag{4.81}$$

and

$$S_k \triangleq \begin{bmatrix} S_1(k) & S_2(k) \\ S_2(k) & S_4(k) \end{bmatrix} \tag{4.82}$$

Now use Equations 4.17 and 4.25 (and Exercise 4.1!) to see that

$$V_{k+1} = \begin{bmatrix} \frac{1+k(p_0/r)}{1+(k+1)(p_0/r)} & 0 \\ 0 & 1 \end{bmatrix} \begin{bmatrix} (k+1)\tau^2 + V_1(k) + V_2(k)\tau & V_2(k)+\tau \\ (k+1)\tau & 1 \end{bmatrix} \tag{4.83}$$

Thus, for large k the first factor is I and then $V_1(k)$, $V_2(k)$, and $V_3(k)$ increase with k.

Finally, by Equations 4.18 and 4.26

$$\begin{aligned} S_k^- &= \begin{bmatrix} 1 & 0 \\ 0 & 0 \end{bmatrix} S_k \begin{bmatrix} 1 & 0 \\ 0 & 0 \end{bmatrix} + \begin{bmatrix} 0 & \tau \\ 0 & 1 \end{bmatrix} U_k \begin{bmatrix} 0 & 0 \\ \tau & 1 \end{bmatrix} \\ &\quad + \begin{bmatrix} 1 & 0 \\ 0 & 0 \end{bmatrix} V_k \begin{bmatrix} 0 & 0 \\ \tau & 1 \end{bmatrix} + \begin{bmatrix} 0 & \tau \\ 0 & 1 \end{bmatrix} V_k^T \begin{bmatrix} 1 & 0 \\ 0 & 0 \end{bmatrix} \\ &= \begin{bmatrix} S_1(k) + 2V_2(k)\tau + \tau^2 & V_2(k)+\tau \\ V_2(k)+\tau & 1 \end{bmatrix} \end{aligned} \tag{4.84}$$

$$\begin{aligned} S_{k+1} &= \begin{bmatrix} \frac{1+k(p_0/r)}{1+(k+1)(p_0/r)} & 0 \\ 0 & 1 \end{bmatrix} \begin{bmatrix} S_1(k) + 2V_2(k)\tau + \tau^2 & V_2(k)+\tau \\ V_2(k)+\tau & 1 \end{bmatrix} \\ &\quad \times \begin{bmatrix} \frac{1+k(p_0/r)}{1+(k+1)(p_0/r)} & 0 \\ 0 & 1 \end{bmatrix} + \begin{bmatrix} \frac{p_0^2/r}{[1+(k+1)(p_0/r)]^2} & 0 \\ 0 & 0 \end{bmatrix} \end{aligned} \tag{4.85}$$

Again, for large k the first and third factors of the first term in Equation 4.85 tend to I, and $S_1(k)$ then increases with k. Note that $S_4(k)$ does not increase but remains at $S_4(0) = 1$. The actual error associated with the unmodeled state $x_2'(k)$ does not increase; the effect here of the unmodeled unstable dynamics is to provide an unaccounted for "input" to error component, one which makes it increase.

It turns out that it is not very difficult to correct the divergence in this example. The problem is that the Kalman gain K_k goes to zero with k, and so the filter disregards the measurements for large k. All we need to do is prevent this from happening! We give two methods for doing this in the next subsections. □

4.1.3 Fictitious Process Noise Injection

We discuss two methods for preventing divergence: *fictitious process noise injection* and *exponential data weighting*. Both methods work by preventing the predicted error covariance P_k, and hence the Kalman gain K_k from going to zero with k.

To prevent divergence, we can ensure that *in the model* all states are sufficiently excited by the process noise. Thus, we must ensure that $(A, G\sqrt{Q})$ is reachable, even though in the plant there may be no process noise ($w'_k = 0$). If $(A, G\sqrt{Q})$ is reachable, then as was seen in Section 2.5, the detectability of (A, H) guarantees that P_k approaches a strictly positive definite (i.e., nonzero) steady-state value P.

At this point, it is worthwhile to look back at the radar tracking problem in Example 2.8. In that example the actual plant was nonlinear. To apply the Kalman filter in Table 2.1, we assumed a model of simplified linear structure. To prevent divergence, although we were not yet speaking in those terms, we added fictitious process noise (maneuver noise) in the model to compensate for the unmodeled nonlinear dynamics.

It is easy to see that adding process noise to the model in Example 4.1 will correct the divergence problem, for this will make the model the same as the plant!

The next example shows how to use fictitious process noise injection to correct to divergence in Example 4.2.

Example 4.3 Prevention of Divergence by Process Noise Injection

The plant in Example 4.2 is a random ramp with no process noise. The model assumed for filter design was a random bias with no process noise. It was found that while the predicted error covariance p_k tended to zero with k, the actual error correlation S_k increased with k.

We can use the simplified random bias model to design a first-order nondivergent filter for this second-order plant by adding fictitious process noise w_k to the model. Thus, assume a model of the form

$$x_{k+1} = x_k + w_k \tag{4.86}$$

$$z_k = x_k + v_k \tag{4.87}$$

$w_k \sim (0, q)$, $v_k \sim (0, r)$, $x_0(\bar{x}_0, p_0)$.

According to Table 2.1, the predicted error covariance is given by

$$p_{k+1}^- = p_k + q \tag{4.88}$$

$$p_k = \frac{p_k^-}{1 + (p_k^-/r)} \tag{4.89}$$

With nonzero q it is not easy to find a closed-form expression for the error covariance like the one we gave in Example 4.2; however, p_k is sketched in Figure 4.5 for $p_0 = 10$, $r = 10$, and $q = 0.1$. This figure was made using a simple program for Equations 4.88 and 4.89, like the one in Figure 4.2. Note that p_k does not go to zero. We can find its steady-state value (cf., Example 2.1) by noting that at steady state Equations 4.88 and 4.89 can be combined with $p_{k+1}^- = p_k^- \triangleq p^-$ to yield

$$p^- = \frac{rp^-}{r + p^-} + q$$

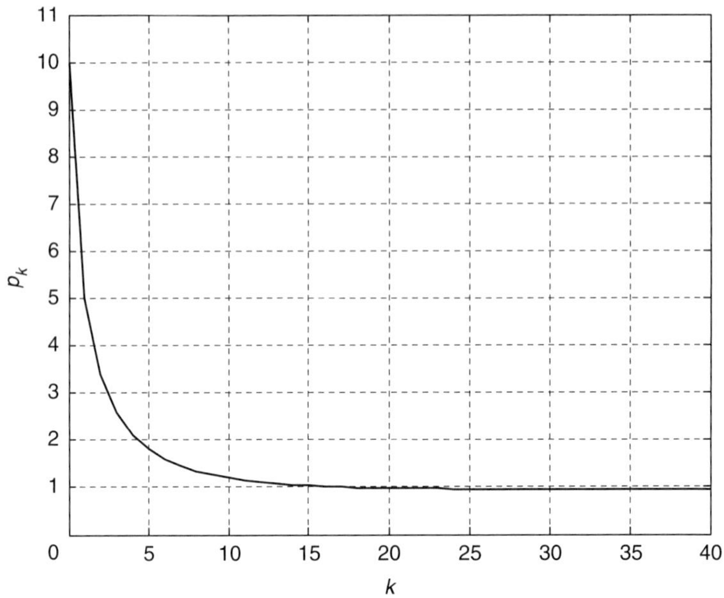

FIGURE 4.5
Predicted error covariance with fictitious model process noise.

or

$$(p^-)^2 - qp^- - qr = 0 \tag{4.90}$$

whose positive solution is

$$p^- = \frac{q}{2}\left(1 + \sqrt{1 + \frac{4r}{q}}\right) \tag{4.91}$$

Hence the steady-state value of p_k is

$$p = \frac{q}{2}\left(\sqrt{1 + \frac{4r}{q}} - 1\right) \tag{4.92}$$

The Kalman gain is

$$K_k = \frac{p_k}{r} \tag{4.93}$$

which reaches a nonzero, steady-state value of

$$K = \frac{q}{2r}\left(\sqrt{1 + \frac{4r}{q}} - 1\right) \tag{4.94}$$

The steady-state Kalman gain is plotted as a function of q/r in Figure 4.6. Note that it varies from 0 to 1 as q/r increases.

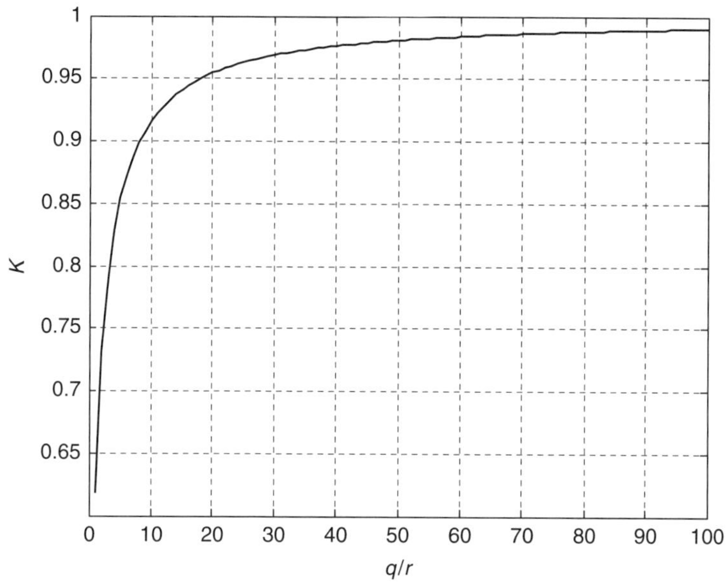

FIGURE 4.6
Steady-state Kalman gain as a function of q/r.

Now suppose the actual plant is the random ramp

$$x'_{k+1} = \begin{bmatrix} 1 & T \\ 0 & 1 \end{bmatrix} x'_k \tag{4.95}$$

$$z'_k = \begin{bmatrix} 1 & 0 \end{bmatrix} x'_k + v'_k \tag{4.96}$$

$v'_k \sim (0, r)\, x'_0 = (\bar{x}'_0, \bar{P}_0)$.

If we apply the above Kalman filter to this plant, the model mismatch matrices are as in Example 4.2 and the actual error system becomes

$$\tilde{x}^-_{k+1} = \begin{bmatrix} 1 - K_k & 0 \\ 0 & 0 \end{bmatrix} \tilde{x}^-_k + \begin{bmatrix} 0 & T \\ 0 & 1 \end{bmatrix} x'_k + \begin{bmatrix} K_k \\ 0 \end{bmatrix} v'_k \tag{4.97}$$

with gain K_k given by (Equation 4.93). The difference between this and Equation 4.77 in Example 4.2 is that, now that fictitious process noise has been added to the model, K_k does not go to zero. This means that as k becomes large the measurements are not discounted. In fact, Equation 4.97 approaches the steady-state error system

$$\tilde{x}^-_{k+1} = \begin{bmatrix} 1 - K & 0 \\ 0 & 0 \end{bmatrix} \tilde{x}^-_k + \begin{bmatrix} 0 & T \\ 0 & 1 \end{bmatrix} x'_k + \begin{bmatrix} K \\ 0 \end{bmatrix} v'_k \tag{4.98}$$

Two effects are of interest here. First, if $q \neq 0$, then the plant measurement noise v'_k continues to excite the error as $k \to \infty$, that the limiting value of the first component $\tilde{x}^-_1(k)$ is nonzero even if plant state component $x_2(k)$ is zero.

Fictitious process noise injection into the model has resulted in suboptimal behavior of the filter. This degradation is compensated for by the second effect: if q is correctly chosen, the error remains bounded even in the face of the unmodeled plant dynamics. To see this, note that since in the model (A, H) is detectable and $(A, G\sqrt{Q})$ is reachable, $A(I - KH) = (1 - K)$ is always stable (i.e., $|1 - K| < 1$ for all q, so the error in Equation 4.98 remains bounded. As q increased, $(1-K)$ tends to zero (see Figure 4.6). In the extreme, if we select q so that $(1 - K) \simeq 0$, then the error is given in steady state by

$$\tilde{x}_1^-(k) = \tau x_2'(k) + v_k' \tag{4.99}$$

$$\tilde{x}_2^-(k) = x_2'(k) \tag{4.100}$$

which is bounded since, according to Figure 4.4, $x_2'(k)$ has a constant value of $x_2'(0)$.

This bounded behavior of $\tilde{x}_k^-$ induced by the fictitious process noise can also be seen by examining the actual error correlation S_k. From Equations 4.16 through 4.18 and Equations 4.24 through 4.26 it is apparent that we still have (if $\bar{P}_0 = I$)

$$U_k = \begin{bmatrix} 1 + k^2\tau^2 & k\tau \\ k\tau & 1 \end{bmatrix} \tag{4.101}$$

Now, however,

$$V_{k+1} = \begin{bmatrix} 1 - K_{k+1} & 0 \\ 0 & 1 \end{bmatrix} \begin{bmatrix} (k+1)\tau^2 + V_1(k) + V_2(k)\tau & V_2(k) + \tau \\ (k+1)\tau & 1 \end{bmatrix} \tag{4.102}$$

and

$$\begin{aligned} S_{k+1} = &\begin{bmatrix} 1 - K_{k+1} & 0 \\ 0 & 1 \end{bmatrix} \begin{bmatrix} S_1(k) + 2V_2(k)\tau + \tau^2 & V_2(k) + \tau \\ V_2(k) + \tau & 1 \end{bmatrix} \begin{bmatrix} 1 - K_{k+1} & 0 \\ 0 & 1 \end{bmatrix} \\ &+ \begin{bmatrix} K_{k+1}^2 r & 0 \\ 0 & 0 \end{bmatrix} \end{aligned} \tag{4.103}$$

To make the point, if we choose q very large so that $K \simeq 1$, then in steady state there results

$$V_k = \begin{bmatrix} 0 & 0 \\ k\tau & 1 \end{bmatrix} \tag{4.104}$$

$$S_k = \begin{bmatrix} r & 0 \\ 0 & 1 \end{bmatrix} \tag{4.105}$$

so that S_k is bounded. ▯

4.1.4 Exponential Data Weighting

Another way to prevent divergence by keeping the filter from discounting measurements for large k is to use *exponential data weighting.* Additional

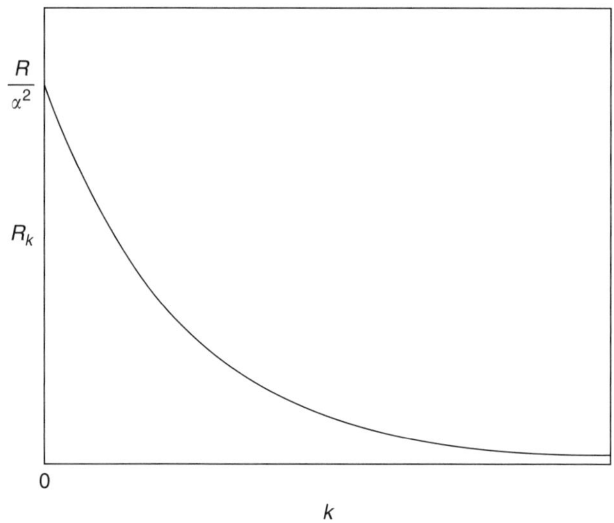

FIGURE 4.7
Exponential data weighting.

references for this material are Anderson (1973), Anderson and Moore (1979), and Gelb (1974).

Let us set the model covariance matrices equal to

$$R_k = R\alpha^{-2(k+1)} \tag{4.106}$$

$$Q_k = Q\alpha^{-2(k+1)} \tag{4.107}$$

for some $\alpha \geq 1$ and constant matrices Q and R. Since $\alpha \geq 1$, this means that as time k increases, the process and measurement noise covariances decrease, so that we are giving more credibility to the recent data by decreasing the noise covariance exponentially with k (see Figure 4.7).

According to Table 2.3, the Kalman filter is now given as follows. The error covariance develops by

$$P_{k+1}^- = A\big[P_k^- - P_k^- H^T\big(HP_k^- H^T + R\alpha^{-2k-2}\big)^{-1} HP_k^-\big]A^T + GQG^T\alpha^{-2k-2} \tag{4.108}$$

or

$$P_{k+1}^-\alpha^{2(k+1)} = \alpha^2 A\left[P_k^-\alpha^{2k} - P_k^-\alpha^{2k}H^T\left(HP_k^-\alpha^{2k}H^T + \frac{R}{\alpha^2}\right)^{-1} HP_k^-\alpha^{2k}\right]A^T + GQG^T$$

By defining the *weighted covariance*

$$P_k^\alpha \triangleq P_k^-\alpha^{2k} \tag{4.109}$$

this becomes

$$P_{k+1}^{\alpha} = \alpha^2 A \left[P_k^{\alpha} - P_k^{\alpha} H^T \left(H P_k^{\alpha} H^T + \frac{R}{\alpha^2} \right)^{-1} H P_k^{\alpha} \right] A^T + GQG^T \quad (4.110)$$

The initial condition is $P_0^{\alpha} = P_0$.

The Kalman gain is

$$K_k = P_k^- H^T \left(H P_k^- H^T + R\alpha^{-2k-2} \right)^{-1}$$

which can be written in terms of P_k^{α} as

$$K_k = P_k^{\alpha} H^T \left(H P_k^{\alpha} H^T + \frac{R}{\alpha^2} \right)^{-1} \quad (4.111)$$

Finally, the estimate update equation is

$$\hat{x}_{k+1}^- = A\hat{x}_k^- + Bu_k + AK_k(z_k - H\hat{x}_k^-) \quad (4.112)$$

or

$$\hat{x}_{k+1}^- = A(I - K_k H)\hat{x}_k^- + Bu_k + AK_k z_k \quad (4.113)$$

This filter has time-varying noise covariance Q_k and R_k, however, since they are exponentially varying we see from Equations 4.110 and 4.111 that we can compute P_k^{α} and K_k by considering the equivalent *time-invariant* model

$$x_{k+1} = \alpha A x_k + Bu_k + Gw_k \quad (4.114a)$$

$$z_k = Hx_k + v_k \quad (4.114b)$$

$w_k \sim (0, Q), v_k \sim (0, R/\alpha^2)$. The error covariance update equation for this model that is given in Table 2.3 is identical to Equation 4.110.

While P_k^{α} is the error covariance for the time-invariant model (Equation 4.114), it is *not* the error covariance for the original model (Equation 4.1) with time-varying covariances (Equations 4.106 and 4.107). By using the sensitivity relations Equations 4.16 through 4.18 and Equations 4.24 through 4.26 it can be shown (Anderson 1973) that P_k^{α} provides an upper bound for the actual error covariance.

The error covariance P_k^- with no data weighting reaches a bounded steady-state value if (A, H) is detectable. The covariance P_k^{α} reaches such a value if $(\alpha A, H)$ is detectable. But detectability of (A, H) is equivalent to detectability of $(\alpha A, H)$, so that if the plant (Equation 4.1) is detectable then P_k^{α} is bounded.

To investigate the stability of the filter (Equations 4.110, 4.111, and 4.113), note the following. If $(\alpha A, G\sqrt{Q})$ is stabilizable, then the filter designed from Table 2.3 for the equivalent model (4.1-51) is asymptotically stable for large k. Although the covariance for this filter is given by Equation 4.110 and the

Kalman gain by Equation 4.111, the estimate update is not given by Equation 4.113. It is given, from Table 2.3, by

$$\hat{x}_{k+1}^- = \alpha A(I - K_k H)\hat{x}_k^- + Bu_k + \alpha A K_k z_k \tag{4.115}$$

The stabilizability property for Equation 4.114 guarantees that for large k the poles of $\alpha A(I - K_k H)$ are inside $|z| = 1$. This means that the poles of $A(I - K_k H)$ are inside

$$|z| = \frac{1}{\alpha} < 1 \tag{4.116}$$

so that in the limit, Equation 4.113 has at least exponential stability with terms decaying like α^{-k}.

Reachability of $(A, G\sqrt{Q})$ [i.e., of the model 4.1] is equivalent to reachability of $(\alpha A, G\sqrt{Q})$ (i.e., of the equivalent model 4.114). Thus, by using exponential data weighting, reachability of $(A, G\sqrt{Q})$ guarantees a *stability margin* of α^{-1} for the filter Equation 4.113.

We can show that exponential data weighting is equivalent to fictitious process noise injection. To wit, write Equation 4.110 as

$$P_{k+1}^\alpha = A\left[P_k^\alpha - P_k^\alpha H^T\left(HP_k^\alpha H^T + \frac{R}{\alpha^2}\right)^{-1} HP_k^\alpha\right]A^T + Q_k' \tag{4.117}$$

where

$$Q_k' = GQG^T + (\alpha^2 - 1)A\left[P_k^\alpha - P_k^\alpha H^T\left(HP_k^\alpha H^T + \frac{R}{\alpha^2}\right)^{-1} HP_k^\alpha\right]A^T \tag{4.118}$$

Evidently, Q_k' corresponds to a process noise whose covariance is increased over the original Q.

If $\alpha \neq 1$, then the actual error covariance is increased since the filter is suboptimal. Compare with the discussion of (Equation 4.98) in Example 4.3. This means that P_k^1 (i.e., P_k^α for $\alpha = 1$) provides a lower bound for the actual error covariance, while P_k^α (for the α actually used) provides an upper bound.

The next example shows that if $\alpha > 1$, then P_k^α reaches a nonzero steady-state value. Thus, exponential data weighting can be used to prevent divergence. It works by preventing the discounting of data for large k as much as fictitious process noise injection prevented divergence in Example 4.3.

Example 4.4 Steady-State Kalman Gain with Exponential Data Weighting

Let us again consider the scalar system of Example 4.1,

$$x_{k+1} = x_k \tag{4.119}$$

$$z_k = x_k + v_k \tag{4.120}$$

$v_k \sim (0, r)$, $x_0 \sim (\bar{x}_0, p_0)$. In Examples 4.1 and 4.2, we showed that the predicted error covariance p_k went to zero with k, which resulted in divergence problems. We showed in Example 4.3 how to prevent divergence by keeping p_k from going to zero with fictitious process noise. In this example, we show how to prevent p_k from going to zero with exponential data weighting.

If we use the Kalman filter with exponential data weighting, then Equation 4.110 becomes

$$\begin{aligned} p_{k+1}^\alpha &= \alpha^2 \left[p_k^\alpha - \frac{(p_k^\alpha)^2}{p_k^\alpha + r/\alpha^2} \right] \\ &= \left(\frac{p_k^\alpha r/\alpha^2}{p_k^\alpha + r/\alpha^2} \right) \alpha^2 \end{aligned} \tag{4.121}$$

This can be written (cf. Equation 2.64) as

$$\frac{\alpha^2}{p_{k+1}^\alpha} = \frac{1}{p_k^\alpha} + \frac{\alpha^2}{r} \tag{4.122}$$

We can more easily solve this for the information matrix

$$l_k \triangleq (p_k^\alpha)^{-1} \tag{4.123}$$

Equation 4.122 can be written as the linear difference equation in l_k,

$$l_{k+1} = \alpha^{-2} l_k + \frac{1}{r} \tag{4.124}$$

The solution to this (for $\alpha > 1$) is

$$l_k = \alpha^{-2k} l_0 + \sum_{i=0}^{k-1} \alpha^{-2(k-i-1)} \left(\frac{1}{r} \right)$$

or,

$$\begin{aligned} l_k &= \alpha^{-2k} l_0 + \frac{\alpha^{-2(k-1)}}{r} \sum_{i=0}^{k-1} \alpha^{2i} \\ &= \alpha^{-2k} l_0 + \frac{\alpha^{-2(k-1)}}{r} \left(\frac{1 - \alpha^{2k}}{1 - \alpha^2} \right) \end{aligned}$$

so

$$l_k = \alpha^{-2k} l_0 + \frac{1}{r} \left(\frac{1 - \alpha^{-2k}}{1 - \alpha^{-2}} \right) \tag{4.125}$$

In terms of p_k^α,

$$p_k^\alpha = \left[\alpha^{-2k} p_0^{-1} + \frac{1}{r} \left(\frac{1 - \alpha^{-2k}}{1 - \alpha^{-2}} \right) \right]^{-1} \tag{4.126}$$

or

$$p_k^\alpha = \frac{p_0 \alpha^{2k}}{1 + (p_0/r)\alpha^{2k}[(1 - \alpha^{-2k})/(1 - \alpha^{-2})]} \tag{4.127}$$

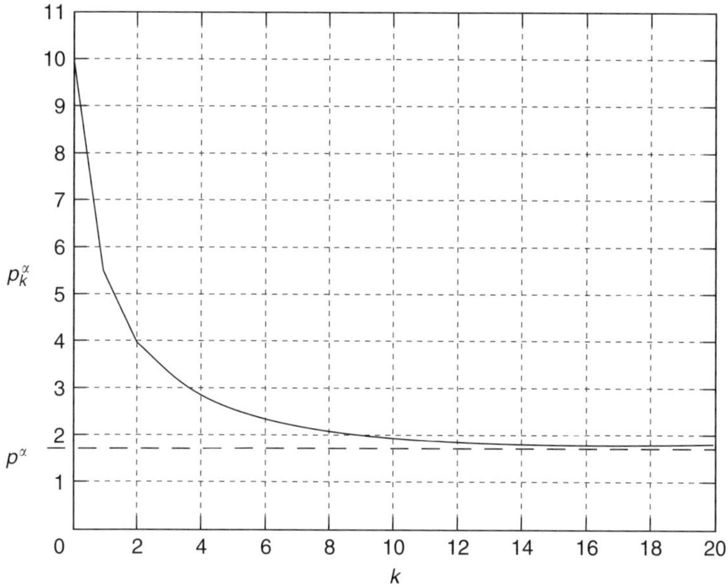

FIGURE 4.8
Covariance matrix with exponential data weighting.

This can be profitably compared to Equation 4.32 in Example 4.1. Figure 4.8 shows p_k^α for $p_0 = 10, r = 10$, and $\alpha = 1.1$.

The steady-state value of p_k^α is easily seen by Equation 4.126 to be

$$p^\alpha = r\left(1 - \frac{1}{\alpha^2}\right) \tag{4.128}$$

so that the steady-state Kalman gain is

$$K = \frac{p^\alpha}{p^\alpha + r/\alpha^2} = \left(1 - \frac{1}{\alpha^2}\right) \tag{4.129}$$

This is nonzero if $\alpha > 1$, which shows that if exponential data weighting is used, data are not discounted as k becomes large. ▯

EXERCISE 4.4 Expression for Covariance with Exponential Data Weighting

a. Suppose that in Equation 4.1, $A = I, G = 0$, and H is time-varying so that $z_k = H_k x_k + v_k$. Then show that the solution to Equation 4.110 can be written explicitly as

$$(P_k^\alpha)^{-1} = \alpha^{-2k}(P_0)^{-1} + \sum_{i=0}^{k-1} H_{i+1}^T R^{-1} H_{i+1} \alpha^{-2(k-i-1)} \tag{4.130}$$

b. Suppose that H_k does decrease with k and that $\alpha = 1$ corresponding to no exponential data weighting. Show that in this case P_k^α approaches a limiting value of zero for all P_0 as k increases.

c. Suppose that $H_k = H \neq 0$ is constant and $\alpha > 1$. Then show that P_k^α approaches a bounded but nonzero positive semidefinite limit, which is independent of P_0.

4.2 Reduced-Order Filters and Decoupling

The dominant factor in determining the computational complexity of the Kalman filter is the dimension n of the model state vector. The number of computations per iteration is on the order of n^3. Any reduction in the number of states will therefore pay off directly in terms of increased computation time. In this section we discuss methods of reducing the dimension of the state vector in the Kalman filter.

4.2.1 Decoupling and Parallel Processing

If the model is a block diagonal of the form

$$\begin{bmatrix} \dot{x}_1 \\ \dot{x}_2 \end{bmatrix} = \begin{bmatrix} A_1 & 0 \\ 0 & A_2 \end{bmatrix} \begin{bmatrix} x_1 \\ x_2 \end{bmatrix} + \begin{bmatrix} w_1 \\ w_2 \end{bmatrix} \tag{4.131a}$$

$$\begin{bmatrix} z_1 \\ z_2 \end{bmatrix} = \begin{bmatrix} H_1 & 0 \\ 0 & H_2 \end{bmatrix} \begin{bmatrix} x_1 \\ x_2 \end{bmatrix} + \begin{bmatrix} v_1 \\ v_2 \end{bmatrix} \tag{4.131b}$$

$w_1 \sim (0, Q_1), w_2 \sim (0, Q_2), v_1 \sim (0, R_1), v_2 \sim (0, R_2), P(0) = \text{diag}[P_1(0), P_2(0)]$, where $x_1 \in R^{n_1}$ and $x_2 \in R^{n_2}$, then the filtering problem decouples into two separate and unrelated problems. A Kalman filter can be designed for each subsystem $(A_i, G_i, H_i, Q_i, R_i)$. Each filter has the form

$$\dot{\hat{x}}_i = A_i\hat{x}_i + K_i(z_i - H_i\hat{x}_i) \tag{4.132}$$

and they can be implemented in a parallel fashion on separate microprocessors.

We have seen an example of this parallel filter implementation in the α–β radar tracker of Example 2.9. In that case, the range and azimuth filters are independent.

The effect of this decoupling into independent filters is to reduce the number of operations from the order of $(n_1+n_2)^3$ to the order of $(n_1^3+n_2^3)$. If $n = 8$, for example, and if we have two 4×4 subsystems, then the number of operations is reduced from about 512 to about 128.

If model matrices A, H, Q, and R are block diagonals, then using decoupled filters, each of which is optimal for its subsystem, results in an *optimal* overall filter. (See the problems.) If A is not block diagonal it may still be desirable in some instances to implement suboptimal decoupled filters by ignoring the off-diagonal terms.

Example 4.5 Suboptimal Filters for Weakly Coupled Systems

Let the second-order plant be

$$\begin{bmatrix} \dot{x}_1 \\ \dot{x}_2 \end{bmatrix} = \begin{bmatrix} a_1 & \varepsilon \\ 0 & a_2 \end{bmatrix} \begin{bmatrix} x_1 \\ x_2 \end{bmatrix} + \begin{bmatrix} w_1 \\ w_2 \end{bmatrix} \tag{4.133}$$

$$\begin{bmatrix} z_1 \\ z_2 \end{bmatrix} = \begin{bmatrix} 1 & 0 \\ 0 & 1 \end{bmatrix} \begin{bmatrix} x_1 \\ x_2 \end{bmatrix} + \begin{bmatrix} v_1 \\ v_2 \end{bmatrix} \tag{4.134}$$

with $w_i(t) \sim (0, q_i)$, $v_i(t) \sim (0, r_i)$, and $x_i(0) \sim (\overline{x_i(0)}, 0)$ uncorrelated (Gelb 1974).

a. Optimal filter

To find the optimal second-order filter, we must solve the second-order Riccati equation. If we let

$$P = \begin{bmatrix} p_1 & p_{12} \\ p_{12} & p_2 \end{bmatrix} \tag{4.135}$$

then the Riccati equation can be written as the three scalar equations.

$$\dot{p}_1 = -\frac{p_1^2}{r_1} - \frac{p_{12}^2}{r_2} + 2a_1 p_1 + 2\varepsilon p_{12} + q_1 \tag{4.136a}$$

$$\dot{p}_{12} = -\frac{p_1 p_{12}}{r_1} - \frac{p_{12} p_2}{r_2} (a_1 + a_2) p_{12} + \varepsilon p_2 \tag{4.136b}$$

$$\dot{p}_2 = -\frac{p_{12}^2}{r_1} - \frac{p_2^2}{r_2} + 2a_2 p_2 + q_2 \tag{4.136c}$$

The Kalman gain is then $PH^T R^{-1}$ or

$$K = \begin{bmatrix} p_1/r_1 & p_{12}/r_2 \\ p_{12}/r_1 & p_2/r_2 \end{bmatrix} \tag{4.137}$$

and the (second-order) optimal filter is

$$\begin{bmatrix} \dot{\hat{x}}_1 \\ \dot{\hat{x}}_2 \end{bmatrix} = \begin{bmatrix} a_1 & \varepsilon \\ 0 & a_2 \end{bmatrix} \begin{bmatrix} \hat{x}_1 \\ \hat{x}_2 \end{bmatrix} + \begin{bmatrix} p_1/r_1 & p_{12}/r_2 \\ p_{12}/r_1 & p_2/r_2 \end{bmatrix} \begin{bmatrix} z_1 - \hat{x}_1 \\ z_2 - \hat{x}_2 \end{bmatrix} \tag{4.138}$$

b. Optimal filter for $\varepsilon = 0$

If $\varepsilon = 0$, then Equation 4.133 decouples into two separate systems. Writing a Riccati equation for each of these first-order systems we get

$$\dot{p}_1 = -\frac{p_1^2}{r_1} + 2a_1 p_1 + q_1 \tag{4.139a}$$

$$\dot{p}_2 = -\frac{p_2^2}{r_2} + 2a_2 p_2 + q_2 \tag{4.139b}$$

The Kalman gains are

$$K_1 = \frac{p_1}{r_1} \tag{4.140a}$$

$$K_2 = \frac{p_2}{r_2} \tag{4.140b}$$

and the optimal filter in case $\varepsilon = 0$ is composed of the two separate filters

$$\dot{\hat{x}}_1 = a_1\hat{x}_1 + \frac{p_1}{r_1}(z_1 - \hat{x}_1) \tag{4.141a}$$

$$\dot{\hat{x}}_2 = a_2\hat{x}_2 + \frac{p_2}{r_2}(z_2 - \hat{x}_2) \tag{4.141b}$$

Note that, since $p_{12}(0) = 0$, if $\varepsilon \to 0$, then the filter Equations (Equation 4.136) and the filter (Equation 4.138) tend to Equations 4.139 and 4.141.

c. Suboptimal filter for $\varepsilon \neq 0$

The filter in part a is optimal for any ε. If $\varepsilon = 0$, the filter in part b is optimal.

If ε is small but nonzero, it is worth asking whether a state estimate based on the simplified filter 4.139 and 4.141 would be accurate enough for our purposes. To answer this question, the optimal error covariances (Equation 4.136) must first be computed.

MATLAB code is shown in Figure 4.9 to implement Equation 4.136. The outputs are shown in Figure 4.10 for several values of ε. The steady-state variances are indicated. ▯

```
% Example 4.5
% Optimal error covariance using Runge-Kutta integrator
function f4p2d1()
   global eps;
   eps_set=[0 0.3 0.7 1.0 1.5 2.0]; % Different values for epsilon
   for j=1:6
        eps=eps_set(j);
        [t p] = ode45(@errcov422,[0 3],[0 0 0]); % Runge-Kutta Integrator
        subplot(3,2,j);
        plot(t,p(:,1),'k-',t,p(:,2),'k--',t,p(:,3),'k-.');% Plot p
        axis([0 3 -0.1 0.5]); grid on; xlabel('t (secs)')
        title(strcat('\epsilon = ',num2str(eps)));
   end
return

% Differential Equations
function dp=errcov422(t,p)
    global eps
    dp = zeros(3,1);
    a1=-2; a2=-1; q1=1; q2=1; r1=1; r2=1;
    p1=p(1); p12=p(2); p2=p(3);
    dp(1)=-p1^2/r1-p12^2/r2+2*a1*p1+2*eps*p12+q1;
    dp(2)=-p1*p12/r1-p12*p2/r2+(a1+a2)*p12+eps*p2;
    dp(3)=-p12^2/r1-p2^2/r2+2*a2*p2+q2;
return
```

FIGURE 4.9
Subroutine for Runge–Kutta to find optimal error covariance.

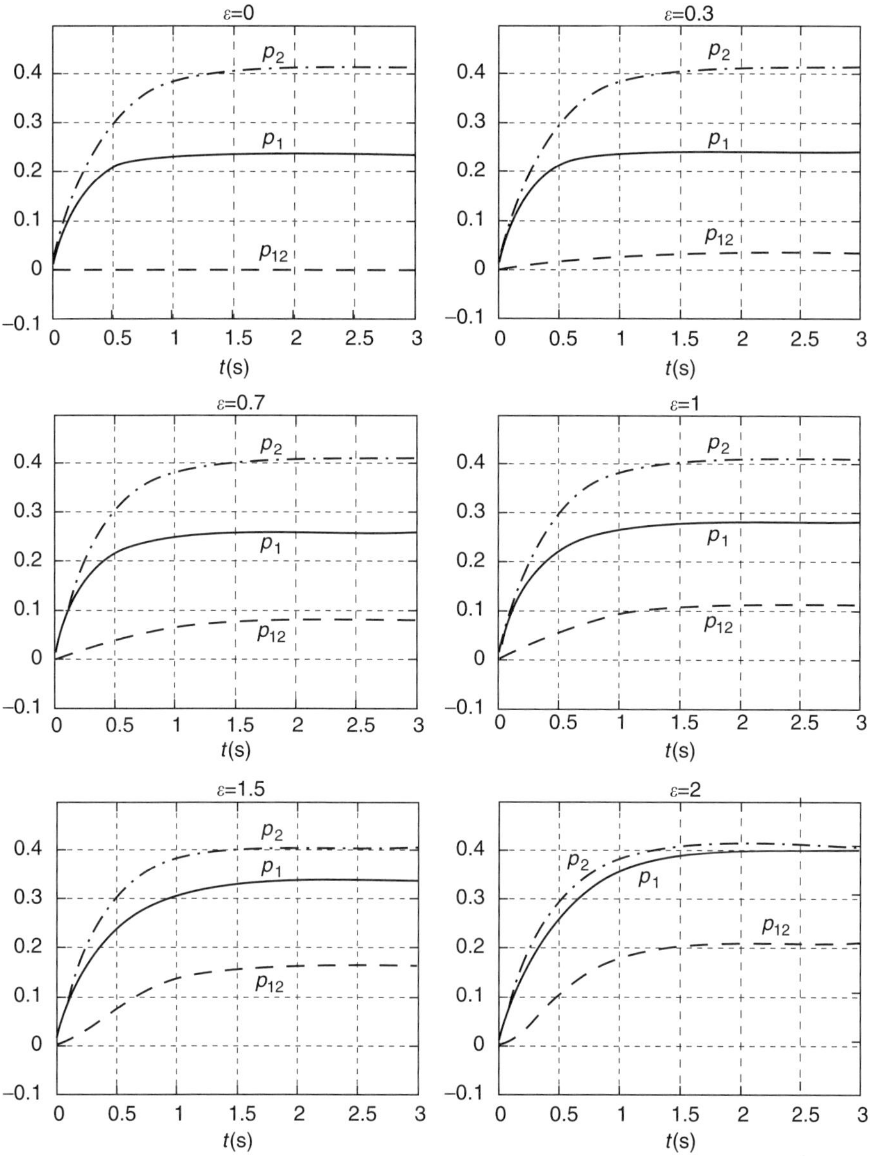

FIGURE 4.10
Optimal error variances.

Now suppose that we use the decoupled filter 4.139 and 4.141 to estimate the plant state in Equation 4.133. This is suboptimal when $\varepsilon \neq 0$. For the sake of discussion, let $\varepsilon = 0.7$ so that the curves in Figure 4.10 represent the *optimal* error variances, which are obtained by using the filter 4.136 and 4.138. The curves for $\varepsilon = 0$ in the figure are the *predicted* error variances using the suboptimal filter 4.139 and 4.141. It is clear that we do not yet have enough information to judge the performance of our suboptimal filter on the plant (Equation 4.133) when $\varepsilon = 0.7$. If $\hat{x} = [\hat{x}_1 \ \hat{x}_2]^T$ is the estimate provided by the suboptimal filter 4.139 and 4.141 and $x'(t)$ is the state of the plant (1), then the *actual* error is

$$\tilde{x}(t) = x'(t) - \hat{x}(t) \tag{4.142}$$

The *actual* error correlation $S(t) = E[\tilde{x}(t)\tilde{x}^T(t)]$ must be examined to determine the adequacy of our suboptimal filter. Note that we must deal with *three* error covariances to do a complete performance analysis.

According to Exercise 4.2, $S(t)$ is found as follows. The model mismatch matrices are

$$\Delta A = \begin{bmatrix} a_1 & \varepsilon \\ 0 & a_2 \end{bmatrix} - \begin{bmatrix} a_1 & 0 \\ 0 & a_2 \end{bmatrix} = \begin{bmatrix} 0 & \varepsilon \\ 0 & 0 \end{bmatrix} \tag{4.143}$$

and

$$\Delta H = I - I = 0$$

$S(t)$ is found by using Equation 4.57. Let

$$U = \begin{bmatrix} U_1 & U_{12} \\ U_{12} & U_2 \end{bmatrix}, \qquad V = \begin{bmatrix} V_1 & V_{12} \\ V_{21} & V_2 \end{bmatrix}, \qquad S = \begin{bmatrix} s_1 & s_{12} \\ s_{12} & s_2 \end{bmatrix} \tag{4.144}$$

[Note that correlations $U(t)$ and $S(t)$ are symmetric while cross-correlation $V(t)$ is not.] Then we obtain the scalar equations

$$\dot{U}_1 = 2a_1U_1 + 2\varepsilon U_{12} + q_1 \tag{4.145a}$$

$$\dot{U}_{12} = (a_1 + a_2)U_{12} + \varepsilon U_2 \tag{4.145b}$$

$$\dot{U}_2 = 2a_2U_2 + q_2 \tag{4.145c}$$

$$\dot{V}_1 = (2a_1 - K_1)V_{12} + \varepsilon U_{12} \tag{4.146a}$$

$$\dot{V}_{12} = (a_1 + a_2 - K_1)V_{12} + \varepsilon U_2 \tag{4.146b}$$

$$\dot{V}_{21} = (a_1 + a_2 - K_2)V_{21} + \varepsilon V_2 \tag{4.146c}$$

$$\dot{V}_2 = (2a_2 - K_2)V_2 + q_2 \tag{4.146d}$$

$$\dot{s}_1 = 2(a_1 - K_1)s_1 + r_1K_1^2 + 2\varepsilon V_{12} + q_1 \tag{4.147a}$$

$$\dot{s}_{12} = (a_1 + a_2 - K_1 - K_2)s_{12} + \varepsilon V_2 \tag{4.147b}$$

$$\dot{s}_2 = 2(a_2 - K_2)s_2 + r_2K_2^2 + q_2 \tag{4.147c}$$

In these equations, K_1 and K_2 are the gains actually used to find $\hat{x}$. In this case they are the suboptimal values in Equation 4.140. Note that if $\varepsilon = 0$ then Equation 4.147 reduces to Equation 4.136, which is equivalent to Equation 4.139, since $p_{12}(0) = 0$.

These equations are rather tedious, but it is straightforward to integrate them by using a Runge–Kutta integrator, as shown in Figure 4.11. Note that

```
% Code to find the predicted and actual error covariances using
% Runge-Kutta Integrator
function f4p2d3()
   global eps;
   eps_set=[0.3 0.7 1.0 1.5];% Different values for epsilon
   for j=1:4
         eps=eps_set(j);
         [t s] = ode45(@s422,[0 3],[0 0 0 0 0 0 0 0 0 0 0 0 0]);% Runge-Kutta
Integrator
         subplot(2,2,j);
         plot(t,s(:,11),'k-',t,s(:,12),'k--',t,s(:,13),'k--',t,s(:,1),'k-.');
         axis([0 3 -0.1 0.5]); grid on; xlabel('t (secs)')
         title(strcat('\epsilon = ',num2str(eps)));
      end
return

% Function for actual error covariance
function ds=s422(t,s)

   ds = zeros(12,1);
   a1=-2; a2=-1; q1=1; q2=1; r1=1; r2=1;
   global eps

   % suboptimal (predicted) error covariance
   p1=s(1); p12=s(2); p2=s(3);
   ds(1)=-p1^2/r1-p12^2/r2+2*a1*p1+2*eps*p12+q1;
   ds(2)=-p1*p12/r1-p12*p2/r2+(a1+a2)*p12+eps*p2;
   ds(3)=-p12^2/r1-p2^2/r2+2*a2*p2+q2;

   % Suboptimal Gains
   K1=p1/r1;
   K2=p2/r2;

   % Plant state correlations
   u1=s(4); u12=s(5); u2=s(6);
   ds(4)=2*a1*u1+2*eps*u12+q1;
   ds(5)=(a1+a2)*u12+eps*u2;
   ds(6)=2*a2*u2+q2;

   % Actual error/plant state crosscorrelations
   v1=s(7); v12=s(8); v21=s(9); v2=s(10);
   ds(7)=(2*a1-K1)*v12+eps*u12;
   ds(8)=(a1+a2-K1)*v12+eps*u2;
   ds(9)=(a1+a2-K2)*v21+eps*v2;
   ds(10)=(2*a2-K2)*v2+q2;

   % Actual error correlations
   s1=s(11); s12=s(12); s2=s(13);
   ds(11)=2*(a1-K1)*s1+r1*K1^2+2*eps*v12+q1;
   ds(12)=(a1+a2-K1-K2)*s12+eps*v2;
   ds(13)=2*(a2-K2)*s2+r2*K2^2+q2;
return
```

FIGURE 4.11
Subroutine to find actual error covariance.

we have been forced to include Equation 4.139 to find the suboptimal gains K_1 and K_2. Figure 4.12 shows the predicted error variances $P(t)$ from Equation 4.139 and the actual error variances $S(t)$ from Equation 4.147 for several values of ε. The actual variances in this figure should be compared with the optimal variances in Figure 4.10 to decide for a given ε, if the suboptimal filter has a satisfactory performance.

Note that Equations 4.137 and 4.147c with gains as in Equation 4.140 are identical. Thus the predicted and actual error variances for the second plant state are the same. They are independent of ε, and they are both greater than the optimal error variance (Equation 4.136c), which decreases with ε.

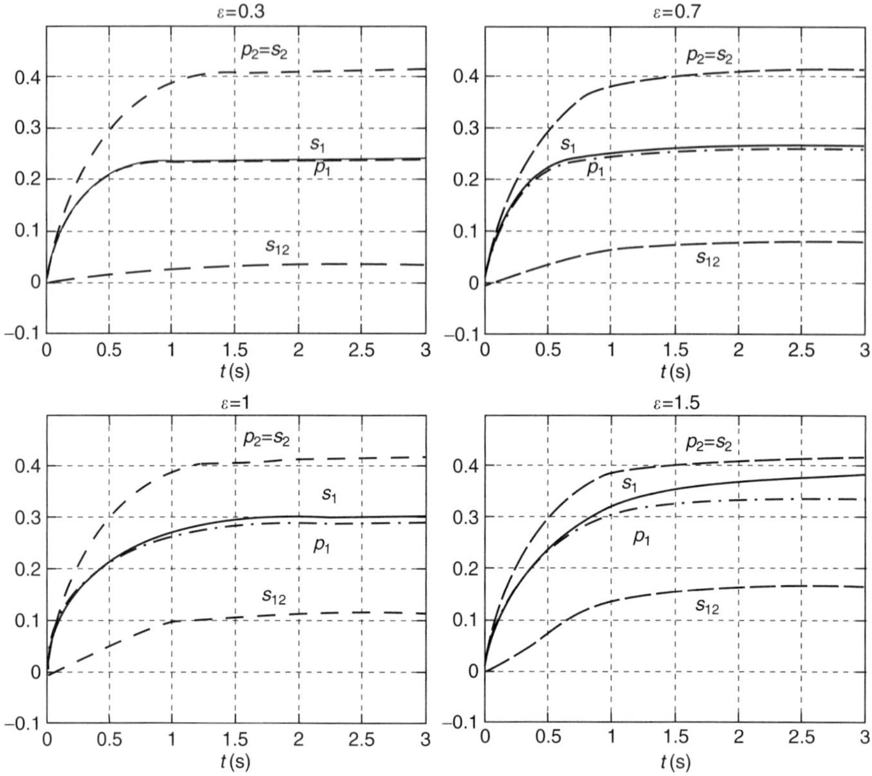

FIGURE 4.12
Predicted $[p(t)]$ and actual $[s(t)]$ error variances.

The predicted error variance (Equation 4.139a) and actual error correlation (Equation 4.147a) for the first plant state are not the same. As ε increases, the actual correlation becomes significantly larger than the predicted variance. The actual variance is again larger than the optimal error variance given by Equation 4.136a, though our prediction says it should be smaller.

4.2.2 Reduced-Order Filters

We discuss here two topics: elimination of a portion of the state vector by substitution and ignoring of a portion of the state vector.

Suppose first that the system is of the form

$$\begin{bmatrix} x_{k+1} \\ \xi_{k+1} \end{bmatrix} = \begin{bmatrix} A_1 & 0 \\ A_{21} & A_2 \end{bmatrix} \begin{bmatrix} x_k \\ \xi_k \end{bmatrix} + \begin{bmatrix} w_k \\ \eta_k \end{bmatrix} \tag{4.148a}$$

$$z_k = \begin{bmatrix} H_1 & H_2 \end{bmatrix} \begin{bmatrix} x_k \\ \xi_k \end{bmatrix} \tag{4.148b}$$

with $w_k \sim (0, Q_1), \eta_k \sim (0, Q_2)$ white and uncorrelated with each other, and with $x_0 \sim (\bar{x}_0, P_0)$. An estimate of $x_k \in R^{n_1}$ is desired, but the value of $\xi_k \in R^{n_2}$ is unimportant to us and an estimate for this portion of the state is not required. Suppose that H_2 has full column rank so that a left inverse H_2^+ exists,

$$H_2^+ H_2 = I \tag{4.149}$$

(Note that $H_2^+ = (H_2^T H_2)^{-1} H_2^T$.)

We can reduce this from an (n_1+n_2)-order problem to an n_1-order problem by substituting for ξ_k as follows (Simon and Stubberud 1969).

Define the *derived measurement* (cf., Section 2.7) as

$$z'_{k+1} = z_{k+1} - H_2 A_2 H_2^+ z_k \tag{4.150}$$

Then we have

$$z'_{k+1} = H_1(A_1 x_k + w_k) + H_2(A_{21} x_k + A_2 \xi_k + \eta_k) - H_2 A_2 H_2^+ H_1 x_k - H_2 A_2 \xi_k$$

This becomes

$$z'_{k+1} = H' x_k + v_k \tag{4.151}$$

where the new measurement matrix is

$$H' = H_1 A_1 + H_2 A_{21} - H_2 A_2 H_2^+ H_1 \tag{4.152}$$

and the new measurement noise is

$$v_k = H_1 w_k + H_2 \eta_k \tag{4.153}$$

The new measurement noise is white noise with zero mean and a covariance of

$$R \triangleq \overline{v_k v_k^T} = H_1 Q_1 H_1^T + H_2 Q_2 H_2^T \tag{4.154}$$

Now the system model can be reduced to

$$x_{k+1} = A_1 x_k + w_k \tag{4.155a}$$

$$z'_{k+1} = H' x_k + v_k \tag{4.155b}$$

with $w_k \sim (0, Q_1), v_k \sim (0, R)$, and $x_0 \sim (\bar{x}_0, P_0)$. The noises w_k and v_k are correlated with cross-correlation

$$\overline{v_j w_k^T} = \overline{(H_1 w_j + H_2 \eta_j) w_k^T} = H_1 Q \delta_{jk} \tag{4.156}$$

so that for optimality the modified Kalman gain of Section 2.7 should be used.

A Kalman filter of order n_1 designed using this reduced-order model will provide the *optimal* estimate $\hat{x}_k$ for the $(n_1 + n_2)$-order plant (Equation 4.148).

EXERCISE 4.5 Kinematic Modeling for Continuous Systems

As a variation on the above procedure, consider the plant

$$\dot{x} = A_1 x + A_{12}\xi + w \tag{4.157a}$$

$$\dot{\xi} = f(x, \xi, u, \eta) \tag{4.157b}$$

where $w(t) \sim (0, Q_1), \eta(t) \sim (0, Q_2)$ are white noise processes uncorrelated with each other and with the initial states. The function $f(\cdot)$ is in general nonlinear and $u(t)$ is a deterministic input. See Kortüm (1976).

In the context of mechanical systems, Equation 4.157a is the set of linearized kinematic equations describing the behavior of the angles and angular velocities. Equation 4.157b represents the nonlinear dynamic equations of motion, the Euler equations.

Suppose measurements are available of the form

$$z_1 = H_1 x + v_1 \tag{4.158a}$$

$$z_2 = H_2\xi + v_2 \tag{4.158b}$$

where $v_1(t) \sim (0, R_1)$ and $v_2(t) \sim (0, R_2)$ are white noises uncorrelated with each other and with $w(t), \eta(t)$, and the initial states. Assume that H_2 has a left inverse H_2^+. This corresponds to having good information about the dynamic state components ξ.

We desire an estimate of $x(t)$ but not of $\xi(t)$.

a. Eliminate $\xi(t)$ to show that

$$\dot{x} = A_1 x + A_{12}H_2^+ z_2 + (w - A_{12}H_2^+ v_2) \tag{4.159}$$

where $A_{12}H_2^+ z_2(t)$ can be treated as a known deterministic input.

b. Find the spectral density matrix of the new process noise

$$w'(t) = w(t) - A_{12}H_2^+ v_2(t) \tag{4.160}$$

c. A Kalman filter can now be applied to the reduced model 4.158a and 4.159. Write down the Riccati equation, the Kalman gain, and the filter equation.

d. Why is this filter suboptimal? (*Note:* This procedure is called "kinematic modeling." The dynamics (Equation 4.157b) can even be unknown.)

Let us now examine the effects of simply ignoring some of the states of a system. To do this we shall use the results of Exercises 4.1 and 4.3.

Let the actual plant be given by

$$x'_{k+1} = \bar{A}x'_k + \bar{G}w'_k \tag{4.161a}$$

$$z'_k = \bar{H}x'_k + v'_k \tag{4.161b}$$

with $w'_k \sim (0, \bar{Q})$, $v'_k \sim (0, \bar{R})$ white and uncorrelated with each other and with $x'_0 \sim [E(x'_0), \bar{P}_0]$. Let the plant state x'_k be an n_p vector. (The following results can easily be modified to incorporate a deterministic input.)

To design a filter for this plant, suppose that we assume the model

$$x_{k+1} = Ax_k + Gw_k \tag{4.162a}$$

$$z_k = Hx_k + v_k \tag{4.162b}$$

with $w_k \sim (0, Q)$, $v_k \sim (0, R)$ white and uncorrelated with each other and with $x_0 \sim [E(x_0), P_0]$. Let the model state x_k be an n_m vector.

The object here is to obtain filters or reduced order by assuming a model of order n_m less than n_p, the order of the plant. Let us simplify things by supposing that to find a reduced-order model we just select certain components of the plant state x'_k to incorporate in Equation 4.162 discarding the remaining components. Then we have

$$x_k = Tx'_k \tag{4.163}$$

with the state transformation given without loss of generality by

$$T = \begin{bmatrix} I_{n_m} & 0 \end{bmatrix} \tag{4.164}$$

A right inverse of T is given by

$$T^+ = T^T = \begin{bmatrix} I_{n_m} \\ 0 \end{bmatrix} \tag{4.165}$$

In this case, the estimate $\hat{x}_k$ provided by a filter based on the model will be a (suboptimal) estimate for the portion of the plant state retained in the model.

According to Exercise 4.1 the correlation matrix S_k of the actual error

$$\tilde{x}_k = x'_k - T^T \hat{x}_k \tag{4.166}$$

is found as follows. Let U_k be the correlation of plant state vector x'_k and V_k the cross-correlation of $\tilde{x}_k$ with x'_k. Define the model mismatch matrices as

$$\Delta A = \bar{A} - T^T AT \tag{4.167a}$$

$$\Delta H = \bar{H} - HT \tag{4.167b}$$

Then, to find S_k perform the time update

$$U_{k+1} = \bar{A}U_k\bar{A}^T + \bar{G}\bar{Q}\bar{G}^T \tag{4.168a}$$

$$V_{k+1}^- = \Delta AU_k\bar{A}^T + T^T ATV_k\bar{A}^T + \bar{G}\bar{Q}\bar{G}^T \tag{4.168b}$$

$$\begin{aligned} S_{k+1}^- &= T^T ATS_kT^T A^T T + \bar{G}\bar{Q}\bar{G}^T + \Delta AU_k\Delta A^T \\ &\quad + \left(T^T ATV_k\Delta A^T + \Delta AV_k^T T^T A^T T\right) \end{aligned} \tag{4.168c}$$

followed by the measurement update

$$U_k = U_k \tag{4.169a}$$

$$V_k = \left(I - T^T K_k H T\right) V_k^- - T^T K_k \Delta H U_k \tag{4.169b}$$

$$\begin{aligned} S_k = {} & \left(I - T^T K_k H T\right) S_k^- \left(I - T^T K_k H T\right)^T \\ & + T^T K_k \left(\bar{R} + \Delta H U_k \Delta H^T\right) K_k^T T - \left(I - T^T K_k H T\right) V_k^- \Delta H^T K_k^T T \\ & + T^T K_k \Delta H (V_k^-)^T \left(I - T^T K_k H T\right)^T \end{aligned} \tag{4.169c}$$

(Note that these equations are not as messy as they seem, since the multiplications by T and T^T simply correspond to "padding" a matrix with zero blocks!)

The dynamics of the *a priori* actual error ($\tilde{x}_k^- = x_k' - T^T \hat{x}_k^-$) are described by the error system

$$\begin{aligned} \tilde{x}_{k+1}^- = {} & T^T A (I - K_k H) T \tilde{x}_k^- + \left(\Delta A - T^T A K_k \Delta H\right) x_k' \\ & + \bar{G} w_k' - T^T A K_k v_k' \end{aligned} \tag{4.170}$$

The continuous counterparts to these equations are given in Exercise 4.3.

To carry out a *performance analysis* of the effects of disregarding the bottom portion of the plant state vector, it is only necessary to proceed as follows:

1. Compute the optimal error covariance using the Riccati equation for the full plant (Equation 4.161).
2. Design a Kalman filter for the reduced model (Equation 4.162).
3. Use the filter gain K_k found in step 2 in Equations 4.168 and 4.169 to compute the actual error covariance S_k resulting when the reduced filter in step 2 is applied to the full plant.
4. If S_k is unacceptably greater than the optimal covariance from step 1, select a different set of plant states to retain in the model and return to step 2.

Basically, we should retain in the model all the states we are directly interested in estimating and all states strongly coupled with them, and all unstable states.

An illustration of an analysis for a reduced-order filter is given in Example 4.2. In practice we would not fine error covariances analytically, as we did in that example. Instead we would implement the equations for S_k or $S(t)$ in a subroutine and make some plots for comparison with the optimal error covariance.

A special case of reduced-order modeling occurs when the plant has the form

$$\begin{bmatrix} \dot{x}' \\ \dot{\xi}' \end{bmatrix} = \begin{bmatrix} A_1 & A_{12} \\ 0 & A_2 \end{bmatrix} \begin{bmatrix} x' \\ \xi' \end{bmatrix} + \begin{bmatrix} w' \\ \eta' \end{bmatrix} \tag{4.171a}$$

$$z' = \bar{H} x' + v' \tag{4.171b}$$

with $w'(t) \sim (0, Q_1)$, $\eta'(t) \sim (0, Q_2)$, and $v'(t) \sim (0, R)$ white and uncorrelated with each other and with the initial states. An estimate of $x'(t)$ is required, but we are not interested in $\xi'(t)$.

This system could arise, for example, when the subsystem of $\xi'(t)$ is a shaping filter for a portion of the process noise (Sections 2.7 and 3.6). See Huddle (1967), Huddle and Wismer (1968), and Darmon (1976).

The optimal Kalman gain is of the form

$$K(t) = \begin{bmatrix} K_1(t) \\ K_2(t) \end{bmatrix} \tag{4.172}$$

To design a reduced-order filter, assume the reduced-order model

$$\dot{x} = A_1 x + w \tag{4.173a}$$

$$z = Hx + v \tag{4.173b}$$

with $w(t) \sim (0, Q_1), v(t) \sim (0, R)$ white and uncorrelated. (In the case where $\xi'(t)$ is the state of a shaping filter, this corresponds to assuming the process noise is white when it is not.) Then $x(t)$ is related to the plant state by the transformation 4.164.

Discounting the lower portion of the plant state is equivalent to restricting the filter gain to the form

$$K(t) = \begin{bmatrix} K_1(t) \\ 0 \end{bmatrix} \tag{4.174}$$

which results in a suboptimal but simplified filter. Such a design has been called a *Kalman–Schmidt* filter (Schmidt 1966).

We could analyze the actual error by using the results of Exercise 4.3. It is quite enlightening, however, to use instead another approach which gives the same answers in a simplified form.

Let the error covariance for the plant $\bar{P}$ be partitioned conformably with $x'(t), \xi'(t)$ as

$$\bar{P} = \begin{bmatrix} P_1 & P_{12} \\ P_{12}^T & P_2 \end{bmatrix} \tag{4.175}$$

Applying a filter with gain of $K(t)$ to the full plant (Equation 4.171), the error covariance satisfies the full Riccati equation

$$\dot{\bar{P}} = (\bar{A} - K\bar{H})\bar{P} + \bar{P}(\bar{A} - K\bar{H})^T + \bar{Q} + K\bar{R}K^T \tag{4.176}$$

Let us now restrict the gain $K(t)$ to have the suboptimal form in Equation 4.174. Then by using the partitioning in Equations 4.171 and 4.175, and the fact that

$$\bar{H} = \begin{bmatrix} H & 0 \end{bmatrix} \tag{4.177}$$

we can write Equation 4.176 as

$$\dot{P}_1 (A_1 - K_1 H)P_1 + P_1(A_1 - K_1 H)^T + A_{12}P_{12}^T + P_{12}A_{12}^T + K_1 R K_1^T + Q_1 \tag{4.178a}$$

$$\dot{P}_{12} = (A_1 - K_1 H)P_{12} + A_{12}P_2 + P_{12}A_2^T \tag{4.178b}$$

$$\dot{P}_2 = A_2 P_2 + P_2 A_2^T + Q_2 \tag{4.178c}$$

The interpretation of these *Kalman–Schmidt* equations is as follows. $P_1(t)$ is the actual error covariance for the estimate of $x'(t)$ using the suboptimal gain (Equation 4.174). $P_2(t)$ is the actual error covariance for the estimate of $\xi'(t)$. Since no correction is applied to this estimate based on the measurements, it is the same as the covariance of $\xi'(t)$ itself (see Section 1.1). $P_{12}(t)$ is the cross-covariance between the estimation error in $x'(t)$ and the state portion $\xi'(t)$.

The predicted error covariance based on the model 4.173 only is given by

$$\dot{P} = (A_1 - K_1H)P + P(A_1 - K_1H)^T + Q_1 + K_1RK_1^T \tag{4.179}$$

Define the degradation in performance incurred due to using the suboptimal gain as

$$E = P_1 - P \tag{4.180}$$

Then we have, by subtracting Equation 4.179 from Equation 4.178a

$$\dot{E} = (A_1 - K_1H)E + E(A_1 - K_1H)^T + A_{12}P_{12}^T + P_{12}A_{12}^T \tag{4.181}$$

This is the error increase induced by the unmodeled states $\xi'(t)$. In some cases this may be small enough to warrant using the reduced-order filter implied by Equation 4.174.

The gain $K_1(t)$ is usually selected to be the optimal (Kalman) gain based on the reduced-order model,

$$K_1 = PH^TR^{-1} \tag{4.182}$$

EXERCISE 4.6 Alternative Derivation of Kalman–Schmidt Error Covariance Equations

Equation 4.178 can also be derived using Exercise 4.3.

a. Show that the model mismatch matrices are

$$\Delta A = \begin{bmatrix} 0 & A_{12} \\ 0 & A_2 \end{bmatrix}, \qquad \Delta H = 0 \tag{4.183}$$

b. Write down the equations of Exercise 4.3 in terms of the subblocks of the partitioned plant matrices, also partitioning the correlation matrices as

$$U = \begin{bmatrix} U_1 & U_{12} \\ U_{12}^T & U_2 \end{bmatrix}, \quad V = \begin{bmatrix} V_1 & V_{12} \\ V_{21} & V_2 \end{bmatrix}, \quad S = \begin{bmatrix} S_1 & S_{12} \\ S_{12}^T & S_2 \end{bmatrix} \tag{4.184}$$

c. Show that, for the correct choice of initial conditions,

$$S_{12} = V_{12} \tag{4.185}$$

$$S_2 = U_2 = V_2 \tag{4.186}$$

so that the simplified equations (Equation 4.179) are recovered.

4.3 Using Suboptimal Gains

The Kalman filter is time-varying even when the system is time-invariant because the gain K depends on time. In the discrete filter this means that K_k must either be precomputed and stored as a sequence, or that the Riccati equation must be solved for each k during the measurement run. In the continuous filter implemented on an analog computer, it means that continuously variable gains must be used. In the continuous filter implemented on a digital computer, it means that a time-dependent gain must be used in the filter routine that is called by the integration algorithm.

In some application it is desirable to approximate the optimal Kalman gain by suboptimal gain that is easier to implement. One common choice is to use the steady-state gain K_∞; for this gain the Kalman filter is just the Wiener filter.

To examine the effects of using suboptimal gains, let us summarize some results from Sections 2.5 and 3.4.

The discrete Kalman filter is given by

$$\hat{x}_{k+1}^- = A(I - K_k H)\hat{x}_k^- + Bu_k + AK_k z_k \tag{4.187}$$

The error system is

$$\tilde{x}_{k+1}^- = A(I - K_k H)\tilde{x}_k^- + Gw_k - AK_k v_k \tag{4.188}$$

and the associated error covariance is given by

$$P_{k+1}^- = A(I - K_k H)P_k^-(I - K_k H)^T A^T + AK_k R K_k^T A^T + GQG^T \tag{4.189}$$

If the gain is selected as the optimal Kalman value

$$K_k = P_k H^T (HP_k^- H^T + R)^{-1} \tag{4.190}$$

then Equation 4.189 reduces to the Riccati equation in Table 2.3. With a little bit of thought, however, it is clear that Equation 4.189 is the error covariance of filter for *any* fixed gain sequence K_k. This is because, once K_k has been selected, Equation 4.188 is a stochastic system with a fixed plant matrix. The error covariance, therefore, propagates according to the *Lyapunov* (i.e., not Riccati Equation 4.189). This equation can, therefore, be used to analyze the effects of selecting a suboptimal K_k.

If the steady-state Kalman gain K_∞ is used for all k, we obtain the (time-invariant) Wiener filter which is Equation 4.187 with K_k replaced by K_∞. The Wiener filter has an associated error covariance of Equation 4.189 with K_k replaced by K_∞.

The continuous estimate for any gain $K(t)$ is given by

$$\dot{\hat{x}} = [A - K(t)H]\hat{x} + Bu + K(t)z \tag{4.191}$$

The error system is

$$\dot{\tilde{x}} = [A - K(t)H]\tilde{x} + Gw - K(t)v \tag{4.192}$$

with associated error covariance

$$\dot{P} = (A - KH)P + P(A - KH)^T + KRK^T + GQG^T \tag{4.193}$$

If the optimal gain

$$K = PH^T R^{-1} \tag{4.194}$$

is used in Equation 4.191 then Equation 4.193 reduces to the Riccati equation in Table 3.1. It is important to realize, however, that Equation 4.193 describes the error covariance associated with *any* gain $K(t)$ used in the filter (Equation 4.191). If $K(t)$ is a given gain function, then Equation 4.193 is a Lyapunov equation, not a Riccati equation.

If the steady-state Kalman gain K_∞ is used, the result is the time-invariant Wiener filter (Equation 4.191), with $K(t)$ replaced by K_∞. This filter has an associated error covariance of Equation 4.193 with $K(t)$ replaced by K_∞.

The next example shows how Equation 4.189 and 4.193 are used to analyze the effects of using simplified suboptimal filter gains.

Example 4.6 Suboptimal Filter Gains for Continuous Wiener Process

Let the system model be

$$\dot{x} = w \tag{4.195}$$

$$z = x + v \tag{4.196}$$

with $w(t) \sim (0, q)$ and $v(t) \sim (0, r)$ white and uncorrelated with $x(0) \sim (0, p_0)$.

A low-pass filter for estimating $x(t)$ from the data $z(t)$ is given by

$$\dot{\hat{x}} = K(z - \hat{x}) = -K\hat{x} + Kz \tag{4.197}$$

Let us discuss several ways to select the filter gain $K(t)$.

a. Kalman filter

To find the optimal gain $K(t)$ we must solve the Riccati equation

$$\dot{p} = -\frac{p^2}{r} + q \tag{4.198}$$

According to Example 3.3 part *d*,

$$p(t) = \sqrt{qr} + \frac{\sqrt{qr}}{ce^{t/\tau} - (1/2)} \tag{4.199}$$

where

$$c = \frac{1}{2}\frac{p_0 + \sqrt{qr}}{p_0 - \sqrt{qr}} \tag{4.200}$$

$$\tau = \frac{1}{2}\sqrt{\frac{r}{q}} \tag{4.201}$$

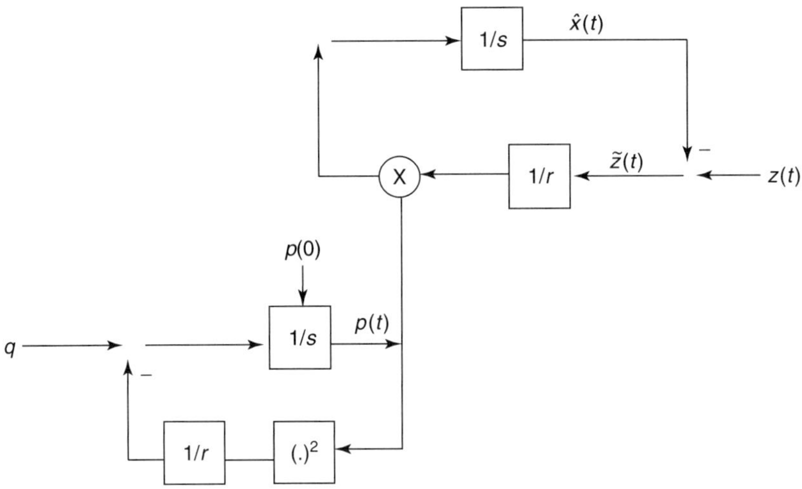

FIGURE 4.13
Optimal filter.

Hence $K = p/r$ or

$$K(t) = \sqrt{\frac{q}{r}}\left(1 + \frac{1}{ce^{t/\tau} - (1/2)}\right) \tag{4.202}$$

This time-varying gain must be implemented to get the optimal filter. One way to do this is shown in Figure 4.13.

b. Time-invariant filter

To simplify the implementation, let us use the Wiener filter. The steady-state covariance and gain are

$$p_\infty = \sqrt{qr} \tag{4.203}$$

$$K_\infty = \sqrt{\frac{q}{r}} \tag{4.204}$$

so the Wiener filter is

$$\dot{\hat{x}} = -\sqrt{\frac{q}{r}}\hat{x} + \sqrt{\frac{q}{r}}z \tag{4.205}$$

An implementation is shown in Figure 4.14.

To compare the performances of the Kalman and Wiener filters, we need to find the Wiener filter error covariance. This can be done by using Equation 4.193 with $K = K_\infty$, which becomes

$$\dot{p} = -2K_\infty p + rK_\infty^2 + q \tag{4.206}$$

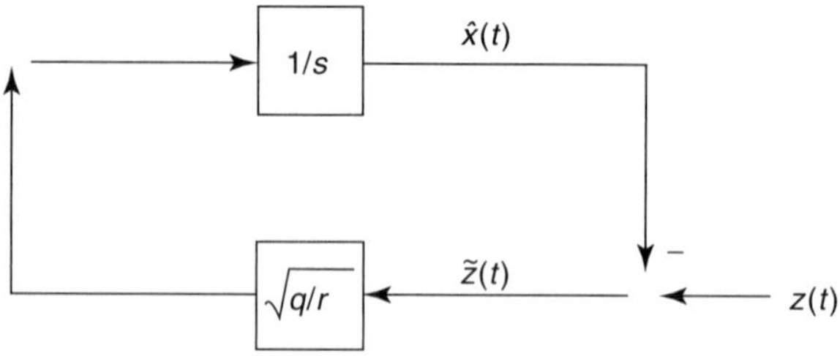

FIGURE 4.14
Suboptimal time-invariant filter.

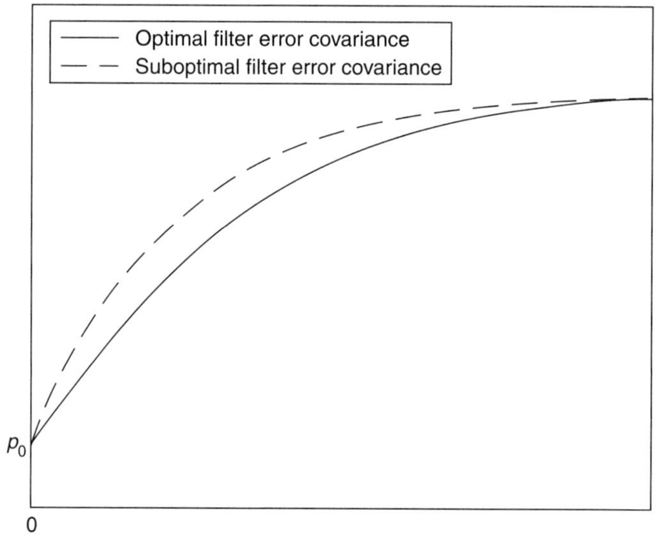

FIGURE 4.15
Error covariances for Kalman and Wiener filters.

or

$$\dot{p} = -2\sqrt{\frac{q}{r}}p + 2q \tag{4.207}$$

The solution to Equation 4.207 is

$$p(t) = \mathrm{e}^{-t/\tau}p_0 + \int_0^t \mathrm{e}^{-(t-\lambda)/\tau}2q\,\mathrm{d}\lambda$$

or

$$p(t) = \sqrt{qr} + \big(p_0 - \sqrt{qr}\big)\mathrm{e}^{-t/\tau} \tag{4.208}$$

The optimal and suboptimal error covariances 4.199 and 4.208 are sketched in Figure 4.15. Note that their steady-state values are identical. By comparing the two error covariances, we can now decide if the simplified structure of the suboptimal filter justifies its degraded performance in our particular application.

The curves in Figure 4.15 were obtained by writing MATLAB code to implement Equations 4.198 and 4.207. □

4.4 Scalar Measurement Updating

If the measurement z_k is a scalar, then the discrete Kalman filter measurement update is considerably simplified. In this case we have, from Table 2.2,

$$K_k = \frac{P_k^- H^T}{HP_k^- H^T + R} \tag{4.209a}$$

$$P_k = (I - K_k H)P_k^- \tag{4.209b}$$

$$\hat{x}_k = \hat{x}_k^- + K_k(z_k - H\hat{x}_k^-) \tag{4.209c}$$

where no matrix inversions are needed, since $(HP_k^- H^T + R)$ is a scalar.

In general, $z_k \in R^p$ is not a scalar, but we can always perform the single p-vector measurement update with p sequential scalar updates by *prewhitening* the data. This results in a dramatic reduction in computing time if p is large. (In a vector update, the number of operations depends on p^3. In sequential scalar updates, it depends on p.)

To do this, we proceed as in Section 1.4. The measurement equation is

$$z_k = Hx_k + v_k \tag{4.210}$$

$v_k \sim (0, R)$. Factor R into its square roots U:

$$R = UU^T \tag{4.211}$$

To save operations, it is convenient to require that U be upper triangular, which we can always do. We can find U by the *upper triangular Cholesky decomposition* in Bierman (1977). Now invert the triangular matrix U. Then a decomposition of R^{-1} is

$$R^{-1} = U^{-T}U^{-1} \tag{4.212}$$

As in Section 1.4 we now define new data, measurement matrix, and noise by

$$z_k^w = U^{-1}z_k \tag{4.213}$$

$$H^w = U^{-1}H \tag{4.214}$$

$$v_k^w = U^{-1}v_k \tag{4.215}$$

Then Equation 4.210 becomes

$$z_k^w = H^w x_k + v_k^w \tag{4.216}$$

Since $v_k^w \sim (0, I)$, the components of z_k^w are statistically uncorrelated so that z_k^w is a *white* data vector. The operation (4.213) corresponds to *prefiltering* the data by U^{-1} prior to using it in the Kalman filter.

Suppose the data have been prewhitened. Then according to the results in Section 1.4, at each time R we can incorporate *each of the p components of* z_k *separately* by a scalar update like Equation 4.209. To do this, in Equation 4.209 we replace R by 1, z_k by the ith component of the whitened data z_k^w, and H by the ith row of H^w. Then we cycle through Equation 4.209 for $i = 1, \ldots, p$. This procedure is repeated at each time k.

If the noise covariance is time-varying (R_k), then we need to do the Cholesky decomposition $R_k = U_k U_k^T$ at each time k to get $H_k^w = U_k^{-1} H$ at each time. In the time-invariant case we only need to factor R and find H^w at the beginning of the estimation run.

A stabilized (e.g., Equation 2.59) version of scalar measurement update (Equation 4.209) is given in Bierman (1977). The algorithm uses symmetry to minimize the number of operations.

Scalar sequential processing of the data when $R = I$ should be compared to our discussion in Section 4.2 on breaking the optimal filter into smaller "subfilters," when the plant matrices A, Q, R, and P_0 are block diagonals.

Problems

Section 4.1

1. Divergence in the Continuous Kalman Filter

(a) Let the model be the random constant

$$\dot{x} = 0 \tag{4.217}$$

$$z = x + v \tag{4.218}$$

with $v \sim (0, 1)$ white and uncorrelated with $x(0) \sim (0, 1)$. Show that the optimal estimate is given by

$$\dot{\hat{x}} = \frac{1}{t+1}(z - \hat{x}) \tag{4.219}$$

with predicted error variance

$$p(t) = \frac{1}{t+1} \tag{4.220}$$

(b) Show that the state transition matrix of (4.219) is

$$\phi(t, t_0) = \frac{1}{t - t_0 + 1} \tag{4.221}$$

(Hint: Solve $(\mathrm{d}/\mathrm{d}t)\phi(t,0) = [-1/(t+1)]\,\phi(t,0)$, with $\phi(0,0) = 1$.) Hence find the estimate $\hat{x}(t)$ if the data are $z(t) = tu_{-1}(t)$.

(c) Now suppose that the actual plant is the Wiener process

$$\dot{x} = w \tag{4.222}$$

$$z = x + v \tag{4.223}$$

with $w \sim (0, \varepsilon)$ and $v \sim (0, 1)$ white and uncorrelated with $x(0) \sim (0, 1)$. Show that if the filter (Equation 4.219) is used then the actual error variance is

$$s(t) = \frac{1}{t+1} + \frac{\varepsilon}{3}(t+1) - \frac{\varepsilon}{3(t+1)^2} \tag{4.224}$$

which is unbounded for any $\varepsilon \neq 0$ (Anderson 1973).

2. Inaccurate Modeling of Newton's System as Constant Unknown. The model used to design a filter for an unknown x is

$$\dot{x} = 0$$
$$z = x + v$$

$v(t) \sim (0, r)$. In fact, the unknown has dynamic described by

$$\dot{x} = \varepsilon x_2$$
$$\dot{x}_2 = 0$$

for a small ε.

(a) Find Kalman gain and predicted error covariance.

(b) Write down the actual error dynamics.

(c) Use Exercise 4.3 to determine the actual error covariance.

(d) Select another first-order model than can be used to design a filter in which divergence does not occur. For this model, repeat (a) through (c).

3. Exponential Data Weighting. Use exponential data weighting to correct the filter divergence experienced in

(a) Example 4.2.

(b) Problem 1.

(c) Problem 2.

4. Filter with a Margin of Stability. Redo Example 2.12 if the steady-state filter must be guaranteed to have poles inside $|z| = 0.98$.

5. Filter with a Margin of Stability. Redo Problem 10 if the steady-state filter must be guaranteed to have poles to the left of $s = -0.1$.

Section 4.2

6. Decoupled Filter

(a) Let the model be decoupled so that $A = \text{diag}[A_1, A_2], Q = \text{diag}[Q_1, Q_2]$, $R = \text{diag}[R_1, R_2]$, and $P(0) = \text{diag}[P_1(0), P_2(0)]$. Show that $P(t)$ is block diagonal for all t, and that the optimal filter decomposes into two independent subfilters.

(b) Repeat (a) for the discrete-time filter.

7. Block Triangular Filter Structure. Suppose the system is block triangular of the form

$$\begin{bmatrix} \dot{x}_1 \\ \dot{x}_2 \end{bmatrix} = \begin{bmatrix} A_1 & A_{12} \\ 0 & A_2 \end{bmatrix} \begin{bmatrix} x_1 \\ x_2 \end{bmatrix} + \begin{bmatrix} w_1 \\ w_2 \end{bmatrix} \tag{4.225a}$$

$$\begin{bmatrix} z_1 \\ z_2 \end{bmatrix} = \begin{bmatrix} H_1 & H_{12} \\ 0 & H_2 \end{bmatrix} \begin{bmatrix} x_1 \\ x_2 \end{bmatrix} + \begin{bmatrix} v_1 \\ v_2 \end{bmatrix} \tag{4.225b}$$

$w_1(t) \sim (0, Q_1)$, $w_2(t) \sim (0, Q_2)$, $v_1(t) \sim (0, R_1)$, $v_2(t) \sim (0, R_2)$, $x_1(0) \sim [\overline{x_1(0)}, P_1(0)]$, $x_2(0) \sim [\overline{x_2(0)}, P_2(0)]$. Then subsystem 2 is not influenced by subsystem 1, but drives it. In this case we might try the simplified filter structure

$$\dot{\hat{x}}_2 = A_2\hat{x}_2 + K_2(z_2 - H_2\hat{x}_2) \tag{4.226a}$$

$$\dot{\hat{x}}_1 = A_1\hat{x}_1 + A_{12}\hat{x}_2 + K_1(z_1 - H_1\hat{x}_1 - H_{12}\hat{x}_2) \tag{4.226b}$$

where Equation 4.226a does not depend on Equation 4.226b but only drives it.

(a) Write the Riccati equation in terms of the submatrices.

(b) Is the filter 4.226 ever optimal?

8. Effect of Mismodeling of Process Noise. Let

$$\dot{x} = \varepsilon w \tag{4.227}$$

$$z = x + v \tag{4.228}$$

where $x(0) \sim (0, 1)$ and $v(t) \sim (0, 1)$ is white and uncorrelated with the nonwhite process noise which has spectral density of

$$\Phi_w(\omega) = \frac{1}{\omega^2 + 1} \tag{4.229}$$

(a) Find a shaping filter for $w(t)$ and augment the state equation.

(b) Design a filter for Equations 4.227 and 4.228 assuming incorrectly that $w(t)$ is white $(0, 1)$. Find error covariance and optimal gain.

(c) Making the same erroneous assumption as in (b), find K_∞ and use this suboptimal gain in a filter for $x(t)$. Find suboptimal error variance. Assume $\varepsilon = 0.1$ if simplification is desired.

(d) Now we use this K_∞ on the actual overall plant found in (a). Solve Equation 4.178 for the actual error variance $p_1(t)$.

(e) Write the optimal Kalman filter equations for the full plant in (a). Find optimal steady-state error covariance and compare to the steady-state value of $p_1(t)$ from (d).

Section 4.3

9. Suboptimal Filtering. In Problem 12:

(a) Find the steady-state Kalman gain K_∞ if $r_k = r$ is constant.

(b) Redo Problem 12 (a), determining for $k = 0$ through 5 the estimate and (suboptimal) error covariance if K_∞ is used for the data processing instead of the optimal K_k.

10. Suboptimal Filtering. Let

$$\dot{x} = 0.1x + w$$
$$z = x + v$$

with $w(t) \sim (0, 2)$, $v(t) \sim (0, 1)$ uncorrelated white noise processes and $x(0) \sim (0, 1)$.

(a) Find optimal error covariance and Kalman gain.

(b) Find steady-state Kalman gain K_∞. Now use a suboptimal filter

$$\dot{\hat{x}} = (A - KH)\hat{x} + Kz \tag{4.230}$$

with $K = K_\infty$. Find the resulting error covariance.

(c) Now a constant feedback gain K is selected in Equation 4.227 to make the filter stable with a pole at $s = -1$. Find this gain and compute the error covariance resulting when this suboptimal filter is used on the data.

(d) Plot the three error covariances on the same graph.

11. Suboptimal Filtering. For Newton's system in Example 3.6, two suboptimal filters are proposed.

(a) A filter is built using K_∞. Find the resulting suboptimal error covariance.

(b) Use Ackermann's formula to find a filter gain that places both the filter poles at $s = -\sqrt{\gamma}$. Compute the error covariance of this suboptimal filter.

Section 4.4

12. Scalar Measurement Updating

$$x_{k+1} = \begin{bmatrix} 1 & 0.5 \\ 0 & 1 \end{bmatrix} x_k + \begin{bmatrix} 0 \\ 1 \end{bmatrix} u_k + w_k$$

$$z_k = \begin{bmatrix} 1 & 1 \\ 1 & -1 \end{bmatrix} x_k + v_k$$

where

$$Q = \begin{bmatrix} 1 & -0.5 \\ -0.5 & 1 \end{bmatrix}, \qquad R = \begin{bmatrix} 2 & 1 \\ 1 & 1 \end{bmatrix}, \qquad x_0(0) \sim (0, I)$$

The input is $u_k = u_{-1}(k)$ and the data are

$$z_1 = \begin{bmatrix} 1 \\ -1 \end{bmatrix}, \qquad z_2 = \begin{bmatrix} 2.5 \\ -1.5 \end{bmatrix}, \qquad z_3 = \begin{bmatrix} 4.5 \\ -1.5 \end{bmatrix}$$

(a) Determine $\hat{x}_k$ and P_k for $k = 0, 1, 2, 3$.

(b) Repeat using prewhitening of the data and scalar measurement updating.

5

Estimation for Nonlinear Systems

We have derived the optimal estimator for linear stochastic systems with Gaussian statistics. It is known as the Kalman filter, and is itself a linear system with a gain matrix that depends on the solution to a matrix quadratic equation.

The optimal estimation problem for nonlinear systems is in general very complicated, and only in a few special cases do algorithms exist that are easy to implement or understand.

In this chapter, we find time and measurement updates for the entire hyperstate (i.e., the conditional probability density function of the unknown given the data). We then use these to find update equations for the estimate and error covariance that are in general not computationally realizable. It is demonstrated that these reduce to the Kalman filter in the linear Gaussian case.

To obtain a computationally viable algorithm for general nonlinear systems, we make some approximations that result in the *extended Kalman filter* (EKF).

5.1 Update of the Hyperstate

In this section, general equations for time and measurement updates of the hyperstate $f_{x/Z}$, are derived where x is the unknown and Z is all of the available data. Since most practical applications today use sampled time signals, we deal with the case of discrete measurements.

First discrete systems, and then continuous systems are considered.

5.1.1 Discrete Systems

Consider a general dynamical discrete-time system with state x_k. The system is assumed to be Markov, so that the conditional PDF of x_{k+1} given the entire previous history of the system is simply f_{x_{k+1}/x_k}, the conditional PDF given the previous state. There are prescribed measurements z_k of the unknown state x_k. The measurement process has no memory and z_k depends only on x_k, so that the measurement is completely described probabilistically by the conditional PDF f_{z_k/x_k}. This situation is shown in Figure 5.1, where $f_{x_k/x_{k-1}}$ and f_{z_k/x_k} are assumed known.

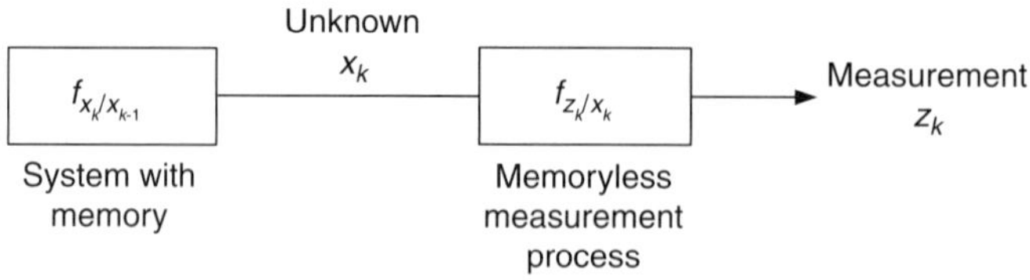

FIGURE 5.1
Characterization of general nonlinear stochastic system using conditional PDFs.

As an illustration of such a characterization of systems using PDFs, consider the special case of linear Gaussian measurements $z_k \in R^p$,

$$z_k = Hx_k + v_k \tag{5.1}$$

where $v_k \sim N(0, R)$, or

$$f_{v_k}(v_k) = \frac{1}{\sqrt{(2\pi)^p |R|}} e^{-(1/2)v_k^T R^{-1} v_k} \tag{5.2}$$

This situations can be alternatively represented by specifying the conditional PDF

$$f_{z_k x_k}(z_k/x_k) = f_{v_k}(z_k - Hx_k) = \frac{1}{\sqrt{(2\pi)^p |R|}} e^{-(1/2)(z_k - Hx_k)^T R^{-1}(z_k - Hx_k)} \tag{5.3}$$

The characterizations 5.1–5.3 are equivalent.

The optimal mean-square estimate of x_k is the conditional mean given all previous data (Section 1.1). Let us represent the data through time k by

$$Z_k = \{z_1, z_2, \ldots, z_k\} \tag{5.4}$$

It is desired to find a recursive optimal estimation procedure by finding an expression for the optimal estimate $\overline{x_{k+1}/Z_{k+1}}$ in terms of $\overline{x_k/Z_k}$, where overbar denotes expected value. To determine an error bound we would also like an expression for the error covariance $P_{x_{k+1}/Z_{k+1}}$ in terms of P_{x_k/Z_k}. Both of these expressions can in principle be determined from a recursive expression for the conditional PDF $f_{x_{k+1}/Z_{k+1}}$ in terms of f_{x_k/Z_k}. The quantity f_{x_k/Z_k} is the *hyperstate* of the system.

We can use the PDF chain rule and its integral the Chapman–Kolmogorov equation (Papoulis 1965) to find quite easily a recursion for the hyperstate. (See Sorenson 1970a and Bryson and Ho 1975.)

According to the chain rule we have

$$\begin{aligned} f_{x_{k+1}, z_{k+1}/Z_k} &= f_{x_{k+1}/z_{k+1}, Z_k} f_{z_{k+1}/Z_k} \\ &= f_{z_{k+1}/x_{k+1}, Z_k} f_{x_{k+1}/Z_k} \end{aligned} \tag{5.5}$$

Rearranging and taking into account the memoryless nature of the measurement process, there results the difference equation

$$f_{x_{k+1}/Z_{k+1}} = \frac{f_{z_{k+1}/x_{k+1}}}{f_{z_{k+1}/Z_k}} f_{x_{k+1}/Z_k} \tag{5.6}$$

This represents the *measurement update* of the hyperstate.

To find the time update from f_{x_k/Z_k} to f_{x_{k+1}/Z_k}, use the Chapman–Kolmogorov equation to write

$$f_{x_{k+1}/Z_k} = \int f_{x_{k+1}/x_k,Z_k} f_{x_k/Z_k} \mathrm{d}x_k$$

By the Markov system assumption, we obtain the time update

$$f_{x_{k+1}/Z_k} = \int f_{x_{k+1}/x_k} f_{x_k/Z_k} \mathrm{d}x_k \tag{5.7}$$

which is a prediction of f_{x_{k+1}/Z_k} in terms of f_{x_k/Z_k}.

Combining the time and measurement updates yields the desired recursion for the hyperstate

$$f_{x_{k+1}/Z_{k+1}} = \frac{\int f_{z_{k+1}/x_{k+1}} f_{x_{k+1}/x_k} f_{x_k/Z_k} \mathrm{d}x_k}{f_{z_{k+1}/Z_k}} \tag{5.8}$$

According to the Chapman–Kolmogorov equation, the normalizing constant can be expressed as

$$f_{z_{k+1}/Z_k} = \int f_{z_{k+1}/x_{k+1},x_k,Z_k} f_{x_{k+1}/x_k,Z_k} f_{x_k/Z_k} \mathrm{d}x_k \mathrm{d}x_{k+1}$$

Taking into account *both* the memorylessness of the measurements and the Markov property results in

$$f_{z_{k+1}/Z_k} = \int f_{z_{k+1}/x_{k+1}} f_{x_{k+1}/x_k} f_{x_k/Z_k} \mathrm{d}x_k \mathrm{d}x_{k+1} \tag{5.9}$$

The hyperstate recursion then becomes

$$f_{x_{k+1}/Z_{k+1}} = \frac{\int f_{z_{k+1}/x_{k+1}} f_{x_{k+1}/x_k} f_{x_k/Z_k} \mathrm{d}x_k}{\int f_{z_{k+1}/x_{k+1}} f_{x_{k+1}/x_k} f_{x_k/Z_k} \mathrm{d}x_k \mathrm{d}x_{k+1}} \tag{5.10}$$

The next example shows the power of this recursion by demonstrating that it yields the Kalman filter as a special case!

Example 5.1 Derivation of Discrete Kalman Filter from Hyperstate Recursion

Let the system be given by

$$x_{k+1} = Ax_k + Gw_k \tag{5.11}$$

$$z_k = Hx_k + v_k \tag{5.12}$$

with $w_k \sim N(0, Q)$ and $v_k \sim N(0, R)$ white and uncorrelated with each other and with $x_0 \sim N(\bar{x}_0, P_0)$. Suppose $x_k \in R^n$, $z_k \in R^p$.

a. Time update

Assume that $x_k/Z_k \sim N(\hat{x}_k, P_k)$, or

$$f_{x_k/Z_k} = \frac{1}{\sqrt{(2\pi)^n |P_k|}} e^{-(1/2)(x_k-\hat{x}_k)^T P_k^{-1}(x_k-\hat{x}_k)} \tag{5.13}$$

We could use $f_{x_{k+1}/x_k} = f_{w_k}[G^{-1}(x_{k+1} - Ax_k)]$ and perform the integration in Equation 5.7. (If G is singular we must use characteristic functions. See Sorenson 1970a,b.) However, since x_{k+1} is a sum of Gaussian random vectors (RVs) (see the problems at the end of this chapter), we know that

$$x_{k+1}/Z_k \sim N(A\hat{x}_k, AP_kA^T + GQG^T) \tag{5.14}$$

or

$$f_{x_{k+1}/Z_k} = \frac{1}{\sqrt{(2\pi)^n |P_{k+1}^-|}} e^{-(1/2)(x_{k+1}-\hat{x}_k^-)^T (P_{k+1}^-)^{-1}(x_{k+1}-\hat{x}_{k+1}^-)} \tag{5.15}$$

where the predicted estimate and error covariance are

$$\hat{x}_{k+1}^- = A\hat{x}_k \tag{5.16}$$

$$P_{k+1}^- = AP_kA^T + GQG^T \tag{5.17}$$

b. Measurement update

We have *a priori*

$$x_{k+1}/Z_k \sim N\left(\hat{x}_{k+1}^-, P_{k+1}^-\right) \tag{5.18}$$

and we must use Equation 5.6 to determine the PDF of the *a posteriori* RV x_{k+1}/Z_{k+1}.

First, let us find the normalizing constant. We could evaluate the expression (Equation 5.9). However, since z_{k+1} is a sum of Gaussian RVs, we know that

$$z_{k+1}/Z_k \sim N\left(H\hat{x}_{k+1}^-, HP_{k+1}^-H^T + R\right) \tag{5.19}$$

We can also say (see Equation 5.3)

$$z_{k+1}/x_{k+1} \sim N(Hx_{k+1}, R) \tag{5.20}$$

Using Equations 5.18 through 5.20 in Equation 5.6

$$\begin{aligned} f_{x_{k+1}/Z_{k+1}} = {} & \frac{\sqrt{(2\pi)^p |HP_{k+1}^-H^T + R|}}{\sqrt{(2\pi)^p|R|}\sqrt{(2\pi)^n|P_{k+1}^-|}} \\ & \times \exp\left\{-\frac{1}{2}[(z_{k+1} - Hx_{k+1})^T R^{-1}(z_{k+1} - Hx_{k+1}) \right. \\ & + (x_{k+1} - \hat{x}_{k+1}^-)^T (P_{k+1}^-)^{-1}(x_{k+1} - \hat{x}_{k+1}^-) \\ & \left. - (z_{k+1} - H\hat{x}_{k+1}^-)^T \left(HP_{k+1}^-H^T + R\right)^{-1}(z_{k+1} - H\hat{x}_{k+1}^-)]\right\} \end{aligned} \tag{5.21}$$

After completing the square in the exponent and some slightly tedious work, we realize that this can be written as

$$f_{x_{k+1}/Z_{k+1}} = \frac{1}{\sqrt{(2\pi)^n |P_{k+1}|}} e^{-(1/2)(x_{k+1}-\hat{x}_{k+1})^T P_{k+1}^{-1}(x_{k+1}-\hat{x}_{k+1})} \tag{5.22}$$

where the updated estimate and error covariance are

$$\hat{x}_{k+1} = \hat{x}_{k+1}^- + P_{k+1} H^T R^{-1}(z_{k+1} - H\hat{x}_{k+1}^-) \tag{5.23}$$

$$P_{k+1} = P_{k+1}^- - P_{k+1}^- H^T \left(HP_{k+1}^- H^T + R\right)^{-1} HP_{k+1}^- \tag{5.24}$$

(See Example 1.2.)

Equations 5.16, 5.17, 5.23, and 5.24 are exactly the Kalman filter equations.

A simple alternative derivation of Equations 5.23 and 5.24 is worth mentioning. Equation 5.6 can be written as

$$f_{x_{k+1}/z_{k+1}} = \frac{f_{z_{k+1}/x_{k+1}} f_{x_{k+1}}}{f_{z_{k+1}}} \tag{5.25}$$

where all PDFs are interpreted as being conditioned on the previous data Z_k. This is Bayes' rule.

Now, consider the augmented RV

$$\begin{bmatrix} x_{k+1} \\ z_{k+1} \end{bmatrix} \sim N\left(\begin{bmatrix} \hat{x}_{k+1}^- \\ H\hat{x}_{k+1}^- \end{bmatrix}, \begin{bmatrix} P_{k+1}^- & P_{k+1}^- H^T \\ HP_{k+1}^- & HP_{k+1}^- H^T + R \end{bmatrix} \right) \tag{5.26}$$

with means and covariances conditioned on Z_k. We want to find an alternate to Equation 5.25 for determining the conditional mean and covariance of x_{k+1} given the new data z_{k+1} (and all previous data; this is implicit in Equation 5.18). But this is exactly the problem we solved in Example 1.3, which yields Equations 5.23 and 5.24 immediately. □

5.1.2 Continuous Systems

Now let us discuss a nonlinear continuous-time system with discrete measurements. A full analysis requires the use of *Itô calculus*. We shall avoid this by ignoring the mathematical niceties and taking some liberties with notation to make clear the essential ideas. For a rigorous treatment, see Jazwinski (1970).

Let there be prescribed nonlinear time-varying system

$$\dot{x} = a(x, t) + g(x, t)w \tag{5.27}$$

with discrete measurements at times t_k given by

$$z_k = h[x(t_k), k] + v_k \tag{5.28}$$

with $w(t) \sim (0, Q)$ and $v_k \sim (0, R)$ uncorrelated white noises. Let $x(t) \in R^n$.

The hyperstate measurement update can easily be found by using the same techniques as in the discrete case. It is

$$f_{x(t_k)/Z_k} = \frac{f_{z_k/x(t_k)}}{f_{z_k/Z_{k-1}}} f_{x(t_k)/Z_{k-1}} \tag{5.29}$$

with normalization constant

$$f_{z_k/Z_{k-1}} = \int f_{z_k/x} f_x/Z_{k-1} \mathrm{d}x \tag{5.30}$$

The time update is not so easy to discover, and it is here that the Itô calculus is needed (Jazwinski 1970). We merely state the result that between measurements the hyperstate satisfies the *Fokker–Planck* (or Kolmogorov) diffusion equation

$$\frac{\partial}{\partial t} f_{x(t)/Z_k} = -\text{trace}\,\frac{\partial f_{x(t)/Z_k} a}{\partial x} + \frac{1}{2}\sum_{i,j=1}^{n} \frac{\partial^2 f_{x(t)/Z_k}[gQg^T]_{ij}}{\partial x_i \partial x_j}; \quad t_k \le t < t_{k+1} \tag{5.31}$$

where $[M]_{ij}$ is the (i, j)th element of a matrix M. This is very similar to the heat equation.

The Fokker–Planck equation has only been solved in a few simple cases, one of which is the following.

Example 5.2 Fokker–Planck Equation for Wiener Process

Let

$$\dot{x}(t) = w(t) \tag{5.32}$$

with $w(t) \sim N(0, 1)$ white. Suppose $x(0)$ is initially zero with probability 1, that is,

$$f_{x(t)}(x, 0) = \delta(x) \tag{5.33}$$

a. Mean and variance

To find the mean and variance of $x(t)$, we can use Example 3.1 to see that

$$\dot{\bar{x}}(t) = 0, \qquad \bar{x}(0) = 0 \tag{5.34}$$

so

$$\bar{x}(t) = 0 \tag{5.35}$$

Furthermore, the Lyapunov equation is

$$\dot{p}(t) = 1, \qquad p(0) = 0 \tag{5.36}$$

so

$$p(t) = t \tag{5.37}$$

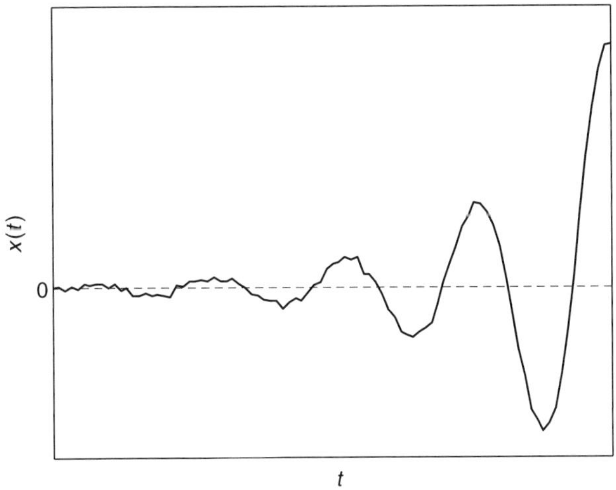

FIGURE 5.2
Sample function of a Wiener process.

Thus, the variance of a Wiener process increases linearly with t, which expresses our increasing uncertainty regarding $x(t)$.

A sample function of the Wiener process $\{x(t)\}$ is shown in Figure 5.2.

b. Fokker–Planck equation

In this simple case, the Fokker–Planck equation for the unconditional PDF $f_{x(t)}(x, t)$ is

$$\frac{\partial f_{x(t)}}{\partial t} = \frac{1}{2}\frac{\partial^2 f_{x(t)}(x,t)}{\partial x^2}; \quad t \geq 0 \tag{5.38}$$

with $f_{x(t)}(x, 0) = \delta(x)$.

Since the white noise $w(t)$ is normal, so is $x(t)$. Thus, according to Equations 5.35 and 5.37

$$f_{x(t)}(x,t) = \frac{1}{\sqrt{2\pi t}} e^{-x^2/2t} \tag{5.39}$$

It is straightforward to verify that Equation 5.39 satisfies Equation 5.38.

A sketch of $f_{x(t)}(x, t)$ for several values of t is shown in Figure 5.3. □

5.2 General Update of Mean and Covariance

Here, we discuss the exact time and measurement updates for the mean and covariance of the nonlinear continuous system with discrete measurements (Equations 5.27 and 5.28). First, we derive a useful preliminary result.

To simplify the appearance of our equations as far as possible, let us define the following shorthand notation for the *a posteriori* and *a priori* conditional

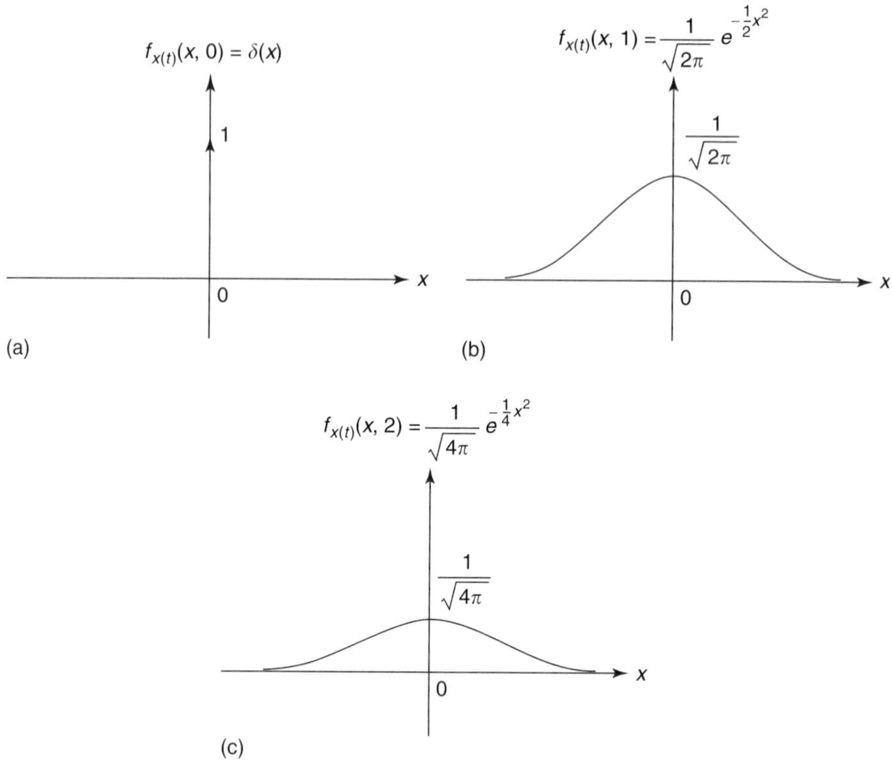

FIGURE 5.3
Propagation of PDF for a Wiener process. (a) $t = 0$. (b) $t = 1$. (c) $t = 2$.

expected values:

$$\hat{x}(t) = \varepsilon[x(t)] \triangleq E[x(t)/(z_k; t_k \leq t)] \tag{5.40a}$$

$$\hat{x}^-(t) = \varepsilon^-[x(t)] \triangleq E[x(t)/(z_k; t_k < t)] \tag{5.40b}$$

Thus, $\hat{x}(t_k)$ incorporates the measurements $z_k, z_{k-1}, \ldots$, while $\hat{x}^-(t_k)$ incorporates only $z_{k-1}, z_{k-2}, \ldots$. (The only reason we introduce $\varepsilon(\cdot)$ is to avoid using very large carats over entire expressions!) Note that the carat has the same meaning here as it has in the past; it simply denotes the optimal estimate.

5.2.1 Time Update

Let $\phi[x(t)]$ be an arbitrary twice differentiable scalar function of $x \in R^n$. To determine a time update between measurements for $\phi[x(t)]$ due to Equation 5.27, we proceed as follows (Jazwinski 1970). By definition

$$\hat{\phi}[x(t)] = \int \phi(x) f_{x/Z_k} \mathrm{d}x; \quad t_k \leq t \tag{5.41}$$

Therefore,

$$\frac{\mathrm{d}\hat{\phi}[x(t)]}{\mathrm{d}t} = \int \phi(x)\frac{\partial f_{x/Z_k}}{\partial t}\mathrm{d}x \tag{5.42}$$

Taking into account the Fokker–Planck equation, this becomes

$$\frac{\mathrm{d}\hat{\phi}[x(t)]}{\mathrm{d}t} = -\int \mathrm{trace}\left[\phi(x)\frac{\partial fa}{\partial x}\right]\mathrm{d}x + \frac{1}{2}\int \phi(x)\sum_{i,j=1}^{n}\frac{\partial^2 f[gQg^T]_{ij}}{\partial x_i \partial x_j}\mathrm{d}x$$

where $f \triangleq f_{x(t)/Z_k}$. Integrating by parts, assuming f is well behaved at infinity, and using some matrix identities, this yields (see the problems at the end of this chapter)

$$\frac{\mathrm{d}\hat{\phi}[x(t)]}{\mathrm{d}t} = \int \phi_x^T a f_{x/Z_k}\mathrm{d}x + \frac{1}{2}\int \mathrm{trace}(gQg^T\phi_{xx})f_{x/Z_k}\mathrm{d}x \tag{5.43}$$

By definition, this is

$$\frac{\mathrm{d}\hat{\phi}[x(t)]}{\mathrm{d}t} = \varepsilon\left(\phi_x^T a\right) + \frac{1}{2}\,\mathrm{trace}\,\varepsilon(gQg^T\phi_{xx}) \tag{5.44}$$

This is an important preliminary result which we can use to determine exact time updates for the mean and covariance for the nonlinear system (Equation 5.27).

To wit, first let $\phi(x_i) = x_i$, the ith component of $x \in R^n$. Then, gradient $\phi_{x_i} = 1$ and Hessian $\phi_{x_i x_i} = 0$. Therefore,

$$\frac{\mathrm{d}\hat{x}_i}{\mathrm{d}t} = \varepsilon(a_i) = \hat{a}_i \tag{5.45}$$

with $a_i(x,t)$ the ith component of $a(x,t)$. Therefore, a time update between measurements for the conditional mean is provided by

$$\dot{\hat{x}} = \hat{a}(x,t) \tag{5.46}$$

To find a time update for the conditional covariance

$$P(t) \triangleq P_{x(t)/Z_k} = \varepsilon[(x-\hat{x})(x-\hat{x})^T] = \varepsilon(xx^T) - \hat{x}\hat{x}^T \tag{5.47}$$

we first let $\phi(y) = x_i x_j$, with $y = [x_i \quad x_j]^T$, a temporary auxiliary random vector. Then, gradient $\phi_y = [x_j \quad x_i]^T$ and Hessian $\phi_{yy} = \left[\begin{smallmatrix} 0 & 1 \\ 1 & 0 \end{smallmatrix}\right]$. Therefore, by Equation 5.44

$$\begin{aligned}\frac{\mathrm{d}}{\mathrm{d}t}\varepsilon(x_i x_j) &= \varepsilon(a_i x_j + x_i a_j) + \frac{1}{2}([gQg^T]_{ij} + [gQg^T]_{ji}) \\ &= \varepsilon(a_i x_j + x_i a_j) + \varepsilon([gQg^T]_{ij})\end{aligned} \tag{5.48}$$

where we have used the symmetry of gQg^T. (To show this, first find an expression for $\dot{y}$.) Now,

$$\frac{\mathrm{d}}{\mathrm{d}t}(\hat{x}_i\hat{x}_j) = \hat{x}_i\frac{\mathrm{d}}{\mathrm{d}t}\hat{x}_j + \hat{x}_j\frac{\mathrm{d}}{\mathrm{d}t}\hat{x}_i$$

so by Equation 5.45

$$\frac{\mathrm{d}}{\mathrm{d}t}(\hat{x}_i\hat{x}_j) = \hat{x}_i\hat{a}_j + \hat{a}_i\hat{x}_j \tag{5.49}$$

By combining Equations 5.47 through 5.49 and employing some bookkeeping to keep track of indices, we finally achieve the desired error covariance time update

$$\dot{P} = [\varepsilon(xa^T) - \hat{x}\hat{a}^T] + [\varepsilon(ax^T) - \hat{a}\hat{x}^T] + \varepsilon(gQg^T) \tag{5.50}$$

This componentwise derivation is tedious, but it is required because differentiating a matrix with respect to a vector yields a three-dimensional object (a third-order tensor).

Equations 5.46 and 5.50 are the exact estimate and error covariance time updates between measurements for the nonlinear system (Equation 5.27). They are in general unsolvable, since they are *not* ordinary differential equations. To see this, note that Equation 5.46 is really

$$\frac{\mathrm{d}\hat{x}}{\mathrm{d}t} = \int a(x,t) f_{x(t)/Z_k}\mathrm{d}x; \quad t_k \le t \tag{5.51}$$

which depends on *all* the conditional moments of $x(t)$. Note the important point that, in general,

$$\hat{a}(x,t) \neq a(\hat{x},t) \tag{5.52}$$

In the linear case

$$\dot{x} = a(x,t) + g(x,t)w = Ax + Gw \tag{5.53}$$

we do have $\hat{a}(x,t) = \varepsilon(Ax) = A\hat{x} = a(\hat{x},t)$, so the mean update is the familiar

$$\dot{\hat{x}} = A\hat{x} \tag{5.54}$$

Furthermore, in the linear case, Equation 5.50 reduces to the familiar Lyapunov equation since

$$\begin{aligned}\dot{P} &= [\varepsilon(xx^TA^T) - \hat{x}\hat{x}^TA^T] + [\varepsilon(Axx^T) - A\hat{x}\hat{x}^T] + GQG^T \\ &= PA^T + AP + GQG^T \end{aligned} \tag{5.55}$$

5.2.2 Measurement Update

It is quite easy to determine how an arbitrary *matrix* function $\Phi[x(t)]$ is updated at a measurement due to Equation 5.28. Multiplying the PDF update

(Equation 5.29) by $\Phi[x(t)]$ and integrating with respect to $x(t_k)$, there results

$$\varepsilon\{\Phi[x(t_k)]\} = \frac{\varepsilon^-\{\Phi[x(t_k)]f_{z_k/x(t_k)}\}}{\varepsilon^-[f_{z_k/x(t_k)}]} \tag{5.56}$$

From this it follows immediately that the exact measurement updates for the estimate and error covariance are given by

$$\hat{x}(t_k) = \frac{\varepsilon^-[x(t_k)f_{z_k/x(t_k)}]}{\varepsilon^-[f_{z_k/x(t_k)}]} \tag{5.57}$$

and

$$P(t_k) = \frac{\varepsilon^-[x(t_k)x^T(t_k)f_{z_k/x(t_k)}]}{\varepsilon^-[f_{z_k/x(t_k)}]} - \hat{x}(t_k)\hat{x}^T(t_k) \tag{5.58}$$

Unfortunately, these expressions are of no use algorithmically. Furthermore, judging by Example 5.1 part *b* which was for the linear Gaussian case, any attempt to use them to derive more convenient measurement updates for $\hat{x}$ and P will quickly become embroiled in some lengthy manipulations. Consequently, we select an alternate approach. For further discussion of Equations 5.57 and 5.58, see Jazwinski (1970).

5.2.3 Linear Measurement Update

To find more convenient expressions for measurement updating of the estimate and error covariance in the nonlinear case, we shall restrict the estimate to be a linear function of the data (Gelb 1974).

Define the *a priori* and *a posteriori* estimation errors as

$$\tilde{x}_k^- \triangleq x(t_k) - \hat{x}^-(t_k) \tag{5.59}$$

$$\tilde{x}_k \triangleq x(t_k) - \hat{x}(t_k) \tag{5.60}$$

Then the *a priori* and *a posteriori* error covariance are

$$P^-(t_k) \triangleq E[\tilde{x}_k^-(\tilde{x}_k^-)^T] \tag{5.61}$$

$$P(t_k) \triangleq E[\tilde{x}_k\tilde{x}_k^T] \tag{5.62}$$

For notational ease, define

$$x_k \triangleq x(t_k) \tag{5.63}$$

and suppress the k argument of $h(\cdot,\cdot)$. The nonlinear measurement is then written as

$$z_k = h(x_k) + v_k \tag{5.64}$$

with white noise $v_k \sim (0, R)$ uncorrelated with $x(t)$.

To make this problem tractable, let us now restrict the *a posteriori* estimate to be a linear function of the data, so that

$$\hat{x}_k = b_k + K_k z_k \tag{5.65}$$

for some deterministic b_k and K_k which are yet to be found. Then the estimation error update is

$$\tilde{x}_k = x_k - b_k - K_k[h(x_k) + v_k] = -b_k - K_k h(x_k) - K_k v_k + \tilde{x}_k^- + \hat{x}_k^- \tag{5.66}$$

It is desired that $\hat{x}_k$ be unbiased if $\hat{x}_k^-$ is, so by taking expected values

$$0 = -b_k - K_k \hat{h}^-(x_k) + \hat{x}_k^- \tag{5.67}$$

Solving for b_k and substituting into Equation 5.65 yields the familiar form

$$\hat{x}_k = \hat{x}_k^- + K_k[z_k - \hat{h}^-(x_k)] \tag{5.68}$$

Taking Equation 5.67 into account, Equation 5.66 becomes

$$\tilde{x}_k = \tilde{x}_k^- - K_k[h(x_k) - \hat{h}^-(x_k)] - K_k v_k \tag{5.69}$$

Using this, an expression for the error covariance $P(t_k)$ in terms of the as yet undetermined K_k can now be written as

$$\begin{aligned} P(t_k) = P^-(t_k) &+ K_k E\{[h(x_k) - \hat{h}^-(x_k)][h(x_k) - \hat{h}^-(x_k)]^T\}K_k^T \\ &- K_k E\{[h(x_k) - \hat{h}^-(x_k)](\tilde{x}_k^-)^T\} \\ &- E\{\tilde{x}_k^-[h(x_k) - \hat{h}^-(x_k)]^T\}K_k^T + K_k R K_k^T \end{aligned} \tag{5.70}$$

The gain K_k is now selected to minimize $P(t_k)$. Differentiating $P(t_k)$ with respect to K_k and solving for K_k results in the "Kalman-like" gain

$$\begin{aligned} K_k = &E\{(x_k - \hat{x}_k^-)[h(x_k) - \hat{h}^-(x_k)]^T\} \\ &\times (E\{[h(x_k) - \hat{h}^-(x_k)][h(x_k) - \hat{h}^-(x_k)]^T\} + R)^{-1} \end{aligned} \tag{5.71}$$

Substituting this into Equation 5.70 and simplifying yields

$$P(t_k) = P^-(t_k) - K_k E\{[h(x_k) - \hat{h}^-(x_k)](x_k - \hat{x}_k^-)]^T\} \tag{5.72}$$

The complete linear estimate update due to nonlinear measurements is given by Equations 5.68, 5.71, and 5.72. These equations are computationally intractable because they are not ordinary difference equations. They depend on all conditional moments of $x(t)$ to compute $\hat{h}^-[x(t_k)]$. In the next subsection, we make first-order approximations to derive a computationally viable algorithm known as the *EKF*.

It can quickly be demonstrated that in the case of linear measurements

$$z_k = Hx_k + v_k \tag{5.73}$$

our equations reduce to the Kalman filter measurement update. In general $\hat{h}^-[x(t_k)] \neq h[\hat{x}^-(t_k)]$, however, in the linear case

$$\hat{h}^-[x(t_k)] = \varepsilon^-(Hx_k) = H\hat{x}_k^- = h[\hat{x}^-(t_k)] \tag{5.74}$$

Substituting this into Equations 5.68, 5.71, and 5.72 yields the measurement update in Table 2.2.

5.3 Extended Kalman Filter

Let us discuss continuous nonlinear systems of the form

$$\dot{x} = a(x,t) + G(t)w \tag{5.75a}$$

with measurements at discrete times t_k given by

$$z_k = h[x(t_k), k] + v_k \tag{5.75b}$$

where $w(t) \sim (0, Q)$ and $v_k \sim (0, R)$ are white noise processes uncorrelated with each other and with $x(0) \sim (\tilde{x}_0, P_0)$. For simplicity, we assume the process noise matrix $G(t)$ is independent of $x(t)$. The case $g[x(t), t]$ is covered in the problems.

Exact time and measurement updates for the hyperstate $f_{x(t)/Z_k}$, where $Z_k = \{z_k, z_{k-1}, \ldots,\}$ is all the data available through time t_k, are given by Equations 5.31 and 5.29, respectively. Using these equations, we derived the exact time and measurement updates for the estimate (Equations 5.46 and 5.57) and error covariance (Equations 5.50 and 5.58). To make the measurement updates assume a more convenient form, we next restricted the estimate to a linear dependence on the data z_k. This resulted in the measurement updates (Equation 5.68) for the estimate and (Equations 5.61 and 5.71) for the error covariance.

All these update equations are in general intractable, since they depend on the entire hyperstate, or equivalently on all the moments of $x(t)$. We demonstrated at several points that in the linear Gaussian case, the Kalman filter was recovered.

What we would like to do now is to make some approximations to $a(x,t)$ and $h(x,k)$ to obtain a computationally viable algorithm for filtering in nonlinear systems. We shall see that the result is an "extended" version of the Kalman filter. See Gelb (1974).

5.3.1 Approximate Time Update

The exact time update for nonlinear $a(x,t)$ is given by Equations 5.46 and 5.50. To find a time update that can be conveniently programmed on a computer, expand $a(x,t)$ in a Taylor series about $\hat{x}(t)$, the current state estimate

$$a(x,t) = a(\hat{x},t) + \left.\frac{\partial a}{\partial x}\right|_{x=\hat{x}} (x - \hat{x}) + \cdots \tag{5.76}$$

Taking conditional expectations yields

$$\hat{a}(x,t) = a(\hat{x},t) + O(2) \tag{5.77}$$

where $O(2)$ denotes terms of order 2. A first-order approximation to $\hat{a}(x,t)$ is, therefore, $a(\hat{x},t)$. Update 5.46 is thus replaced by

$$\dot{\hat{x}} = a(\hat{x},t) \tag{5.78}$$

To determine a first-order approximation to Equation 5.50, substitute for $a(x,t)$ using the first two terms of Equation 5.76 and then take the indicated expected values. Representing the Jacobian as

$$A(x,t) \triangleq \frac{\partial a(x,t)}{\partial x} \tag{5.79}$$

this results in

$$\dot{P} = A(\hat{x},t)P + PA^T(\hat{x},t) + GQG^T \tag{5.80}$$

Equations 5.78 and 5.80 represent an approximate, computationally feasible time update for the estimate and error covariance. The estimate simply propagates according to the nonlinear dynamics, and the error covariance propagates like that of a linear system with plant matrix $A(\hat{x},t)$.

Note that Jacobian $A(\cdot,t)$ is evaluated for each t at the *current estimate*, which is provided by Equation 5.78, so that there is coupling between Equations 5.78 and 5.80. To eliminate this coupling, it is possible to introduce a further approximation and solve not Equation 5.80 but instead

$$\dot{P}(t) = A[\hat{x}(t_k),t]P(t) + P(t)A^T[\hat{x}(t_k),t] + G(t)QG^T(t); \quad t_k \le t < t_{k+1} \tag{5.81}$$

In this equation, the Jacobian is evaluated once using the estimate $\hat{x}(t_k)$ after updating at t_k to include z_k. This is used as the plant matrix for the time propagation over the entire interval until the next measurement time t_{k+1}.

We shall soon present an example demonstrating how to accomplish the time update (Equations 5.78 and 5.80) very easily on a digital computer using a Runge–Kutta integrator.

5.3.2 Approximate Measurement Update

The optimal linear measurement update for nonlinear measurement matrix $h[x(t_k),k]$ was found to be given by Equations 5.68, 5.71, and 5.72. To find a measurement update that can be conveniently programmed, expand $h(x_k)$ in a Taylor series about $\hat{x}_k^-$, the *a priori* estimate at time t_k:

$$h(x_k) = h(\hat{x}_k^-) + \left.\frac{\partial h}{\partial x}\right|_{x=\hat{x}_k^-} (x_k - \hat{x}_k^-) + \cdots \tag{5.82}$$

Represent the Jacobian as

$$H(x,k) = \frac{\partial h(x,k)}{\partial x} \tag{5.83}$$

To find a first-order approximate update, substitute for $h(x_k)$ in Equations 5.68, 5.71, and 5.72 using the first two terms of Equation 5.82. The result is

$$\hat{x}_k = \hat{x}_k^- + K_k[z_k - h(\hat{x}_k^-)] \tag{5.84}$$

$$K_k = P^-(t_k)H^T(\hat{x}_k^-)[H(\hat{x}_k^-)P^-(t_k)H^T(\hat{x}_k^-) + R]^{-1} \tag{5.85}$$

$$P(t_k) = [I - K_k H(\hat{x}_k^-)]P^-(t_k) \tag{5.86}$$

These equations represent an approximate, computationally feasible linear measurement update for the estimate and the error covariance. The residual is computed using the nonlinear measurement function $h(\cdot, k)$ evaluated at the *a priori* estimate $\hat{x}_k^-$. The error covariance is found using the Jacobian matrix $H(\hat{x}_k^-)$.

5.3.3 The Extended Kalman Filter

The approximate time and measurement updates we have derived are collected for reference in Table 5.1. They comprise what is known as the EKF for a continuous system with discrete measurements. If Q and R are time varying, it is only necessary to use their values $Q(t)$ and R_k in the filter. A deterministic input $u(t)$ is included in the equations for completeness.

An important advantage of the continuous–discrete EKF is that the optimal estimate is available continuously at all times, including times *between* the measurement times t_k. Note that if $a(x, u, t)$ and $h(x, k)$ are linear, then the EKF reverts to the continuous–discrete Kalman filter in Section 3.7.

Given the continuous nature of dynamical systems and the requirement of microprocessors for discrete data, the continuous–discrete EKF is usually the most useful formulation for modern control purposes. The EKF approximation can also be applied to continuous systems with *continuous* measurements. For the sake of completeness, the continuous–continuous EKF is given in Table 5.2. A discrete–discrete Kalman filter could also be written down very easily.

Since the time and measurement updates depend on Jacobians evaluated *at the current estimate*, the error covariance and Kalman gain K_k cannot be computed off-line for the EKF. They must be computed in real time as the data become available.

In some cases, the nominal trajectory $x_N(t)$ of the state $x(t)$ is known *a priori*. An example is a robot arm for welding a car door which always moves along the same path. In this case, the EKF equations can be used with $\hat{x}(t)$ replaced by $x_N(t)$. Since $x_N(t)$ is known beforehand, the error covariance and K_k can now be computed off-line before the measurements are taken. This procedure of linearizing about a known nominal trajectory results in what is called the *linearized Kalman filter*. If a controller is employed to keep the state

TABLE 5.1
Continuous–Discrete Extended Kalman Filter

System model and measurement model

$$\dot{x} = a(x, u, t) + G(t)w$$
$$z_k = h[x(t_k), k] + v_k$$
$$x(0) \sim (\bar{x}_0, P_0), w(t) \sim (0, Q), v_k \sim (0, R)$$

Assumptions

$\{w(t)\}$ and $\{v_k\}$ are white noise processes uncorrelated with $x(0)$ and with each other.

Initialization

$$P(0) = P_0, \hat{x}(0) = \bar{x}_0$$

Time update

Estimate $\quad \dot{\hat{x}} = a(\hat{x}, u, t)$

Error covariance $\quad \dot{P} = A(\hat{x}, t)P + PA^T(\hat{x}, t) + GQG^T$

Measurement update

Kalman gain

$$K_k = P^-(t_k)H^T(\hat{x}_k^-)[H(\hat{x}_k^-)P^-(t_k)H^T(\hat{x}_k^-) + R]^{-1}$$

Error covariance $\quad P(t_k) = [I - K_k H(\hat{x}_k^-)P^-(t_k)]$

Estimate $\quad \hat{x}_k = \hat{x}_k^- + K_k[z_k - h(\hat{x}_k^-, k)]$

Jacobians

$$A(x, t) = \frac{\partial a(x, u, t)}{\partial x}$$
$$H(x) = \frac{\partial h(x, k)}{\partial x}$$

$x(t)$ on the nominal trajectory $x_N(t)$, the linearized Kalman filter performs quite well.

Higher-order approximations to the optimal nonlinear updates can also be derived by retaining higher-order terms in the Taylor series expansions. For details, see Jazwinski (1970) and Gelb (1974).

It should be realized that the measurements times t_k need not be equally spaced. The time update is performed over any interval during which no data are available. When data become available, a measurement update is performed. This means that the cases of intermittent or missing measurements, and pure prediction in the absence of data can easily be dealt with using the EKF.

A block diagram of a software implementation of the continuous–discrete EKF is shown in Figure 3.14. The implementation of the continuous–continuous Kalman filter is even easier. The time update function is used to

TABLE 5.2
Continuous–Continuous Extended Kalman Filter

System model and measurement model

$$\dot{x} = a(x, u, t) + G(t)w$$
$$z = h(x, t) + v$$
$$x(0) \sim (\bar{x}_0, P_0), w(t) \sim (0, Q), v(t) \sim (0, R)$$

Assumptions

$\{w(t)\}$ and $\{v(t)\}$ are white noise processes uncorrelated with $x(0)$ and with each other.

Initialization

$$P(0) = P_0, \hat{x}(0) = \bar{x}_0$$

Estimate update

$$\dot{\hat{x}} = a(\hat{x}, u, t) + K[z - h(\hat{x})]$$

Error covariance update

$$\dot{P} = A(\hat{x}, t)P + PA^T(\hat{x}, t) + GQG^T - PH^T(\hat{x}, t)R^{-1}H(\hat{x}, t)P$$

Kalman gain

$$K = PH^T(\hat{x}, t)R^{-1}$$

Jacobians

$$A(x, t) = \frac{\partial a(x, u, t)}{\partial x}$$
$$H(x, t) = \frac{\partial h(x, t)}{\partial x}$$

integrate the estimate and error covariance dynamics simultaneously. The vector X in the subroutine argument will contain the components of both $\hat{x}(t)$ and $P(t)$. The next examples illustrate how to write code to use the EKF. They also make the point that the filter implementation can be considerably simplified by doing some preliminary analysis!

Example 5.3 FM Demodulation

a. System and measurement

Let message signal $s(t)$ be normal with zero mean and variance σ^2. Suppose, it has a first-order Butterworth spectrum with a bandwidth of α (cf., Anderson and Moore 1979),

$$\Phi_s(\omega) = \frac{2\alpha\sigma^2}{\omega^2 + \alpha^2} \tag{5.87}$$

The FM modulated signal that is transmitted is

$$y(t) = \sqrt{2}\sigma \sin[\omega_c t + \theta(t)] \tag{5.88}$$

where ω_c is the carrier frequency and the phase is related to the message by

$$\theta(t) = \int_0^t s(\tau)\mathrm{d}\tau \tag{5.89}$$

It is desired to use the EKF to estimate $s(t)$ from samples of the received signal, which is $z(t) = y(t) + v(t)$ with $v(t) \sim (0, r)$, a white noise added by the communication channel.

Defining state $x = [s \quad \theta]^T$, the processes 5.87 and 5.89 can be modeled as (see Figure 2.18).

$$\dot{x} = \begin{bmatrix} -\alpha & 0 \\ 1 & 0 \end{bmatrix} x + \begin{bmatrix} 1 \\ 0 \end{bmatrix} w \triangleq Ax + Gw \tag{5.90}$$

with white noise $w(t) \sim (0, 2\alpha\sigma^2)$. The discrete nonlinear time-varying measurements are

$$z_k = \sqrt{2}\sigma \sin(\omega_c Tk + Cx_k) + v_k \triangleq h(x_k, k) + v_k \tag{5.91}$$

with $C \triangleq [0 \quad 1]$ and T the sampling period.

b. Time update

The time update is straightforward since the dynamics (Equation 5.90) are linear. Therefore, the error covariance propagates according to

$$\dot{P} = AP + PA^T + GQG^T \tag{5.92}$$

Letting

$$P \triangleq \begin{bmatrix} p_1 & p_2 \\ p_2 & p_4 \end{bmatrix} \tag{5.93}$$

we can write Equation 5.92 as the set of linear scalar equations

$$\dot{p}_1 = -2\alpha p_1 + 2\alpha\sigma^2 \tag{5.94a}$$

$$\dot{p}_2 = -\alpha p_2 + p_1 \tag{5.94b}$$

$$\dot{p}_4 = 2p_2 \tag{5.94c}$$

It is quite easy to solve these analytically; however, we shall deliberately refrain from doing so to make the point that an analytical solution is not required to implement the Kalman filter.

The estimate propagates by

$$\dot{\hat{x}} = A\hat{x} \tag{5.95}$$

or

$$\dot{\hat{s}} = -\alpha \hat{s} \tag{5.96a}$$

$$\dot{\hat{\theta}} = \hat{s} \tag{5.96b}$$

c. Measurement update

The measurement is nonlinear, so the EKF update in Table 5.1 is required here. The measurement Jacobian is

$$H(x) = \frac{\partial h(x,k)}{\partial x} = \sqrt{2}\sigma C \cos(\omega_c T_k + Cx) \tag{5.97}$$

Letting $H(\hat{x}_k^-) \triangleq [h_1 \quad h_2]$, we have

$$h_1 = 0 \tag{5.98a}$$

$$h_2 = \sqrt{2}\sigma \cos(\omega_c T_k + \hat{\theta}_k^-) \tag{5.98b}$$

The updated Kalman gain is found as follows. Let

$$\delta \triangleq H(\hat{x}_k^-) P^-(t_k) H^T(\hat{x}_k^-) + \frac{r}{T} \tag{5.99}$$

Defining

$$P^-(t_k) \triangleq \begin{bmatrix} p_1^- & p_2^- \\ p_2^- & p_4^- \end{bmatrix} \tag{5.100}$$

After simplifying, we have

$$\delta = 2\sigma^2 p_4^- \cos^2(\omega_c T_k + \hat{\theta}_k^-) + r = h_2^2 p_4^- + \frac{r}{T} \tag{5.101}$$

Hence,

$$K_k \triangleq \begin{bmatrix} k_1 \\ k_2 \end{bmatrix} = \begin{bmatrix} p_2^- h_2/\delta \\ p_4^- h_2/\delta \end{bmatrix} \tag{5.102}$$

The update covariance is given by

$$P(t_k) = \begin{bmatrix} 1 & -k_1 h_2 \\ 0 & 1 - k_2 h_2 \end{bmatrix} P^-(t_k) \tag{5.103}$$

or

$$p_1 = p_1^- - k_1 h_2 p_2^- = p_2^- - \frac{h_2^2 (p_2^-)^2}{\delta} \tag{5.104a}$$

$$p_2 = (1 - k_2 h_2) p_2^- = \frac{r p_2^-}{\delta T} \tag{5.104b}$$

$$p_4 = (1 - k_2 h_2) p_4^- = \frac{r p_4^-}{\delta T} \tag{5.104c}$$

Finally, the estimate update is (omitting the time subscript k)

$$\hat{s} = \hat{s}^- + k_1[z - \sqrt{2}\sigma \sin(\omega_c T_k + \hat{\theta}^-)] \tag{5.105a}$$

$$\hat{\theta} = \hat{\theta}^- + k_2[z - \sqrt{2}\sigma \sin(\omega_c T_k + \hat{\theta}^-)] \tag{5.105b}$$

d. Filter implementation

The time update (Equations 5.94 and 5.96) is implemented between measurement with an integration routine such as MATLAB's ODE45. The measurement update (Equations 5.98, 5.101, 5.102, 5.104, and 5.105) is performed whenever data become available. Note that it is *not* necessary for measurements to be taken at regular intervals. For irregular or intermittent measurements at times t_k, it is only necessary to change the time dependence T_k in Equations 5.98b and 5.105 to t_k. Note also that the EKF yields the optimal estimate at all times, even between the measurement times t_k.

The preliminary analysis we performed resulted in scalar updates, which are easier to program and faster to run than matrix updates. MATLAB implementations of the time update and measurement update are given in Figure 5.4. The vector X in that figure contains both the estimates and the error covariance entries. We have avoided using two sets of error covariances, one for $P(t_k)$ and one for $P^-(t_k)$, by paying attention to the order of operations. Thus, in the program Equation 5.104a must be performed before Equation 5.104b since it needs the *a priori* value p_2^-, which is updated in Equation 5.104b.

We would expect the error covariance to behave much like the error covariance in Example 3.9. □

Example 5.4 Satellite Orbit Estimation

a. Satellite equations of motion

The equations of motion of a satellite in a planar orbit about a point mass are

$$\ddot{r} = r\dot{\theta}^2 - \frac{\mu}{r^2} + w_r \tag{5.106}$$

$$\ddot{\theta} = \frac{-2\dot{r}\dot{\theta}}{r} + \frac{1}{r}w_\theta \tag{5.107}$$

where r is the radial distance of the satellite from the mass and θ its angle from a reference point on the orbit (usually perigee, the closest point of the satellite's orbit to the attracting mass M). White process noises $w_r(t) \sim (0, q_r)$ and $w_\theta(t) \sim (0, q_\theta)$ represent disturbance accelerations. The gravitational constant of the point mass M is $\mu \triangleq GM$, with G the universal gravitation constant. See Maybeck (1979), Bryson and Ho (1975), and Jazwinski (1970).

```
% Example 5.3
% FM Demodulation


function f5p3d1()


    % px(1:3)=Error covariances, px(4:5)=State estimates
    % Sampling period T=0.1 sec, Measurement every 1 sec.
    px=[10 10 10 10 10];
    i=0; wc=1; r=1, del=0.1; del_mu=1;
    global sig; sig=1;
    rnd=randn(1000,1);
    for t=0:del:10


        % Time Update
        t1=t+del;
        [td px] = ode45(@diffeqns,[t t1],px);
        px=px(end,:);


        % Measurement Update
        if (mod(t1,del_mu)==0)
            i=i+1;
            h=[0 sqrt(2)*sig*cos(wc*t1+px(5))];
            d=2*sig^2*px(3)*cos(wc*t1+px(5))+r/del;
            % Kalman Gain
            k=[px(2)*h(2)/d ; px(3)*h(2)/d];
            % Error covariances
            px(1)=px(1)-k(1)*h(2)*px(2);
            px(2)=(1-k(2)*h(2))*px(2);
            px(3)=(1-k(2)*h(2))*px(3);
            % Available measurement zk
            z=sqrt(2)*sig*sin(wc*t1+px(5))+rnd(i);
            % State estimate after measurement update
            px(4)=px(4)+k(1)*(z-sqrt(2)*sig*sin(wc*t1+px(5)));
            px(5)=px(5)+k(2)*(z-sqrt(2)*sig*sin(wc*t1+px(5)));
        end


        % Plotting
        subplot(3,1,1); plot(t1,px(1),'k.'); hold on; ylabel('p_1'); grid on;
        subplot(3,1,2); plot(t1,px(2),'k.'); hold on; ylabel('p_2'); grid on;
        subplot(3,1,3); plot(t1,px(3),'k.'); hold on; ylabel('p_4'); grid on;
        xlabel('Time (Secs.)');


    end
return


function dpx=diffeqns(t,px)
    global sig;
    a=1;
    % Differential equations for Error covariance
    dpx(1)=-2*a*px(1)+2*a*sig^2;
    dpx(2)=-a*px(2)+px(1);
    dpx(3)=2*px(2);
    % Differential equations for states
    dpx(4)=-a*px(4);
    dpx(5)=px(4);
    dpx=dpx';
return
```

FIGURE 5.4
Time update and measurement update subroutines for FM demodulation.

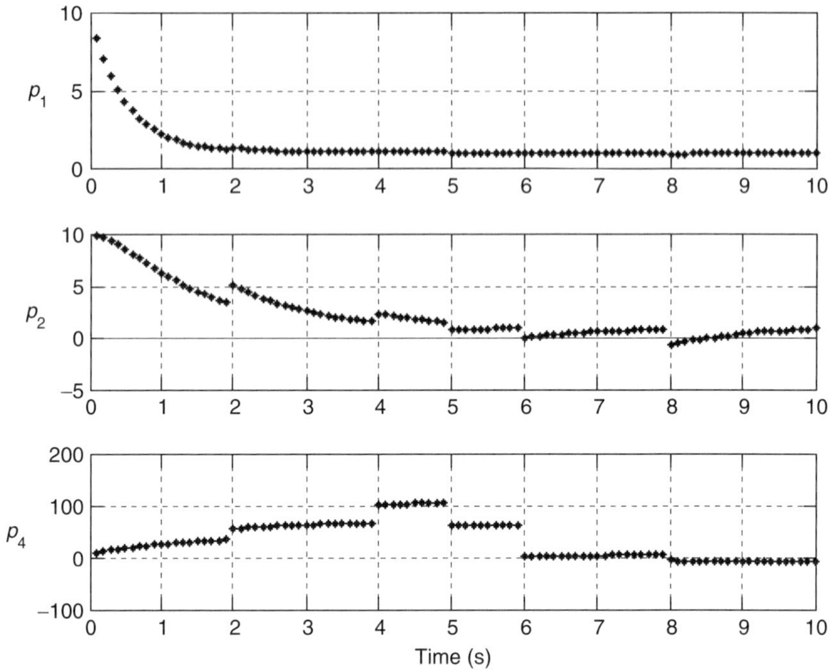

FIGURE 5.4
(Continued.)

b. Orbit estimation

An elliptic orbit in a plane is completely specified if we know

a, the semimajor axis

e, the eccentricity

t_p, the time of perigee passage

θ_p, the angle between perigee and a reference axis.

Within a specified orbit, the satellite's position is completely specified if we know $\theta(t)$. The variables a, e, θ_p, and $\theta(t)$ are the *Keplerian orbital elements*.

The orbital elements of a satellite can be determined from its state $x \triangleq [r\ \dot{r}\ \theta\ \dot{\theta}]^T$. For example,

$$a^2 = \frac{2}{r} - \frac{\dot{r}^2}{\mu} \tag{5.108}$$

$$e^2 = \left(1 - \frac{r}{2}\right)^2 + \frac{(r\dot{r})^2}{a\mu} \tag{5.109}$$

Another useful parameter is the period

$$T = 2\pi\sqrt{\frac{a^3}{\mu}} \tag{5.110}$$

The problem of orbit estimation is thus equivalent to the problem of determining the instantaneous state $x(t)$ of the satellite, which we can do using the continuous–discrete EKF.

c. Time update

The estimate for $x = [r \quad \dot{r} \quad \theta \quad \dot{\theta}]^T$ propagates between measurements according to the deterministic (noise-free) version of Equations 5.106 and 5.107.

$$\frac{\mathrm{d}\hat{r}}{\mathrm{d}t} = \hat{\dot{r}} \tag{5.111a}$$

$$\frac{\mathrm{d}\hat{\dot{r}}}{\mathrm{d}t} = \hat{r}\hat{\dot{\theta}}^2 - \frac{\mu}{\hat{r}^2} \tag{5.111b}$$

$$\frac{\mathrm{d}\hat{\theta}}{\mathrm{d}t} = \hat{\dot{\theta}} \tag{5.111c}$$

$$\frac{\mathrm{d}\hat{\dot{\theta}}}{\mathrm{d}t} = \frac{-2\hat{\dot{\theta}}\hat{\dot{r}}}{\hat{r}} \tag{5.111d}$$

To determine the error covariance time update we require the Jacobian

$$A(x,t) = \begin{bmatrix} 0 & 1 & 0 & 0 \\ \dot{\theta}^2 + 2\mu/r^3 & 0 & 0 & 2r\dot{\theta} \\ 0 & 0 & 0 & 1 \\ 2\dot{r}\dot{\theta}/r^2 & -2\dot{\theta}/r & 0 & -2\dot{r}/r \end{bmatrix} \tag{5.112}$$

In addition, the process noise matrix is

$$g(x,t) = \begin{bmatrix} 0 & 0 \\ 1 & 0 \\ 0 & 0 \\ 0 & 1/r \end{bmatrix} \tag{5.113}$$

which is a function of the state, so we must use (see the problems at the end of this chapter)

$$\begin{aligned} \dot{P} &= A(\hat{x},t)P + PA^T(\hat{x},t) + g(\hat{x},t)Qg^T(\hat{x},t) \\ &+ \frac{1}{2}\sum_{i,j=1}^{n} \left.\frac{\partial g}{\partial x_i}\right|_{x=\hat{x}} Q \left.\frac{\partial g^T}{\partial x_j}\right|_{x=\hat{x}} p_{ij}(t) \end{aligned} \tag{5.114}$$

where $x \in R^n$ and p_{ij} is the (i,j)th element of P. We could approximate by omitting the last term, but we shall include it for completeness since it presents no real complication. The only nonzero term in the double sum is

$$\frac{1}{2}\frac{\partial g}{\partial r}\bigg|_{x=\hat{x}} Q \frac{\partial g^T}{\partial r}\bigg|_{x=\hat{x}} p_{11}(t) = \frac{1}{2}\begin{bmatrix} 0 & 0 & 0 & 0 \\ 0 & 0 & 0 & 0 \\ 0 & 0 & 0 & 0 \\ 0 & 0 & 0 & p_{11}q_\theta/\hat{r}^4 \end{bmatrix} \tag{5.115}$$

By substituting Equations 5.112 and 5.115 in Equations 5.114 and simplifying, 10 scalar differential equations are obtained for the $n(n+1)/2$ distinct components of P. Defining auxiliary variables

$$a_1 \triangleq \hat{\dot{\theta}}^2 + \frac{2\mu}{\hat{r}^3} \tag{5.116a}$$

$$a_2 \triangleq \frac{2\hat{\dot{r}}\hat{\dot{\theta}}}{\hat{r}^2} \tag{5.116b}$$

$$a_3 \triangleq \frac{-2\hat{\dot{\theta}}}{\hat{r}} \tag{5.116c}$$

$$a_4 \triangleq -2\hat{r}\hat{\dot{\theta}} \tag{5.116d}$$

$$a_5 \triangleq \frac{-2\hat{\dot{r}}}{\hat{r}} \tag{5.116e}$$

we have

$$\dot{p}_{11} = 2p_{12} \tag{5.117a}$$

$$\dot{p}_{12} = a_1 p_{11} + p_{22} + a_4 p_{14} \tag{5.117b}$$

$$\dot{p}_{13} = p_{14} + p_{23} \tag{5.117c}$$

$$\dot{p}_{14} = a_2 p_{11} + a_3 p_{12} + a_5 p_{14} + p_{24} \tag{5.117d}$$

$$\dot{p}_{22} = 2a_1 p_{12} + 2a_4 p_{24} + q_r \tag{5.117e}$$

$$\dot{p}_{23} = a_1 p_{13} + a_4 p_{34} + p_{24} \tag{5.117f}$$

$$\dot{p}_{24} = a_2 p_{12} + a_3 p_{22} + a_5 p_{24} + a_1 p_{14} + a_4 p_{44} \tag{5.117g}$$

$$\dot{p}_{33} = 2p_{34} \tag{5.117h}$$

$$\dot{p}_{34} = a_2 p_{13} + a_3 p_{23} + a_5 p_{34} + p_{44} \tag{5.117i}$$

$$\dot{p}_{44} = 2a_2 p_{14} + 2a_3 p_{24} + 2a_5 p_{44} + \frac{q_\theta}{\hat{r}^2} + \frac{p_{11}q_\theta}{2\hat{r}^4} \tag{5.117j}$$

The time update equations 5.111, 5.116, and 5.117 can easily be programmed. It is hoped that we are making the point that, even for fairly complicated nonlinear systems, some preliminary analysis makes the EKF very straightforward to implement!

d. Measurement update

We have deliberately made no mention yet of measurements. The time update is dependent only on the satellite dynamics, and whatever measurements are used, the subroutine time update routine does not change.

Several measurement schemes can be used, some of which combine measurements by both ground tracking stations and instruments on board the satellite. We shall discuss a simple scheme in which range r and range rate $\dot{r}$ are measured by a station on the Earth's surface. Thus, the data are

$$z_k = \begin{bmatrix} 1 & 0 & 0 & 0 \\ 0 & 1 & 0 & 0 \end{bmatrix} x_k + \begin{bmatrix} v_1(k) \\ v_2(k) \end{bmatrix} \tag{5.118}$$

with $v_1(k) \sim (0, \sigma_r^2)$ and $v_2(k) \sim (0, \sigma_{\dot{r}}^2)$. Suppose data are taken at times t_k, so that $x_k \triangleq x(t_k)$. Suppose also that r and $\dot{r}$ are determined by independent means (e.g., radar ranges and doppler range rates) so that $v_1(k)$ and $v_2(k)$ are uncorrelated.

The measurements are linear, so the EKF measurement update in Table 5.1 reduces to the regular discrete Kalman filter update. This means that the stabilized measurement update software in Bierman (1977), for example, can be used. Since

$$R = \begin{bmatrix} \sigma_r^2 & 0 \\ 0 & \sigma_{\dot{r}}^2 \end{bmatrix} \tag{5.119}$$

is diagonal, scalar updates can be performed with no prewhitening of the data. Thus, at each time t_k we would perform a two-step update to include separately the two independent measurements

$$z_1(k) = \begin{bmatrix} 1 & 0 & 0 & 0 \end{bmatrix} x_k + v_1(k) \tag{5.120}$$

and

$$z_2(k) = \begin{bmatrix} 0 & 1 & 0 & 0 \end{bmatrix} x_k + v_2(k) \tag{5.121}$$

e. Discussion

The sample times t_k need not be uniform. We propagate through time by integrating the differential equations until data are available, at which point we call a measurement update routine to incorporate the data. If, at a given t_k, only one of the measurements (Equations 5.120 and 5.121) is available, we can include it individually with a call only to the appropriate scalar measurement update routine.

These examples clearly demonstrate the power and convenience of the EKF. It is a scheme that can be tailored to each application, and its modular structure can be exploited by using modular programming techniques that make it easy to modify the measurement scheme with a minimum of software redevelopment. It does not depend on a uniform data arrival rate. □

5.4 Application to Adaptive Sampling

In this section, we present an in-depth discussion of the application of EKF to localization of mobile robotic platforms while concurrently sampling an unknown physical variable of interest.

The use of robots in environmental monitoring applications requires distributed sensor systems optimized for effective estimation of relevant models. Here, we consider mobile sensing nodes (robots) characterized by sensing or measurement noise and localization uncertainty, and navigation in a region with a parameterized field model. We discuss a problem of great importance to the robotic deployment of such nodes, namely, adaptive sampling by repositioning to estimate a parametric distributed variable field models in the presence of localization uncertainty.

5.4.1 Mobile Robot Localization in Sampling

Recently, there has been renewed interest in using mobile robots as sensor-carrying platforms to perform hazardous tasks, such as searching for harmful biological and chemical agents, search and rescue in disaster areas, or environmental mapping and monitoring. (See, for instance, the work of Mataric *et al.* 2002, Kumar *et al.* 2003, and Popa *et al.* 2004).

Developing robust deployment algorithms for mobile sensor units requires simultaneous consideration of several optimization problems that have traditionally addressed separately. One problem is related to the quality and usefulness of the collected sensor information (e.g., choosing optimal locations in space where environmental samples are taken to yield the maximum information), another is related to the robot team localization and navigation (e.g., how do we achieve accurate sampling location).

The EKF has been used extensively as a localization tool for mobile robots, including the so-called simultaneous localization and mapping algorithms (SLAM) (see, for instance, Fenwick *et al.* 2002). In this context, the EKF has been used as a means to accomplish sensor fusion from different inertial sensors, global positioning units, and other sensor measurements.

While multiple vehicle localization and sensor fusion are classic problems in robotics, the problem of distributed field variable estimation is typically relevant to charting and prediction in oceanography and meteorology. The typical inverse modeling approach involves finding a partial differential equation and corresponding boundary conditions to express the evolution of the field variables in the 4D position–time space, based on a set of observation samples. In this context, measurement uncertainty has also been addressed using EKF estimation (Bennet 2002).

5.4.2 The Combined Adaptive Sampling Problem

We now describe the problem of adaptive sampling, e.g., the problem of reconstructing a parameterized field model from observations taken using mobile robots. In this section, we assume that *a parametric form of a measured field* is known, as might be the case, e.g., with a bottom profile or systematic variations in temperature or salinity. As an example, consider navigation in a shallow region (lake or coastal region) where smooth changes in depth provide a linear field model measured as altitude by the vehicle. We will use the EKF to integrate the field estimation with localization to improve these estimates.

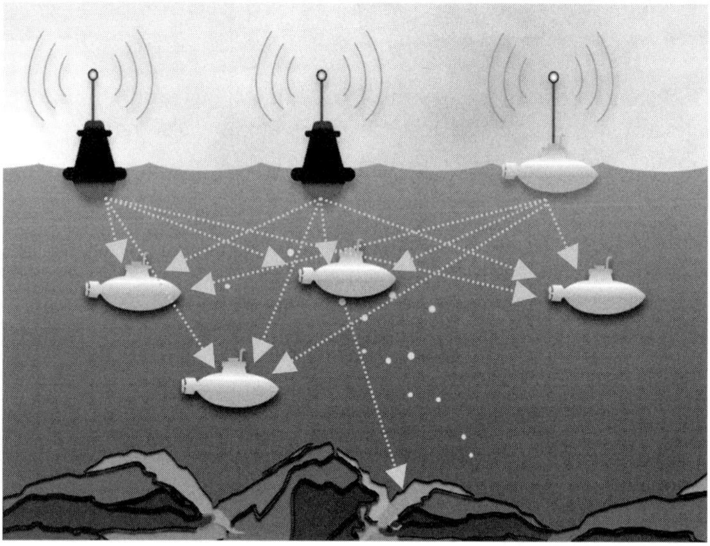

FIGURE 5.5
Cooperative mapping, localization, and sampling of a field variable using multiple autonomous vehicles.

Given uncertainty models for the field measurements and the location where the samples are taken, determining optimal sampling points can be based on minimizing information measures. An example of such measures is the state covariance estimates of the Kalman filter. In the case of simple distributions, such as linear ones, we can also compute a closed-form solution of the information gain without the need of a Kalman filter. The sampling objective is to determine the unknown coefficients of the hyperplane that describes the field variable, given a known localization uncertainty described by the robot dynamic model.

We are coordinating N autonomous vehicles (AVs) to sample a space–time field distribution, subject to a set of constraints (Figure 5.5). Localization uncertainties are relevant here since position estimates of robots are often inaccurate due to navigational errors, especially in GPS-denied environments. We can reduce localization uncertainty by building accurate models of distributed fields and vice versa. For instance, if a robot is sampling an unknown field, but its location is accurately known, a distributed parameter field model can be constructed by taking repeated field samples. Later on, this field model can be used to reduce the localization error. We pose the adaptive sampling (AS) problem both qualitatively and qualitatively as follows (see Popa *et al.* 2004 and Popa *et al.* 2006):

> *Given a parametric field variable $Z = Z(A, X)$ depending on an unknown parameter vector A and position vector X, and a robotic vehicle navigating the field with localization uncertainty, what*

sampling locations $X_{i,1\leq i\leq n}$ should be chosen such that the uncertainty in estimating the unknown parameter vector is minimized?

Let $X_i[k]$ denote the 3D position of the ith AV at sample number k. The ith vehicle motion is typically governed by kinematics that is nonlinear or nonholonomic, and describes the state evolution as

$$X_i[k+1] = X_i[k] + h(X_i[k], u_i[k]) + Gw_i[k] \tag{5.122}$$

where u_i is the control input to the vehicle and w_i the process noise. At each sample point, the ith vehicle measurement model is written as

$$g(X_i[k], A[k], Y_i[k] + \nu_i[k]) = 0 \tag{5.123}$$

where g is a nonlinear function of the vehicle position, $X_i[k]$, measurements of a noisy field variable $Y_i[k]$, with a white noise component with zero mean, and covariance matrix Λ_i. The vector A is a set of known or unknown coefficients describing the field variable dependence with the sample location. If the set of coefficients is unknown but constant, the vector A will be added to the overall system state, and its evolution will be governed by $A[k+1] = A[k]$. The simultaneous sampling and navigation estimation problem reduces to estimating the overall state vector $X[k] = (X_1[k]; \ldots X_N[k], A)$, along with its covariance matrix $P[k]$.

The adaptive sampling algorithm will then seek to sample at a new location such as to minimize the variance of the field parameter and the RV location estimates. In the next sections, we describe how we can represent uncertainty to guide the adaptive sampling process. A solution to determine the covariance of the overall state uncertainty is available using the EKF. The effectiveness of EKF algorithm is also compared to direct calculation of the coefficient variance through closed-form formulas.

5.4.3 Closed-Form Estimation for a Linear Field without Localization Uncertainty

For the AS problem with a single vehicle, let us first assume that the position of the robot is accurately known, and that we are estimating a field distribution that is also known to be linear in its parameters. For instance, the field can represent dissolved oxygen in a body of water that varies linearly with the location of the sample point in the (x, y) direction, and quadratically with the sample point depth (z). Or it can represent a general field distribution that has been approximated through a finite number of coefficients using a polynomial or a radial basis function expansion. The linear-in-parameters form allows us to compute a closed-form solution for the information measure used by the sampling algorithm. Since there is no uncertainty in the vehicle localization, the unknown field coefficient covariance can be calculated directly using a simple least-square estimation. After n measurements taken at location x_i, the field model depends linearly on the coefficients a_j via position-dependent functions, and we can directly estimate the unknown coefficients from the least-square solution:

$$
\begin{aligned}
Z_1 &= a_o + a_1 g_1(x_1) + \cdots + a_m g_m(x_1)\\
Z_2 &= a_o + a_1 g_1(x_2) + \cdots + a_m g_m(x_2)\\
&\vdots\\
Z_n &= a_o + a_1 g_1(x_n) + \cdots + a_m g_m(x_n),
\end{aligned}
$$

$$\hat{A}_n = \begin{pmatrix} 1 & g_i(x_j) \end{pmatrix}^{+}_{i \le m, j \le n} \begin{pmatrix} x_1 \\ \vdots \\ x_n \end{pmatrix} = M_n^{+} \begin{pmatrix} Z_1 \\ \vdots \\ Z_n \end{pmatrix} \tag{5.124}$$

In Equation 5.124, $g_i(x)$ are nonlinear basis functions that depend on the sample location x, while the field variables Z_i represents the field model at location x_i. Because the pseudoinverse of the matrix $M_n = (1 \; g_i(x_j))_{i \le m, j \le n}$ can be directly calculated from $M_n^+ = (M_n^T M_n)^{-1} M_n^T$, we obtain a closed-form solution for the covariance of the unknown parameters a_j:

$$\hat{A}_n = \left(\sum_{j=1}^{n} (1 \ldots g_i(x_j) \ldots g_m(x_j)) \begin{pmatrix} 1 \\ \vdots \\ g_i(x_j) \\ \vdots \\ g_m(x_j) \end{pmatrix} \right)^{-1} \sum_{j=1}^{n} Z_j \begin{pmatrix} 1 \\ \vdots \\ g_i(x_j) \\ \vdots \\ g_m(x_j) \end{pmatrix} \tag{5.125}$$

The covariance matrix of $\hat{A}_n$ can now be related directly to the (constant) measurement uncertainty as $\mathrm{var}(\hat{A}_n) = \mathrm{var}(z_i)(M_n^T M_n)^{-1}$, and the adaptive sampling algorithm will move the vehicle from location x_n to x_{n+1}, such that the following p-norm is maximized over the search space Θ:

$$m(x) = \left\| \left(M_n^T \begin{pmatrix} 1 \\ \vdots \\ g_i(x) \\ \vdots \\ g_m(x) \end{pmatrix} \right) \begin{pmatrix} & M_n & \\ 1 & \ldots g_i(x) \ldots & g_m(x) \end{pmatrix} \right\|_p, \; m(x_{n+1}) \ge m(x), (\forall) x \in \Theta \tag{5.126}$$

Example 5.5 Sampling of a 1D and 2D Linear Fields

This example is similar to the problem of estimating the slope and intercept of a line fit from Exercise 2.10. Specifically, we have $m = 1$ and $g_i(x) = x$, for the 1D case and $m = 2$ and $g_i(x) = (x, y)$ for the 2D case. We consider two uncertainty norms, $p = 2$ and $p = \infty$.

We simulate sets of Monte Carlo measurement runs for the sampling problem in both 1D and 2D. In each case, a nominal linear model was imposed, and measurement samples were generated using Gaussian random numbers with a set covariance of 10% of the nominal field value. The averaged results over 1000-field-sample instantiations were compared to one another using the 2-norm, the infinity norm, and the default raster scan (e.g., sequential scanning). To determine the next best sampling location, we use the closed-form parameter variance formulas. Figures 5.6 and 5.7, compare the results of simulation sets for the 1D and 2D cases, respectively. While the resulting sampling sequences are different for each sampling set, the value of the estimation residuals are similar in the adaptive sampling cases, and much smaller than the estimate residuals obtained in the default raster-scan sampling sequence. As clearly seen in Figure 5.8, the field prediction error (e.g., the difference between actual value and the estimated one in the unsampled points) decreases much faster if we sample adaptively. ▯

Example 5.6 Sampling of a 2D Gaussian Field

For a 2D spatially distributed field variable, a common basis is the RBF distribution. If the Gaussian centers and variance are known, then the unknown RBF coefficients a_j can be estimated using Equations 5.124–5.126. If, however, the centers and variances of the Gaussians are unknown, then the field to be estimated becomes nonlinear-in-parameters. For instance, for $m=2$, an example of such a field is given by the measurement equation:

$$z = a_0 + a_1 e^{-\frac{(x-x_1)^2+(y-y_1)^2}{2\sigma_1^2}} + a_2 e^{-\frac{(x-x_2)^2+(y-y_2)^2}{2\sigma_2^2}} \tag{5.127}$$

and we are estimating $a_0, a_1, a_2, x_1, y_1, x_2, y_2, \sigma_1, \sigma_2$. Instead of the linear least-square solution (Equation 5.126), we can a use a numerical nonlinear-least-square solution, but it is difficult to obtain a closed-form measure of uncertainty for the whole parameter vector.

First, on a 2D 100×100 sampling grid we assume that the centers and covariances of two Gaussians $g_1(x,y), g_2(x,y)$ are known, and they are located at $(30,30), (65,45), \sigma_1 = \sigma_2 = 10$. The resulting sampling sequences generated are shown in Figures 5.9(a) through 5.9(d).

The second sampling sequence is obtained for a Gaussian field assuming that the location of the centers are not known, but the variance is known. In this case, we use a numerical constrained optimization solver from MATLAB (fmincon) to solve the variance minimization problem defining the next sampling step. Assuming nominal values of $a_0=1, a_1=4, a_2=-5.5, (x_1,y_1)=(30,30), (x_2,y_2)=(35,65)$, we first divide the Gaussian center's search space into a $N \times N$ grid. We constrain our nonlinear optimization routine to searching inside all grid pairs where the centers are located (to avoid local minima). The optimal location of the grid center moves according to the graphs shown in Figure 5.10 if $N=2$. The figure also shows the convergence of the three linear field coefficients to their actual values. ▯

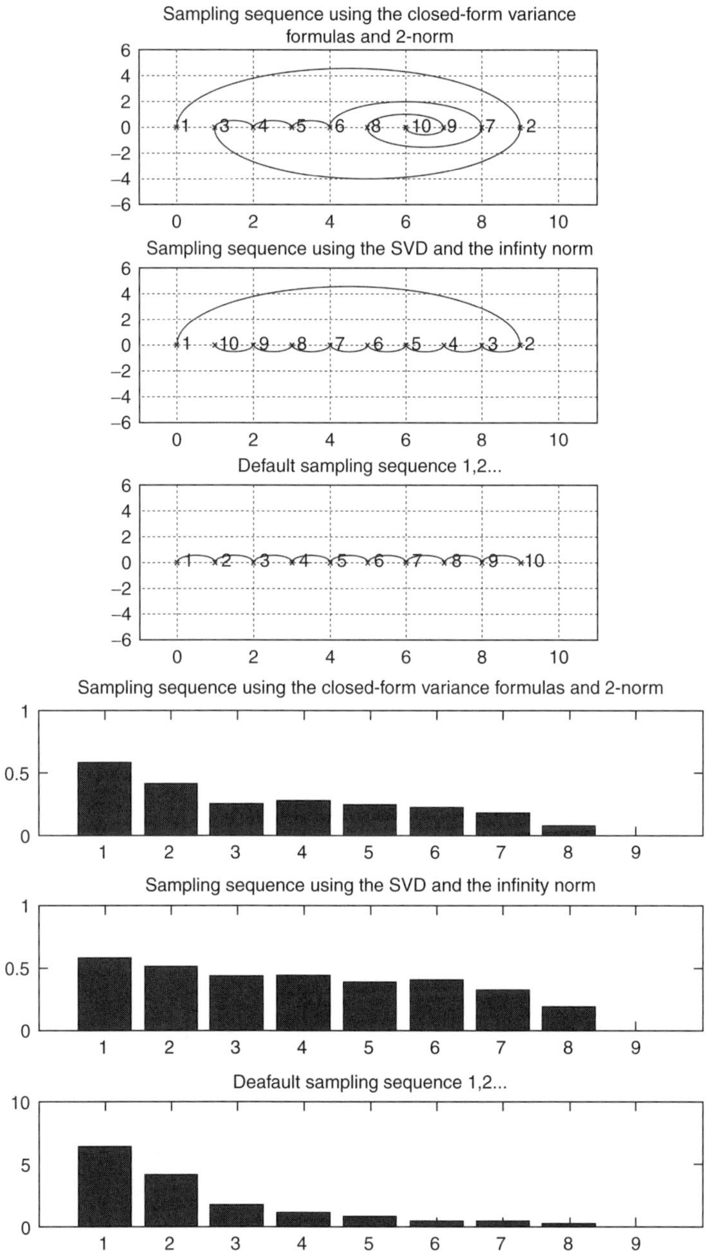

FIGURE 5.6

1D problem: different sampling paths using information measures and default raster scanning (top), and corresponding parameter residuals for different sampling paths using information measures and the default raster scan (bottom).

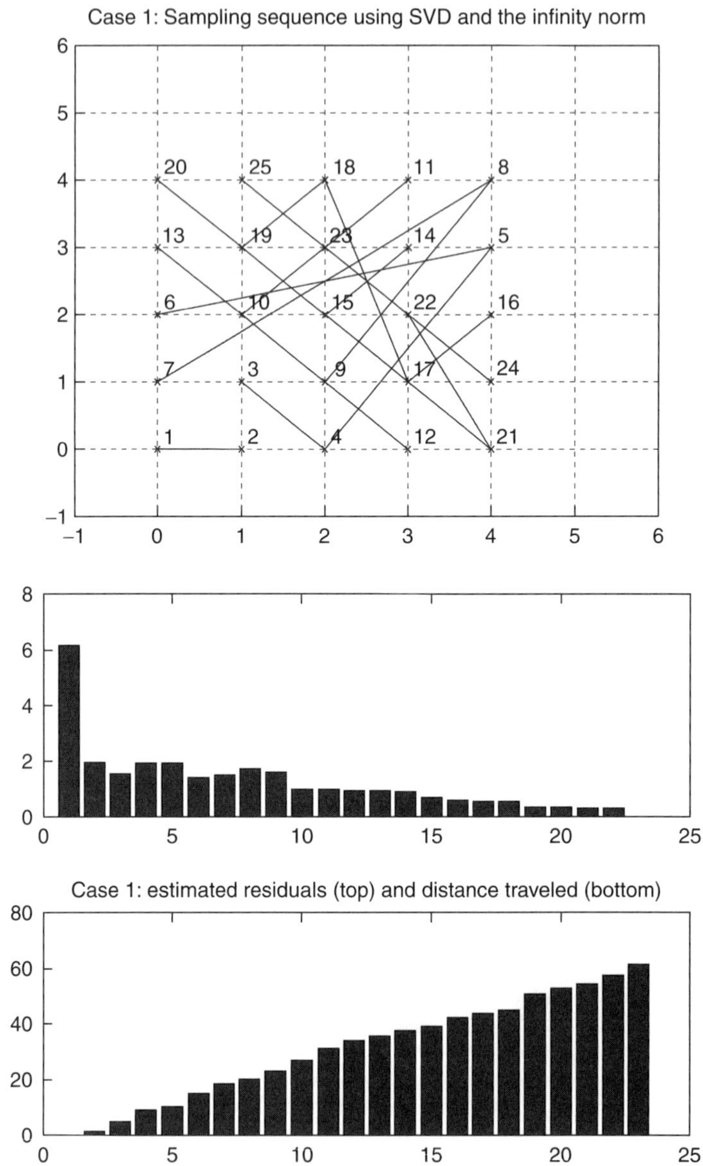

FIGURE 5.7
2D problem: adaptive sampling path using $p=2$ (top) and corresponding residuals and distance traveled plots (bottom).

5.4.4 Closed-Form Estimation for a Linear Field with Localization Uncertainty

Previously, we used a numerical "closed-form" calculation of the parameter variance, in the absence of location state uncertainty. In the linear case, the

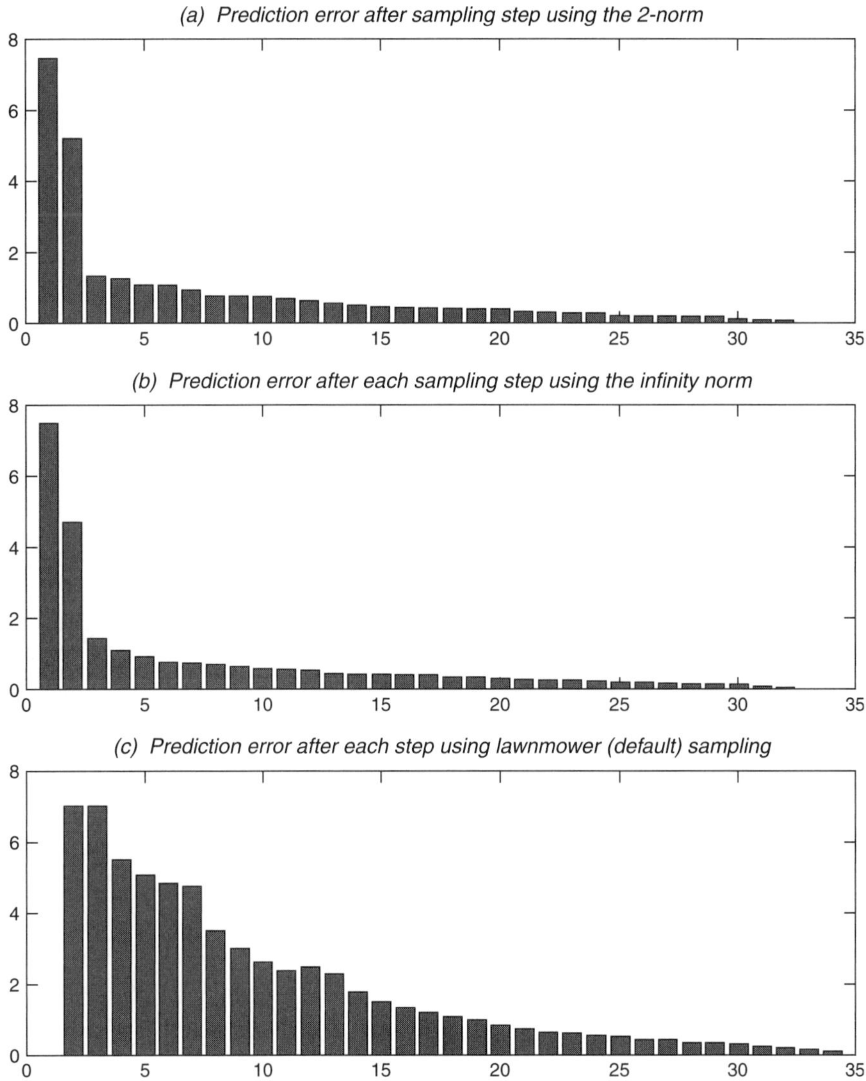

FIGURE 5.8
2D problem: comparison of prediction error residuals for 2 AS algorithms and the default raster scan. Clearly, sampling to maximize information gain is much better than a default lawn mower or raster-scan sampling sequence.

parameter variance calculation relies on calculating the pseudoinverse of a matrix whose entries are assumed to be known. Example 5.6 shows that if the field is nonlinear, the AS algorithm requires numerical optimizations that may suffer from convergence difficulties. In addition, the entries of the optimization function depend on the position where measurements are taken, and in the context of mobile robots, this location is not known exactly.

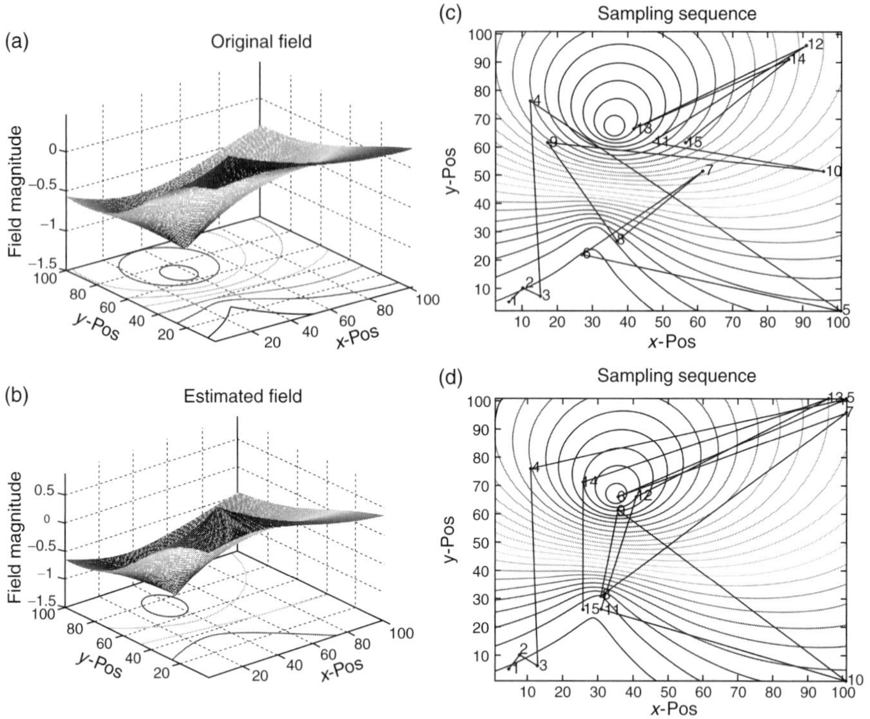

FIGURE 5.9
Sampling sequence for a Gaussian field distribution (original (a), estimated (b)) is generated to minimize error variance using the infinity norm (c), and the 2-norm (d). First 15 sampling locations are shown.

Let us consider a very simple one-dimensional sampling example, to which we add localization uncertainty. We set $m=2$ and $g(x)=x$ to obtain an algebraic closed-form solution of the parameter variance. If the set of measurements is given by

$$\begin{aligned} z_1 &= a_1 + a_2 x_1 \\ z_2 &= a_1 + a_2 x_2 \\ &\vdots \\ z_n &= a_1 + a_2 x_n \end{aligned} \tag{5.128}$$

$$\begin{bmatrix} z_1 \\ z_2 \\ \vdots \\ z_n \end{bmatrix} = \begin{bmatrix} 1 & x_1 \\ 1 & x_2 \\ & \vdots \\ 1 & x_n \end{bmatrix} \begin{bmatrix} a_1 \\ a_2 \end{bmatrix} \tag{5.129}$$

$$Z_n = M_n A$$

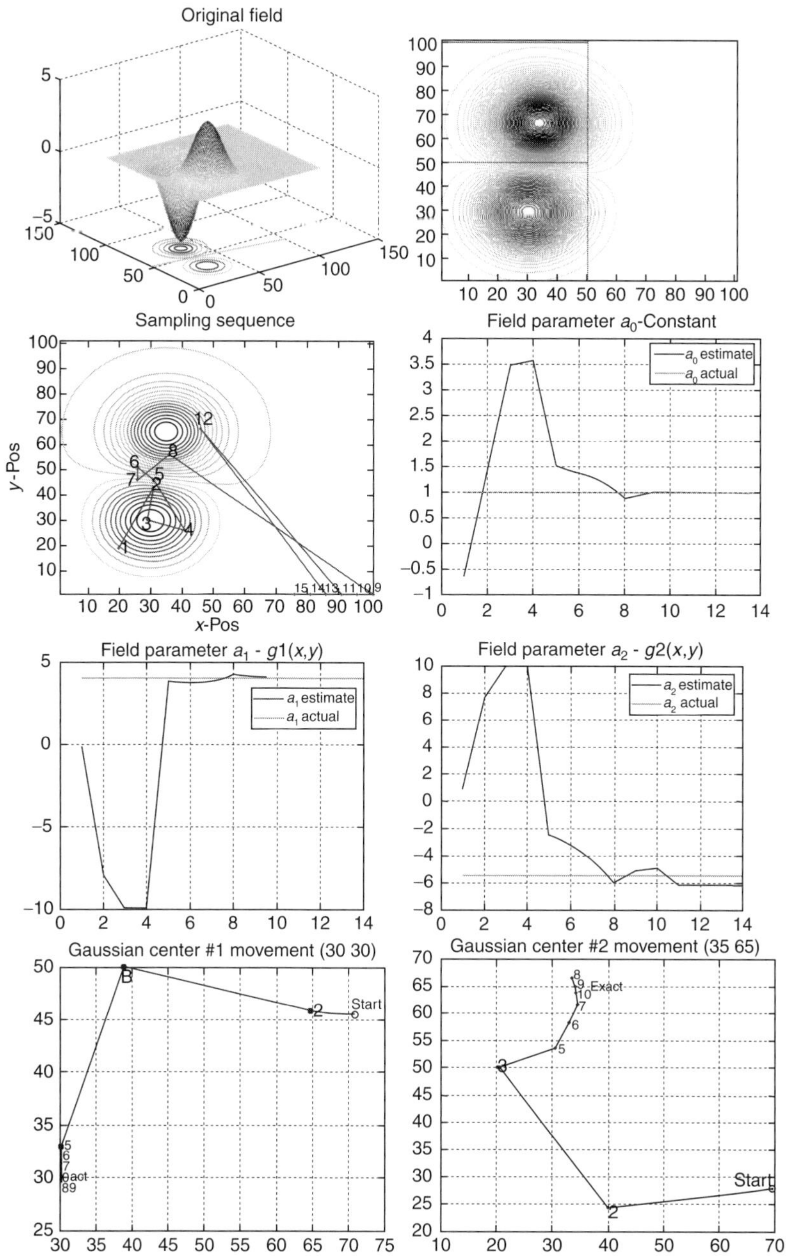

FIGURE 5.10
Sampling sequence for a Gaussian field distribution where the Gaussian centers are unknown, generated using MATLAB's fmincon and the 2-norm of the linear field parameters. Convergence of all parameters, including the Gaussian centers to their true values is observed, but one must use a branch-and-bound search routine on a 2×2 grid for initial search conditions.

in which Z_n is the measurement vector after n samples, M_n the basis function matrix, and A the vector of unknown parameters. The least-square estimate of A after n measurements will be

$$\hat{A} = M_n^+ Z_n = (M_n^T M_n)^{-1} M_n^T Z_n \tag{5.130}$$

$$\hat{A} = \left\{ \begin{bmatrix} 1 & 1 & \dots & 1 \\ x_1 & x_2 & \dots & x_n \end{bmatrix} \begin{bmatrix} 1 & x_1 \\ 1 & x_2 \\ \vdots & \\ 1 & x_n \end{bmatrix} \right\}^{-1} \begin{bmatrix} 1 & 1 & \dots & 1 \\ x_1 & x_2 & \dots & x_n \end{bmatrix} \begin{bmatrix} z_1 \\ z_2 \\ \vdots \\ z_n \end{bmatrix} \tag{5.131}$$

$$\begin{aligned} \hat{A} &= \frac{1}{n\sum_{i=1}^{n} x_i^2 - \left(\sum_{i=1}^{n} x_i\right)^2} \begin{bmatrix} \sum_{i=1}^{n} z_i \sum_{i=1}^{n} x_i^2 - \left(\sum_{i=1}^{n} x_i z_i\right) \sum_{i=1}^{n} x_i \\ -\sum_{i=1}^{n} z_i \sum_{i=1}^{n} x_i + n\left(\sum_{i=1}^{n} x_i z_i\right) \end{bmatrix} \\ &= \frac{1}{f_0(x)} \begin{bmatrix} f_1(x,z) \\ f_2(x,z) \end{bmatrix} = \begin{bmatrix} F_1(x,z) \\ F_2(x,z) \end{bmatrix} \end{aligned} \tag{5.132}$$

The variance of $\hat{A}$ can now be calculated based on the variance of (x, z) through Jacobians:

$$\text{var}(\hat{A}\,) = \left(\frac{\partial \hat{A}}{\partial(x,z)}\right) \text{var}(x,z) \left(\frac{\partial \hat{A}}{\partial(x,z)}\right)^T \tag{5.133}$$

where the Jacobian is given by

$$\frac{\partial \hat{A}}{\partial(x,z)} = \begin{bmatrix} \dfrac{\partial F_1}{\partial x_i} & \dfrac{\partial F_1}{\partial z_i} \\ \dfrac{\partial F_2}{\partial x_i} & \dfrac{\partial F_2}{\partial z_i} \end{bmatrix} \tag{5.134}$$

and

$$\begin{aligned} \frac{\partial F_1}{\partial x_j} &= \frac{f_0 \frac{\partial f_1}{\partial x_j} - f_1 \frac{\partial f_0}{\partial x_j}}{f_0^2}, \quad \frac{\partial F_2}{\partial x_j} = \frac{f_0 \frac{\partial f_2}{\partial x_j} - f_2 \frac{\partial f_0}{\partial x_j}}{f_0^2} \\ \frac{\partial f_1}{\partial x_j} &= \left(\sum_{i=1}^{n} z_i\right) 2x_j - \left(\sum_{i=1}^{n} x_i z_i\right) - \left(\sum_{i=1}^{n} x_i\right) z_j \\ \frac{\partial f_0}{\partial x_j} &= 2n x_j - 2\sum_{i=1}^{n} x_i, \quad \frac{\partial f_2}{\partial x_j} = -\left(\sum_{i=1}^{n} z_i\right) + n z_j \end{aligned} \tag{5.135}$$

Similarly,

$$\begin{aligned} \frac{\partial F_1}{\partial z_j} &= \frac{1}{f_0} \frac{\partial f_1}{\partial z_j} = \frac{1}{f_0} \left\{ \left(\sum_{i=1}^{n} x_i^2\right) - \left(\sum_{i=1}^{n} x_i\right) x_j \right\} \\ \frac{\partial F_2}{\partial z_j} &= \frac{1}{f_0} \frac{\partial f_2}{\partial z_j} = \frac{1}{f_0} \left\{ -\left(\sum_{i=1}^{n} x_i\right) + n x_j \right\} \end{aligned} \tag{5.136}$$

Even though we obtained closed-form algebraic solutions for the unknown parameter covariance, we can see that introducing localization uncertainty to our sampling problem makes it hard to obtain closed-form solutions for a large number of parameters or for nonlinear fields. Using closed-form variance solutions to determine optimal sampling locations is only feasible if no localization uncertainty is considered.

5.4.5 Adaptive Sampling Using the Extended Kalman Filter

Owing to difficulties in estimating the combined parametric and localization uncertainty using closed-form solutions, we employ an iterative estimation approach to select optimal sampling points. Let x_k denote the 3D position of one of the mobile robots at sample number k. A model for the mobile node is usually nonlinear, and describes the state evolution as

$$x_{k+1} = x_k + f(x_k, u_k) + w_k \tag{5.137}$$

where u_k is the control input to the vehicle and w_k the state measurement noise, assumed to be white, with zero mean, and covariance matrix Q_k^1. Additional navigation measurements are then given by

$$y_k = h(x_k) + \xi_k \tag{5.138}$$

The field measurements depend on the state as well as a set of parameters which are unknown:

$$z_k = g(x_k, a_k) + \zeta_k \tag{5.139}$$

If the set of unknown coefficient is constant or slow time-varying, we add the vector a to the overall system state, with an evolution governed by $a_{k+1} = a_k + v_k$. The noise covariances are given by $E[\xi_k \xi_k^T] = R_k^1$, $E[\varsigma_k \varsigma_k^T] = R_k^2$, and $E[\nu_k \nu_k^T] = Q_k^2$.

A simultaneous sampling and navigation estimation problem formulated using the EKF, using an overall state vector $X_k = (x_k, a_k)$, a measurement vector $Z_k = (y_k, z_k)$, using Jacobians $F_X(\cdot) \equiv \frac{\partial F}{\partial X}(.)$, $H_X(.) \equiv \frac{\partial H}{\partial X}(.)$ is as follows:

System equations:

$$X_{k+1} = F(X_k, u_k) + p_k, \qquad Z_k = H(X_k) + n_k \tag{5.140}$$

$$E[p_k p_k^T] = \begin{pmatrix} Q_k^1 & 0 \\ 0 & Q_k^2 \end{pmatrix} = Q_k, \qquad E[n_k n_k^T] = \begin{pmatrix} R_k^1 & 0 \\ 0 & R_k^2 \end{pmatrix} = R_k \tag{5.141}$$

Time update:

$$\hat{X}_{k+1}^- = F(\hat{X}_k, u_k) \tag{5.142}$$

$$\hat{Z}_{k+1}^- = H(\hat{X}_{k+1}^-) \tag{5.143}$$

$$P_{k+1}^- = F_X(\hat{X}_k, u_k) P_k F_X^T(\hat{X}_k, u_k) + Q \tag{5.144}$$

Measurement update:

$$(P_{k+1})^{-1} = (P_{k+1}^{-})^{-1} + H_X^T(X_{k+1}^{-})R^{-1}H_X^T(X_{k+1}^{-}) \tag{5.145}$$

or

$$\begin{aligned} P_{k+1} = P_{k+1}^{-} - P_{k+1}^{-}H_X(X_{k+1}^{-})\big(H_X^T(X_{k+1}^{-})P_{k+1}^{-}H_X(X_{k+1}^{-}) + R\big)^{-1} \\ \times H_X^T(X_{k+1}^{-})P_{k+1}^{-} \end{aligned} \tag{5.146}$$

$$\hat{X}_{k+1} = \hat{X}_{k+1}^{-} + P_{k+1}H_X^T(\hat{X}_{k+1}^{-})R^{-1}(Z_{k+1} - \hat{Z}_{k+1}^{-}) \tag{5.147}$$

The EKF provides a recursive method for computing the state covariance which can be utilized as an information measure. The sampling algorithm will choose locations in space to sample next such that the covariance norm is minimized. Note that in the time and measurement updates, *no measurement* is needed to be taken for the covariance updates. The measurement Z_{k+1} is *only* needed for the estimate measurement update (Equation 5.147).

This means that Equations 5.144 and 5.146 are covariance updates available *before* the measurement is taken at time $k+1$. They can be used to predict the usefulness of the measurement at time $k+1$ before it is ever taken.

Also, they show the effect not only of the measurement Z_{k+1} in reducing the error covariance, they also show the cost of moving to the estimated position $\hat{X}_{k+1}^{-}$ needed to take the measurement Z_{k+1} in terms of uncertainty injected due to the motion uncertainty (e.g., covariance increase in Equation 5.144). This aspect may avoid long-distance movements where we are jumping to far corners of the map to take the next measurements, and has added benefits in that the EKF linear approximation becomes less accurate as we move long distances. To choose the next sampling location, we use the following algorithm:

EKF—Adaptive Sampling Algorithm:

Evaluate Equations 5.142 *through* 5.146 *over the search space* Θ *of possible next states and select the next measurement position that yields the smallest norm of* P_{k+1}. *Then move to that position and take the measurement* Z_{k+1} *and do Equation* 5.147.

Example 5.7 EKF Estimation for a Parameter-Linear Field

Instead of using the closed-form numerical or algebraic solutions from the previous sections, sampling measures can be obtained for parameter-linear fields by using the EKF. First, if there is no localization uncertainty, the state and output equations can be written as

$$\begin{aligned} A_{k+1} &= A_k \\ Z_k &= G_k A_k + \nu_k \end{aligned} \tag{5.148}$$

where

$$G_k = (1 \ldots g_i(X_k) \ldots g_m(X_k)), E\big[\nu_k \nu_k^T\big] = R \tag{5.149}$$

Because of the simple state update equation, the error covariance update can be reduced to

$$A_o \sim (\hat{A}_o, P_{A_o}), P_o = P_{A_0}, P_{k+1}^{-1} = P_k^{-1} + G_{k+1}^T R^{-1} G_{k+1}$$
$$\hat{A}_{k+1} = \hat{A}_k + P_{k+1} G_{k+1}^T R^{-1} (Z_{k+1} - G_{k+1} \hat{A}_k) \tag{5.150}$$

The error covariance is similar to the least-squares solution, and can be directly calculated by

$$P_k = \left(P_0^{-1} + \sum_{j=1}^{k} G_j^T R^{-1} G_j \right)^{-1} \tag{5.151}$$

Because the field has a linear in the parameter form, we don't have to use the Taylor series expansion in the EKF. If however, the parameterized field is nonlinear, such as in the case of RBF basis with unknown centers or variance, the recursive equations need to be set up using the Taylor series expansion of H.

A set of simulations was performed for the 1D case sampling case (e.g., a linear field with two unknowns—intercept and slope). We considered data sets generated under Gaussian noise assumptions, using nominal coefficient values $a_0 = 2$ and $a_1 = 0.5$ (intercept and slope), and measurement noise covariance of 0.5, state measurement noise covariance of 0.1, and state transition noise covariance of 0.1. The convergence of the coefficient estimates to their nominal values is shown in Figures 5.11a through 5.11d for two different sampling sequences. For Figures 5.11a and 5.11c, the sampling sequence is the default one $(1, 2, \ldots)$, while Figures 5.11b and 5.11d were obtained by minimizing the error covariance, e.g., the adaptive sampling algorithm moves the vehicle from location X_n to X_{n+1}, such that the following infinity norm is maximized over the search space Θ:

$$m(X) = \|P_n\|_\infty, m(X_{n+1}) \geq m(X), (\forall)X \in \Theta \tag{5.152}$$

Note that the slope estimate of the second sampling sequence convergences faster to its nominal value. The optimal sampling sequences are very similar to the ones obtained in Figure 5.6 using the closed-form parameter variance. □

5.4.6 Simultaneous Localization and Sampling Using a Mobile Robot

AS algorithms direct the robot to sample a spatiotemporally distributed field variable to estimate it. In the previous example, the localization of the robot was not explicitly included in the EKF equations. The advantage of using the EKF framework is its ability to combine robot navigation and sensor measurement uncertainty. Accounting for localization errors is important in environments where global positioning estimates are not available (e.g., GPS-denied environments), such as underwater or inside buildings.

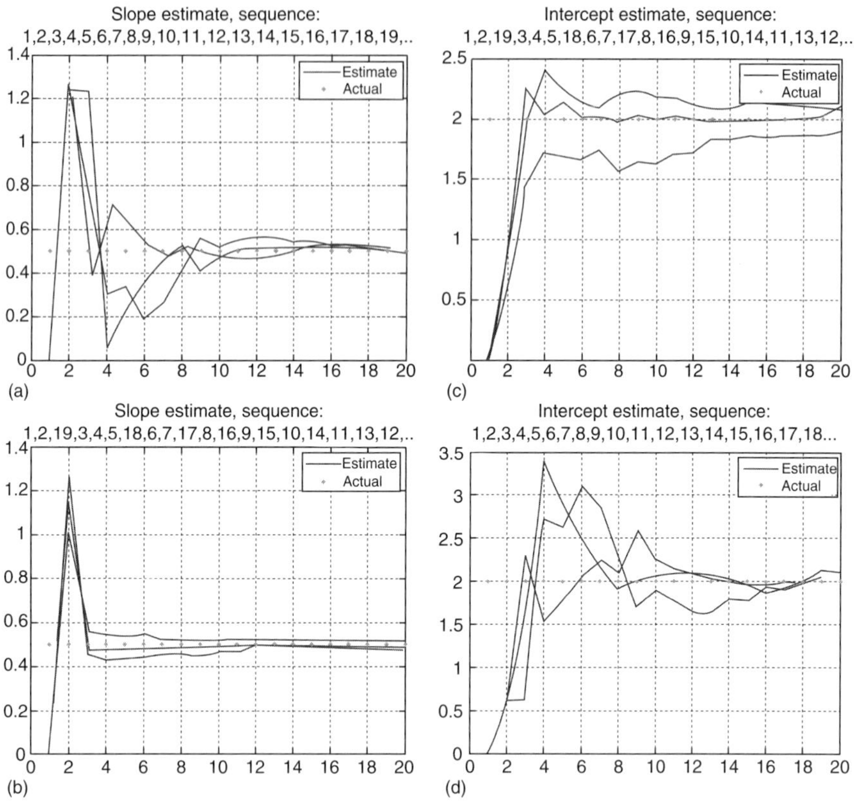

FIGURE 5.11
The convergence of slope and intercept in a field estimation sampling problem for the default (a,c) and optimal (b,d) 1D sampling sequence using the infinity norm. The field parameter error covariance decreases much faster for the adaptive sampling sequence.

Assuming that the field variable distribution is time stationary and linear in parameters, the combined nonlinear vehicle dynamics and sensor model with Gaussian noise assumptions can be represented as in Equations 5.137 through 5.139.

For a differential drive robot (such as the ARRI-Bot shown in Figure 5.12), localization uncertainties arise from model parameter errors and the use of dead-reckoning with inertial navigation sensors (e.g., wheel encoders).

$$\hat{X}_1(k) = \hat{X}_1(k-1) + \frac{K_{\text{drv}-R}\Delta\phi_R r_R + K_{\text{drv}-L}\Delta\phi_L r_L}{2}\cos(\hat{X}_3(k-1))$$

$$\hat{X}_2(k) = \hat{X}_2(k-1) + \frac{K_{\text{drv}-R}\Delta\phi_R r_R + K_{\text{drv}-L}\Delta\phi_L r_L}{2}\sin(\hat{X}_3(k-1))$$

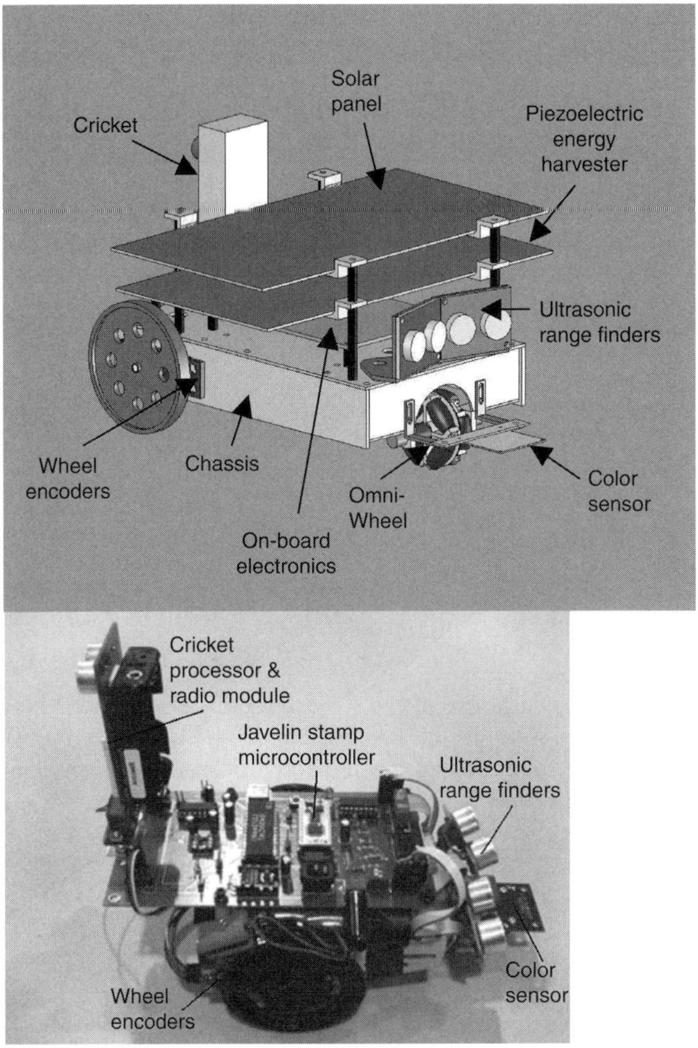

FIGURE 5.12
Schematic diagram of the differential-drive ARRI-bot V-2.

$$\hat{X}_3(k) = \hat{X}_3(k-1) + \frac{K_{\text{turn}-R}\Delta\phi_R r_R - K_{\text{turn}-L}\Delta\phi_L r_L}{L_b}$$
$$X = \begin{bmatrix}\hat{X}_1 & \hat{X}_2 & \hat{X}_3\end{bmatrix} = \begin{bmatrix}x & y & \theta\end{bmatrix} \tag{5.153}$$

which include nominal physical dimensions (lengths—L_b, wheel radii—r_l, r_R), the wheel encoder counts $\Delta\phi_L, \Delta\phi_R$, and the actuator drive constants $K_{\text{drv}}, K_{\text{turn}}$. The states of the robot are the position and orientation, $X = \begin{bmatrix}x & y & \theta\end{bmatrix}^T$.

In the case of an m-dimensional linear-in-parameters field distribution,

$$g(X_k, A_k) = a_{ko} + \sum_{i=1}^{m} g_i(X_k)a_{ki} = \begin{pmatrix}1 & g_1(X_k)^T\end{pmatrix} A_k \tag{5.154}$$

where (A_k) is the parameter vector and g_i a nonlinear basis for the field. The nonlinear EKF equations can be represented by

$$\begin{aligned}
\begin{pmatrix} X_{k+1} \\ A_{k+1} \end{pmatrix} &= \begin{pmatrix} X_k \\ A_k \end{pmatrix} + \begin{pmatrix} I_3 \\ 0 \end{pmatrix} U_k + \begin{pmatrix} \omega_k \\ 0 \end{pmatrix} = \begin{pmatrix} X_k \\ A_k \end{pmatrix} + B_k U_k + \vartheta_k \\
Z_k &= \begin{pmatrix}1 & g_1(X_k)^T\end{pmatrix} A_k + \nu_k \\
X_{k+1}^{DR} &= X_k^{DR} + h\left(X_k^{DR}, \bar{u}_k\right)
\end{aligned} \tag{5.155}$$

where,

$$E\left[\vartheta_k \vartheta_k^T\right] = Q = \begin{pmatrix} Q_1 & 0 \\ 0 & 0 \end{pmatrix}, \quad E\left[\lambda_k \lambda_k^T\right] = R \tag{5.156}$$

are the white noise covariances of the state and output, and X_k^{DR} is the dead-reckoning estimate for the robot position computed after a command vector $\bar{u}_k$ is applied to the robot to cause a predicted position change equivalent to U_k. The nonlinear Kalman filter update equations become

$$\begin{aligned}
P_{k+1}^- &= P_k + Q, \quad \begin{pmatrix} \hat{X}_{k+1}^- \\ \hat{A}_{k+1}^- \end{pmatrix} = \begin{pmatrix} \hat{X}_k \\ \hat{A}_k \end{pmatrix} + B_k\left(X_{k+1}^{DR} - X_k^{DR}\right) \\
H_k &= \frac{\partial g_k}{\partial(\hat{X}_k^-, \hat{A}_k^-)} = \left(\hat{A}_k^{-T} \begin{pmatrix} 0 \\ \frac{\partial g_1(\hat{X}_k^-)}{\partial \hat{X}_k^-} \end{pmatrix} \begin{pmatrix}1 & g_1(\hat{X}_k^-)^T\end{pmatrix} \right) \\
P_{k+1} &= \left((P_{k+1}^-)^{-1} + H_{k+1}^T R^{-1} H_{k+1}\right)^{-1} \\
\begin{pmatrix} \hat{X}_{k+1} \\ \hat{A}_{k+1} \end{pmatrix} &= \begin{pmatrix} \hat{X}_{k+1}^- \\ \hat{A}_{k+1}^- \end{pmatrix} + P_{k+1} H_{k+1}^T R^{-1} \left[Z_{k+1} - g_{k+1}(\hat{X}_k^-, \hat{A}_k^-)\right]
\end{aligned} \tag{5.157}$$

The dead-reckoning estimates are calculated on the vehicle and used to correct the vehicle position estimate at the end of the vehicle transition. Note that the command vector $\bar{u}_k$ is determined using a chosen path planner, which for the case of our mobile robot nodes is a simple holonomic planner. The AS algorithm selects best next-grid closest sampling point by searching over possible $\bar{U}_k$'s, such that the 2-norm of the covariance matrix P_{k+1} is minimized.

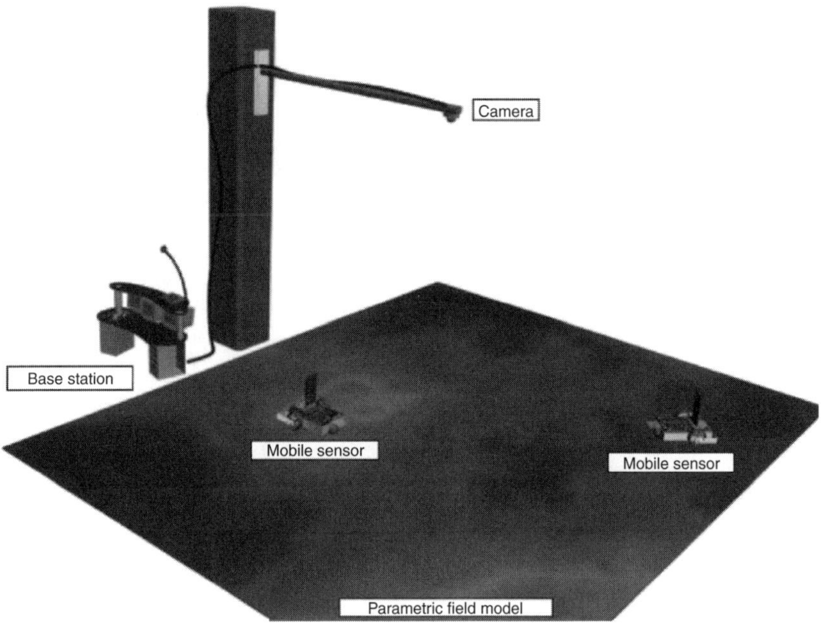

FIGURE 5.13
Illustration of adaptive sampling of a color field using ARRI-Bots.

Example 5.8 Sampling a Color Field with Mobile Robots

The ARRI-Bots equipped with a color sensor, are used to reconstruct a color field printed on the lab floor as shown in Figure 5.13. The field is distributed over a search space of $3.15\,\text{m} \times 2.25\,\text{m}$. A wide-angle overhead camera system gives the true robot location and is used to compare it with the estimated robot location. The camera at a height of 10.5 m encompasses the entire area in its field of view. Three separate linear field models, one for each primary color, are used to generate a linear RGB color field with nine parameters:

$$R = r_0 + r_1 x + r_2 y, \quad G = g_0 + g_1 x + g_2 y, \quad B = b_0 + b_1 x + b_2 y \tag{5.158}$$

Nominal values for the field shown in Figure 5.13 are

$$\begin{aligned} &r_0 = 0.2307, \quad r_1 = 0.0012, \quad r_2 = -0.00048, \quad g_0 = 0, \quad g_1 = 0.0002 \\ &g_2 = 0.0018, \quad b_0 = 1.0, \quad b_1 = -0.00078, \quad b_2 = -0.001 \end{aligned} \tag{5.159}$$

Based on this field, we test the effectiveness of the sampling algorithm by using the color sensor measurement for field estimation as well as localization. Figures 5.14 and 5.15 show the resulting field distribution assuming ever-increasing values of color sensor measurement uncertainty in the EKF model. The sensor measurement uncertainty is increased from 0.01 to 0.45, and the norm of error in field parameter estimates is compared for different

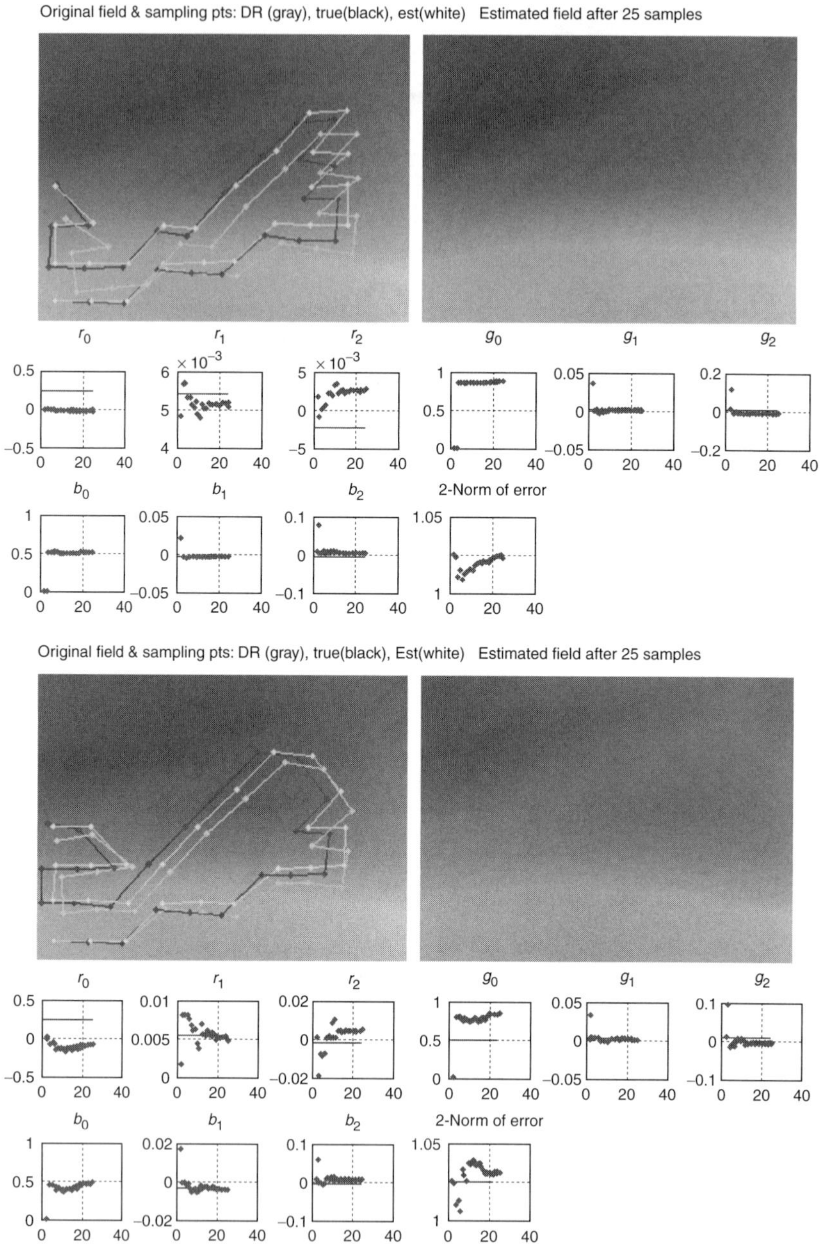

FIGURE 5.14
Uncertainty of (0.01, 0.01, 0.01) (top) and (0.1, 0.1, 0.1) (bottom) in color sensor measurement: Field estimate converges and localization error is very small (after 25 samples).

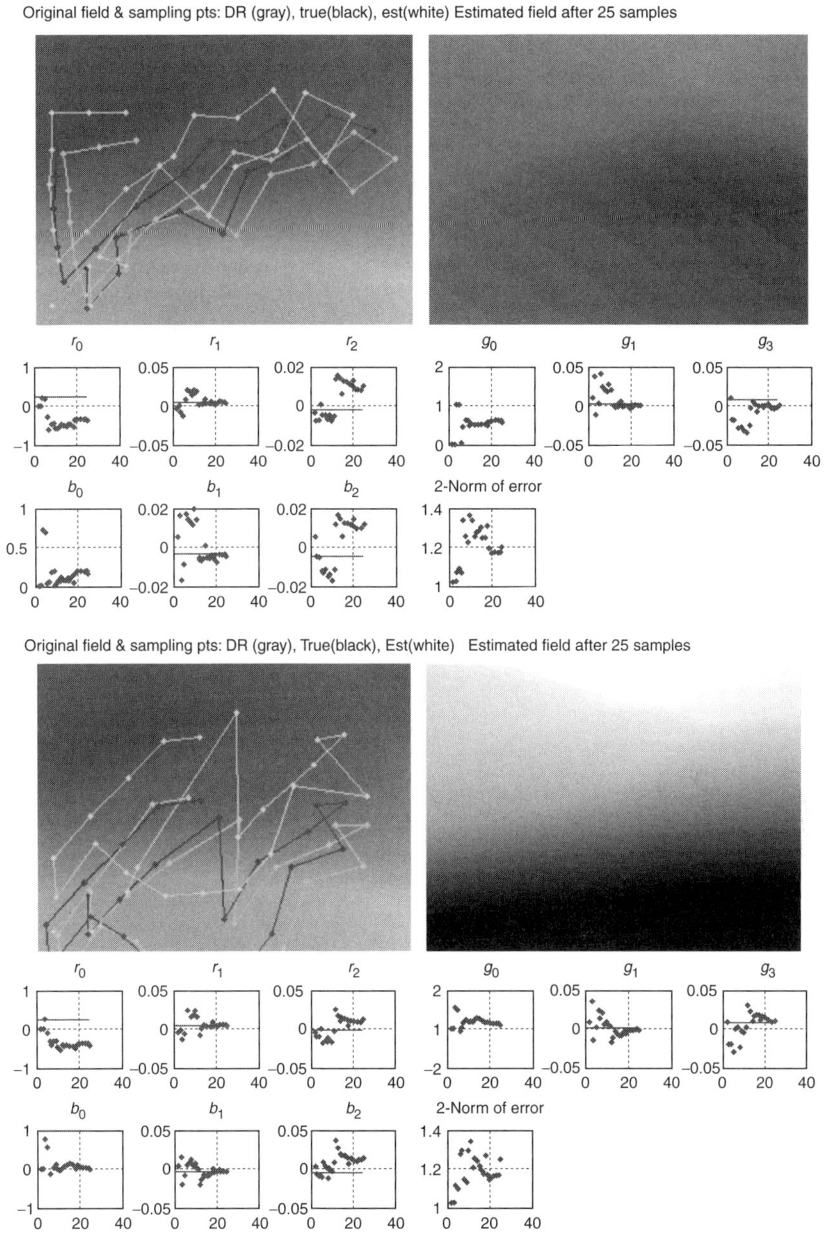

FIGURE 5.15
Uncertainty of (0.35, 0.35, 0.35) (top) and (0.45, 0.45, 0.45) (bottom) in color sensor measurement: Field estimates are less accurate and the localization error is large.

sensor measurements uncertainties. As expected, since only the color sensor measurement is used for localization, the increase in the color sensor noise adversely affects both the estimation of the field, as well as the localization of the robot.

In Figures 5.14 and 5.15, the lines in gray color show the dead-reckoning robot location calculated from a differential drive kinematics without using geometric and encoder uncertainties. The location estimate based on the color sensor measurement EKF is shown in white, while the "true" robot position calculated from differential drive kinematics with added geometric and encoder uncertainties is plotted in black. The "true" sampling locations correspond to measurements of position using the overhead camera. As expected, the dead-reckoning error increases as the robot covers more distance. It is also clear that the location estimate is close to the true location for small error in color sensor measurement, and that large discrepancies in both the estimated field and robot position result in large measurement uncertainties (Figure 5.16). □

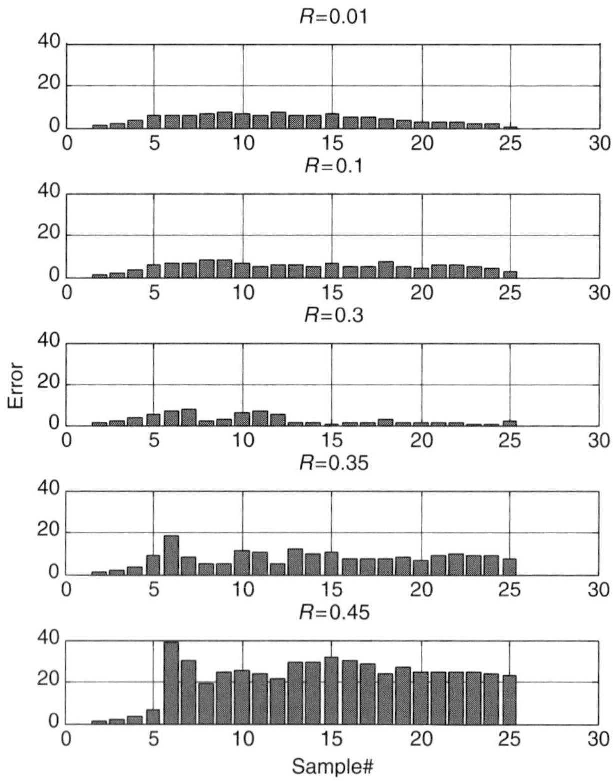

FIGURE 5.16
Difference between the true and estimated robot location versus the number of samples for different sensor measurement errors.

Problems

Section 5.1

1. Sum of Random Variables. Let Z be a linear combination of RVs X and Y so that

$$Z = AX + BY$$

with A and B deterministic matrices. Show that the mean and covariance of Z are

$$\bar{Z} = A\bar{X} + B\bar{Y}$$
$$P_Z = AP_XA^T + AP_{XY}B^T + BP_{YX}A^T + BP_YB^T$$

with P_X, P_Y the covariances of X and Y, and P_{XY} the cross-covariance.

2. Fokker–Planck Equation for Wiener Process. Verify that $f_{x(t)}$ in Example 5.2 satisfies the Fokker–Planck equation.

3. Fokker–Planck Equation for Markov Process. Let

$$\dot{x} = ax + w$$

with $w(t) \sim N(0, q)$ white and independent of $x(0) \sim N(\bar{x}_0, p_0)$.

(a) Determine the mean and covariance, and hence the unconditional PDF $f(x, t)$ of $x(t)$.

(b) Write the Fokker–Planck equation and verify that the PDF satisfies it.

(c) Now, measurements are taken every T s so that

$$z = x + v$$

with $v(t) \sim N(0, r)$ independent of $w(t)$ and $x(0)$. Determine the measurement update of $f(x, t^-)$ to $f(x, t)$ at $t = kT$.

(d) Between measurements, the PDF of $x(t)$ satisfies the Fokker–Planck equation. Sketch the covariance of $x(t)$ over several measurement periods.

4. Let

$$x_{k+1} = x_k + w_k$$
$$z_k = x_k + v_k$$

with $w_k \sim N(0, 1)$ and $v_k \sim N(0, 1)$ white and independent of each other and of $x_0 \sim N(\bar{x}_0, 2)$. Use Equations 5.6 and 5.7 to find time and measurement updates for the hyperstate.

5. Measurement Update for Uniform Noise. Repeat Problem 4 if v_k is zero mean white noise uniformly distributed on $[-\frac{1}{2}, \frac{1}{2}]$.

6. Update with Nonlinear Measurement. Repeat Problem 4 if

$$z_k = \ln\left(\frac{1}{x_k}\right) + v_k$$

with v_k white noise with PDF

$$f_{v_k}(v) = \begin{cases} e^{-v}, & v \geq 0 \\ 0, & v < 0 \end{cases}$$

and w_k uniform on $[0, 1]$.

Section 5.2

7. Prove Equation 5.43.

8. Time Update of Moments. Let

$$\dot{x} = ax + w$$

with $w(t) \sim (0, q)$ white and $x(0) \sim (\bar{x}_0, p_0)$. Use Equation 5.44 to compute analytically the following.

(a) $\hat{x}(t)$.

(b) $\varepsilon[x^2(t)]$ and the covariance $p(t)$.

(c) $\varepsilon[x^3(t)]$. Suppose $\varepsilon[x^3(0)] = d$ is given.

9. Nonlinear System Dynamics. Let

$$\dot{x} = ax^2 + w$$

with $w(t) \sim (0, q)$ and $x(0) \sim (\bar{x}_0, p_0)$. Find differential equations for $\hat{x}(t)$ and the covariance $p(t)$. How can these be solved?

10. Nonlinear Measurements. Let

$$\dot{x} = w$$

$$z_k = \ln\left(\frac{1}{x_k}\right) + v$$

with $w(t)$ uniform on $[0,\ 1]$ and $v(t)$ independent white noise and

$$f_v(v) = \begin{cases} e^{-v}, & v \geq 0 \\ 0, & v < 0 \end{cases}$$

The measurements are taken every T s. Go as far as you can in determining $\hat{x}(t)$ and $p(t)$ using the linearized measurement updates (Equations 5.68 and 5.72).

Section 5.3

11. Nonlinear Process Noise-Weighting Matrix. Suppose that a nonlinear system is described by

$$\dot{x}(t) = a(x,t) + g(x,t)w \tag{5.160}$$

with $w(t) \sim (0, Q)$. Show that to a first-order approximation

$$\varepsilon(gQg^T) = g(\hat{x},t)Qg^T(\hat{x},t) + \frac{1}{2}\sum_{i,j=1}^{n} \left.\frac{\partial g}{\partial x_i}\right|_{x=\hat{x}} Q \left.\frac{\partial g^T}{\partial x_j}\right|_{x=\hat{x}} p_{ij}(t) \tag{5.161}$$

where $p_{ij}(t)$ is element (i, j) of $P(t)$.

12. EKF. Let

$$\dot{x} = x^2 + w$$
$$z = \frac{1}{x} + v$$

with $w(t) \sim (0, 1)$ and $v(t) \sim (0, 1)$ independent white noises, and $x(0) \sim (1, 1)$.

(a) Write the continuous–discrete EKF equations if measurements are taken every second.

(b) Solve for and sketch the EKF error covariance.

(c) Write MATLAB code to implement the EKF.

13. Repeat Problem 12 using the continuous–continuous EKF.

14. EKF for Pendulum. The equation of motion of a simple pendulum of length l is

$$\ddot{\theta} = -\frac{g}{l}\sin\theta + w$$

The horizontal displacement of the bob is measured so that

$$Z = l(1 - \cos\theta) + v$$

Suppose that $w(t) \sim N(0, 0.02)$, $v(t) \sim N(0, 1)$, and $\dot{\theta}(0) \sim (0, 0.02)$.

(a) Write the EKF equations if data are taken every $T = (\pi/5)\sqrt{l/g}$ s. You can let $g/l = 1$ if you desire.

(b) Solve for and sketch $p(t)$ over several measurement periods.

15. EKF for Rocket Dynamics. The equations of motion of a rocket are

$$\begin{aligned}\dot{r} &= w \\ \frac{\mathrm{d}}{\mathrm{d}t}\dot{m} &= 0 \\ \dot{v} &= -\frac{vw}{r} + \frac{\dot{m}V_e \cos\phi}{m_0 + \dot{m}t} + n_1 \\ \dot{w} &= \frac{v^2}{r} - \frac{\mu}{r^2} + \frac{\dot{m}V_e \sin\phi}{m_0 + \dot{m}t} + n_2\end{aligned}$$

where $\phi(t)=$ thrust direction angle, $\gamma(t)=$ flight path angle, total velocity $V(t)$ has radial component $w(t)$ and tangential component $v(t)$, $\mu = GM$ is the gravitational constant of the attracting center, $V_e=$ constant exhaust velocity, $m_0 =$ initial mass, $m(t) =$ rocket mass, and $r(t) =$ radius from the attracting mass M. Independent white process noises $n_i(t) \sim (0, q_i)$ have been added to compensate for unmodeled dynamics. Take the state as $x = [v \quad w \quad r \quad \dot{m}]^T$.

Suppose that $\dot{m}$ and r are measured every second so that

$$\begin{aligned}z_k^1 &= \dot{m}_k + v_k^1 \\ z_k^2 &= r_k + v_k^2\end{aligned}$$

where $v_k^1 \sim (0, r_1), v_k^2 \sim (0, r_2)$.

(a) Find the Jacobian and write the EKF error covariance propagation equation.

(b) Write MATLAB code to implement the continuous–discrete EKF.

(c) Write MATLAB code to implement the continuous–continuous EKF.

16. Spacecraft Reentry Trajectory Estimation. The state of a spacecraft in the 3D reentry problem is

$$x = \begin{bmatrix} v & \gamma & \psi & h & \phi & \theta \end{bmatrix}^T \tag{5.162}$$

where v is velocity magnitude, γ the flight-path angle measured from the local horizontal, ψ the heading angle measured from east toward north in the horizon plane, h the altitude above the Earth's surface, ϕ the latitude, and θ the longitude (Jazwinski 1970). If m the (constant) vehicle mass and r its distance from the Earth's center, the state equations are

$$\dot{v} = \frac{F_{xv}}{mv} - \frac{\mu}{r^2}\sin\gamma + \Delta\dot{v}_{\text{rot}} \tag{5.163}$$

$$\dot{\gamma} = \frac{F_{zv}}{mv} + \frac{v\cos\gamma}{r} - \frac{\mu}{vr^2}\cos\gamma + \Delta\dot{\gamma}_{\text{rot}} \tag{5.164}$$

$$\dot{\psi} = \frac{F_{yv}}{mv} - \frac{v}{r}\cos\gamma\cos\psi\tan\phi + \Delta\dot{\psi}_{\text{rot}} \tag{5.165}$$

$$\dot{h} = v\sin\gamma \tag{5.166}$$

$$\dot{\phi} = \frac{v}{r} \cos\gamma \sin\psi \tag{5.167}$$

$$\dot{\theta} = \frac{v}{r\cos\phi} \cos\gamma \cos\psi \tag{5.168}$$

The Earth's gravitational constant is $\mu = GM$, with G the universal gravitation constant and M the mass of the Earth. F_{xv}, F_{zv}, and F_{yv} are aerodynamic forces, and the last terms in Equations 5.163 through 5.165 are due to the Earth's rotation. Assume a nonrotating Earth so that these terms are zero.

The aerodynamic forces are not well known since they depend on winds, air density, temperature, and so on. To make the problem tractable, assume that they are uncorrelated white noise processes with covariances of σ_x^2, σ_z^2, and σ_y^2.

We intend to apply the EKF to estimate the state x and find the error covariance.

(a) Find the time update equations. Write MATLAB code to implement it.

(b) Suppose we make discrete measurements of h, ϕ, and θ from a ground-tracking station, and of v using instruments abroad the vehicle. Assume all measurements are independent, and add white measurement noise to each one. Write four separate sets of measurement update equations, one to incorporate each of these pieces of data. Write separate measurement update subroutines for each of these measurements. Then, whenever one of these measurements becomes available, we can simply call the appropriate subroutine to incorporate it into the state estimate.

(c) Write the time update function to implement the continuous–continuous EKF.

17. Radar Tracking in $x - y$ Coordinates. An aircraft is flying straight and in level with a constant velocity V in an easterly direction as shown in Figure 5.17. It is tracked by a radar that provides measurements of range r and azimuth θ every 5 s.

(a) Let the state x be $[x_1 \quad \dot{x}_1 \quad x_2 \quad \dot{x}_2]$. Write the state equations, adding acceleration disturbances $w_1(t) \sim (0, q_1)$ and $w_2(t) \sim (0, q_2)$ to the northerly and easterly motion, respectively.

(b) Write the (nonlinear) measurement equation for data points $z_k = [r(kT) \quad \theta(kT)]^T$, adding independent measurement noises $v_1(k) \sim (0, \sigma_r^2)$, $v_2(k) \sim (0, \sigma_\theta^2)$.

Compare to Example 2.8. In r–θ coordinates with straight-line motion by the target we get nonlinear state equations with linear measurements. In x–y coordinates with straight-line target motion we get linear equations with nonlinear measurements!

(c) Find the EKF time update for the estimate and error covariance. Simplify by preliminary analysis and write a MATLAB function to implement the time update.

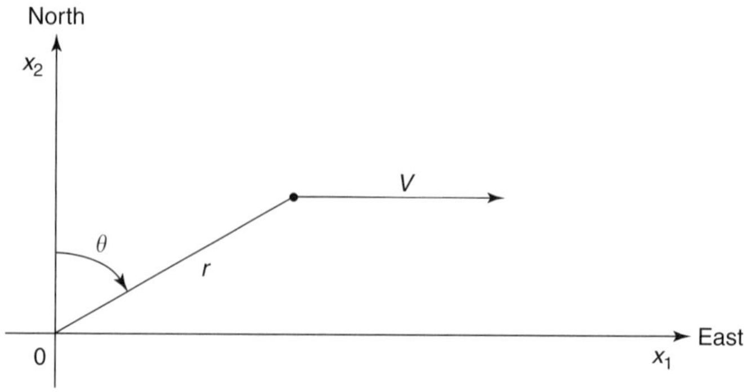

FIGURE 5.17
Radar tracking problem.

(d) Find the EKF measurement update, simplify, and implement in MATLAB a measurement update function. Since the measurements of r and θ are assumed independent, two scalar updates can be performed at each sample time t_k to incorporate separately the effects of the two measurements. Write update subroutines for each measurement.

(e) (Optional.) Simulate a tracking run in MATLAB. (You will find that one of the problems with simulations is generating the data z_k.)

18. Converting Nonlinear Systems to Linear Systems by Using Two Sets of Coordinates. Consider the radar tracking situation in Figure 5.17. To obtain linear state equations we can use x–y coordinates. To obtain linear measurements we must use r–θ coordinates.

Explore the possibility of formulating the continuous-time Kalman filter with the time update in x–y coordinates and the measurement update in r–θ coordinates. You will have to find a state-space transformation to apply to the state estimate before and after each measurement. It will also be necessary to determine how a state-space transformation affects the error covariance.

19. Tracking a Body in Free Fall. The state of a body falling vertically through the atmosphere is $x = [h \quad \dot{h} \quad \beta]$, where h is its height above the Earth's surface and β its *ballistic coefficient.* The ballistic coefficient is included as a state because it is not well known and so it must be estimated.

The state equations are

$$\frac{\mathrm{d}\dot{h}}{\mathrm{d}t} = D - g \tag{5.169a}$$

$$\frac{\mathrm{d}\beta}{\mathrm{d}t} = w(t) \tag{5.169b}$$

where $g = 9.8\ \mathrm{m/s^2}$ is acceleration due to gravity, $w(t) \sim (0, q)$ the white noise,

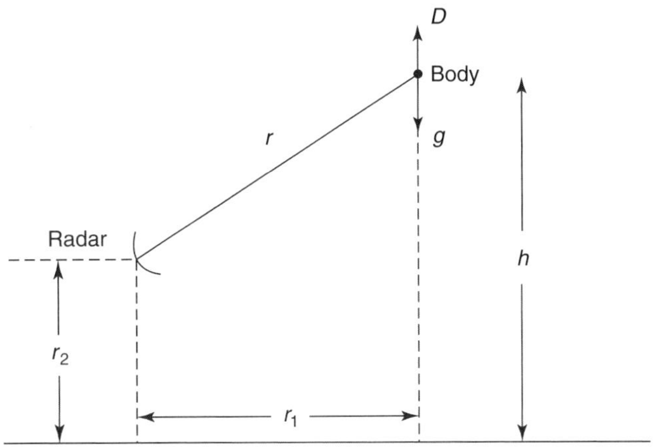

FIGURE 5.18
Tracking a free-falling body.

and drag is given by

$$D = \frac{\rho \dot{h}^2}{2\beta} \tag{5.170}$$

Atmospheric density is

$$\rho = \rho_0 e^{-\mathrm{h}/\mathrm{c}} \tag{5.171}$$

with $\rho_0 = 1220\ \mathrm{g/m^3}$ being the density at sea level and $c = 10263\,\mathrm{m}$ a decay constant. Equation 5.171 is good to about 6000 m in the troposphere.

Suppose range measurements are taken every second as in Figure 5.18. Thus,

$$z_k = \sqrt{r_1^2 + (h_k - r_2)^2 + v_k} \tag{5.172}$$

with $h_k \triangleq h(kT)$ and $v_k \sim (0, \sigma_r^2)$. (See Gelb 1974.)
It is desired to use the EKF to estimate the state.

(a) Write down the time update equations and simplify them by preliminary analysis. Write a time update function in MATLAB to update the estimate and covariance.

(b) Write the measurement update and simplify by preliminary analysis. Write a measurement update function in MATLAB to implement this.

(c) (Optional.) Run a simulation using $q = 1000\,\mathrm{g^2/m^2\,s^6}$, $h(0) \sim N(10{,}000\,\mathrm{m}, 500\,\mathrm{m}^2)$, $\dot{h}(0) \sim N(-500\,\mathrm{m/s}, 100\,\mathrm{m^2/s^2})$, $\beta(0) \sim N(6 \times 10^7\,\mathrm{g/m\,s^2}, 2 \times 10^{13}\mathrm{g^2/m^2\,s^4})$. Use initial error variances of $p_{11}(0) = 50\,m^2$, $p_{22}(0) = 200\,\mathrm{m^2/s^2}$, $p_{33}(0) = 2 \times 10^{12}\,\mathrm{g^2/m^2\,s^4}$. Let $\sigma_r^2 = 5\,\mathrm{m^2/Hz}$, and the radar tracker distances be $r_1 = 1000\,\mathrm{m}$, $r_2 = 500\,\mathrm{m}$. The measured data are taken every second beginning at $t_1 = 1\mathrm{s}$, and are given in meters by $z_k = 9055, 8560, 7963, 7467, 7000, 6378, 5885, 5400, 4928, 4503$.

(d) The acceleration due to gravity is not a constant but varies as

$$g = g_0 \left(\frac{R}{R+h} \right)^2 \tag{5.173}$$

where R is the radius of the Earth and $g_0 = 9.8$ m/s^2. Is it worth accounting for this variation in this situation? How would we account for this if we wanted to?

20. Biological Treatment of Waste Sewage. An industrial plant's effluent waste is fed into a pool of bacteria that transforms the waste into nonpolluting forms. If the bacteria concentrations is x_1 and the pollutant concentration is x_2, a state model is given by (Luenberger 1979)

$$\dot{x}_1 = \frac{ax_1x_2}{x_2 + K} + w_1 \tag{5.174a}$$

$$\dot{x}_2 = -\frac{bx_1x_2}{x_2 + K} + w_2 \tag{5.174b}$$

with $w_1(t) \sim N(0, q_1), w_2(t) \sim N(0, q_2)$ uncorrelated white noises. Let measurements of x_2 be taken every hour, and suppose they are affected by the bacteria concentration so that

$$z_k = x_2(k) + cx_1^2(k) + v_k \tag{5.175}$$

with noise $v_k \sim N(0, r)$.

(a) Determine the EKF time update and simplify.

(b) Determine the EKF measurement update and simplify.

(c) Write MATLAB code to implement the EKF.

Part II

Robust Estimation

6

Robust Kalman Filter

In Chapter 2, we studied the optimal filtering of a system whose dynamics is known exactly. In practice, there inevitably exists uncertainty in the system model either due to modeling error or due to parameter drifting of the system. This chapter discusses the design of a filter that has certain performance robustness against possible system parameter uncertainties. In particular, we are interested in the so-called robust Kalman filter which is concerned with the design of a fixed filter for a family of plants under consideration such that the filtering error covariance is of minimal upper bound (Jain 1975; Petersen and McFarlane 1991; Xie *et al.* 1994).

6.1 Systems with Modeling Uncertainties

Consider the following dynamical system:

$$x_{k+1} = A_\delta x_k + G w_k \tag{6.1}$$

$$z_k = H_\delta x_k + v_k \tag{6.2}$$

$$s_k = L x_k \tag{6.3}$$

where $x_k \in R^n$ is the state vector, $z_k \in R^p$ the measurement, $s_k \in R^r$ the signal to be estimated, $w_k \in R^l$ the noise input, and $v_k \in R^p$ the measurement noise. The initial state x_0 of the system is a random vector of zero mean and covariance $P_{x_0} > 0$ and is independent of w_k and v_k for any $k \geq 0$.

We shall adopt the following assumptions for the process and measurement noises.

Assumption 6.1 *For all integers k and $j \geq 0$,*

(i) $E(w_k) = 0, \;\; E(w_k w_j^T) = Q\delta_{kj}, \;\; Q \geq 0$

(ii) $E(v_k) = 0, \;\; E(v_k v_j^T) = R\delta_{kj}, \;\; R > 0$

(iii) $E(w_k v_j^T) = 0$

where δ_{kj} is the Kronecker Delta.

We allow the uncertainty to be time-varying and assume that the matrices A_δ and H_δ have the form (Xie *et al.* 1991; Fu *et al.* 1992):

$$\begin{bmatrix} A_\delta \\ H_\delta \end{bmatrix} = \begin{bmatrix} A \\ H \end{bmatrix} + \begin{bmatrix} D_1 \\ D_2 \end{bmatrix} \Delta_k F \tag{6.4}$$

where $\Delta_k \in R^{\alpha \times \beta}$ is an unknown real time-varying matrix satisfying

$$\Delta_k^T \Delta_k \leq I, \quad \forall\, k \geq 0 \tag{6.5}$$

and D_1, D_2, and F are known real constant matrices of appropriate dimensions that specify how the elements of the system matrices A_δ and H_δ are affected by the uncertainty in Δ_k. This type of uncertainty is commonly referred to as norm-bounded uncertainty, which can be used to describe the variations of system parameters over known ranges.

Example 6.1

Consider a wireless fading channel described by a second-order auto-regressive model:

$$h(k) = d_1 h(k-1) + d_2 h(k-2) + w(k)$$

where d_1 and d_2 are the transition coefficients of the channel often estimated through the use of training sequences and they suffer from uncertainty. Denote $x_1(k) = h(k-1)$ and $x_2(k) = h(k-2)$. Then,

$$\begin{bmatrix} x_1(k+1) \\ x_2(k+1) \end{bmatrix} = \begin{bmatrix} d_1 & d_2 \\ 1 & 0 \end{bmatrix} \begin{bmatrix} x_1(k) \\ x_2(k) \end{bmatrix} + \begin{bmatrix} 1 \\ 0 \end{bmatrix} w(k) \tag{6.6}$$

Let $d_1 \in [0.4, 0.6]$ and $d_2 \in [-0.2, -0.4]$. Then, Equation 6.6 can be put into the model 6.1 with uncertainty of the form in Equations 6.4 and 6.5, and

$$A = \begin{bmatrix} 0.5 & -0.3 \\ 1 & 0 \end{bmatrix}, \qquad D_1 = \begin{bmatrix} 0.1 & 0.1 \\ 0 & 0 \end{bmatrix}, \qquad \Delta_k = \begin{bmatrix} \delta_1 & 0 \\ 0 & \delta_2 \end{bmatrix}, \qquad F = I$$

□

We also make the following assumption concerning system parameters.

Assumption 6.2 $rank\begin{bmatrix} A & D_1 & GQ^{\frac{1}{2}} \end{bmatrix} = n.$

We are concerned with the design of a filter for the uncertain system 6.1 through 6.3 to achieve an optimal guaranteed filtering performance regardless of the uncertainty characterized in Equation 6.5. More specifically, we would like to design a robust Kalman *a priori* filter of the form

$$\hat{x}_{k+1} = \hat{A}_k \hat{x}_k + K_k z_k, \quad \hat{x}_0 = 0 \tag{6.7}$$

$$\hat{s}_k = \hat{L}_k \hat{x}_k \tag{6.8}$$

where $\hat{A}_k$, K_k, and $\hat{L}_k$ are time-varying matrices to be determined so that the variance of the filtering error $s_k - \hat{s}_k$ is guaranteed to be smaller than a certain bound for all uncertainty matrices Δ_k satisfying Equation 6.5.

Similar to the Kalman filtering for systems without uncertainty, it can be shown that the optimal estimate of a linear combination of the state is the same as the linear combination of the optimal estimate of the state (Shaked and de Souza 1995). Therefore, in the sequel, we shall set $\hat{L}_k = L$ and design a filter so that the filtering error dynamics satisfies

$$E\left[(x_k - \hat{x}_k)(x_k - \hat{x}_k)^T\right] \leq P_k \tag{6.9}$$

with P_k being an optimized upper bound of the filtering error covariance. In this case,

$$E\left[(s_k - \hat{s}_k)(s_k - \hat{s}_k)^T\right] \leq S_k = LP_kL^T$$

6.2 Robust Finite Horizon Kalman *a Priori* Filter

In terms of system 6.1 through 6.3 with uncertainty of Equation 6.5 and filter (Equations 6.7 and 6.8), the state-space equations for the filtering error $\tilde{s}_k = s_k - \hat{s}_k$ are as follows:

$$\xi_{k+1} = (\bar{A}_k + \bar{D}_k\Delta_k\bar{F})\xi_k + \bar{G}_k\eta_k \tag{6.10}$$

$$\tilde{s}_k = \bar{L}\xi_k \tag{6.11}$$

where $\xi_k = \begin{bmatrix} e_k^T & \hat{x}_k^T \end{bmatrix}^T$ with $e_k = x_k - \hat{x}_k$ and $\xi_0 = \begin{bmatrix} x_0^T & 0 \end{bmatrix}^T$, $\eta_k = \begin{bmatrix} w_k^T & v_k^T \end{bmatrix}^T$ and

$$\bar{A}_k = \begin{bmatrix} A - K_kH & A - \hat{A}_k - K_kH \\ K_kH & \hat{A}_k + K_kH \end{bmatrix}, \qquad \bar{G}_k = \begin{bmatrix} G & -K_k \\ 0 & K_k \end{bmatrix} \tag{6.12}$$

$$\bar{D}_k = \begin{bmatrix} D_1 - K_kD_2 \\ K_kD_2 \end{bmatrix}, \qquad \bar{F} = \begin{bmatrix} F & F \end{bmatrix}, \qquad \bar{L} = \begin{bmatrix} L & 0 \end{bmatrix} \tag{6.13}$$

Unlike a system without uncertainty, the above filtering error is also dependent on the system state.

For a given Δ_k, the covariance matrix Σ_k of ξ_k, i.e., $\Sigma_k = E(\xi_k\xi_k^T)$, satisfies the following Lyapunov equation:

$$(\bar{A}_k + \bar{D}_k\Delta_k\bar{F})\Sigma_k(\bar{A}_k + \bar{D}_k\Delta_k\bar{F})^T - \Sigma_{k+1} + \bar{G}_k\bar{Q}\bar{G}_k^T = 0 \tag{6.14}$$

where $\bar{Q} = \left[\begin{smallmatrix} Q & 0 \\ 0 & R \end{smallmatrix}\right]$ is the covariance matrix of η_k.

Note, however, that the covariance matrix Σ_k is dependent on Δ_k which is unknown. Therefore, in robust Kalman filtering, we are concerned with finding an upper bound, $\bar{\Sigma}_k$, of Σ_k for all uncertainties. Thus, we shall replace the above Lyapunov equation by the Lyapunov inequality

$$(\bar{A}_k + \bar{D}_k\Delta_k\bar{F})\bar{\Sigma}_k(\bar{A}_k + \bar{D}_k\Delta_k\bar{F})^T - \bar{\Sigma}_{k+1} + \bar{G}_k\bar{Q}\bar{G}_k^T \leq 0 \tag{6.15}$$

which is to be satisfied for all admissible uncertainties characterized in Equation 6.5. It is easy to show that $\Sigma_k \leq \bar{\Sigma}_k$ for all Δ_k if $\bar{\Sigma}_k$ satisfies Equation 6.15. In fact, if Equation 6.15 is satisfied for some $\bar{\Sigma}_k \geq 0$, there exists $\Xi_k \geq 0$ such that

$$(\bar{A}_k + \bar{D}_k\Delta_k\bar{F})\bar{\Sigma}_k(\bar{A}_k + \bar{D}_k\Delta_k\bar{F})^T - \bar{\Sigma}_{k+1} + \bar{G}_k\bar{Q}\bar{G}_k^T + \Xi_k = 0 \tag{6.16}$$

Subtracting Equation 6.16 from Equation 6.14 leads to

$$(\bar{A}_k + \bar{D}_k\Delta_k\bar{F})(\Sigma_k - \bar{\Sigma}_k)(\bar{A}_k + \bar{D}_k\Delta_k\bar{F})^T - (\Sigma_{k+1} - \bar{\Sigma}_{k+1}) - \Xi_k = 0$$

or

$$(\Sigma_{k+1} - \bar{\Sigma}_{k+1}) = (\bar{A}_k + \bar{D}_k\Delta_k\bar{F})(\Sigma_k - \bar{\Sigma}_k)(\bar{A}_k + \bar{D}_k\Delta_k\bar{F})^T - \Xi_k$$

With $\bar{\Sigma}_0 = \Sigma_0 = diag\{P_{x_0}, 0\}$ and $\Xi_k \geq 0$, the above implies $\Sigma_k - \bar{\Sigma}_k \leq 0$, i.e., $\Sigma_k \leq \bar{\Sigma}_k$.

We focus on the design of a robust *a priori* filter as defined below.

DEFINITION 6.1 *The filter (Equations 6.7 and 6.8) is said to be a robust* a priori *filter if for some $\varepsilon_k > 0$, there exists a bounded $\tilde{\Sigma}_k = \tilde{\Sigma}_k^T \geq 0$ that satisfies the following difference Riccati equation (DRE):*

$$\begin{aligned}\tilde{\Sigma}_{k+1} = {} & \bar{A}_k\tilde{\Sigma}_k\bar{A}_k^T + \bar{A}_k\tilde{\Sigma}_k\bar{F}^T(\varepsilon_k^{-1}I - \bar{F}\tilde{\Sigma}_k\bar{F}^T)^{-1}\bar{F}\tilde{\Sigma}_k\bar{A}_k^T \\ & + \varepsilon_k^{-1}\bar{D}_k\bar{D}_k^T + \bar{G}_k\bar{Q}\bar{G}_k^T\end{aligned} \tag{6.17}$$

such that $I - \varepsilon_k\bar{F}\tilde{\Sigma}_k\bar{F}^T > 0$, where $\tilde{\Sigma}_0 = \Sigma_0 = \begin{bmatrix} P_{x_0} & 0 \\ 0 & 0 \end{bmatrix}$.

It is easy to show that a robust *a priori* filter guarantees that $\Sigma_k \leq \bar{\Sigma}_k \leq \tilde{\Sigma}_k$ for any admissible uncertainty in Equation 6.5. The proof is left as an exercise (see Problem 1). Thus, for a given robust filter, an upper bound for the filtering error variance can be given by

$$E\left[(s_k - \hat{s}_k)^T(s_k - \hat{s}_k)\right] \leq trace\left(L\tilde{\Sigma}_{1,1,k}L^T\right) \tag{6.18}$$

where $\tilde{\Sigma}_{1,1,k}$ is the (1,1)-block of $\tilde{\Sigma}_k$ and has the dimension $n \times n$.

Before presenting a solution to the optimal robust *a priori* filtering, we introduce the following two DREs:

$$X_{k+1} = AX_kA^T + AX_kF^T\left(\frac{1}{\varepsilon_k}I - FX_kF^T\right)^{-1} FX_kA^T + \frac{1}{\varepsilon_k}D_1D_1^T + GQG^T \tag{6.19}$$

$$\begin{aligned} P_{k+1} = AY_kA^T &- \left(AY_kH^T + \frac{1}{\varepsilon_k}D_1D_2^T\right)\left(R_{\varepsilon_k} + HY_kH^T\right)^{-1} \\ &\times \left(AY_kH^T + \frac{1}{\varepsilon_k}D_1D_2^T\right)^T + \frac{1}{\varepsilon_k}D_1D_1^T + GQG^T \end{aligned} \tag{6.20}$$

where $X_0 = P_0 = P_{x_0}$, $\varepsilon_k > 0$ is a scaling parameter, $Y_k^{-1} = P_k^{-1} - \varepsilon_k F^T F$, and $R_{\varepsilon_k} = R + \frac{1}{\varepsilon_k}D_2D_2^T$.

REMARK 6.1 The DRE 6.20 can be rewritten as (see Problem 2)

$$\begin{aligned} P_{k+1} = AP_kA^T &- \left(AP_k\tilde{H}_k^T + \tilde{D}_{1,k}\tilde{D}_{2,k}^T\right)\left(\tilde{H}_kP_k\tilde{H}_k^T + \tilde{R}_k\right)^{-1} \\ &\times \left(AP_k\tilde{H}_k^T + \tilde{D}_{1,k}\tilde{D}_{2,k}^T\right)^T + \tilde{D}_{1,k}\tilde{D}_{1,k}^T \end{aligned} \tag{6.21}$$

where

$$\tilde{H}_k = \sqrt{\varepsilon_k}\begin{bmatrix} H \\ F \end{bmatrix}, \qquad \tilde{D}_{1,k} = \begin{bmatrix} GQ^{\frac{1}{2}} & \frac{1}{\sqrt{\varepsilon_k}}D_1 & 0 \end{bmatrix} \tag{6.22}$$

$$\tilde{D}_{2,k} = \begin{bmatrix} \bar{D}_{2,k} \\ 0 \end{bmatrix}, \qquad \tilde{R}_k = \begin{bmatrix} \bar{D}_{2,k}\bar{D}_{2,k}^T & 0 \\ 0 & -I \end{bmatrix}, \qquad \bar{D}_{2,k} = \begin{bmatrix} 0 & D_2 & \sqrt{\varepsilon_k}R^{\frac{1}{2}} \end{bmatrix} \tag{6.23}$$

∎

LEMMA 6.1 (Zhu *et al.* 2002)
Consider the system (Equations 6.1 through 6.3) satisfying Assumptions 6.1 and 6.2. Then, the following hold:

(i) If for a given horizon $N > 0$, scaling parameter $\varepsilon_k > 0$, and filter (Equations 6.7 and 6.8), the DRE (Equation 6.17) has a bounded solution $\tilde{\Sigma}_k$ over $[0, N]$ such that $I - \varepsilon_k\bar{F}\tilde{\Sigma}_k\bar{F}^T > 0$, then there exists a bounded solution $X_k = X_k^T > 0$ to the DRE (Equation 6.19) satisfying $X_k^{-1} - \varepsilon_k F^T F > 0$ for the same ε_k. Further, $X_k = \tilde{\Sigma}_{11,k} + \tilde{\Sigma}_{12,k} + \tilde{\Sigma}_{21,k} + \tilde{\Sigma}_{22,k}$.

(ii) If for a given horizon $N > 0$, scaling parameter $\varepsilon_k > 0$ and filter (Equations 6.7 and 6.8), the DRE (Equation 6.19) has a bounded solution X_k over $[0, N]$ satisfying $X_k^{-1} - \varepsilon_k F^T F > 0$, then there exists $P_k = P_k^T > 0$ satisfying $P_k^{-1} - \varepsilon_k F^T F > 0$ in Equation 6.20 for the same ε_k. Furthermore, $X_k \geq P_k > 0$.

The proof of the lemma is left as an exercise (See Problem 3).

THEOREM 6.1 (Zhu *et al.* 2002)
Consider the uncertain system 6.1 through 6.3 with uncertainty of Equation 6.5 and satisfying Assumptions 6.1 and 6.2. There exists a robust a priori *filter for the system that minimizes the bound on the filtering error variance in Equation 6.18 if and only if for some* $\varepsilon_k > 0$ *there exists a solution* $X_k = X_k^T > 0$ *over [0, N] to the DRE 6.19 such that* $X_k^{-1} - \varepsilon_k F^T F > 0$.

Under this condition, a robust a priori *filter with an optimized upper bound of filtering error covariance is given by*

$$\hat{x}_{k+1} = (A + A_{e,k})\hat{x}_k + K_k\left[z_k - (H + H_{e,k})\hat{x}_k\right], \quad \hat{x}_0 = 0 \tag{6.24}$$

$$\hat{s}_k = L\hat{x}_k \tag{6.25}$$

where

$$A_{e,k} = \varepsilon_k A P_k F^T \left(I - \varepsilon_k F P_k F^T\right)^{-1} F \tag{6.26}$$

$$H_{e,k} = \varepsilon_k H P_k F^T \left(I - \varepsilon_k F P_k F^T\right)^{-1} F \tag{6.27}$$

$$K_k = \left(A Y_k H^T + \frac{1}{\varepsilon_k} D_1 D_2^T\right)\left(R_{\varepsilon_k} + H Y_k H^T\right)^{-1} \tag{6.28}$$

$$R_{\varepsilon_k} = R + \frac{1}{\varepsilon_k} D_2 D_2^T \tag{6.29}$$

and $P_k = P_k^T > 0$ *is a solution of DRE 6.20 or 6.21 over [0, N] satisfying* $Y_k^{-1} = P_k^{-1} - \varepsilon_k F^T F > 0$.

Moreover, the optimized covariance bound of the filtering error, $x_k - \hat{x}_k$, *is* P_k.

The proof of Theorem 6.1 can be found at the end of this chapter.

Note that Equation 6.19 can be rewritten as

$$\begin{aligned} X_{k+1} &= \begin{bmatrix} A & D_1 & GQ^{\frac{1}{2}} \end{bmatrix} diag\{X_k, \varepsilon_k^{-1} I, I\} \begin{bmatrix} A & D_1 & GQ^{\frac{1}{2}} \end{bmatrix}^T \\ &\quad + A X_k F^T \left(\varepsilon_k^{-1} I - F X_k F^T\right)^{-1} F X_k A^T \end{aligned}$$

It is clear that under Assumption 6.2, $X_k > 0$ and $I - \varepsilon_k F X_k F^T > 0$ imply that $X_{k+1} > 0$. Further, in view of Lemma 6.1, the existence of $P_k > 0$ such that $P_k^{-1} - \varepsilon F^T F > 0$ is guaranteed.

REMARK 6.2 Note from Theorem 6.1 that the filter parameters (Equations 6.26 through 6.28) are only dependent on the solution of the DRE 6.20 and have nothing to do with the solution of the DRE 6.19. However, the existence of a solution to Equation 6.19 is needed to ensure that the derived filter is indeed a robust *a priori* filter. That is to say, the filter (Equations 6.26 through 6.28) computed based on the solution of Equation 6.20 may not guarantee the filtering error covariance bound P_k if the DRE Equation 6.19 does not admit a solution $X_k > 0$.

The matrices $A_{e,k}$ and $H_{e,k}$ of the filter can be viewed respectively as correction matrices on A and H in consideration of the norm-bounded uncertainty in

A_δ and H_δ. We observe from Equation 6.20 and (Equations 6.26 through 6.28) that when the parameter uncertainty in system 6.1 and 6.2 disappears, the DRE 6.20 and the robust filter of Theorem 6.1 reduce respectively to those in the standard finite horizon Kalman *a priori* filter for the nominal system of Equations 6.1 and 6.2 as discussed in Chapter 2. Further, Equation 6.19 becomes

$$X_{k+1} = AX_kA^T + GQG^T$$

whose existence of a solution is guaranteed and hence becomes redundant. ∎

From Theorem 6.1, it is clear that the optimal robust *a priori* filtering performance at time k depends not only on system data but also on the scaling parameters ε_0, $\varepsilon_1, \ldots, \varepsilon_{k-1}$. This is a major difference from the optimal estimation of systems without uncertainty where the optimal estimate at time k can be obtained from the optimal estimate at $k-1$ by a one-step recursion. This latter property, however, is no longer valid for the robust Kalman filtering in general except for a scalar system (see Problem 4). From Theorem 6.1, to obtain an optimal robust estimate at time k, an optimization with respect to the scaling parameters $\varepsilon_0, \varepsilon_1, \ldots, \varepsilon_{k-1}$ needs to be carried out, which may be costly with the increase of k.

To reduce the computational complexity, instead of optimizing the cost over all the k scaling parameters, we may optimize $trace(P_k)$ over a fixed length of scaling parameters $\varepsilon_{k-1}, \varepsilon_{k-2}, \ldots, \varepsilon_{k-d}$, where d is called a window (Fu *et al.* 2001). This can be done as follows. Given the computed suboptimal X_{k-d} and P_{k-d}. First, for any $k-d \le i \le k-1$ it follows that given $X_i > 0$ and $P_i > 0$, the solution $P_{i+1} > 0$ of Equation 6.20 exists if $0 < \varepsilon_i < \bar{\varepsilon}_i = \|FX_iF^T\|^{-1}$. Hence, we can grid the interval $(0, \bar{\varepsilon}_i)$ and for each ε_k from the interval calculate the corresponding X_{i+1} and P_{i+1}. The optimization of $trace(P_k)$ over $\varepsilon_{k-1}, \varepsilon_{k-2}, \ldots, \varepsilon_{k-d}$ will then involve the calculation of X_i and P_i of a tree structure. The optimal solution will be the one with minimal $trace(P_k)$. The performance of the robust filter and its computational cost will depend on the size of window, d. The larger the d the better the performance. However, a larger d also implies a higher computational cost.

For $d = 1$, given P_k, let $\bar{\varepsilon}_k$ be the supremum of ε_k such that $\varepsilon_k FP_kF^T < I$, the DREs 6.19 and 6.20 admit positive definite solutions for any $\varepsilon \in (0, \bar{\varepsilon}_k)$. The scaling parameter ε_k could be exploited to minimize $trace(P_{k+1})$, which is a convex function of ε_k (Xie and Soh 1995).

6.3 Robust Stationary Kalman *a Priori* Filter

We now discuss the design of robust stationary *a priori* filter. Unlike the standard Kalman filtering in Chapter 2 where the dynamic system is allowed to be unstable, the uncertain system for robust Kalman filtering is required to be stable when discussing stationary filtering. This is because in the presence of parametric uncertainty it is impossible to isolate the filtering error from the system state unlike the standard Kalman filtering case. That is to say, the filtering error is driven by the system state when there exists parameter uncertainty in the system. This can be clearly seen from Equations 6.10 and 6.11.

We assume that the system 6.1 is quadratically stable (Petersen and McFarlane 1994). The definition is as follows.

DEFINITION 6.2 *The system 6.1 is said to be quadratically stable if there exists a symmetric positive definite matrix P such that*

$$A_\delta^T P A_\delta - P < 0 \tag{6.30}$$

for all admissible uncertainties Δ_k in 6.5.

We note that the notion of quadratic stability is only sufficient for the asymptotic stability of an uncertain system. This basically means that there exists a single quadratic Lyapunov function, $x^T Px$, for all admissible uncertainties. It is a powerful machinery in dealing with parameter uncertainty, especially time-varying parameter uncertainty.

We consider a linear time-invariant filter (Equations 6.7 and 6.8) where $\hat{A}_k$, K_k, and $\hat{L}_k$ are now constant matrices denoted by $\hat{A}$, K, and $\hat{L}$, respectively. Recall that $\hat{L} = L$.

We introduce the notion of stabilizing solution of an algebraic Riccati equation (ARE) as follows.

DEFINITION 6.3 *Consider the ARE*

$$APA^T - P + APH^T(R - HPH^T)^{-1}HPA^T + Q = 0$$

where A, H, R, and Q are known real matrices of appropriate dimensions with Q and R being symmetric. Then a real matrix $P = P^T$ is called a stabilizing solution of the ARE if it satisfies the ARE such that the matrix $A + APH^T (R - HPH^T)^{-1}H$ is asymptotically stable, i.e., all its eigenvalues lie in the open unit disk.

It is a known fact that the stabilizing solution of an ARE is unique.

We introduce the definition of robust stationary *a priori* filter, similar to the finite horizon case.

DEFINITION 6.4 *The filter (Equations 6.7 and 6.8) is said to be a robust stationary* a priori *filter if for some $\varepsilon > 0$, there exists a stabilizing solution $\tilde{\Sigma} = \tilde{\Sigma}^T \geq 0$ to the ARE:*

$$\tilde{\Sigma} = \bar{A}\tilde{\Sigma}\bar{A}^T + \bar{A}\tilde{\Sigma}\bar{F}^T(\varepsilon^{-1}I - \bar{F}\tilde{\Sigma}\bar{F}^T)^{-1}\bar{F}\tilde{\Sigma}\bar{A}^T + \varepsilon^{-1}\bar{D}\bar{D}^T + \bar{G}\bar{Q}\bar{G}^T \tag{6.31}$$

such that $\varepsilon^{-1}I - \bar{F}\Sigma\bar{F}^T > 0$, where $\bar{A}, \bar{D}, \bar{F}$, and $\bar{G}$ are as in Equations 6.12 and 6.13 by dropping the subscript k since they are time-invariant now.

Define two AREs related to the optimal robust stationary *a priori* filter design below.

$$AXA^T - X + AXF^T\left(\frac{1}{\varepsilon}I - FXF^T\right)^{-1} FXA^T + \frac{1}{\varepsilon}D_1D_1^T + GQG^T = 0 \tag{6.32}$$

$$APA^T - P - (AP\tilde{H}^T + \tilde{D}_1\tilde{D}_2^T)(\tilde{H}S\tilde{H}^T + \tilde{R})^{-1}(AP\tilde{H}^T + \tilde{D}_1\tilde{D}_2^T)^T + \tilde{D}_1\tilde{D}_1^T = 0 \tag{6.33}$$

where $\tilde{H}$, $\tilde{D}_1$, $\tilde{D}_2$, and $\tilde{R}$ are as given in Equations 6.22 and 6.23 by dropping the subscript k and $\varepsilon > 0$ is a scaling parameter.

The next theorem provides a solution to the optimal robust *a priori* filtering for the uncertain system (Equations 6.1 through 6.3) in the stationary case.

THEOREM 6.2

Consider the uncertain quadratically stable system 6.1 through 6.3 satisfying Assumption 6.1. Then there exists a robust a priori filter with an optimized upper bound of filtering error covariance if and only if for some $\varepsilon > 0$, there exists a stabilizing solution $P = P^T \geq 0$ to the ARE 6.32 such that $I - \varepsilon FPF^T > 0$. Under this condition, a robust a priori filter that optimizes the upper bound of filtering error covariance is given by

$$\hat{x}_{k+1} = (A + A_e)\hat{x}_k + K\left[z_k - (H + H_e)\hat{x}_k\right] \tag{6.34}$$

where

$$A_e = \varepsilon APF^T\left(I - \varepsilon FPF^T\right)^{-1}F$$
$$H_e = \varepsilon HPF^T\left(I - \varepsilon FPF^T\right)^{-1}F$$
$$K = \left(AYH^T + \frac{1}{\varepsilon}D_1D_2^T\right)\left(R_\varepsilon + HYH^T\right)^{-1}$$
$$Y = P + \varepsilon P\left(I - \varepsilon FPF^T\right)^{-1}P$$

with $P = P^T \geq 0$ the stabilizing solution of the ARE 6.33 satisfying $I - \varepsilon FPF^T > 0$ and $R_\varepsilon = R + \frac{1}{\varepsilon}D_2D_2^T$. Moreover, the optimized filtering error covariance bound is P.

We note again that the ARE 6.32 has nothing to do with the computation of the optimal robust filter. However, it is needed to ensure that the computed filter using the ARE 6.33 is indeed a robust *a priori* filter. In fact, the existence of the stabilizing solution of the ARE 6.32 implies the quadratic stability of the system 6.1.

The filtering error variance $trace(P)$ is a convex function of the scaling parameter $\varepsilon > 0$. The optimal robust filter can be designed by a simple one-dimensional search of ε.

Example 6.2

Consider the following scalar system.

$$x_{k+1} = (a + \delta a)x_k + gw_k$$
$$z_k = hx_k + v_k$$

where $|\delta a| \leq d_1$ and w_k and v_k are uncorrelated white noises with zero means and covariances q and r, respectively. In this case, D_1, D_2, and F in Equation 6.4 can be given respectively by $D_1 = d_1$, $D_2 = 0$, and $F = 1$.

Associated with the system, the DRE 6.19 and DRE 6.20 become

$$X_{k+1} = \frac{a^2 X_k}{1 - \varepsilon_k X_k} + \varepsilon_k^{-1} d_1^2 + g^2 q$$

and

$$P_{k+1} = \frac{a^2 r P_k}{r(1 - \varepsilon_k P_k) + h^2 P_k} + \varepsilon_k^{-1} d_1^2 + g^2 q$$

with $\varepsilon_k P_k < 1$. Note from Lemma 6.1 that $1 - \varepsilon_k P_k > 0$ when $\varepsilon_k X_k < 1$. Hence, $P_{k+1} > 0$ as long as $P_k > 0$ and $\varepsilon_k < X_k^{-1}$.

Let

$$a = 0.5, \quad h = 1, \quad d_1 = 0.2, \quad g = 1, \quad P_{x_0} = 1, \quad q = 1, \quad r = 1$$

The above DREs become

$$X_{k+1} = \frac{0.25 X_k}{1 - \varepsilon_k X_k} + 0.04\varepsilon_k^{-1} + 1$$

and

$$P_{k+1} = \frac{0.25 P_k}{1 + (1 - \varepsilon_k) P_k} + 0.04\varepsilon_k^{-1} + 1$$

respectively.

Setting $\varepsilon_k = \varepsilon = 0.205$, the evolutions of X_k and P_k are shown in Figure 6.1. It is clear that $P_k \leq X_k$ and P_k converges after a number of iterations to $P_\infty = 1.3584$ while X_k takes a longer time to converge to $X_\infty = 2.1826$.

The robust *a priori* filter is given by Equation 6.24 with

$$A_{e,k} = \frac{0.5\varepsilon P_k}{1 - \varepsilon P_k}, \quad H_{e,k} = \frac{\varepsilon P_k}{1 - \varepsilon P_k}$$
$$K_k = \frac{0.5 Y_k}{1 + Y_k}, \quad Y_k = \frac{P_k}{1 - \varepsilon P_k}$$

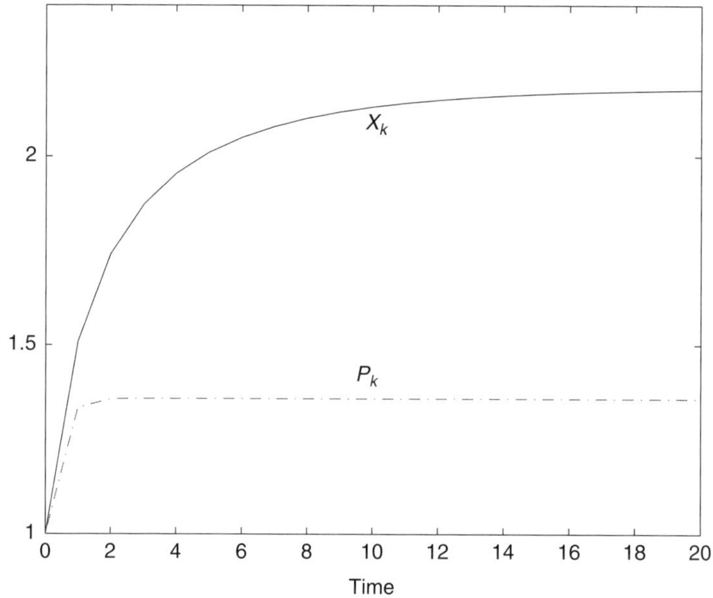

FIGURE 6.1
The evolutions of X_k and P_k.

We now consider the stationary case. It can be verified numerically that there exists a stabilizing solution $X \geq 0$ to the ARE 6.32 such that $1 - \varepsilon X > 0$ if $0 < \varepsilon < 0.21$. For $\varepsilon \in (0, 0.21)$, the larger the ε the smaller the P. Thus, by setting $\varepsilon = 0.205$, we obtain a close to optimal solution $X = 2.1826$ and $P = 1.3584$ that are the final values of the finite horizon case computed earlier. The robust stationary *a priori* filter is given by

$$\hat{x}_{k+1} = 0.2404\hat{x}_k + 0.3265z_k, \quad \hat{s}_k = \hat{x}_k$$

whereas the standard Kalman *a priori* filter without considering the parameter uncertainty is given by

$$\hat{x}_{k+1} = 0.2344\hat{x}_k + 0.2656z_k, \quad \hat{s}_k = \hat{x}_k$$

The filtering error covariance for the robust *a priori* filter in comparison with the standard Kalman *a priori* filter is shown in Figure 6.2. The maximum filtering error covariance for the robust filter is 1.3079, which is below the bound $trace(LPL^T) = 1.3584$. As observed from this simple example, the robust filter performs more consistently over the entire uncertainty range than the standard Kalman filter, which does not take into account possible parameter variation. $\square$

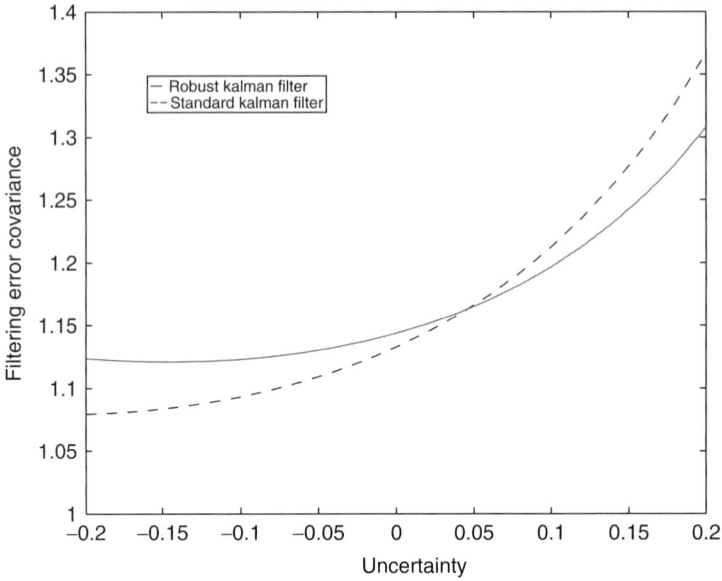

FIGURE 6.2
The comparison between the robust and standard Kalman *a priori* filters when δa varies from -0.2 to 0.2.

6.4 Convergence Analysis

6.4.1 Feasibility and Convergence Analysis

Recall that for systems without uncertainty, the solution to the DRE associated with the standard Kalman *a priori* filter converges to that of the ARE of its corresponding stationary filter under very mild conditions on system matrices (Anderson and Moore 1979). This, however, is no longer the case for the robust filter whose convergence depends on the given initial condition, that is, the covariance matrix of the initial state. For Example 6.2, if we set the initial condition $P_{x_0} = 2.6715$, the solution X_k of the DRE 6.19 becomes negative for some values of k as seen from Figure 6.3.

In this section, we shall study the feasibility and convergence properties of the solutions of the DREs 6.19 and 6.20 with the associated AREs 6.32 and 6.33. That is, we shall find conditions on the initial condition P_{x0} such that the solutions of the DREs 6.19 and 6.20 exist and converge to the stabilizing solutions of the AREs 6.32 and 6.33, respectively, as k approaches ∞. We assume here that $\varepsilon_k = \varepsilon$, a constant.

Note that the existence of finite horizon robust *a priori* filter requires the fulfillment at each step of a suitable matrix inequality (feasibility condition).

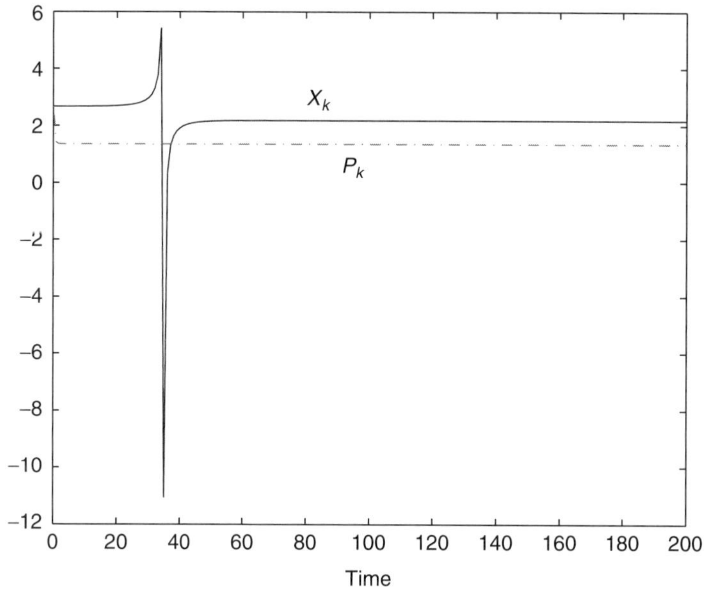

FIGURE 6.3
The solutions X_k and P_k when $P_{x_0} = 2.6715$.

In such case, we are interested in finding conditions (relating to the initial state uncertainty and the scaling parameter ε) under which we can ensure feasibility of the solutions of DREs over an arbitrarily long time interval, and convergence toward the optimal stationary robust filter (Bolzern and Maron 1999; Zhu *et al.* 2000).

DEFINITION 6.5 **(Feasible Solution)** *A real positive definite solution* X_k *(or* P_k*) of* DRE *6.19 (or* DRE *6.20) is termed a "feasible solution of* DRE *6.19 (or* DRE *6.20)" if it satisfies the condition* $X_k^{-1} - \varepsilon F^T F > 0$ *(or* $P_k^{-1} - \varepsilon F^T F > 0$*) at each step* $k \in [0, N]$.

Assumption 6.3 *A and* $A - \sqrt{\varepsilon}\tilde{D}_1 \bar{D}_2^T (\bar{D}_2 \bar{D}_2^T)^{-1}$ *are invertible for some given* $\varepsilon > 0$, *where* $\tilde{D}_1$ *and* $\bar{D}_2$ *are as in Equation 6.23 by dropping the subscript* k, *and ARE 6.32 and ARE 6.33 admit stabilizing and feasible solutions* $X_s > 0$ *and* $P_s > 0$, *respectively.*

Under Assumption 6.3, the feasibility and convergence analysis can be stated as follows: *Given an arbitrarily large N, find suitable conditions on the initial state covariance* P_{x_0} *such that the solutions* X_k *and* P_k *are feasible at every step* $k \in [0, N]$, *and* X_k *and* P_k *converge respectively to the stabilizing solutions* X_s *and* P_s *as* $N \to \infty$ *and* $k \to \infty$.

To state the conditions on P_{x_0} for feasibility and convergence, we introduce two Lyapunov equations. The first one is

$$\hat{A}_1^T Y_1 \hat{A}_1 - Y_1 = -M_{1-} \tag{6.35}$$

where

$$\begin{aligned}
\hat{A}_1 &= A + AX_sF^T\left(\varepsilon^{-1}I - FX_sF^T\right)^{-1}F \\
M_1 &= \hat{A}_1^{-T}\left[G_1 - F^T\left(\varepsilon^{-1}I - FX_sF^T\right)^{-1}F\right]\hat{A}_1^{-1} - G_1 \\
G_1 &= -X_s^{-1} - X_s^{-1}\left(\varepsilon F^TF - X_s^{-1}\right)^{-1}X_s^{-1}
\end{aligned}$$

The definition of M_- is: $M = M_+ + M_-$, where $M_+ \geq 0$ ($M_- \leq 0$) and the nonzero eigenvalues of M_+ (the nonzero eigenvalues of M_-) are the positive (negative) eigenvalues of M. The second Lyapunov equation is

$$\hat{A}_2^T Y_2 \hat{A}_2 - Y_2 = -M_{2-} \tag{6.36}$$

where

$$\begin{aligned}
\hat{A}_2 &= A - \left(AP_s\tilde{H}^T + \tilde{D}_1\tilde{D}_2^T\right)\left(\tilde{H}P_s\tilde{H}^T + \tilde{R}\right)^{-1}\tilde{H} \\
M_2 &= \left(\hat{A}_2^{-1}\right)^T\left[G_2 + \tilde{H}^T\left(\tilde{H}P_s\tilde{H}^T + \tilde{R}\right)^{-1}\tilde{H}\right]\hat{A}_2^{-1} - G_2 \\
G_2 &= -P_s^{-1} - P_s^{-1}\left(\varepsilon F^TF - P_s^{-1}\right)^{-1}P_s^{-1}
\end{aligned}$$

and $\tilde{H}$, $\tilde{R}$ are as defined in Equations 6.22 and 6.23, respectively, by dropping the subscript k. Finally, we let

$$\Xi_i = G_i - Y_i + M_{i-}, \quad i = 1, 2 \tag{6.37}$$

It can be known that Ξ_i ($i = 1, 2$) are positive semi-definite.

The following theorem provides a sufficient condition for ensuring feasibility and convergence of the solutions of DRE 6.19 and DRE 6.20 over $[0, \infty)$.

THEOREM 6.3

Consider the DREs 6.19 and 6.20, and let Assumptions 6.1 through 6.3 hold. Then the solutions X_k of DRE 6.19 and P_k of DRE 6.20 are feasible over $[0, \infty)$, and converge to the stabilizing solutions X_s and P_s, respectively, as $k \to \infty$, if for a sufficiently small scalar $\rho > 0$, the positive initial state P_{x_0} satisfies

$$\begin{bmatrix} P_{x_0} & 0 \\ 0 & P_{x_0} \end{bmatrix} < \begin{bmatrix} X_s + (\Xi_1 + \rho I)^{-1} & 0 \\ 0 & P_s + (\Xi_2 + \rho I)^{-1} \end{bmatrix} \tag{6.38}$$

where Ξ_i ($i = 1, 2$) is defined as in Equation 6.37.

Example 6.3

Consider Example 6.2 again. With $\varepsilon = 0.205$, we know that there exist stabilizing solutions $X_s = 2.1826$ and $P_s = 1.3564$ to Equations 6.32 and 6.33, respectively. It can be computed that

$$\hat{A}_1 = 0.9049, \quad G_1 = 0.3710, \quad M_1 = -0.3710$$

Thus, $M_{1-} = -0.3710$. Solving Lyapunov Equation 6.35 yields that $Y_1 = -2.0474$. Similarly, we can obtain

$$\hat{A}_2 = 0.2404, \quad G_2 = 0.2841, \quad M_2 = 11.2465.$$

In this case, $M_{2-} = 0$ and $Y_2 = 0$. Hence, $\Xi_1 = 2.0474$ and $\Xi_2 = 0.2841$. By applying Theorem 6.3, a sufficient condition for feasibility and convergence of the DREs 6.19 and 6.20 is given by

$$0 < P_{x_0} < 2.6710$$

In fact, this condition is also necessary for the example as observed from Figure 6.3 where $P_{x_0} = 2.6715$ is adopted. □

6.4.2 ε–Switching Strategy

In the last subsection, we learned that, for a given ε, the convergence and feasibility of X_k and P_k depend on the initial state covariance bound P_{x_0} and parameter ε. Since P_{x_0} is usually given *a priori*, it is necessary to study the following problem: *given $P_{x_0} > 0$ and an obtained optimal $\varepsilon_{opt} \in (0, \bar{\varepsilon})$, which minimizes trace$(P)$, is it possible to drive X_k and P_k from $P_{x_0} > 0$ to $X_s(\varepsilon_{opt})$ and $P_s(\varepsilon_{opt})$ by means of a suitable time-varying function ε_k?*

To solve this problem, we shall apply a strategy which varies ε according to a piecewise-constant pattern $\{\varepsilon_i,\ i = 0, 1, \ldots, \bar{n},\ \varepsilon_{\bar{n}} = \varepsilon_{opt}\}$ at switching times $\{T_i,\ i = 0, 1, \ldots, \bar{n}\}$. We determine the sequences $\{\varepsilon_i\}$ and $\{T_i\}$ such that the solutions X_k and P_k with

$$\varepsilon_k = \begin{cases} \varepsilon_i & (T_i \le k < T_{i+1}), \qquad i = 0, 1, \ldots, \bar{n} - 1 \\ \varepsilon_{opt} & (k \ge T_{\bar{n}}) \end{cases}$$

are feasible and converge to $X_s(\varepsilon_{opt})$ and $P_s(\varepsilon_{opt})$, respectively, and the minimum *trace*(P) is then obtained.

We note that if the initial covariance P_{x_0} is sufficiently small and satisfies Equation 6.38, the optimal value ε_{opt} with minimum *trace*$(P_{\varepsilon_{opt}})$ can be computed directly and convergence and feasibility conditions can also be satisfied over $[0, \infty)$. However, if the initial covariance P_{x_0} is very large, we cannot approach ε_{opt} immediately. In fact, we must first select a suitable $\varepsilon_0 \neq \varepsilon_{opt}$ which satisfies Equation 6.38 such that feasibility and convergence are guaranteed with this P_{x_0}. Then ε_i $(i = 1, 2, \ldots, \bar{n})$ could approach ε_{opt} through a finite number of steps occurring at suitable instants. The selection of such instants and the changes in ε_i must satisfy Equation 6.38 and follow the $\varepsilon-$switching algorithm given below.

Step 1. Given an *a priori* initial state $P_0^0 = X_0^0 = P_{x_0}$ and let $i = 0$. In this algorithm, X_0^i and P_0^i $(i = 1, 2, \ldots)$ denote the (initial) state of DREs 6.19 and 6.20 at T_i, respectively. If $P_{x_0} \le P_0^{opt}$, where P_0^{opt} denotes the maximum possible initial state which satisfies Equation 6.38 with $\varepsilon = \varepsilon_{opt}$, then set $\varepsilon_k = \varepsilon_{opt}$ for the subsequent recursions.

Step 2. Let initial state be $X_0 = X_0^i$ and $P_0 = P_0^i$, find an ε_i which satisfies Equation 6.38 and minimizes $|\varepsilon_{opt} - \varepsilon_i|$. Then the stabilizing solutions $X_s(\varepsilon_i)$ and $P_s(\varepsilon_i)$ of AREs 6.32 and 6.33, respectively, can be computed.

Step 3. Let $\varepsilon = \varepsilon_i$, $X_0 = X_0^i$, and $P_0 = P_0^i$, and iteratively compute $X_k(\varepsilon_i)$ of DRE 6.19 and $P_k(\varepsilon_i)$ of DRE 6.20 until they almost approach the stabilizing solutions $X_s(\varepsilon_i)$ and $P_s(\varepsilon_i)$, respectively. We denote $X_k(\varepsilon_i)$ and $P_k(\varepsilon_i)$ at this instant as $\hat{X}_k(\varepsilon_i)$ and $\hat{P}_k(\varepsilon_i)$, respectively.

Step 4. If $\varepsilon_i = \varepsilon_{opt}$, then set $\varepsilon_k = \varepsilon_{opt}$ for the subsequent recursions. Otherwise, go to *Step 5.*

Step 5. Let $\hat{X}_k(\varepsilon_i)$ and $\hat{P}_k(\varepsilon_i)$ be new initial states X_0^{i+1} and P_0^{i+1}, respectively, and let $i = i + 1$, go to *Step 2.*

Example 6.4

Consider the following uncertain discrete-time system:

$$x_{k+1} = \begin{bmatrix} 0 & -0.5 \\ 1 & 1+\delta \end{bmatrix} x_k + \begin{bmatrix} -6 \\ 1 \end{bmatrix} w_k \tag{6.39}$$

$$z_k = \begin{bmatrix} -100 & 10 \end{bmatrix} x_k + v_k \tag{6.40}$$

where δ is an uncertain parameter satisfying $|\delta| \leq 0.3$. Note that the above system is of the form in Equations 6.1 and 6.2 with $D_1 = \left[\begin{smallmatrix} 0 \\ 10 \end{smallmatrix}\right], D_2 = 0, F = [0 \quad 0.03], Q = 1$, and $R = 1$.

We now want to estimate the signal $s_k = Lx_k$ with $L = [1 \quad 0]$. In such case, the filtering error variance is bounded by $tr(LPL^T)$. We apply the stationary filter design procedure in Section 6.3 to the system (Equations 6.39 and 6.40). By applying Theorem 6.2, we obtain $\bar{\varepsilon} = 1.38$. The optimal ε is also equal to $\bar{\varepsilon}$, i.e., $\varepsilon_{opt} = 1.38$. Then,

$$X_s(\varepsilon_{opt}) = \begin{bmatrix} 225.5 & -195.3 \\ -195.3 & 390.4 \end{bmatrix}, \qquad P_s(\varepsilon_{opt}) = \begin{bmatrix} 69.2 & -79.0 \\ -79.0 & 234.1 \end{bmatrix}$$

The corresponding optimal robust *a priori* filter is given by

$$\hat{x}_{k+1} = \bar{A}\hat{x}_k + K\left[z_k - \bar{H}\hat{x}_k\right] \tag{6.41}$$

where

$$\bar{A} = \begin{bmatrix} 0 & -0.71 \\ 1.00 & 1.27 \end{bmatrix}, \qquad \bar{H} = \begin{bmatrix} -100 & 27.94 \end{bmatrix}, \qquad K = \begin{bmatrix} -0.007 \\ 0.005 \end{bmatrix}$$

and the optimal filtering error variance bound is 69.2.

The actual filtering error variances with $\delta = 0$, $\delta = 0.3$, and $\delta = -0.3$ are 51.06, 54.44, and 52.80, respectively.

Now we analyze feasibility and convergence properties. For the finite horizon case, according to Theorem 6.3, when the initial state P_{x_0} satisfies

$$0 < P_{x_0} < \begin{bmatrix} 249.45 & -213.23 \\ -213.23 & 414.38 \end{bmatrix} = \bar{P}_0^{up} \tag{6.42}$$

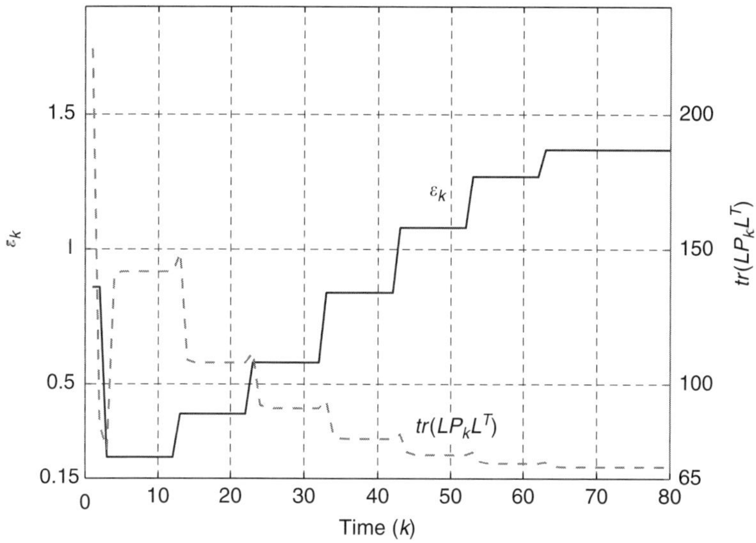

FIGURE 6.4
Time courses of ε-switching strategy—right vertical scale: $tr(LP_kL^T)$; left vertical scale: ε_k.

then X_k and P_k are both feasible and converge to the stabilizing solution $X_s(\varepsilon_{opt})$ and $P_s(\varepsilon_{opt})$, respectively, as $k \to \infty$. If the initial condition does not satisfy Equation 6.42, for example, $P_{x_0} = \bar{P}_0^{up} + diag\{200, 200\} > \bar{P}_0^{up}$, then the feasibility is lost at the first step itself. In such case, we can apply the $\varepsilon-$switching strategy by finding a sequence $\{\varepsilon_i,\ i=0,1,2,\ldots,\bar{n},\ \varepsilon_{\bar{n}} = \varepsilon_{opt} = 1.38\}$, such that X_k and P_k are feasible and converge to $X_s(\varepsilon_{opt})$ and $P_s(\varepsilon_{opt})$ with $trace(LP_{\varepsilon_{opt}}L^T) = 69.2$. Simulation results are shown in Figure 6.4. □

6.5 Linear Matrix Inequality Approach

Here, we only consider the stationary filter design using a linear matrix inequality (LMI) approach (Xie 2005). Recall the system under investigation:

$$x_{k+1} = A_\delta x_k + Gw_k \tag{6.43}$$

$$z_k = H_\delta x_k + v_k \tag{6.44}$$

$$s_k = Lx_k \tag{6.45}$$

where $x_k \in R^n$ is the state vector, $z_k \in R^p$ the measurement, $s_k \in R^r$ the signal to be estimated, $w_k \in R^l$ the noise input and $v_k \in R^p$ the measurement noise. The uncertain matrices A_δ and H_δ are described by

$$\begin{bmatrix} A_\delta \\ H_\delta \end{bmatrix} = \begin{bmatrix} A \\ H \end{bmatrix} + \begin{bmatrix} D_1 \\ D_2 \end{bmatrix} \Delta_k F \tag{6.46}$$

where Δ_k is an unknown real time-varying matrix having a block diagonal structure:

$$\Delta_k = \text{diag}\{\Delta_{1,k}, \Delta_{2,k}, \ldots, \Delta_{l,k}\} \tag{6.47}$$

with $\Delta_{i,k} \in R^{\alpha_i \times \beta_i}$ satisfying $\Delta_{i,k}^T \Delta_{i,k} \leq I$, $i = 1, 2, \ldots, l$. The block diagonal structure offers more flexibility in characterizing uncertainty in practical systems and can lead to a less conservative filter design than the single block uncertainty in general. Observe that the uncertainty in Example 6.1 has the above block diagonal structure.

We consider a linear time-invariant filter of the general form:

$$\hat{x}_{k+1} = \hat{A}\hat{x}_k + Kz_k \tag{6.48}$$

$$\hat{s}_k = \hat{L}\hat{x}_k + Jz_k \tag{6.49}$$

where $\hat{A}, K, \hat{L}$, and J are matrices to be determined. Obviously, if J is set to zero, Equations 6.48 and 6.49 will be *a priori* filter. If $J \neq 0$, the filter is known to be *a posteriori* filter.

Our objective is to design a time-invariant filter of the above form such that for any uncertainty of Equation 6.47 the filtering error dynamic is asymptotically stable and

$$E\left[(s_k - \hat{s}_k)(s_k - \hat{s}_k)^T\right] \leq S \tag{6.50}$$

for a constant matrix $S = S^T \geq 0$. Furthermore, $trace(S)$ is to be minimized.

6.5.1 Robust Filter for Systems with Norm-Bounded Uncertainty

In this subsection we shall address the robust Kalman filtering of systems with norm-bounded uncertainty through the LMI approach.

Denote $\xi_k = [x_k^T \quad \hat{x}_k^T]^T$, $\tilde{s}_k = s_k - \hat{s}_k$, and $\eta_k = [w_k^T \quad v_k^T]^T$. Observe that η_k is a random vector with zero mean and covariance matrix of $\bar{Q} = diag\{Q, R\}$. Then, it follows from Equations 6.1 through 6.3 and 6.47 through 6.49 that

$$\xi_{k+1} = (\bar{A} + \bar{D}_1\Delta_k\bar{F})\xi_k + \bar{G}\eta_k \tag{6.51}$$

$$\tilde{s}_k = (\bar{H} + \bar{D}_2\Delta_k\bar{F})\xi_k + \bar{J}\eta_k \tag{6.52}$$

where

$$\bar{A} = \begin{bmatrix} A & 0 \\ KH & \hat{A} \end{bmatrix}, \qquad \bar{D}_1 = \begin{bmatrix} D_1 \\ KD_2 \end{bmatrix}, \qquad \bar{G} = \begin{bmatrix} G & 0 \\ 0 & K \end{bmatrix} \tag{6.53}$$

$$\bar{F} = \begin{bmatrix} F & 0 \end{bmatrix}, \qquad \bar{J} = -\begin{bmatrix} 0 & J \end{bmatrix} \tag{6.54}$$

$$\bar{H} = \begin{bmatrix} L - JH & -\hat{L} \end{bmatrix}, \qquad \bar{D}_2 = -JD_2 \tag{6.55}$$

In association with the system 6.51, we define the Lyapunov matrix inequality

$$(\bar{A} + \bar{D}_1\Delta_k\bar{F})\Sigma(\bar{A} + \bar{D}_1\Delta_k\bar{F})^T - \Sigma + \bar{G}\bar{Q}\bar{G}^T \leq 0 \tag{6.56}$$

where Σ is a constant matrix that provides an upper bound for the covariance matrix of ξ_k for all admissible uncertainties (Equation 6.47).

In view of Equation 6.52, the design objective is

$$\min trace(S)$$

subject to Equation 6.56 and

$$(\bar{H} + \bar{D}_2\Delta_k\bar{F})\Sigma(\bar{H} + \bar{D}_2\Delta_k\bar{F})^T + \bar{J}\bar{Q}\bar{J}^T \le S \tag{6.57}$$

THEOREM 6.4

Suppose that the system (Equations 6.43 through 6.45) is quadratically stable. Then, there always exists a robust Kalman filter of the form 6.48 and 6.49. Further, the optimal robust Kalman filter can be obtained by solving the following optimization

$$\min_{X,Y,\mathcal{A},\mathcal{K},\mathcal{L},J,\varepsilon_1,\ldots,\varepsilon_l,\varepsilon'_1,\ldots,\varepsilon'_l} trace(S) \tag{6.58}$$

subject to

$$\begin{bmatrix} -X & -Y & XA+\mathcal{K}H & \mathcal{A} & XGQ^{\frac{1}{2}} & \mathcal{K}R^{\frac{1}{2}} & XD_1+\mathcal{K}D_2 \\ * & -Y & YA & YA & YGQ^{\frac{1}{2}} & 0 & YD_1 \\ * & * & -X+F^T\Gamma_2F & -Y+F^T\Gamma_2F & 0 & 0 & 0 \\ * & * & * & -Y+F^T\Gamma_2F & 0 & 0 & 0 \\ * & * & * & * & -I & 0 & 0 \\ * & * & * & * & * & -I & 0 \\ * & * & * & * & * & * & -\Gamma_1 \end{bmatrix} < 0 \tag{6.59}$$

$$\begin{bmatrix} -S & L-JH & L-\mathcal{L} & -JR^{\frac{1}{2}} & -JD_2 \\ * & -X+F^T\Gamma'_2F & -Y+F^T\Gamma'_2F & 0 & 0 \\ * & * & -Y+F^T\Gamma'_2F & 0 & 0 \\ * & * & * & -I & 0 \\ * & * & * & * & -\Gamma'_1 \end{bmatrix} < 0 \tag{6.60}$$

where "$*$" denotes entry that can be deduced from the symmetry property of the matrix and

$$\begin{aligned} \Gamma_1 &= diag\{\varepsilon_1 I_{\alpha_1}, \varepsilon_2 I_{\alpha_2}, \ldots, \varepsilon_l I_{\alpha_l}\} \\ \Gamma_2 &= diag\{\varepsilon_1 I_{\beta_1}, \varepsilon_2 I_{\beta_2}, \ldots, \varepsilon_l I_{\beta_l}\} \\ \Gamma'_1 &= diag\{\varepsilon'_1 I_{\alpha_1}, \varepsilon'_2 I_{\alpha_2}, \ldots, \varepsilon'_l I_{\alpha_l}\} \\ \Gamma'_2 &= diag\{\varepsilon'_1 I_{\beta_1}, \varepsilon'_2 I_{\beta_2}, \ldots, \varepsilon'_l I_{\beta_l}\} \end{aligned}$$

are scaling matrices.

Indeed, given the optimal solution $(X>0, Y>0, \mathcal{A}, \mathcal{K}, \mathcal{L}, J, \varepsilon_1, \ldots, \varepsilon_l, \varepsilon'_1, \ldots, \varepsilon'_l)$, *an optimal robust Kalman filter can be computed as*

$$K = M^{-1}\mathcal{K} \tag{6.61}$$

$$\hat{L} = (\mathcal{L} - JH)Y^{-1}W^{-T} \tag{6.62}$$

$$\hat{A} = M^{-1}(\mathcal{A} - XA - \mathcal{K}H)Y^{-1}W^{-T} \tag{6.63}$$

where M and W are any invertible matrices satisfying

$$MW^T = I - XY^{-1} \tag{6.64}$$

Note that J is a decision variable of the LMIs 6.59 and 6.60 that has been obtained from the optimization.

The proof of the theorem can be found at the end of this chapter.

It is observed that Equations 6.59 and 6.60 are linear in X, Y, $\mathcal{A}$, $\mathcal{K}$, $\mathcal{L}$, J, ε_1, $\varepsilon_2, \ldots, \varepsilon_l, \varepsilon_1'$, $\varepsilon_2', \ldots, \varepsilon_l'$. Therefore, the optimal filter design involves a convex optimization which can be easily solved by using the Matlab LMI Toolbox. Although the Riccati equation approach in the last section provides good insights on the optimal robust filter in terms of its structure and how uncertainty affects the optimal filter, it requires searching for a scaling parameter. In the LMI approach of this section, even with the block diagonal uncertainty the scaling parameters can be obtained together with the optimal filter parameters by convex optimization which is very convenient. Note that Theorem 6.4 also allows to handle both the optimal robust *a priori* filtering ($J = 0$) and *a posteriori* filtering ($J \neq 0$).

It can be known that $I - XY^{-1}$ is invertible if Equations 6.59 and 6.60 hold (see Problem 7). Thus, there exist invertible matrices M and W in Equation 6.64. In practice, for simplicity, we may let $W = I$ and thus $M = I - XY^{-1}$. The choices of M and W do not affect the filter performance and the filter transfer function but will give rise to different state space realizations of the filter. In fact, the transfer function of the filter (Equations 6.48 and 6.49) is given by

$$\begin{aligned}
\mathcal{F}(z) &= \hat{L}(zI - \hat{A})^{-1}K + J \\
&= (\mathcal{L} - JH)Y^{-1}W^{-T}\left[zI - M^{-1}(\mathcal{A} - XA - \mathcal{K}H)Y^{-1}W^{-T}\right]^{-1}M^{-1}\mathcal{K} + J \\
&= (\mathcal{L} - JH)Y^{-1}\left[z(I - XY^{-1}) - (\mathcal{A} - XA - \mathcal{K}H)Y^{-1}\right]^{-1}\mathcal{K} + J
\end{aligned}$$

which is independent of the choices of M and W.

Alternatively, by a dual approach, Equations 6.59 and 6.60 may be replaced respectively by

$$\begin{bmatrix}
-X + F^T\Gamma_2 F & * & * & * & * & * \\
-Y + F^T\Gamma_2 F & -Y + F^T\Gamma_2 F & * & * & * & * \\
XA + \mathcal{K}H & \mathcal{A} & -X & * & * & * \\
YA & YA & -Y & -Y & * & * \\
L - JH & L - \mathcal{L} & 0 & 0 & -I & * \\
0 & 0 & D_1^T X + D_2^T \mathcal{K} & D_1^T Y & -D_2^T J^T & -\Gamma_1
\end{bmatrix} < 0 \tag{6.65}$$

$$\begin{bmatrix}
-S_1 & * & * & * & * \\
-S_{21} & -S_2 & * & * & * \\
XGQ^{\frac{1}{2}} & \mathcal{K}R^{\frac{1}{2}} & -X & * & * \\
YGQ^{\frac{1}{2}} & 0 & -Y & -Y & * \\
0 & -JR^{\frac{1}{2}} & 0 & 0 & -I
\end{bmatrix} < 0 \tag{6.66}$$

where

$$S = \begin{bmatrix} S_1 & S_{21}^T \\ S_{21} & S_2 \end{bmatrix}$$

Equations 6.65 and 6.66 are simpler than Equations 6.59 and 6.60. However, which set of LMIs will lead to a better robust filter is not conclusive as both provide only sufficient conditions and upper bounds of filtering error variance.

6.5.2 Robust Filtering for Systems with Polytopic Uncertainty

Consider again the dynamical system 6.43 through 6.45. A polytopic representation of the uncertainty (Shaked *et al.* 2001; Geromel *et al.* 2002) in the matrices A_δ and H_δ is given by

$$\Omega_P = \left\{ (A_\delta, H_\delta) \mid (A_\delta, H_\delta) = \sum_{i=1}^{N_v} \alpha_i \big(A^{(i)}, H^{(i)}\big), \quad \alpha_i \geq 0, \ \sum_{i=1}^{N_v} \alpha_i = 1 \right\} \tag{6.67}$$

where $(A^{(i)}, H^{(i)})$ is the i-th vertex of the polytope.

Example 6.5

Consider again Example 6.1. The uncertainty in A can be represented by a polytope with four vertices:

$$A^{(1)} = \begin{bmatrix} 0.4 & -0.2 \\ 1 & 0 \end{bmatrix}, \qquad A^{(2)} = \begin{bmatrix} 0.4 & -0.4 \\ 1 & 0 \end{bmatrix}$$

$$A^{(3)} = \begin{bmatrix} 0.6 & -0.2 \\ 1 & 0 \end{bmatrix}, \qquad A^{(4)} = \begin{bmatrix} 0.6 & -0.4 \\ 1 & 0 \end{bmatrix}$$ □

As compared to the norm-bounded uncertainty representation, the polytopic representation of uncertainty is more flexible and gives more accurate description of the uncertainty. This implies that the robust filter designed based on the polytopic uncertainty representation could potentially be less conservative, which will be demonstrated by an example later in this chapter. However, the cost of computing the optimal robust filter could be higher, especially when the number of vertices is large.

First, denote $\xi_k = [x_k^T \ \hat{x}_k^T]^T$ and $\eta_k = [w_k^T \ v_k^T]^T$. It follows from Equations 6.43 through 6.45, 6.48 and 6.49 that

$$\xi_{k+1} = \bar{A}_\delta \xi_k + \bar{G} \eta_k \tag{6.68}$$

$$s_k - \hat{s}_k = \bar{H}_\delta \xi_k + \bar{J} \eta_k \tag{6.69}$$

where

$$\bar{A}_\delta = \begin{bmatrix} A_\delta & 0 \\ K H_\delta & \hat{A} \end{bmatrix}, \qquad \bar{G} = \begin{bmatrix} G & 0 \\ 0 & K \end{bmatrix} \tag{6.70}$$

$$\bar{H}_\delta = \begin{bmatrix} L - J H_\delta & -\hat{L} \end{bmatrix}, \qquad \bar{J} = -\begin{bmatrix} 0 & J \end{bmatrix} \tag{6.71}$$

Based on the discussion in the previous section, for any uncertainty from Ω_P, an upper bound for the covariance of the filtering error, $s_k - \hat{s}_k$, of the system (Equations 6.68 and 6.69) can be computed by the following minimization:

$$\min_{\Sigma_\delta} trace(\bar{H}_\delta \Sigma_\delta \bar{H}_\delta^T + \bar{J}\bar{Q}\bar{J}^T) \tag{6.72}$$

subject to

$$\bar{A}_\delta \Sigma_\delta \bar{A}_\delta^T - \Sigma_\delta + \bar{G}\bar{Q}\bar{G}^T < 0 \tag{6.73}$$

where

$$\bar{Q} = \begin{bmatrix} Q & 0 \\ 0 & R \end{bmatrix}$$

We note that the robust filtering methods discussed thus far are all based on a fixed Lyapunov matrix for all admissible uncertainties in the uncertainty set, which is clearly a source of conservatism. In this subsection, we shall discuss a design method based on a parameter-dependent Lyapunov matrix.

First, if the matrices A_δ and H_δ take their values at their i-th vertex of the polytope, i.e., $A_\delta = A^{(i)}$ and $H_\delta = H^{(i)}$, then with $\Sigma_\delta = \Sigma^{(i)}$ and by applying Schur complements, Equation 6.73 is equivalent to

$$\begin{bmatrix} -\left[\Sigma_1^{(i)}\right]^{-1} & \bar{A}^{(i)} & \bar{G}\bar{Q}^{\frac{1}{2}} \\ \bar{A}^{(i)T} & -\Sigma_1^{(i)} & 0 \\ \bar{Q}^{\frac{1}{2}}\bar{G} & 0 & -I \end{bmatrix} < 0 \tag{6.74}$$

where

$$\Sigma_1^{(i)} = [\Sigma^{(i)}]^{-1} \quad \text{and} \quad \bar{A}^{(i)} = \begin{bmatrix} A^{(i)} & 0 \\ KH^{(i)} & \hat{A} \end{bmatrix}$$

Since for any given matrix Π with appropriate dimension,

$$\left[\Pi^{-1} - \left(\Sigma_1^{(i)}\right)^{-1}\right]^T \Sigma_1^{(i)} \left[\Pi^{-1} - \left(\Sigma_1^{(i)}\right)^{-1}\right] \geq 0$$

i.e., $\Pi^{-1} + \Pi^{-T} - \Pi^{-T}\Sigma_1^{(i)}\Pi^{-1} \leq (\Sigma_1^{(i)})^{-1}$, then Equation 6.74 holds if

$$\begin{bmatrix} -\Pi^{-1} - \Pi^{-T} + \Pi^{-T}\Sigma_1^{(i)}\Pi^{-1} & \bar{A}^{(i)} & \bar{G}\bar{Q}^{\frac{1}{2}} \\ \bar{A}^{(i)T} & -\Sigma_1^{(i)} & 0 \\ \bar{Q}^{\frac{1}{2}}\bar{G} & 0 & -I \end{bmatrix} < 0 \tag{6.75}$$

By multiplying Equation 6.75 from the left and the right by $diag\{\Pi^T, I, I\}$ and its transpose, respectively, we have

$$\begin{bmatrix} -\Pi - \Pi^T + \Sigma_1^{(i)} & \Pi^T\bar{A}^{(i)} & \Pi^T\bar{G}\bar{Q}^{\frac{1}{2}} \\ \bar{A}^{(i)T}\Pi & -\Sigma_1^{(i)} & 0 \\ \bar{Q}^{\frac{1}{2}}\bar{G}^T\Pi & 0 & -I \end{bmatrix} < 0, \qquad i = 1, 2, \ldots, N_v \tag{6.76}$$

Since the matrix Π is independent of the vertices of the polytope, for any uncertainty inside the polytope by taking the linear combination of Equation 6.76 with respect to the corresponding α_i, $i = 1,\ 2, \ldots, N_v$, of the uncertainty we can know that $(\sum_{i=1}^{N_v} \alpha_i \Sigma_1^{(i)})^{-1}$ is a Lyapunov matrix satisfying Equation 6.73. Hence, if our design is based on Equation 6.76, we actually consider a parameter (α_i) dependent Lyapunov matrix. A robust Kalman filter based on this parameter-dependent Lyapunov matrix approach is given in the theorem below.

THEOREM 6.5

Consider the system (Equations 6.43 through 6.45) with polytopic uncertainty Ω_P. An estimate of s_k that minimizes an upper bound of the variance of $s_k - \hat{s}_k$ over the polytope Ω_P exists if the solution set of the LMIs 6.79 and 6.80 below is not empty. In this situation, the optimal robust Kalman filter can be given by Equations 6.48 and 6.49 with

$$\hat{A} = \Psi^{-1}\mathcal{A}, \quad K = \Psi^{-1}\mathcal{K}, \quad \hat{L} = \mathcal{L}, \tag{6.77}$$

where matrices Ψ, $\mathcal{A}$, $\mathcal{K}$, $\mathcal{L}$, and the remaining filter parameter J are solutions to the following convex optimization:

$$\min_{\Psi, X, Y, \mathcal{K}, \mathcal{L}, J, \Lambda_j^{(i)},\ j=1,2,3} \quad trace(S) \tag{6.78}$$

subject to

$$\begin{bmatrix} -S & L - JH^{(i)} - \mathcal{L} & L - JH^{(i)} & -JR^{\frac{1}{2}} \\ * & -\Lambda_1^{(i)} & -\Lambda_2^{(i)} & 0 \\ * & * & -\Lambda_3^{(i)} & 0 \\ * & * & * & -I \end{bmatrix} < 0 \tag{6.79}$$

and

$$\begin{bmatrix} -Y - Y^T + \Lambda_1^{(i)} & -Y^T - X - \Psi^T + \Lambda_2^{(i)} & Y^T A^{(i)} & Y^T A^{(i)} & Y^T G Q^{\frac{1}{2}} & 0 \\ * & -X - X^T + \Lambda_3^{(i)} & X^T A^{(i)} + \mathcal{K} H^{(i)} + \mathcal{A} & X^T A^{(i)} + \mathcal{K} H^{(i)} & X^T G Q^{\frac{1}{2}} & \mathcal{K} R^{\frac{1}{2}} \\ * & * & -\Lambda_1^{(i)} & -\Lambda_2^{(i)} & 0 & 0 \\ * & * & * & -\Lambda_3^{(i)} & 0 & 0 \\ * & * & * & * & -I & 0 \\ * & * & * & * & * & -I \end{bmatrix} < 0 \tag{6.80}$$

for $i = 1, 2, \ldots, N_v$.

The proof can be found at the end of this chapter.

REMARK 6.3 Observe that Equations 6.79 and 6.80 are linear in unknown parameters $\Psi, X, Y, \mathcal{A}, \mathcal{K}, \mathcal{L}, , J, \Lambda_j^{(i)}$, $j = 1, 2, 3$. Hence, the optimization in Equation 6.78 can be solved by convex optimization using the Matlab LMI Toolbox. ▮

Example 6.6

Consider again Example 6.4:

$$x_{k+1} = \begin{bmatrix} 0 & -0.5 \\ 1 & 1+\delta \end{bmatrix} x_k + \begin{bmatrix} -6 \\ 1 \end{bmatrix} w_k \tag{6.81}$$

$$z_k = \begin{bmatrix} -100 & 10 \end{bmatrix} x_k + v_k \tag{6.82}$$

and let

$$s_k = \begin{bmatrix} 1 & 0 \end{bmatrix} x_k \tag{6.83}$$

where w_k and v_k are uncorrelated zero-mean white noise signals with unit variances, respectively. δ is the uncertain parameter satisfying $|\delta| \leq \delta_0$, where δ_0 is a known uncertainty bound. Obviously, the uncertainty in this system can be characterized by a norm-bounded uncertainty.

Consider $\delta_0 = 0.3$. The uncertain system matrix can be put into the form 6.4 with single block uncertainty and

$$A = \begin{bmatrix} 0 & -0.5 \\ 1 & 1 \end{bmatrix}, \qquad D_1 = \begin{bmatrix} 0 \\ 10 \end{bmatrix}, \qquad D_2 = 0, \qquad F = [0 \quad 0.03]$$

Using the results in Theorems 6.2 and 6.4 (applying the LMIs 6.65 and 6.66 instead of 6.59 and 6.60), the optimal robust *a priori* filter (setting $J = 0$) based on the Riccati equation approach and the LMI approach can be obtained. The optimal scaling parameter for the LMI approach is obtained as $\varepsilon = 68.3691$ and the corresponding optimal *a priori* filter is

$$\hat{A} = \begin{bmatrix} -0.0273 & 0.1048 \\ -0.0394 & 0.8775 \end{bmatrix}, \qquad K = \begin{bmatrix} -0.2989 \times 10^{-2} \\ 0.1788 \times 10^{-3} \end{bmatrix}$$

$$\hat{L} = \begin{bmatrix} 0.1169 & -0.2960 \end{bmatrix}$$

Theorem 6.4 also allows us to compute the optimal robust *a posteriori* filter (with $J \neq 0$). By applying Theorem 6.4, we obtain the optimal scaling parameter $\varepsilon = 2.7345$ and the optimal *a posteriori* filter of Equations 6.48 and 6.49 with

$$\hat{A} = \begin{bmatrix} 0.0544 & -23.0172 \\ -164.7602 & 5.4716 \times 10^5 \end{bmatrix}, \qquad K = \begin{bmatrix} -0.6728 \times 10^{-4} \\ 0.1206 \times 10^{-7} \end{bmatrix}$$

$$\hat{L} = \begin{bmatrix} 0.0355 & -50.0789 \end{bmatrix}, \qquad J = -0.0084$$

We can also characterize the uncertainty by a polytope with two vertices:

$$A^{(1)} = \begin{bmatrix} 0 & -0.5 \\ 1 & 0.7 \end{bmatrix}, \qquad A^{(2)} = \begin{bmatrix} 0 & -0.5 \\ 1 & 1.3 \end{bmatrix}$$

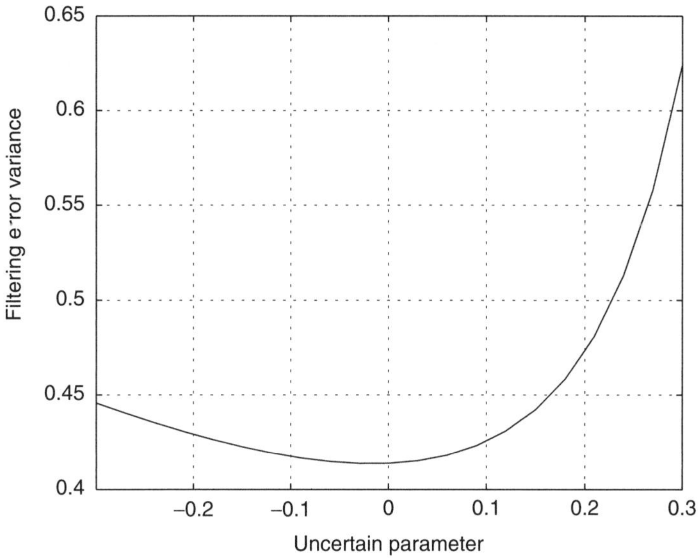

FIGURE 6.5
Actual filtering error variance versus the uncertain parameter δ.

The optimal robust *a priori* filter can be computed by Theorem 6.5 as

$$\hat{A} = \begin{bmatrix} -0.2552 & -0.1441 \\ 1.4361 & 1.0335 \end{bmatrix}, \qquad K = \begin{bmatrix} -0.0113 \\ 0.0107 \end{bmatrix}$$

$$\hat{L} = \begin{bmatrix} 0.4622 & 0.0455 \end{bmatrix}$$

and the optimal robust *a posteriori* filter as

$$\hat{A} = \begin{bmatrix} -0.4103 & -0.3270 \\ 1.3136 & 1.1095 \end{bmatrix}, \qquad K = \begin{bmatrix} -0.0116 \\ 0.0023 \end{bmatrix}$$

$$\hat{L} = \begin{bmatrix} 0.0173 & 0.0146 \end{bmatrix}, \qquad J = -0.0091$$

The actual *a posteriori* filtering error variance versus the uncertain parameter δ is depicted in Figure 6.5, which shows that the worst-case variance is about the same as the bound of 0.62.

A performance comparison is given in Table 6.1. The standard Kalman *a priori* filter is designed based on the nominal system ($\delta = 0$) and without considering parameter uncertainty. It performs poorly when the system parameter deviates from the nominal one. As expected, the optimal robust *a posteriori* filter performs much better than the optimal robust *a priori* filter as the former has access to the current measurement. Also, the polytope uncertainty characterization produces better robust *a priori* and *a posteriori* filters than the norm-bounded one. ∎

TABLE 6.1
Performance of Robust *A Priori* and *A Posteriori* Filters

	$\delta = -0.3$	$\delta = 0$	$\delta = 0.3$	Bound
Standard Kalman *a priori* filter	551.20	36.00	8352.80	
A priori filter by Theorem 6.2	52.8	51.1	54.4	69.20
A priori filter by Theorem 6.4	49.3	52.2	65.9	66.99
A priori filter by Theorem 6.5	49.4	46.18	51.68	52.51
A posteriori filter by Theorem 6.4	0.85	0.77	0.90	1.24
A posteriori filter by Theorem 6.5	0.45	0.41	0.62	0.62

Example 6.7

We consider the discrete-time system in the form of Equations 6.43 through 6.45 with

$$A = \begin{bmatrix} 0.9 & 0.1 + 0.06\alpha \\ 0.01 + 0.05\beta & 0.9 \end{bmatrix}, \qquad G = \begin{bmatrix} 1 & 0 \\ 0 & 1 \end{bmatrix}$$

$$H = \begin{bmatrix} 1 & 0 \end{bmatrix}, \qquad L = [1 \quad 1]$$

where $|\alpha| \leq 1$ and $|\beta| \leq 1$. This is a two-block structured uncertainty that can be described by a four-vertex polytope with vertices:

$$A^{(1)} = \begin{bmatrix} 0.9 & 0.04 \\ -0.04 & 0.9 \end{bmatrix}, \qquad A^{(2)} = \begin{bmatrix} 0.9 & 0.04 \\ 0.06 & 0.9 \end{bmatrix}$$

$$A^{(3)} = \begin{bmatrix} 0.9 & 0.16 \\ 0.06 & 0.9 \end{bmatrix}, \qquad A^{(4)} = \begin{bmatrix} 0.9 & 0.16 \\ -0.04 & 0.9 \end{bmatrix}$$

We first adopt *a priori* filter by letting $J = 0$. Using Theorem 6.5, the optimal filtering error variance is 51.0356 and the optimal robust *a priori* filter is

$$\hat{A} = \begin{bmatrix} -0.3607 & 0.1076 \\ -1.1087 & 0.9402 \end{bmatrix}, \qquad K = \begin{bmatrix} 1.2939 \\ 1.1674 \end{bmatrix}$$

$$\hat{L} = \begin{bmatrix} 1.1617 & 0.3751 \end{bmatrix}$$

When *a posteriori* filter ($J \neq 0$) is considered, we obtain the optimal filter

$$\hat{A} = \begin{bmatrix} -0.9006 & 0.4834 \\ -1.7383 & 1.3157 \end{bmatrix}, \qquad K = \begin{bmatrix} 1.6172 \\ 1.6094 \end{bmatrix}$$

$$\hat{L} = \begin{bmatrix} -0.4981 & 0.5412 \end{bmatrix}, \qquad J = 1.6717$$

which gives the optimal filtering error variance bound of 31.0769. ▯

6.6 Robust Kalman Filtering for Continuous-Time Systems

We consider the uncertain continuous-time system:

$$\dot{x}(t) = [A + \Delta A(t)]\, x(t) + Gw(t) \tag{6.84}$$

$$z(t) = [H + \Delta H(t)]\, x(t) + v(t) \tag{6.85}$$

$$s(t) = Lx(t) \tag{6.86}$$

where $x(t) \in R^n$ is the state vector, $z(t) \in R^p$ the measurement, $s(t) \in R^r$ the signal to be estimated, $w(t) \in R^l$ the noise input, and $v(t) \in R^p$ the measurement noise. The initial state $x(0) = x_0$ of the system is a random vector of zero-mean and covariance $P_{x_0} > 0$, and is independent of $w(t)$ and $v(t)$ for any t. w and v, are uncorrelated white noises with zero means and covariances Q and R, respectively, where $Q \geq 0$ and $R > 0$. $\Delta A(t)$ and $\Delta H(t)$ are unknown matrices representing time-varying uncertainties and are assumed to be of the form

$$\begin{bmatrix} \Delta A(t) \\ \Delta H(t) \end{bmatrix} = \begin{bmatrix} D_1 \\ D_2 \end{bmatrix} \Delta(t) F \tag{6.87}$$

where $\Delta(t) \in R^{\alpha \times \beta}$ is an unknown time-varying matrix satisfying

$$\Delta^T(t)\Delta(t) \leq I \tag{6.88}$$

We shall only discuss the optimal robust stationary filter design in this section. To this end, we make the following assumption.

Assumption 6.4

(i) The system (Equation 6.84) is quadratically stable, that is, there exists a fixed Lyapunov matrix $P > 0$ such that

$$[A + \Delta A(t)]^T P + P\,[A + \Delta A(t)] < 0$$

for any uncertainty satisfying Equation 6.88;

(ii) $\begin{bmatrix} R^{\frac{1}{2}} & D_2 \end{bmatrix}$ is of full row rank.

We note that Assumption 6.4(i) ensures that the filtering error will not go unbounded. The reason for this assumption is the fact that the filtering error dynamic is driven by the system state in the presence of the parameter uncertainty as is in the discrete-time case. Assumption 6.4(ii) basically implies that the filtering problem under consideration is non-singular. Note that when $D_2 = 0$, that is, when there is no uncertainty in the measurement output, Assumption 6.4(ii) reduces to that of $R > 0$, which is a standard assumption of the Kalman filter.

We introduce the stationary filter of the form:

$$\dot{\hat{x}}(t) = \hat{A}\hat{x}(t) + Kz(t), \quad \hat{x}(0) = 0 \tag{6.89}$$

$$\hat{s}(t) = \hat{L}\hat{x}(t) \tag{6.90}$$

where $\hat{s}$ is the estimate of z and $\hat{A}, K$, and $\hat{L}$ are filter matrices to be designed.

Then, the optimal robust Kalman filtering problem is to design the filter (Equations 6.89 and 6.90) such that the filtering error dynamic is asymptotically stable and the upper bound of the filtering error variance $E\left[\tilde{s}^T(t)\tilde{s}(t)\right]$ is minimized, where $\tilde{s}(t) = s(t) - \hat{s}(t)$.

Define the following two AREs:

$$AX + XA^T + \varepsilon XE^T EX + \frac{1}{\varepsilon}D_1 D_1^T + GQG^T = 0 \tag{6.91}$$

and

$$\begin{aligned}&\left(A - \tilde{D}_1\tilde{D}_2^T\tilde{R}^{-1}H\right)P + P\left(A - \tilde{D}_1\tilde{D}_2^T\tilde{R}^{-1}H\right)^T \\ &\quad + P\left(\varepsilon F^T F - H^T\tilde{R}^{-1}H\right)P + \tilde{D}_1\left(I - \tilde{D}_2^T\tilde{R}^{-1}\tilde{D}_2\right)\tilde{D}_1^T = 0\end{aligned} \tag{6.92}$$

where

$$\tilde{D}_1 = \begin{bmatrix} GQ^{\frac{1}{2}} & \frac{1}{\sqrt{\varepsilon}}D_1 & 0 \end{bmatrix}, \qquad \tilde{D}_2 = \begin{bmatrix} 0 & \frac{1}{\sqrt{\varepsilon}}D_2 & R^{\frac{1}{2}} \end{bmatrix} \tag{6.93}$$

$$\tilde{R} = \tilde{D}_2\tilde{D}_2^T = R + \frac{1}{\varepsilon}D_2 D_2^T \tag{6.94}$$

and $\varepsilon > 0$ is a scaling parameter.

THEOREM 6.6

Consider the uncertain system (Equations 6.84 through 6.88) satisfying Assumption 6.4. Then, there exists an asymptotically stable filter (Equations 6.89 and 6.90) that minimizes the upper bound on the asymptotic error variance $E\left[\tilde{s}(t)\tilde{s}^T(t)\right]$ if and only if for some $\varepsilon > 0$ the ARE 6.91 has a stabilizing solution $X = X^T \geq 0$. In this situation, the optimal robust filter is given by

$$\dot{\hat{x}}(t) = \left(A + \varepsilon PF^T F\right)\hat{x}(t) + K\left[z(t) - H\hat{x}(t)\right] \tag{6.95}$$

$$\hat{s}(t) = L\hat{x}(t) \tag{6.96}$$

where

$$K = \left(PH^T + \frac{1}{\varepsilon}D_1 D_2^T\right)\tilde{R}^{-1} \tag{6.97}$$

and $P = P^T \geq 0$ is the stabilizing solution to the ARE 6.92. Moreover, the filter guarantees that asymptotically

$$E\left[\tilde{s}^T(t)\tilde{s}(t)\right] \leq trace(LPL^T)$$

We note that the existence of a stabilizing solution $X = X^T \geq 0$ to Equation 6.91 guarantees that of a stabilizing solution $P = P^T \geq 0$ to Equation 6.92.

Proofs of Theorems

Proof of Theorem 6.1

Proof: (*Necessity*)

Suppose that there exists a robust *a priori* filter for the uncertain system. It follows from Definition 6.1 that there exist $\varepsilon_k > 0$ and a bounded solution $\tilde{\Sigma}_k \geq 0$ to (Equation 6.17) such that $I - \varepsilon_k \bar{F}\tilde{\Sigma}_k \bar{F}^T > 0$ over $[0, N]$. Thus, it is known from Lemma 6.1 that there exist bounded positive definite solutions X_k and P_k to the DREs 6.19 and 6.20 satisfying $X_k^{-1} - \varepsilon_k F^T F > 0$ and $P_k^{-1} - \varepsilon_k F^T F > 0$, respectively, over $[0, N]$.

Next, we shall derive the necessary condition on the filter for optimality of the upper bound on the above error variance, i.e., $trace(\bar{L}\tilde{\Sigma}_k \bar{L}^T)$, $\forall k \in [0, N]$. It will be shown that the filter of Equation 6.24 indeed satisfies this condition.

Note that for $\hat{x}_k$ to be optimal, it should be orthogonal to the estimation error e_k (Shaked and de Souza 1995). Thus, it follows from Lemma 6.1 that $\tilde{\Sigma}_{12,k} = \tilde{\Sigma}_{21,k}^T = 0$ and

$$\tilde{\Sigma}_k = \begin{bmatrix} \tilde{\Sigma}_{11,k} & \tilde{\Sigma}_{12,k} \\ \tilde{\Sigma}_{21,k} & \tilde{\Sigma}_{22,k} \end{bmatrix} = \begin{bmatrix} \tilde{\Sigma}_{11,k} & 0 \\ 0 & X_k - \tilde{\Sigma}_{11,k} \end{bmatrix}$$

where all blocks are $n \times n$ matrices.

By multiplying both sides of Equation 6.17 from the left and from the right by $\begin{bmatrix} I & 0 \end{bmatrix}$ and $\begin{bmatrix} I & 0 \end{bmatrix}^T$, respectively, it follows that $\tilde{\Sigma}_{11,k}$ satisfies

$$\begin{aligned} \tilde{\Sigma}_{11,k+1} &= \begin{bmatrix} I & 0 \end{bmatrix} \tilde{\Sigma}_{k+1} \begin{bmatrix} I & 0 \end{bmatrix}^T \\ &= (A - K_k H)\tilde{\Sigma}_{11,k}(A - K_k H)^T + (A - \hat{A}_k - K_k H)(X_k - \tilde{\Sigma}_{11,k}) \\ &\quad \times (A - \hat{A}_k - K_k H)^T + \left[(A - \hat{A}_k - K_k H)X_k + \hat{A}_k \tilde{\Sigma}_{11,k}\right] F^T \tilde{M}_k^{-1} F \\ &\quad \times \left[(A - \hat{A}_k - K_k H)X_k + \hat{A}_k \tilde{\Sigma}_{11,k}\right]^T \\ &\quad + GQG^T + K_k R K_k^T + \varepsilon_k^{-1}(D_1 - K_k D_2)(D_1 - K_k D_2)^T \end{aligned} \tag{6.98}$$

where $\tilde{M}_k = \varepsilon_k^{-1} I - F X_k F^T > 0$.

Some algebraic manipulations lead to

$$\begin{aligned} \tilde{\Sigma}_{11,k+1} = &\Big\{\hat{A}_k \left[I + (X_k - \tilde{\Sigma}_{11,k})F^T \tilde{M}_k^{-1} F\right] \\ &- (A - K_k H)\big(I + X_k F^T \tilde{M}_k^{-1} F\big)\Big\} \\ &\times \left[I + (X_k - \tilde{\Sigma}_{11,k})F^T \tilde{M}_k^{-1} F\right]^{-1} (X_k - \tilde{\Sigma}_{11,k}) \\ &\times \Big\{\left[I + F^T \tilde{M}_k^{-1} F (X_k - \tilde{\Sigma}_{11,k})\right] \hat{A}_k^T \\ &\quad - (I + X_k F^T \tilde{M}_k^{-1} F)^T (A - K_k H)^T\Big\} + \Xi_k \end{aligned} \tag{6.99}$$

where

$$\begin{aligned}\Xi_k &= (A - K_k H)\tilde{\Sigma}_{11,k}(A - K_k H)^T + GQG^T + K_k R K_k^T \\ &\quad + \frac{1}{\varepsilon_k}(D_1 - K_k D_2)(D_1 - K_k D_2)^T + (A - K_k H)(X_k - \tilde{\Sigma}_{11,k})(A - K_k H)^T \\ &\quad + (A - K_k H)X_k E^T \tilde{M}_k^{-1} F X_k (A - K_k H)^T \\ &\quad - (A - K_k H)(I + X_k F^T \tilde{M}_k^{-1} F)\left[I + (X_k - \tilde{\Sigma}_{11,k})F^T \tilde{M}_k^{-1} F\right]^{-1} \\ &\quad \times (X_k - \tilde{\Sigma}_{11,k})(I + X_k F^T \tilde{M}_k^{-1} F)(A - K_k H)^T\end{aligned}$$

Since $X_k - \tilde{\Sigma}_{11,k} \geq 0$ and $\tilde{M}_k > 0$, to minimize $trace(\tilde{\Sigma}_{11,k}), \forall k \in [0, N]$, in view of Equation 6.99, we therefore need to have

$$\begin{aligned}\hat{A}_k &= (A - K_k H)\left(I + X_k F^T \tilde{M}_k^{-1} F\right)\left[I + (X_k - \tilde{\Sigma}_{11,k})F^T \tilde{M}_k^{-1} F\right]^{-1} \\ &= (A - K_k H) + (A - K_k H)\tilde{\Sigma}_{11,k} F^T \tilde{M}_k^{-1} F\left[I - (\tilde{\Sigma}_{11,k} - X_k)F^T \tilde{M}_k^{-1} F\right]^{-1}\end{aligned} \tag{6.100}$$

By the matrix inversion lemma,

$$\begin{aligned}\tilde{M}_k^{-1} &= (\varepsilon_k^{-1} I - F X_k F^T)^{-1} \\ &= M_k^{-1} - M_k^{-1} F\left[I + (\tilde{\Sigma}_{11,k} - X_k)F^T M_k^{-1} F\right]^{-1}(\tilde{\Sigma}_{11,k} - X_k)F^T M_k^{-1}\end{aligned}$$

where $M_k = \varepsilon_k^{-1} I - F\tilde{\Sigma}_{11,k} F^T > 0$. So we have

$$F^T \tilde{M}_k^{-1} F = F^T M_k^{-1} F\left[I + (\tilde{\Sigma}_{11,k} - X_k)F^T M_k^{-1} F\right]^{-1}$$

and

$$I - (\tilde{\Sigma}_{11,k} - X_k)F^T \tilde{M}_k^{-1} F = \left[I + (\tilde{\Sigma}_{11,k} - X_k)F^T M_k^{-1} F\right]^{-1} \tag{6.101}$$

Thus, it follows from Equations 6.100 and 6.101 that

$$\begin{aligned}\hat{A}_k &= (A - K_k H) + (A - K_k H)\tilde{\Sigma}_{11,k} F^T \tilde{M}_k^{-1} F[I + (\tilde{\Sigma}_{11,k} - X_k)F^T M_k^{-1} F] \\ &= (A - K_k H) + (A - K_k H)\tilde{\Sigma}_{11,k} F^T M_k^{-1} F \\ &= \left[A + \varepsilon_k A\tilde{\Sigma}_{11,k} F^T (I - \varepsilon_k F\tilde{\Sigma}_{11,k} F^T)^{-1} F\right] \\ &\quad - K\left[H + \varepsilon_k H\tilde{\Sigma}_{11,k} F^T (I - \varepsilon_k F\tilde{\Sigma}_{11,k} F^T)^{-1}\right]\end{aligned} \tag{6.102}$$

Next, let us substitute Equation 6.102 into Equation 6.99, we have

$$\tilde{\Sigma}_{11,k+1} = \Xi_k \tag{6.103}$$

After some tedious algebraic manipulations, Equation 6.103 can be simplified as

$$\begin{aligned}\tilde{\Sigma}_{11,k+1} = {} & A\tilde{Y}_k A^T - \left(A\tilde{Y}_k H^T + \frac{1}{\varepsilon_k} D_1 D_2^T\right)\left(R_{\varepsilon_k} + H\tilde{Y}_k H^T\right)^{-1} \\ & \times \left(A\tilde{Y}_k H^T + \frac{1}{\varepsilon_k} D_1 D_2^T\right)^T + \frac{1}{\varepsilon_k} D_1 D_1^T + GQG^T \\ & + \left[\left(R_{\varepsilon_k} + H\tilde{Y}_k H^T\right) K_k - \left(A\tilde{Y}_k H^T + \frac{1}{\varepsilon_k} D_1 D_2^T\right)\right]\left(R_{\varepsilon_k} + H\tilde{Y}_k H^T\right)^{-1} \\ & \times \left[K_k\left(R_{\varepsilon_k} + H\tilde{Y}_k H^T\right) - \left(A\tilde{Y}_k H^T + \frac{1}{\varepsilon_k} D_1 D_2^T\right)\right]^T \end{aligned} \tag{6.104}$$

where R_{ε_k} is positive definite and as defined in Equation 6.29 and $\tilde{Y}_k = \tilde{\Sigma}_{11,k}^{-1} - \varepsilon_k FF^T > 0$.

Because we want to minimize $\tilde{\Sigma}_{11,k}$, $\forall k \in [0, N]$, in view of Equation 6.104, we therefore need to have

$$K_k = \left(A\tilde{Y}_k H^T + \frac{1}{\varepsilon_k} D_1 D_2^T\right)\left(R_{\varepsilon_k} + H\tilde{Y}_k H^T\right)^{-1} \tag{6.105}$$

Now, substituting Equation 6.105 into Equation 6.104, we have

$$\begin{aligned}\tilde{\Sigma}_{11,k+1} = {} & A\tilde{Y}_k A^T - \left(A\tilde{Y}_k H^T + \frac{1}{\varepsilon_k} D_1 D_2^T\right)\left(R_{\varepsilon_k} + H\tilde{Y}_k H^T\right)^{-1} \\ & \times \left(A\tilde{Y}_k H^T + \frac{1}{\varepsilon_k} D_1 D_2^T\right)^T + \frac{1}{\varepsilon_k} D_1 D_1^T + GQG^T \end{aligned} \tag{6.106}$$

It is clear that Equation 6.106 is equivalent to the DRE 6.20 and $\tilde{\Sigma}_{11,k}$ coincides with P_k and $\tilde{Y}_k$ with Y_k.

Further, it is clear that Equation 6.102 implies Equations 6.26 and 6.27, and Equation 6.105 is identical to Equation 6.28.

(*Sufficiency*)

First, it follows from Lemma 6.1 that there exists a bounded solution $P_k = P_k^T > 0$ to the DRE 6.20. Also, $X_k - P_k \geq 0$. Let $\tilde{\Sigma}_k = \begin{bmatrix} P_k & 0 \\ 0 & X_k - P_k \end{bmatrix}$, where X_k and P_k are solutions to the DREs 6.19 and 6.20, respectively. Then, by some straightforward manipulations, it follows from Equations 6.26 through 6.28 that

$$\begin{aligned}& \bar{A}_k \tilde{\Sigma}_k \bar{A}_k^T + \bar{A}_k \tilde{\Sigma}_k \bar{F}^T \left(\frac{I}{\varepsilon_k} - \bar{F}\tilde{\Sigma}_k \bar{F}^T\right)^{-1} \bar{F}\tilde{\Sigma}_k \bar{A}_k^T \\ & + \frac{1}{\varepsilon_k} \bar{D}_k \bar{D}_k^T - \tilde{\Sigma}_{k+1} + \bar{G}_k \bar{Q} \bar{G}_k^T = 0 \end{aligned}$$

Thus, in view of Definition 6.1 and Lemma 6.1, the filter (Equation 6.24) is a robust *a priori* filter with an upper bound of error covariance P_k.

Proof of Theorem 6.4

The proof of the theorem relies on Schur complement which is introduced below.

LEMMA 6.2

Given matrices M, R, and Q of appropriate dimensions, where Q and R are symmetric, then $R>0$ and $Q+M^TR^{-1}M<0$ if and only if $\begin{bmatrix} Q & M^T \\ M & -R \end{bmatrix}<0$, or equivalently, $\begin{bmatrix} -R & M \\ M^T & Q \end{bmatrix}<0$.

LEMMA 6.3

Given matrices D and F, then

$$D\Delta_k F + F^T\Delta_k^T D^T \leq D\Gamma_1 D^T + F^T\Gamma_2^{-1}F$$

for any Δ_k satisfying Equation 6.47, where $\Gamma_1 = diag\{\varepsilon_1 I_{\alpha_1}, \varepsilon_2 I_{\alpha_2}, \ldots, \varepsilon_l I_{\alpha_l}\}$, $\Gamma_2 = diag\{\varepsilon_1 I_{\beta_1}, \varepsilon_2 I_{\beta_2}, \ldots, \varepsilon_l I_{\beta_l}\}$, $\varepsilon_i>0$, $i=1,2,\ldots,l$, are scaling matrices.

Proof of Theorem 6.4:
By applying Lemma 6.2, Equations 6.56 and 6.57 are equivalent to

$$\begin{bmatrix} -\Sigma & (\bar{A}+\bar{D}_1\Delta_k\bar{F})\Sigma & \bar{G}\bar{Q}^{\frac{1}{2}} \\ \Sigma(\bar{A}+\bar{D}_1\Delta_k\bar{F})^T & -\Sigma & 0 \\ \bar{Q}^{\frac{1}{2}}\bar{G}^T & 0 & -I \end{bmatrix} \leq 0 \tag{6.107}$$

$$\begin{bmatrix} -S & (\bar{H}+\bar{D}_2\Delta_k\bar{F}) & \bar{J}\bar{Q}^{\frac{1}{2}} \\ (\bar{H}+\bar{D}_2\Delta_k\bar{F})^T & -\Sigma_1 & 0 \\ \bar{Q}^{\frac{1}{2}}\bar{J}^T & 0 & -I \end{bmatrix} \leq 0 \tag{6.108}$$

respectively, where $\Sigma_1 = \Sigma^{-1}$.

By multiplying Equation 6.107 from the left and the right by $diag\{\Sigma_1, \Sigma_1, I\}$ and its transpose, respectively, we obtain

$$\begin{bmatrix} -\Sigma_1 & \Sigma_1\bar{A} & \Sigma_1\bar{G}\bar{Q}^{\frac{1}{2}} \\ \bar{A}^T\Sigma_1 & -\Sigma_1 & 0 \\ \bar{Q}^{\frac{1}{2}}\bar{G}^T\Sigma_1 & 0 & -I \end{bmatrix} + \begin{bmatrix} \Sigma_1\bar{D}_1 \\ 0 \\ 0 \end{bmatrix}\Delta_k\begin{bmatrix} 0 & \bar{F} & 0 \end{bmatrix} + \begin{bmatrix} 0 \\ \bar{F}^T \\ 0 \end{bmatrix}\Delta_k^T\begin{bmatrix} \bar{D}_1^T\Sigma_1 & 0 & 0 \end{bmatrix} \leq 0$$

By applying Lemma 6.3, the above holds if

$$\begin{bmatrix} -\Sigma_1 & \Sigma_1\bar{A} & \Sigma_1\bar{G}\bar{Q}^{\frac{1}{2}} & \Sigma_1\bar{D}_1 \\ \bar{A}^T\Sigma_1 & -\Sigma_1+\bar{F}^T\Gamma_2\bar{F} & 0 & 0 \\ \bar{Q}^{\frac{1}{2}}\bar{G}\Sigma_1 & 0 & -I & 0 \\ \bar{D}_1^T\Sigma_1 & 0 & 0 & -\Gamma_1 \end{bmatrix} < 0 \tag{6.109}$$

Similarly, Equation 6.108 is satisfied if

$$\begin{bmatrix} -S & \bar{H} & \bar{J}\bar{Q}^{\frac{1}{2}} & \bar{D}_2 \\ \bar{H}^T & -\Sigma_1 + \bar{F}^T\Gamma_2'\bar{F} & 0 & 0 \\ \bar{Q}^{\frac{1}{2}}\bar{J}^T & 0 & -I & 0 \\ \bar{D}_2^T & 0 & 0 & -\Gamma_1' \end{bmatrix} < 0 \tag{6.110}$$

Decompose Σ_1 and Σ_1^{-1} as

$$\Sigma_1 = \begin{bmatrix} X & M \\ M^T & U \end{bmatrix}, \quad \Sigma_1^{-1} = \begin{bmatrix} Y_1 & W \\ W^T & V \end{bmatrix}$$

and denote

$$\Phi = \begin{bmatrix} I & Y_1 \\ 0 & W^T \end{bmatrix} \tag{6.111}$$

$$\mathcal{K} = MK \tag{6.112}$$

$$\mathcal{A} = XA + \mathcal{K}H + M\hat{A}W^TY_1^{-1} \tag{6.113}$$

$$\mathcal{L} = JH + \hat{L}W^TY_1^{-1} \tag{6.114}$$

Then, by multiplying Equation 6.109 from the left by J_1^T and from the right by J_1, where $J_1 = diag\{\Phi, \Phi, I, I\}$, we obtain

$$\begin{bmatrix} -X & -I & XA + \mathcal{K}H & \mathcal{A}Y_1 & XGQ^{\frac{1}{2}} & \mathcal{K}R^{\frac{1}{2}} & XD_1 + \mathcal{K}D_2 \\ * & -Y_1 & A & AY_1 & GQ^{\frac{1}{2}} & 0 & D_1 \\ * & * & -X + F^T\Gamma_2F & -I + F^T\Gamma_2FY_1 & 0 & 0 & 0 \\ * & * & * & -Y_1 + Y_1F^T\Gamma_2FY_1 & 0 & 0 & 0 \\ * & * & * & * & -I & 0 & 0 \\ * & * & * & * & * & -I & 0 \\ * & * & * & * & * & * & -\Gamma_1 \end{bmatrix} < 0. \tag{6.115}$$

Then, Equation 6.59 follows from Equation 6.115 by multiplying its left and right by $diag\{I, Y, I, Y, I, I, I\}$ and its transpose, respectively, where $Y = Y_1^{-1}$.

Similarly, by multiplying Equation 6.110 from the left by J_2^T and from the right by J_2, where $J_2 = diag\{I, \Phi, I, I\}$, we obtain

$$\begin{bmatrix} -S & L - JH & LY_1 - \mathcal{L}Y_1 & -JR^{\frac{1}{2}} & -JD_2 \\ * & -X + F^T\Gamma_2'F & -I + F^T\Gamma_2'FY_1 & 0 & 0 \\ * & * & -Y_1 + Y_1F^T\Gamma_2'FY_1 & 0 & 0 \\ * & * & * & -I & 0 \\ * & * & * & * & -\Gamma_1' \end{bmatrix} < 0 \tag{6.116}$$

which implies Equation 6.60 by multiplying its left and right by $diag\{I, I, Y, I, I\}$ and its transpose, respectively.

Finally, it can be easily shown that $I - XY^{-1}$ is invertible. Thus, there exist nonsingular matrices M and W such that Equation 6.64 holds. Hence, the filter parameters 6.61 through 6.63 follow from Equation 6.111 through 6.114.

Proof of Theorem 6.5

Proof: In view of Equation 6.76, we decompose

$$\Pi = \begin{bmatrix} X & M \\ M_1 & U \end{bmatrix}, \qquad \Pi^{-1} = \begin{bmatrix} Y_1 & W \\ W_1 & V \end{bmatrix}$$

and denote

$$\Phi = \begin{bmatrix} Y_1 & I \\ W_1 & 0 \end{bmatrix}$$

By multiplying Equation 6.76 from the left and the right by $diag\{\Phi^T, \Phi^T, I\}$ and its transpose, respectively, it is obtained that

$$\begin{bmatrix} -Y_1 - Y_1^T + \Sigma_{2,11}^{(i)} & -I - Y_1^T X - W_1^T M_1 + \Sigma_{2,12}^{(i)} & A^{(i)} Y_1 & A^{(i)} & GQ^{\frac{1}{2}} & 0 \\ * & -X - X^T + \Sigma_{2,22}^{(i)} & X^T A^{(i)} Y_1 + M_1^T \bar{K} & X^T A^{(i)} + M_1^T K H^{(i)} & X^T G Q^{\frac{1}{2}} & M_1^T K R^{\frac{1}{2}} \\ * & * & -\Sigma_{2,11}^{(i)} & -\Sigma_{2,12}^{(i)} & 0 & 0 \\ * & * & * & -\Sigma_{2,22}^{(i)} & 0 & 0 \\ * & * & * & * & -I & 0 \\ * & * & * & * & * & -I \end{bmatrix} < 0 \quad (6.117)$$

where

$$\bar{K} = K H^{(i)} Y_1 + \hat{A} W_1, \qquad \begin{bmatrix} \Sigma_{2,11}^{(i)} & \Sigma_{2,12}^{(i)} \\ \Sigma_{2,12}^{(i)T} & \Sigma_{2,22}^{(i)} \end{bmatrix} = \Sigma_2^{(i)} = \Phi^T \Sigma_1^{(i)} \Phi$$

Next, denote

$$Y = Y_1^{-1}, \qquad \Psi = M_1^T W_1 Y, \qquad \mathcal{A} = M_1^T \hat{A} W_1 Y, \qquad \mathcal{K} = M_1^T K \quad (6.118)$$

Multiplying Equation 6.117 from the left and the right by $diag\{Y^T, I, Y^T, I, I, I\}$ and its transpose, Equation 6.80 follows, where

$$\begin{bmatrix} \Lambda_1^{(i)} & \Lambda_2^{(i)} \\ \Lambda_2^{(i)T} & \Lambda_3^{(i)} \end{bmatrix} = \begin{bmatrix} Y^T & 0 \\ 0 & I \end{bmatrix} \Sigma_2^{(i)} \begin{bmatrix} Y & 0 \\ 0 & I \end{bmatrix}$$

By making use of Schur complements, the minimization in Equation 6.72 can be equivalent to

$$\min trace(S)$$

subject to

$$\begin{bmatrix} -S & \bar{H}_\delta & \bar{J}\bar{Q}^{\frac{1}{2}} \\ \bar{H}_\delta^T & -\Sigma_\delta^{-1} & 0 \\ \bar{Q}^{\frac{1}{2}}\bar{J}^T & 0 & -I \end{bmatrix} \leq 0 \tag{6.119}$$

The corresponding matrix inequality of Equation 6.119 at the i-th vertex is

$$\begin{bmatrix} -S & \bar{H}^{(i)} & \bar{J}\bar{Q}^{\frac{1}{2}} \\ \bar{H}^{(i)T} & -\Sigma_1^{(i)} & 0 \\ \bar{Q}^{\frac{1}{2}}\bar{J}^T & 0 & -I \end{bmatrix} \leq 0 \tag{6.120}$$

By multiplying Equation 6.120 from the left and the right by $diag\{I, \Phi^T, I\}$ and its transpose, respectively, it can be known that Equation 6.120 is equivalent to

$$\begin{bmatrix} -S & LY_1 - JH^{(i)}Y_1 - \hat{L}W_1 & L - JH^{(i)} & -JR^{\frac{1}{2}} \\ * & -\Sigma_{2,11}^{(i)} & -\Sigma_{2,12}^{(i)} & 0 \\ * & * & -\Sigma_{2,22}^{(i)} & 0 \\ * & * & * & -I \end{bmatrix} < 0$$

Equation 6.79 then follows by multiplying the above from the left and right by $diag\{I, Y^T, I, I\}$ and its transpose, respectively, taking into consideration Equation 6.118 and denoting $\mathcal{L} = \hat{L}W_1Y$.

We now show that Ψ is invertible. Note that if Equation 6.80 holds, then

$$Z = \begin{bmatrix} -Y - Y^T + \Lambda_1^{(i)} & -Y^T - X - \Psi^T + \Lambda_2^{(i)} \\ -Y - X^T - \Psi + \Lambda_2^{(i)T} & -X - X^T + \Lambda_3^{(i)} \end{bmatrix} < 0$$

which implies

$$[I\ -I]Z\begin{bmatrix} I \\ -I \end{bmatrix} = [I\ -I]\Lambda^{(i)}\begin{bmatrix} I \\ -I \end{bmatrix} + \Psi + \Psi^T < 0$$

Since $\Lambda^{(i)} > 0$, it follows that $\Psi + \Psi^T < 0$. Hence, Ψ is invertible.

Finally, from Equation 6.118 the filter transfer function is given by

$$\begin{aligned} \mathcal{F}(z) &= \mathcal{L}Y^{-1}W_1^{-1}\left(zI - M_1^{-T}\mathcal{A}Y^{-1}W_1^{-1}\right)^{-1}M_1^{-T}\mathcal{K} + J \\ &= \mathcal{L}(z\Psi - \mathcal{A})^{-1}\mathcal{K} + J \\ &= \hat{L}(zI - \hat{A})^{-1}K + J \end{aligned}$$

Problems

1. Show that if Equation 6.17 holds for some $\varepsilon_k > 0$ and $\tilde{\Sigma}_k \geq 0$ satisfying $I - \varepsilon_k \bar{F}\tilde{\Sigma}_k \bar{F}^T > 0$, then Equation 6.15 is satisfied for all Δ_k satisfying $\Delta_k^T \Delta_k \leq I$.

2. Show that the DRE 6.20 is equivalent to the DRE 6.21.

3. Prove Lemma 6.1.

4. Consider the robust filter in Theorem 6.1. For a scalar system, show that the optimal estimate at time instant k can be obtained from the optimal estimate at time instant $k-1$ by a simple optimization over ε_{k-1}. Further, show that the optimization is convex.

5. Consider the scalar system

$$\begin{aligned} x_{k+1} &= ax_k + gw_k \\ z_k &= (h + \delta h)x_k + v_k \end{aligned}$$

where $|\delta h| \leq \bar{h}$. Derive an optimal robust *a priori* filter. Compare the robust *a priori* filter with the Kalman *a priori* filter for $a = 0.5,\ g = 1, h = 1$, and $\bar{h} = 0.3$.

6. Associated with the system (Equations 6.51 and 6.52), the robust Kalman filtering may be formulated as

$$\min trace(S)$$

subject to

$$(\bar{A} + \bar{D}_1\Delta_k\bar{F})^T\Sigma(\bar{A} + \bar{D}_1\Delta_k\bar{F}) - \Sigma + (\bar{H} + \bar{D}_2\Delta_k\bar{F})^T(\bar{H} + \bar{D}_2\Delta_k\bar{F}) \leq 0$$

and

$$\bar{Q}^{\frac{1}{2}}\bar{G}^T\Sigma\bar{G}\bar{Q}^{\frac{1}{2}} + \bar{Q}^{\frac{1}{2}}\bar{J}^T\bar{J}\bar{Q}^{\frac{1}{2}} \leq S$$

Show that the above can be implied by Equations 6.65 and 6.66, respectively.

7. Show that the matrix $I - XY^{-1}$ is invertible if Equations 6.59 and 6.60 hold.

8. Consider the uncertain system

$$\begin{aligned} x_{k+1} &= \begin{bmatrix} -0.5 & 0 \\ 1 & 0.3 + \delta \end{bmatrix} x_k + \begin{bmatrix} 1 \\ 0 \end{bmatrix} w_k \\ z_k &= \begin{bmatrix} 0 & 1 \end{bmatrix} x_k + v_k \end{aligned}$$

where δ is an uncertain parameter satisfying $|\delta| \leq 0.2$ and w_k and v_k are uncorrelated white noises with zero means and unit variances. Using Matlab, find a stationary robust *a priori* filter via the Riccati equation approach and the LMI approach and compare their performance with the standard Kalman *a priori* filter.

9. Consider the uncertain continuous-time system (6.84 through 6.88). Given that the AREs 6.91 and 6.92 exist the stabilizing solutions, show that the solution of Equation 6.92 provides an upper bound for the filtering error covariance.

10. Consider the system

$$\begin{aligned} \dot{x}(t) &= (-1+\delta)x(t) \\ z(t) &= x(t) + v(t) \end{aligned}$$

where $|\delta| \leq 0.5$ is an uncertain parameter and $v(t)$ is a white noise with zero mean and unit variance. Find the optimal stationary robust Kalman filter and compare its performance with the standard Kalman filter.

7

$\mathcal{H}_\infty$ *Filtering of Continuous-Time Systems*

This chapter deals with the design of estimators with an $\mathcal{H}_\infty$ performance measure and where the only uncertainty in the system model is in the form of bounded energy noise signals, i.e., there is no uncertainty in the matrices of the system state–space model. Recall that in the Kalman filtering, the system and measurement noises are assumed to be white noises with known statistics, and the filter is designed to minimize the variance or RMS power of the estimation error. In the $\mathcal{H}_\infty$ filtering, the noise inputs are deterministic signals and are required to be energy-bounded. Other than that no other knowledge of the noises is needed. The design objective of the $\mathcal{H}_\infty$ filter is to ensure that the energy gain from the noise inputs to the estimation error is less than a certain level (Kailath *et al.* 1999; Green and Limebeer 1995; Zhou *et al.* 1995).

We consider both the finite-horizon and stationary $\mathcal{H}_\infty$ estimation problems. For the finite-horizon filtering, we consider two situations for the system initial state: one corresponds to the case when the initial state is unknown, while in the other the initial state is assumed to be zero. The latter case will also correspond to the situation where we are concerned with steady-state filtering, i.e., under the assumption that measurements have been continuing for a sufficiently long time so that the effect of the initial condition has disappeared. For the stationary $\mathcal{H}_\infty$ filtering, in addition to the well-known Riccati equation approach, the linear matrix inequality (LMI) approach is also introduced.

7.1 $\mathcal{H}_\infty$ Filtering Problem

Consider linear systems described by a state-space model of the form

$$\dot{x}(t) = Ax(t) + Bw(t), \quad x(0) = x_0 \tag{7.1}$$

$$z(t) = Hx(t) + Dw(t) \tag{7.2}$$

$$s(t) = Lx(t) \tag{7.3}$$

where $x(t) \in R^n$ is the state, x_0 an unknown initial state, $w(t) \in R^l$ the noise signal, $z(t) \in R^p$ the measurement, and $s(t) \in R^r$ a linear combination of state variables to be estimated over the horizon $[0, T]$, where $T > 0$, using the

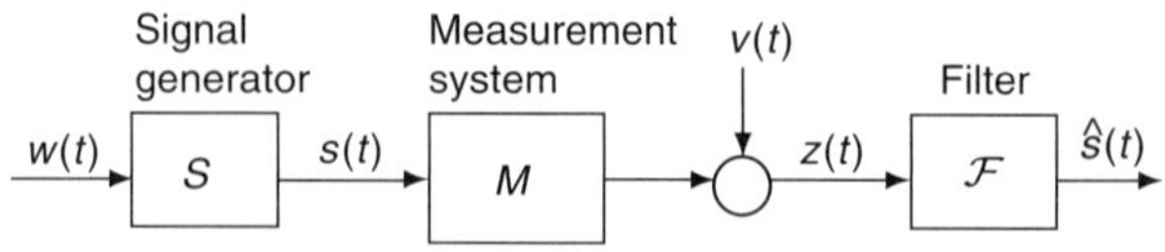

FIGURE 7.1
Signal generating mechanism.

measurements $\{z(\tau),\ 0 \le \tau \le t\}$. The noise w is assumed to be an arbitrary signal in $\mathcal{L}_2[0,T]$, which is the set of energy-bounded signals, and A, B, H, D, and L are known real constant matrices with appropriate dimensions. When the matrix L is the identity, the state $x(t)$ is to be estimated.

For the sake of technical simplification, we shall adopt the following assumption:

Assumption 7.1

$R \triangleq DD^T > 0$

Note that Assumption 7.1 is similar to the standard assumption in Kalman filtering where all the components of the measurement vector are assumed to be corrupted by noise.

We observe that the case where the input and measurement noise signals are different as in Equations 6.1 through 6.2 is a particular case of Equations 7.1 and 7.2 by setting $B = \begin{bmatrix} G & 0 \end{bmatrix}$ and $D = \begin{bmatrix} 0 & I \end{bmatrix}$.

The above signal estimation setting is quite general and encompasses a number of typical situations that arise in the areas of control engineering and signal processing. For example, consider the filtering problem with a signal-generating mechanism as shown in Figure 7.1. The signals $w(t)$ and $v(t)$ are energy-bounded noise sources and $s(t)$ is the signal to be estimated using the noisy measurements, $\{z(\tau),\ 0 \le \tau \le t\}$. Both the signal generator and measurement subsystem are described by linear state–space models with the signal generator model being strictly proper and the measurement system is assumed square. In this situation, it is easy to see that this filtering problem can be recast as a filtering problem similar to those analyzed in this chapter.

Here, we are concerned with obtaining an estimate $\hat{s}(t)$ of $s(t)$ over a horizon $[0,T]$ via a linear causal filter $\mathcal{F}$ using the measurements $\mathcal{Z}_t = \{z(\tau),\ 0 \le \tau \le t\}$, and where no *a priori* estimate of the initial state of Equation 7.1 is assumed. The filter is required to provide a uniformly small estimation error, $\tilde{s}(t) = s(t) - \hat{s}(t),\ \forall\ t \in [0,T]$, for all $w \in \mathcal{L}_2[0,T]$ and $x_0 \in R^n$. We shall consider the following worst-case performance measure:

$$J(P_{x_0}, T) \triangleq \sup \left\{ \left[\frac{\|s - \hat{s}\|^2_{[0,T]}}{\|w\|^2_{[0,T]} + x_0^T P_{x_0}^{-1} x_0} \right]^{1/2} \right\} \tag{7.4}$$

where $\|x\|^2_{[0,T]} = \int_0^T x^T(t)x(t)\mathrm{d}t$ for an integrable function $x(t)$, $P_{x_0} = P_{x_0}^T > 0$ is a given weighting matrix for the initial state and the supremum is taken over all $(w, x_0) \in \mathcal{L}_2[0,T] \oplus R^n$ such that $\|w\|^2_{[0,T]} + x_0^T P_{x_0}^{-1} x_0 \neq 0$. The weighting matrix P_{x_0} is a measure of the uncertainty in x_0 relative to the uncertainty in w. A "small" value of P_{x_0} indicates that the initial state is likely to be very close to zero.

The above index of performance is indeed the induced norm of the operator from the "noise" signal, (w, x_0), to the estimation error, $s - \hat{s}$. Thus, $J(P_{x_0}, T)$ can be interpreted as the worst-case gain from (w, x_0) to $s - \hat{s}$.

In the case where the initial state x_0 is known to be zero, the performance measure of Equation 7.4 is replaced by

$$J_0(T) \triangleq \sup_{0 \neq w \in \mathcal{L}_2[0,T]} \left\{ \frac{\|s - \hat{s}\|_{[0,T]}}{\|w\|_{[0,T]}} \right\} \tag{7.5}$$

In Equation 7.5, T is allowed to be ∞. We observe that the index of performance of Equation 7.5 can be viewed as the limit of Equation 7.4 as the largest eigenvalue of weighting matrix P_{x_0} tends to 0. The reason for this is because for such matrix P_{x_0}, the initial state x_0 will be forced to be zero. Also, note that the performance measure $J_0(\infty)$ coincides with $\|G_{\tilde{s}w}(s)\|_\infty$, where $G_{\tilde{s}w}(s)$ is the transfer function matrix from w to the estimation error, $\tilde{s} = s - \hat{s}$.

The $\mathcal{H}_\infty$ filtering problems are stated as follows:

Given a prescribed level of "noise" attenuation $\gamma > 0$ and an initial state weighting matrix $P_{x_0} = P_{x_0}^T > 0$, find a linear causal filter $\mathcal{F}$ such that

- *In the finite horizon case, $J(P_{x_0}, T) < \gamma$, or alternatively, $J_0(T) < \gamma$ when $x_0 = 0$.*
- *In the stationary case, the filtering error dynamic is asymptotically stable, and under zero initial condition $\|G_{\tilde{s}w}(s)\|_\infty < \gamma$.*

The above estimation problems will be referred to as *standard $\mathcal{H}_\infty$ filtering*, and the resulting filters, which are commonly known as *$\mathcal{H}_\infty$ suboptimal filters*, are said to achieve a level of noise attenuation γ. These filters are designed to guarantee that the worst-case energy gain from (w, x_0) to the estimation error $s - \hat{s}$ is less than γ. We note that in the stationary case, an $\mathcal{H}_\infty$ filter ensures that the "maximum gain" in frequency from the noise source to the estimation error is bounded by γ. An $\mathcal{H}_\infty$ *optimal filter* is the one which minimizes either $J(P_{x_0}, T)$ (alternatively, $J_0(T)$) or $\|G_{\tilde{s}w}(s)\|_\infty$, i.e., it minimizes the worst-case energy gain from the noise input to the estimation error.

Unlike the traditional minimum variance filtering approach, e.g., the celebrated Kalman filtering method, $\mathcal{H}_\infty$ filtering treats the noise signals as deterministic disturbances and thus no *a priori* knowledge of the noise statistics is required, except the boundedness assumption of its energy or average power. This makes the $\mathcal{H}_\infty$ filtering approach suitable for applications where little knowledge about the noise statistics is available.

7.1.1 Relationship with Two-Person Zero-Sum Game

The standard $\mathcal{H}_\infty$ filtering problem can be interpreted as a minimax problem where the estimator strategy $\hat{s}(\cdot)$ is determined such that the payoff function

$$\mathcal{J}_T \triangleq \|s - \hat{s}\|^2_{[0,T]} - \gamma^2 \left(\|w\|^2_{[0,T]} + x_0^T P_{x_0}^{-1} x_0\right) \tag{7.6}$$

satisfies $\mathcal{J}_T < 0$ for all $w \in \mathcal{L}_2[0,T]$ and $x_0 \in R^n$ subject to the constraint $\|w\|^2_{[0,T]} + x_0^T P_{x_0}^{-1} x_0 \neq 0$.

We can view the $\mathcal{H}_\infty$ estimation problem as a dynamic, two-person, zero-sum game (Yaesh and Shaked 1992). In this game the first player, say a statistician, is required to find an estimation strategy $\hat{s}(t) = \mathcal{F}(\mathcal{Z}_t)$, so that the cost function $\mathcal{J}_T$ will be minimized. The statistician opponent, say nature, is looking for the worst-possible noise signal and initial state that comply with the given measurements and will maximize $\mathcal{J}_T$. The $\mathcal{H}_\infty$ estimator will correspond to the statistician's strategy, which is indeed designed to handle the worst-possible x_0 and w.

In the remainder of this section, we introduce a bounded real lemma for the system

$$\dot{x}(t) = Ax(t) + Bw(t), \quad x(0) = x_0 \tag{7.7}$$

$$s(t) = Lx(t) \tag{7.8}$$

where $x \in R^n$, $w \in R^l$, and $s \in R^r$ are, respectively, the state, the input, and the output of the system. We also consider the cost function $\mathcal{J}_T$ of Equation 7.6 (setting $\hat{s} \equiv 0$) for the system 7.7 and 7.8.

LEMMA 7.1

Consider the system 7.7 and 7.8 and the cost function (Equation 7.6) with $\hat{s} \equiv 0$. Then, we have the following:

(a) For given $T > 0$ and P_{x_0} positive definite, $\mathcal{J}_T < 0$ for any nonzero (x_0, w) with x_0 bounded and $w \in \mathcal{L}_2[0,T]$ if and only if the differential Riccati equation (DRE)

$$\dot{P} = AP + PA^T + \gamma^{-2} PL^T LP + BB^T, \quad P(0) = P_{x_0} \tag{7.9}$$

has a solution $P(t)$ over $[0,T]$. Note that if the solution $P(t)$ exists, $P(t) > 0$ for any $t \in [0,T]$.

(b) When $T \to \infty$, for a given $P_{x_0} = P_{x_0}^T > 0$, $\mathcal{J}_\infty < 0$ for any nonzero (x_0, w) with x_0 bounded and $w \in \mathcal{L}_2[0,\infty)$ if and only if the DRE 7.9 has a solution $P(t)$ over $[0,\infty)$ such that the following unforced linear time-varying system

$$\dot{p}(t) = \left[A + \gamma^{-2} P(t) L^T L\right] p(t)$$

is exponentially stable.

(c) If $x(0) = 0$ and $T \to \infty$, then $\mathcal{J}_\infty < 0$ if and only if the algebraic Riccati equation (ARE)

$$AP + PA^T + \gamma^{-2} PL^T LP + BB^T = 0 \tag{7.10}$$

has a stabilizing solution $P = P^T \geq 0$ *or the algebraic Riccati inequality*

$$A\bar{P} + \bar{P}A^T + \gamma^{-2}\bar{P}L^T L\bar{P} + BB^T < 0 \tag{7.11}$$

has a positive definite solution $\bar{P}$. *A stabilizing solution means that* P *is such that* $A + \gamma^{-2}PL^T L$ *is asymptotically stable. The stabilizing solution is unique. Note that* $\mathcal{J}_\infty < 0$ *is equivalent to* $\|G(s)\|_\infty < \gamma$, *where* $G(s)$ *is the transfer function from* w *to* s.

7.2 Finite Horizon $\mathcal{H}_\infty$ Linear Filter

We first note that the standard $\mathcal{H}_\infty$ filtering problem can be solved via several different approaches, including interpolation theory approach (Fu 1991), game theoretic approach (Yaesh 1992), and Riccati equation approach (Nagpal and Khargonekar 1991). In the sequel, we will first present a solution to the standard finite horizon $\mathcal{H}_\infty$ filtering problem using a Riccati equation approach. To this end, we introduce the following DRE:

$$\begin{aligned}\dot{P}(t) = {} & \left(A - BD^T R^{-1}H\right)P(t) + P(t)\left(A - BD^T R^{-1}H\right)^T \\ & + P(t)\left(\gamma^{-2}L^T L - H^T R^{-1}H\right)P(t) + B\left(I - D^T R^{-1}D\right)B^T\end{aligned} \tag{7.12}$$

It can be easily shown that the existence of a bounded solution $P(t)$ to Equation 7.12 over $[0, \tau]$, $\forall \tau > 0$, with initial condition $P(0) > 0$ (respectively, $P(0) \geq 0$), implies that $P(t)$ is symmetric and positive definite (respectively, positive semidefinite) over $[0, \tau]$. Observe that when $BD^T = 0$, i.e., the process noise is independent of the measurement noise, Equation 7.12 becomes

$$\dot{P}(t) = AP(t) + P(t)A^T + P(t)\left(\gamma^{-2}L^T L - H^T R^{-1}H\right)P(t) + BB^T \tag{7.13}$$

THEOREM 7.1

Consider the system 7.1 through 7.3 satisfying Assumption 7.1 and let $\gamma > 0$ *be a given constant. Then we have the following results:*

(a) Assume that x_0 *is unknown and let* $P_{x_0} = P_{x_0}^T > 0$ *be a given initial state weighting matrix. Then, there exists a linear causal filter such that* $J(P_{x_0}, T) < \gamma$ *if and only if there exists a bounded solution* $P(t)$ *to the DRE 7.12 over* $[0, T]$ *with initial condition* $P(0) = P_{x_0}$.

(b) Assuming that $x_0 = 0$, *there exists a linear causal filter such that* $J_0(T) < \gamma$ *if and only if there exists a bounded solution* $P(t)$ *to the DRE 7.12 over* $[0, T]$ *with initial condition* $P(0) = 0$.

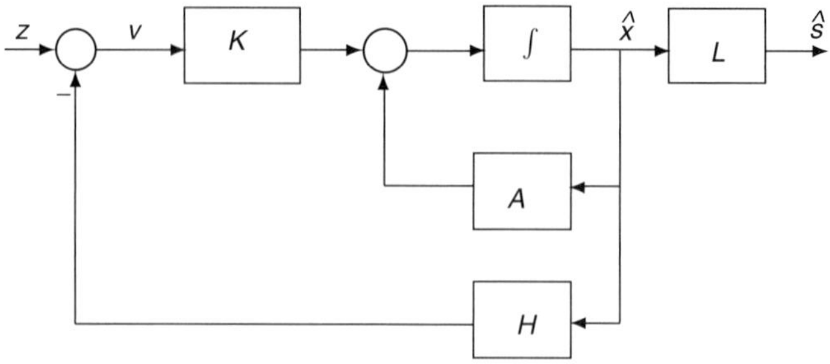

FIGURE 7.2
The central $\mathcal{H}_\infty$ filter.

(c) *When condition (a) or (b) is satisfied, a suitable (central) $\mathcal{H}_\infty$ filter for both the above cases is given by*

$$\dot{\hat{x}}(t) = A\hat{x}(t) + K(t)\left[z(t) - H\hat{x}(t)\right], \quad \hat{x}(0) = 0 \tag{7.14}$$

$$\hat{s}(t) = L\hat{x}(t) \tag{7.15}$$

where the filter gain matrix $K(t)$ satisfies

$$K(t) = \left[P(t)H^T + BD^T\right]R^{-1} \tag{7.16}$$

The block diagram of the central $\mathcal{H}_\infty$ filter is shown in Figure 7.2.

Proof of Sufficiency:

We first consider the proof of the sufficiency of (a). Let $P(t)$ be a bounded solution to Equation 7.12 over $[0, T]$ with initial condition $P(0) = P_{x_0}$. In view of Equation 7.16, it is easy to verify that the DRE 7.12 can be rewritten in the form

$$\begin{aligned}\dot{P}(t) &= [A - K(t)H]\,P(t) + P(t)\,[A - K(t)H]^T + \gamma^{-2}P(t)L^TLP(t)\\ &\quad + [B - K(t)D]\,[B - K(t)D]^T, \quad P(0) = P_{x_0}\end{aligned} \tag{7.17}$$

Next, letting that $e(t) \triangleq x(t) - \hat{x}(t)$ and considering Equations 7.1 through 7.3 and Equations 7.14 and 7.15, we have that the estimation error, $s(t) - \hat{s}(t)$, that can be described by the following state-space model:

$$\begin{aligned}\dot{e}(t) &= [A - K(t)H]e(t) + [B - K(t)D]w(t), \quad e(0) = x_0\\ s(t) - \hat{s}(t) &= Le(t)\end{aligned}$$

Hence, in view of Lemma 7.1, Equation 7.17 implies that $J(P_{x_0}, T) < \gamma$.
The proof of the sufficiency of (b) is similar.

Proof of Necessity:

(a) Since $J(P_{x_0}, T) < \gamma$, it follows that for any $\tau \in [0, T]$

$$\|w\|_{[0,\tau]}^2 + x_0^T P_{x_0}^{-1} x_0 - \gamma^{-2}\|s - \hat{s}\|_{[0,\tau]}^2 > 0 \quad (7.18)$$

whenever $\|w\|_{[0,\tau]}^2 + x_0^T P_{x_0}^{-1} x_0 \neq 0$.

Now, define the Hamiltonian system

$$\begin{bmatrix} \dot{\xi}(t) \\ \dot{q}(t) \end{bmatrix} = \begin{bmatrix} -(A - BD^T R^{-1} H)^T & -(\gamma^{-2} L^T L - H^T R^{-1} H) \\ B(I - D^T R^{-1} D) B^T & A - BD^T R^{-1} H \end{bmatrix} \begin{bmatrix} \xi(t) \\ q(t) \end{bmatrix}$$

$$\begin{bmatrix} \xi(0) \\ q(0) \end{bmatrix} = \begin{bmatrix} P_{x_0}^{-1} x_0 \\ x_0 \end{bmatrix} \quad (7.19)$$

and introduce

$$w_0(t) \triangleq (I - D^T R^{-1} D) B^T \xi(t) - D^T R^{-1} H x(t) \quad (7.20)$$

Note that

$$w_0^T(t) w_0(t) = \xi^T(t) B(I - D^T R^{-1} D) B^T \xi(t) + x^T(t) H^T R^{-1} H x(t) \quad (7.21)$$

We also observe that by setting $w(t) \equiv w_0(t)$, $\forall \tau \in [0, \tau]$ in Equation 7.1, $x(t)$ and $q(t)$ will have the same trajectory. Furthermore, $z(t) = 0$, $\forall\ t \in [0, \tau]$, which implies by the causality that $\hat{s}(t) = 0$ for any $t \in [0, \tau]$. Hence, Equation 7.18 becomes

$$\|w_0\|_{[0,\tau]}^2 + x_0^T P_{x_0}^{-1} x_0 - \gamma^{-2}\|s\|_{[0,\tau]}^2 > 0$$

for any $\tau \in [0, T]$, whenever $x_0 \neq 0$.

In view of Equation 7.19, it follows that for any $t \geq 0$,

$$\frac{\mathrm{d}}{\mathrm{d}t}[q^T(t)\xi(t)] = \xi^T(t) B(I - D^T R^{-1} D) B^T \xi(t) + x^T(t) H^T R^{-1} H x(t) - \gamma^{-2} s^T(t) s(t)$$

Hence, considering Equations 7.19 and 7.21 and integrating the above from 0 to τ, we obtain

$$q^T(\tau)\xi(\tau) = \|w_0\|_{[0,\tau]}^2 + x_0^T P_{x_0}^{-1} x_0 - \gamma^{-2}\|s\|_{[0,\tau]}^2 > 0 \quad (7.22)$$

for any $\tau \in [0, T]$, whenever $x_0 \neq 0$.

Next, denote the transition matrix of Equation 7.19 by

$$\Phi(t, \tau) = \begin{bmatrix} \Phi_{11}(t, \tau) & \Phi_{12}(t, \tau) \\ \Phi_{21}(t, \tau) & \Phi_{22}(t, \tau) \end{bmatrix} \quad (7.23)$$

We now show that the matrix $\Phi_{11}(\tau, 0) P_{x_0}^{-1} + \Phi_{12}(\tau, 0)$ is nonsingular. By contradiction, assume that $\Phi_{11}(\tau, 0) P_{x_0}^{-1} + \Phi_{12}(\tau, 0)$ is singular. Then, there exists a nonzero $\alpha \in R^n$ such that $[\Phi_{11}(\tau, 0) P_{x_0}^{-1} + \Phi_{12}(\tau, 0)]\alpha = 0$. Note from Equation 7.19 that with $x_0 = \alpha$ we conclude that $\xi(\tau) = [\Phi_{11}(\tau,0)P_{x_0}^{-1} + \Phi_{12}(\tau,0)]\alpha = 0$, which contradicts with Equation 7.22. Hence, the matrix $\Phi_{11}(\tau,0)P_{x_0}^{-1} + \Phi_{12}(\tau,0)$ is nonsingular for any $\tau \in [0,T]$.

Now, define the matrix function

$$P(t) = \left[\Phi_{21}(t,0)P_{x_0}^{-1} + \Phi_{22}(t,0)\right]\left[\Phi_{11}(t,0)P_{x_0}^{-1} + \Phi_{12}(t,0)\right]^{-1} \tag{7.24}$$

Since $\Phi_{11}(t,0)P_{x_0}^{-1} + \Phi_{12}(t,0)$ is nonsingular, the above matrix function is well defined and bounded $[0,T]$. Also, note that $P(0) = P_{x_0}$.

It is routine to check that $P(t)$ as defined above satisfies the DRE 7.12. Finally, since $P(0) = P_{x_0} > 0$, $P(t) > 0$ for any $t \in [0,T]$.

(b) Consider the Hamiltonian system of Equation 7.19 but now with initial conditions

$$\begin{bmatrix}\xi^T(0) & q^T(0)\end{bmatrix} = \begin{bmatrix}\xi_0^T & 0\end{bmatrix}$$

where $\xi_0 \in R^n$ is nonzero but otherwise arbitrary.

Similarly to the proof of (a), by choosing $w_0(t)$ as in Equation 7.20, $x(t)$ and $q(t)$ will have the same dynamics and

$$q^T(\tau)\xi(\tau) = \|w_0\|^2_{[0,\tau]} - \gamma^2\|s\|^2_{[0,\tau]} > 0 \tag{7.25}$$

for any $\tau \in [0,T]$, whenever $\xi_0 \neq 0$.

For the transition matrix given in Equation 7.23, it is claimed that $\Phi_{11}(\tau,0)$ is nonsingular. Suppose, contrary to what we want to prove, i.e., there exists some nonzero $\alpha \in R^n$ such that $\Phi_{11}(\tau,0)\alpha = 0$. Since $q(0) = 0$, with $\xi_0 = \alpha$ we have that $\xi(\tau) = \Phi_{11}(\tau,0)\xi_0 + \Phi_{12}(\tau,0)q(0) = 0$, which contradicts with Equation 7.25. Hence, $\Phi_{11}(\tau,0)$ is nonsingular for any $\tau \in [0,T]$. Next, let the matrix function

$$P(t) = \Phi_{12}(t,0)\Phi_{11}^{-1}(t,0)$$

Clearly, $P(t)$ is bounded on $[0,T]$ and $P(0) = 0$. It is straightforward to verify that $P(t)$ is the solution of the DRE 7.12.

$\nabla\nabla\nabla$

REMARK 7.1 It should be observed that the results of Theorem 7.1 still hold in the case where the matrices A, B, H, D, and L of the system 7.1 through 7.3 are bounded, piecewise continuous time-varying. ▮

REMARK 7.2 Note the similarity of the above $\mathcal{H}_\infty$ filtering result with the Kalman filtering. As $\gamma \to \infty$, the Riccati Equation 7.12 becomes the Riccati equation of the finite-horizon Kalman filter for the system 7.1 and 7.2 subject to the assumption that the noise signal $w(t)$ is a zero-mean white process with an identity power spectrum density matrix. Furthermore, the estimator of Equations 7.14 through 7.16 becomes the corresponding Kalman filter for the system 7.1 through 7.3. ▮

Next, we analyze the bound for the variance of the estimation error when the $\mathcal{H}_\infty$ filter is applied to the system 7.1 through 7.3 with the noise, w, being a white noise with unit power spectral density matrix. To this end, we assume that the initial state is a random variable with statistics:

$$E\{x(0)\} = m$$
$$E\left\{[x(0) - m][x(0) - m]^T\right\} = P_{x_0}$$

and is independent of the noise signal, $w(t)$ for any $t \geq 0$.

In order that the $\mathcal{H}_\infty$ filter in Equations 7.14 through 7.16 is unbiased, we need to set that $\hat{x}(0) = m$. By applying the filter, we obtain that the variance of the estimation error, $\tilde{s} = s - \hat{s}$, is given by

$$E\{\tilde{s}(t)\tilde{s}^T(t)\} = L\bar{P}(t)L^T, \quad t \in [0, T]$$

where $\bar{P}(\cdot)$ satisfies

$$\begin{aligned} \dot{\bar{P}}(t) &= [A - K(t)H]\bar{P}(t) + \bar{P}(t)[A - K(t)H]^T \\ &\quad + [B - K(t)D][B - K(t)D]^T, \quad \bar{P}(0) = P_{x_0} \end{aligned} \tag{7.26}$$

Let $\tilde{P}(t) = P(t) - \bar{P}(t)$, where $P(t)$ is a solution of the DRE 7.17. Then, subtracting Equation 7.26 from Equation 7.17 leads to

$$\dot{\tilde{P}}(t) = [A - K(t)H]\tilde{P}(t) + \tilde{P}(t)[A - K(t)H]^T + \gamma^{-2}P(t)L^TLP(t), \quad \tilde{P}(0) = 0 \tag{7.27}$$

Since $\tilde{P}(0) = 0$, it is known that $\tilde{P}(t) \geq 0$ for any $t \in [0, T]$, i.e., $P(t) \geq \bar{P}(t)$. Hence, the central $\mathcal{H}_\infty$ filter will guarantee that

$$E\{\tilde{s}(t)\tilde{s}^T(t)\} = L\bar{P}(t)L^T \leq LP(t)L^T$$

for any $t \in [0, T]$. That is to say, the $\mathcal{H}_\infty$ filter has a predictable bound for its filtering error variance. Note that since the Kalman filtering is optimal in the sense of minimal error variance, we should have that $\hat{P}(t) \leq \bar{P}(t) \leq P(t)$, where $\hat{P}$ is the error variance achieved by the Kalman filter. We observe from Equations 7.12, 7.16, and 7.27 that when $\gamma \to \infty$, $\tilde{P}(t) \to 0$, i.e., $P(t) \to \bar{P}(t) \to \hat{P}(t)$.

7.3 Characterization of All Finite Horizon $\mathcal{H}_\infty$ Linear Filters

A filtered estimate $\hat{s}(t) = \mathcal{F}(\mathcal{Z}_t)$ that achieves the desired suboptimal level of noise attenuation, γ, exists for the system 7.1 through 7.3, on the time interval $[0, T]$, if and only if the DRE 7.12 has a solution on $[0, T]$. When it exists, this

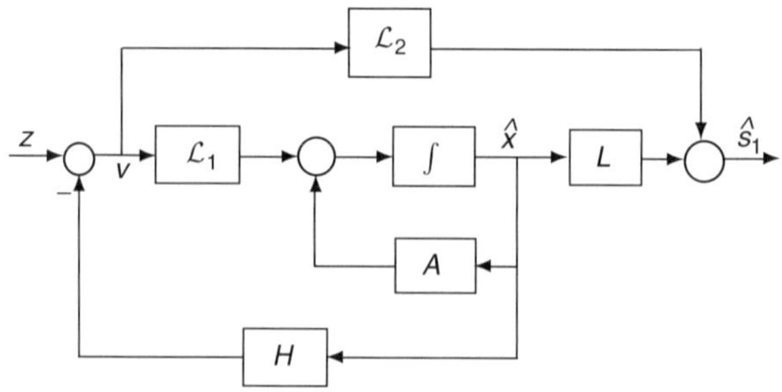

FIGURE 7.3
A parameterization of all linear filters by $\mathcal{L}_1$ and $\mathcal{L}_2$.

estimate is in general not unique. In this section, we shall find all the linear filters that achieve the desired suboptimal $\mathcal{H}_\infty$ performance (Limebeer and Shaked 1991).

To begin with, we note that all linear causal filters can be represented by

$$\dot{\hat{x}} = A\hat{x} + \mathcal{L}_1 v, \quad \hat{x}(0) = 0 \tag{7.28}$$

$$\hat{s}_1 = L\hat{x} + \mathcal{L}_2 v \tag{7.29}$$

$$v = z - H\hat{x} \tag{7.30}$$

where $\mathcal{L}_1$ and $\mathcal{L}_2$ are linear operators. As in the Kalman filtering, v is known as *innovation*.

The block diagram of the above filter is shown in Figure 7.3. Note that the linear operators $\mathcal{L}_1$ and $\mathcal{L}_2$ are in fact dependent. This is to say, only one linear operator may be used in the parameterization of all linear causal filters. To show this, we consider the parameterization given by Figure 7.4, where $P(\cdot)$ is the solution of the DRE 7.12 and $K(\cdot)$ is the gain of the $\mathcal{H}_\infty$ filter given in Equation 7.16, and $\mathcal{U}$ is an operator.

First, we shall show that for any given linear causal operators $\mathcal{L}_1$ and $\mathcal{L}_2$ in Figure 7.3, there exists an operator $\mathcal{U}$ in Figure 7.4 that produces the same estimate for s, i.e., $\hat{s}_1 \equiv \hat{s}$ for the same w. To this end, by setting $\bar{x} = x - \hat{x}$ it follows from Equations 7.1 through 7.3 and 7.28 through 7.30 that

$$\dot{\bar{x}} = (A - \mathcal{L}_1 H)\bar{x} + (B - \mathcal{L}_1 D)w, \quad \bar{x}(0) = x_0 \tag{7.31}$$

$$s - \hat{s}_1 = (L - \mathcal{L}_2 H)\bar{x} - \mathcal{L}_2 D w \tag{7.32}$$

Also, the state-space realization of the filter in Figure 7.4 is as follows:

$$\dot{\hat{x}} = A\hat{x} + K(z - H\hat{x}) - \gamma^{-2} P L^T \eta, \quad \hat{x}(0) = 0 \tag{7.33}$$

$$\hat{s} = L\hat{x} - \eta \tag{7.34}$$

$$\eta = \mathcal{U}(z - H\hat{x}) \tag{7.35}$$

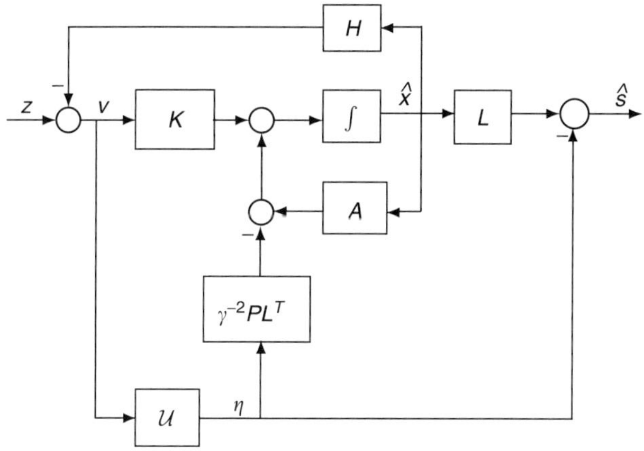

FIGURE 7.4
A parameterization of all linear filters by $\mathcal{U}$.

Define $\tilde{x} = x - \hat{x}$ for the filter in Figure 7.4. Then, the filtering error dynamic for the filter 7.33 through 7.35 is given by

$$\dot{\tilde{x}} = (A - KH)\tilde{x} + (B - KD)w + \gamma^{-2}PL^T\eta, \quad \tilde{x}(0) = x_0 \tag{7.36}$$

$$s - \hat{s} = L\tilde{x} + \eta \tag{7.37}$$

$$v = H\tilde{x} + Dw \tag{7.38}$$

$$\eta = \mathcal{U}(z - H\hat{x}) \tag{7.39}$$

By setting $\hat{s}_1 = \hat{s}$, it follows from Equations 7.32, 7.37, and 7.38 that

$$\begin{aligned} \eta &= (L - \mathcal{L}_2 H)\bar{x} - \mathcal{L}_2 Dw - L\tilde{x} \\ &= (L - \mathcal{L}_2 H)(\bar{x} - \tilde{x}) - \mathcal{L}_2 v \end{aligned} \tag{7.40}$$

Now, subtracting Equation 7.36 from Equation 7.31 yields

$$\begin{aligned} \dot{\bar{x}} - \dot{\tilde{x}} = {} & (A - \mathcal{L}_1 H)\bar{x} + (B - \mathcal{L}_1 D)w \\ & - (A - KH)\tilde{x} - (B - KD)w - \gamma^{-2}PL^T\eta \end{aligned} \tag{7.41}$$

Substituting Equations 7.38 and 7.40, the above leads to

$$\begin{aligned} \dot{\bar{x}} - \dot{\tilde{x}} = {} & \left[A - \mathcal{L}_1 H - \gamma^{-2}PL^T(\mathcal{L}_2 H + L)\right](\bar{x} - \tilde{x}) \\ & + (K + \gamma^{-2}PL^T\mathcal{L}_2 - \mathcal{L}_1)v \end{aligned} \tag{7.42}$$

Thus, the linear causal operator $\mathcal{U} : v \mapsto \eta$ is defined by Equations 7.42 and 7.40, which produces the same estimation error as that of filter 7.28 and 7.29 for given linear operators $\mathcal{L}_1$ and $\mathcal{L}_2$.

Hence, we can concentrate on the linear filter of the form of Equations 7.33 through 7.35 and shall show that

$$\|s-\hat{s}\|^2_{[0,T]} < \gamma^2\left(\|w\|^2_{[0,T]} + x_0^T P_{x_0}^{-1} x_0\right) \tag{7.43}$$

for any nonzero $(w, x_0) \in \mathcal{L}_2[0,T] \oplus R^n$, if and only if $\|\mathcal{U}D\| < \gamma$.

First, we introduce the adjoint system of Equations 7.36 through 7.38 as follows:

$$\dot{\tilde{x}}_a = -(A-KH)^T\tilde{x}_a + L^T w_a + H^T\eta_a, \quad \tilde{x}_a(T)=0 \tag{7.44}$$

$$e_a = -(B-KD)^T\tilde{x}_a + D^T\eta_a \tag{7.45}$$

$$v_a = \gamma^{-2}LP\tilde{x}_a + w_a \tag{7.46}$$

$$\tilde{x}_{a0} = P_{x_0}\tilde{x}_a(0) \tag{7.47}$$

By applying the DRE 7.17 and some routine computations we have that

$$\begin{aligned}
&\gamma^2\|w_a\|^2_{[0,T]} - \left(\|e_a\|^2_{[0,T]} + \tilde{x}_{a0}^T P_{x_0}^{-1}\tilde{x}_{a0}\right)\\
&= \int_0^T \{\gamma^2 w_a^T w_a - [-(B-KD)^T\tilde{x}_a + D^T\eta_a]^T\\
&\qquad \times [-(B-KD)^T\tilde{x}_a + D^T\eta_a]\}\mathrm{d}t - \tilde{x}_a^T(0)P_{x_0}\tilde{x}_a(0)\\
&= \int_0^T \Big[\gamma^2 w_a^T w_a - \tilde{x}_a^T(B-KD)(B-KD)^T\tilde{x}_a + \tilde{x}_a^T(B-KD)D^T\eta_a\\
&\qquad + \eta_a^T D(B-KD)^T\tilde{x}_a - \eta_a^T DD^T\eta_a + \frac{\mathrm{d}}{\mathrm{d}t}\left(\tilde{x}_a^T P\tilde{x}_a\right)\Big]\mathrm{d}t\\
&= \int_0^T [\gamma^2(w_a+\gamma^{-2}LP\tilde{x}_a)^T(w_a+\gamma^{-2}LP\tilde{x}_a) - \eta_a^T DD^T\eta_a]\,\mathrm{d}t\\
&\quad + \int_0^T [\eta_a^T HP\tilde{x}_a + \tilde{x}_a^T PH^T\eta_a + \tilde{x}_a^T(B-KD)D^T\eta_a\\
&\qquad + \eta_a^T D(B-KD)^T\tilde{x}_a]\mathrm{d}t
\end{aligned} \tag{7.48}$$

Note that by using Equation 7.16

$$(B-KD)D^T = -PH^T$$

Thus, it follows from Equation 7.48 that

$$\gamma^2\|w_a\|^2_{[0,T]} - \left(\|e_a\|^2_{[0,T]} + \tilde{x}_{a0}^T P_{x_0}^{-1}\tilde{x}_{a0}\right) = \gamma^2\|v_a\|^2_{[0,T]} - \|D^T\eta_a\|^2_{[0,T]}$$

Since $\eta_a = \mathcal{U}^* v_a$, we thus have that

$$\gamma^2\|w_a\|^2_{[0,T]} - \left(\|e_a\|^2_{[0,T]} + \tilde{x}_{a0}^T P_{x_0}^{-1}\tilde{x}_{a0}\right) = \int_0^T v_a^T(\gamma^2 I - \mathcal{U}DD^T\mathcal{U}^*)v_a\,\mathrm{d}t.$$

Hence,

$$\|e_a\|^2_{[0,T]} + \tilde{x}_{a0}^T P_{x_0}^{-1}\tilde{x}_{a0} < \gamma^2\|w_a\|^2_{[0,T]}$$

if and only if $\|\mathcal{U}D\| < \gamma$. Thus, we conclude that for the original (error) system 7.36 through 7.40.

$$\|s - \hat{s}\|_{[0,T]}^2 + x_0^T P_{x_0}^{-1} x_0 < \gamma^2 \|w\|_{[0,T]}^2$$

for any $(w, x_0) \in \mathcal{L}_2[0,T] \oplus R^n$ if and only if $\|\mathcal{U}D\| < \gamma$.

The above results are summarized in the following theorem:

THEOREM 7.2

Consider the system 7.1 through 7.3 with unknown initial condition x_0 (respectively, $x_0 = 0$). If there exists a bounded solution $P(t)$ with $P(0) = P_{x_0}$ (respectively, $P(0) = 0$) to the DRE 7.12, then all linear causal filters that achieve

$$\|s - \hat{s}\|_{[0,T]}^2 < \gamma^2 \left(\|w\|_{[0,T]}^2 + x_0^T P_{x_0}^{-1} x_0 \right)$$

for all nonzero $(w, x_0) \in \mathcal{L}_2[0,T] \oplus R^n$ are given by Equations 7.33 through 7.35, where K is the $\mathcal{H}_\infty$ filter gain given in Equation 7.16 and $\mathcal{U}$ any linear causal operator satisfying $\|\mathcal{U}D\| < \gamma$.

Note that in Theorem 7.2, when $\mathcal{U}$ is set to be 0, we obtain the $\mathcal{H}_\infty$ filter of Equations 7.14 through 7.16. Thus, the filter given by Equations 7.14 through 7.16 is called *the central $\mathcal{H}_\infty$ filter.*

The parameterization in Theorem 7.2 may be useful when multiobjective performances are required to be achieved by a linear causal filter. For example, when a mixed H_2/H_∞ performance is considered, the free parameter $\mathcal{U}$ in Figure 7.4 may be used to minimize the required H_2 performance.

Finally, we point out that the above result also holds when the system matrices of Equations 7.1 through 7.3 are time-varying, bounded, and continuous.

7.4 Stationary $\mathcal{H}_\infty$ Filter—Riccati Equation Approach

We now analyze the stationary $\mathcal{H}_\infty$ filtering problem. To simplify our analysis, we assume that the system matrices A, B, H, and D are constant. Since the stationary filter is required to be asymptotically stable, we shall make the following assumption for the system 7.1 and 7.2:

Assumption 7.2

(a) (H, A) is detectable

(b) Rank $\begin{bmatrix} A - j\omega I & B \\ H & D \end{bmatrix} = n + p$, *for all* $\omega \in (-\infty, \infty)$

It should be noted that Assumption 7.2 (a) is necessary for the existence of an asymptotically stable filter for the system 7.1 and 7.2. Assumption 7.2 (b) is equivalent to requiring that the transfer function matrix $H(sI - A)^{-1}B + D$

has no transmission zeros on the imaginary axis. We observe that Assumption 7.2 is a standard assumption in stationary Kalman filtering.

A complete solution to the stationary $\mathcal{H}_\infty$ filtering is provided by the next theorem.

THEOREM 7.3

Consider the system 7.1 through 7.3 with $x_0 = 0$ and satisfying Assumptions 7.1 and 7.2. Given a constant $\gamma > 0$, the following statements are equivalent:

(a) There exists a linear causal filter such that the filtering error transfer function $G_{\tilde{s}w}(s)$ is asymptotically stable and $\|G_{\tilde{s}w}(s)\|_\infty < \gamma$.

(b) There exists a stabilizing solution $P = P^T \geq 0$ to the ARE

$$\begin{aligned}&\left(A - BD^T R^{-1} H\right) P + P\left(A - BD^T R^{-1} H\right)^T \\ &\quad + P\left(\gamma^{-2} L^T L - H^T R^{-1} H\right) P + B\left(I - D^T R^{-1} D\right) B^T = 0 \qquad (7.49)\end{aligned}$$

When the above condition (b) is satisfied, we have that a suitable time-invariant $\mathcal{H}_\infty$ filter (central $\mathcal{H}_\infty$ filter) is given by

$$\dot{\hat{x}}(t) = A\hat{x}(t) + K[z(t) - H\hat{x}(t)], \quad \hat{x}(0) = 0 \qquad (7.50)$$

$$\hat{s}(t) = L\hat{x}(t) \qquad (7.51)$$

where the filter gain matrix K is given by

$$K = (PH^T + BD^T)R^{-1} \qquad (7.52)$$

and the matrix $P \geq 0$ is the stabilizing solution of Equation 7.49.

Proof:

(b)→(a): By considering Equation 7.52, it follows that Equation 7.49 can be rewritten as

$$(A-KH)P+P(A-KH)^T+\gamma^{-2}PL^T LP+(B-KD)(B-KD)^T = 0 \qquad (7.53)$$

Since P is the stabilizing solution of Equation 7.49, it is easy to show that $A - KH$ and $A - KH + \gamma^{-2} PL^T L$ are asymptotically stable. In contrast, the transfer function matrix from the disturbance input, w, to the filtering error, $\tilde{s}$, is given by

$$G_{\tilde{s}w}(s) = L[sI - (A - KH)]^{-1}(B - KD)$$

By Lemma 7.1, it is known that $\|G_{\tilde{s}w}(s)\|_\infty < \gamma$.

(a)→(b): Recall from the finite-horizon case that when w is chosen as in Equation 7.20 we have

$$\left\|R^{-\frac{1}{2}} Hx\right\|_2^2 + \left\|(I - D^T R^{-1} D)B^T \xi\right\|_2^2 - \gamma^{-2}\left\|Lx\right\|_2^2 > 0 \qquad (7.54)$$

for any $\xi \in \mathcal{L}_2[0,\infty)$. Then, by Lemma 7.1, we conclude that the algebraic Riccati Equation 7.49 possesses a nonnegative definite stabilizing solution. ∎

Example 7.1

Consider the following second-order system

$$\dot{x}(t) = \begin{bmatrix} 0 & 1 \\ -25 & -7.07 \end{bmatrix} x(t) + \begin{bmatrix} 0.1 \\ 1 \end{bmatrix} w(t) \tag{7.55}$$

$$z(t) = \begin{bmatrix} -100 & 100 \end{bmatrix} x(t) + v(t) \tag{7.56}$$

$$s(t) = \begin{bmatrix} 0 & 10 \end{bmatrix} x(t) \tag{7.57}$$

where the noise sources $w(t)$ and $v(t)$ are assumed to be deterministic and from $\mathcal{L}_2[0,\infty)$. Note that B and D are given by

$$B = \begin{bmatrix} 0.1 & 0 \\ 1 & 0 \end{bmatrix}, \qquad D = \begin{bmatrix} 0 & 1 \end{bmatrix}$$

The problem is to design a stationary $\mathcal{H}_\infty$ filter to attenuate the level of the effect of the noises.

Let $\gamma = 0.3$. By solving the ARE 7.49 using MATLAB, we obtain the stabilizing solution

$$P = \begin{bmatrix} 0.1280 & 0.0958 \\ 0.0958 & 0.0806 \end{bmatrix}$$

and the corresponding $\mathcal{H}_\infty$ filter is given by Equations 7.50 and 7.51 with filter gain

$$K = \begin{bmatrix} -3.2235 \\ -1.5144 \end{bmatrix}$$

The singular value plot of the filtering error transfer function is shown in Figure 7.5. The actual $\mathcal{H}_\infty$ norm of the filtering error transfer function is 0.2996 which is below $\gamma = 0.3$.

The $\mathcal{H}_\infty$ filter gains and actual $\mathcal{H}_\infty$ norms of the filtering error transfer functions for different level of noise attenuation are tabulated in Table 7.1. It is observed that when the γ is reduced, the solution of the ARE is increased and the corresponding filter gain will be higher. Also, the larger the γ the larger the difference between the actual $\mathcal{H}_\infty$ norm and γ. □

Example 7.2

We consider a pitch leveling loop for marine navigation system. Leveling error is a major factor that affects the accuracy of the navigation system. As the ship experiences rotational disturbances due to sea motion, sea disturbances have sinusoidal character and could result in high pitch and roll amplitudes.

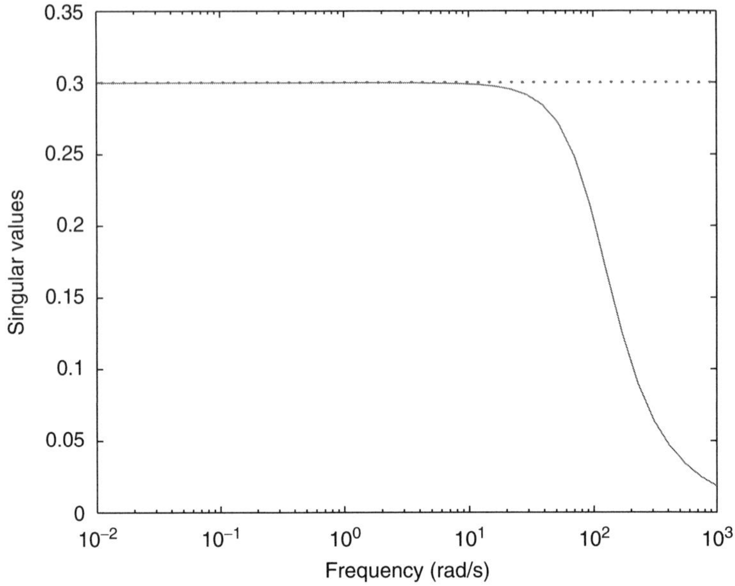

FIGURE 7.5
The singular value plot of the $\mathcal{H}_\infty$ filtering error transfer function and its upper bound γ.

TABLE 7.1
$\mathcal{H}_\infty$ Performance versus γ

γ	$\|G_{\tilde{s}w}(s)\|_\infty$	K	P
0.3	0.2996	$\begin{bmatrix}-3.2235\\-1.5144\end{bmatrix}$	$\begin{bmatrix}0.1280 & 0.0958\\0.0958 & 0.0806\end{bmatrix}$
0.35	0.3389	$\begin{bmatrix}-0.5294\\0.4787\end{bmatrix}$	$\begin{bmatrix}0.0223 & 0.0170\\0.0170 & 0.0218\end{bmatrix}$
0.5	0.4011	$\begin{bmatrix}-0.2321\\0.6815\end{bmatrix}$	$\begin{bmatrix}0.0106 & 0.0083\\0.0083 & 0.0151\end{bmatrix}$
1.00	0.4509	$\begin{bmatrix}-0.1617\\0.7213\end{bmatrix}$	$\begin{bmatrix}0.0078 & 0.0062\\0.0062 & 0.0134\end{bmatrix}$
100	0.4694	$\begin{bmatrix}-0.1463\\0.7286\end{bmatrix}$	$\begin{bmatrix}0.0072 & 0.0057\\0.0057 & 0.0130\end{bmatrix}$

The system model can be represented by

$$\dot{x}(t) = \begin{bmatrix}0 & -g & 0\\0 & 0 & 0\\0 & \alpha & -\alpha\end{bmatrix} x(t) + \begin{bmatrix}\sqrt{q_1} & 0\\0 & \sqrt{q_2}\\0 & 0\end{bmatrix} w(t)$$

$$z(t) = \begin{bmatrix}1 & 0 & 0\end{bmatrix} x(t) + n(t)$$

$$s(t) = \begin{bmatrix}0 & 0 & 1\end{bmatrix} x(t)$$

where $x = \begin{bmatrix} v & \mu & \xi \end{bmatrix}^T$, with v the velocity, μ the direction cosine, and ξ the output of v passing through a low-pass filter:

$$W(s) = \frac{\alpha}{s + \alpha}$$

and n is the measurement noise.

Let $\alpha = 10$ rad/s, $\sqrt{q_1} = 3$ ft/s, $\sqrt{q_2} = 0.1$ rad, and $g = 32.174$ ft/s^2. Thus,

$$B = \begin{bmatrix} 3 & 0 & 0 \\ 0 & 0.1 & 0 \\ 0 & 0 & 0 \end{bmatrix}, \qquad D = \begin{bmatrix} 0 & 0 & 1 \end{bmatrix}$$

By applying Theorem 7.3, we obtain the optimal $\mathcal{H}_\infty$ filtering performance

$$\gamma_{\text{inf}} = 0.096$$

for which the magnitude plots of the $\mathcal{H}_\infty$ filter (solid line) and the Kalman filter (dashed line) are compared in Figure 7.6. It can be observed that the bandwidth of the optimal $\mathcal{H}_\infty$ filter is much higher than that of the Kalman filter. However, the optimal $\mathcal{H}_\infty$ filter gain is

$$K_\infty = \begin{bmatrix} 191.0399 \\ -18.0963 \\ -18.0426 \end{bmatrix}$$

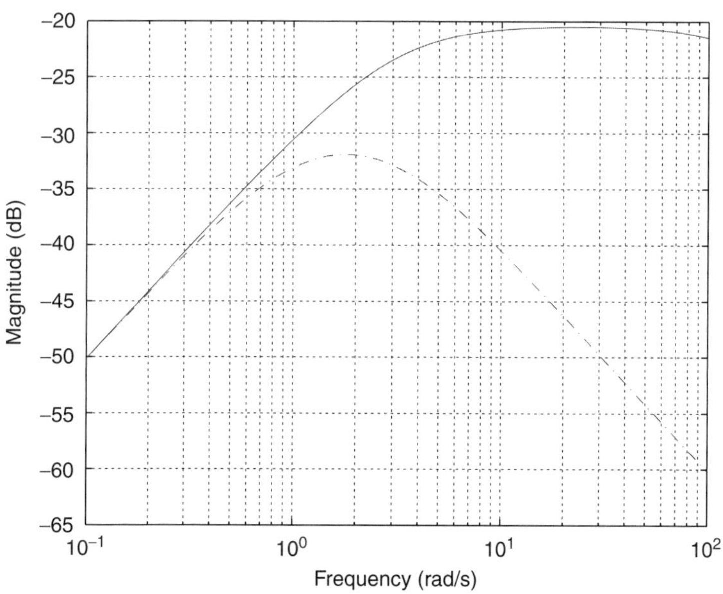

FIGURE 7.6
Comparison of the bandwidths of the optimal $\mathcal{H}_\infty$ filter and the Kalman filter.

which is also much higher than the Kalman filter gain

$$K = \begin{bmatrix} 3.9287 \\ -0.1000 \\ -0.0977 \end{bmatrix}$$

We now take a slightly larger $\gamma = 0.097$. The suboptimal $\mathcal{H}_\infty$ filter 7.50 and 7.51 will have a much small filter gain

$$K_\infty = \begin{bmatrix} 24.0670 \\ -2.0403 \\ -2.0322 \end{bmatrix}$$

but its bandwidth is also significantly reduced as shown in Figure 7.7.

The comparison between the singular value plots of the filtering error transfer functions of the suboptimal $\mathcal{H}_\infty$ filter ($\gamma = 0.097$) and the Kalman filter is shown in Figure 7.8. The actual $\mathcal{H}_\infty$ norm of the Kalman filtering error transfer function is 0.1462, which is much larger than 0.097 given by the $\mathcal{H}_\infty$ filter.

From the singular value plots, it can be known that for disturbances with frequency below 1 rad/s, the $\mathcal{H}_\infty$ filter will result in a much smaller filtering error than the Kalman filter. □

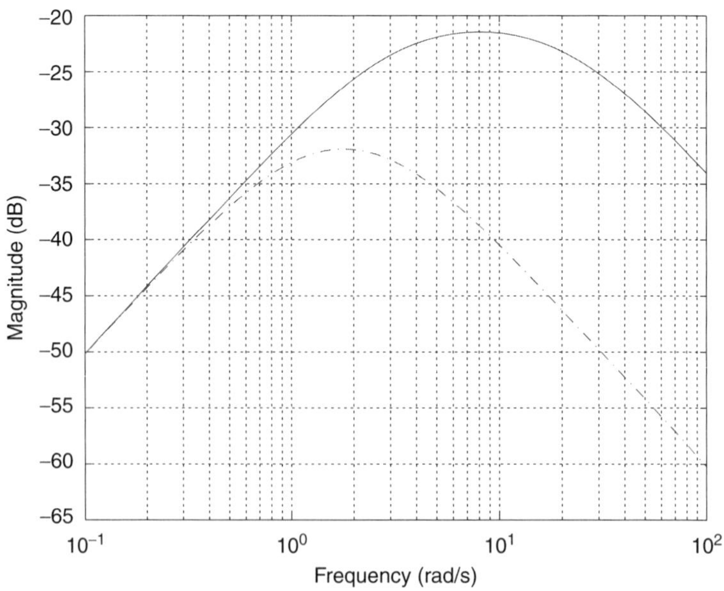

FIGURE 7.7
Comparison of the bandwidths of the suboptimal $\mathcal{H}_\infty$ filter and the Kalman filter.

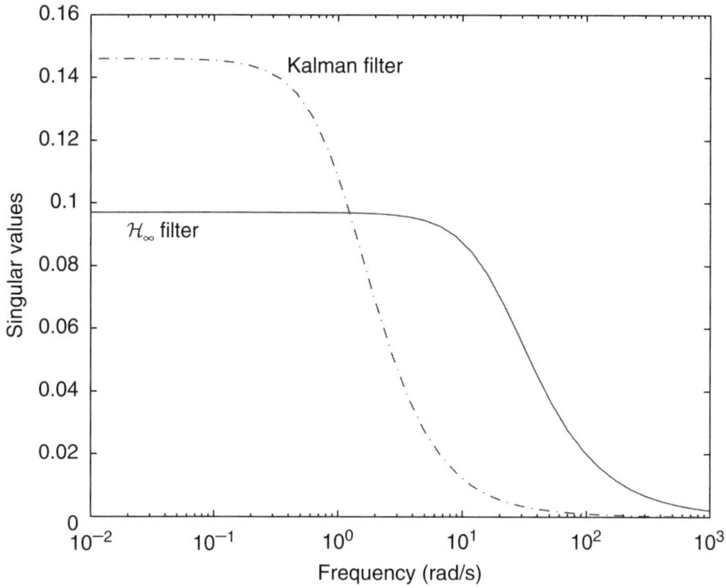

FIGURE 7.8
Comparison between the singular value plot of the $\mathcal{H}_\infty$ filtering error transfer function for $\gamma = 0.097$ and that of the Kalman filter.

7.4.1 Relationship between Guaranteed $\mathcal{H}_\infty$ Norm and Actual $\mathcal{H}_\infty$ Norm

We consider here a special case with $L = H$. That is, the signal to be estimated is the same as the uncorrupted measurement signal. For simplicity of discussions, we adopt the following standard assumption for system 7.1 through 7.3:

Assumption 7.3

(a) (A, B) *is controllable and* (A, H) *is observable*

(b) $BD^T = 0$ *and* $DD^T = I$

The assumption that $BD^T = 0$ essentially implies that the process noise and observation noise are independent of each other.

With the above assumption, Equation 7.49 becomes

$$AP + PA^T - (1 - \gamma^{-2})PH^T HP + BB^T = 0 \tag{7.58}$$

and the filter gain Equation 7.52 becomes

$$K = PH^T$$

From the Kalman filter, we know that if $\gamma > 1$ and with Assumption 7.3(a), Equation 7.58 always admits a stabilizing solution $P = P^T > 0$. Therefore, the minimum achievable $\gamma_{\text{inf}} \leq 1$. For unstable A, Equation 7.58 does not admit a positive definite stabilizing solution for $\gamma_{\text{inf}} < 1$ and hence $\gamma_{\text{inf}} = 1$.

However, the central $\mathcal{H}_\infty$ filter designed with parameter γ generally guarantees a level of attenuation $\bar{\gamma}$ strictly less than γ. The following result gives an estimate of an upper bound of $\bar{\gamma}$ (Bolzern *et al.* 2001):

LEMMA 7.2

Let $\gamma > 1$ and consider the system 7.1 through 7.3 with $L = H$ and satisfying Assumption 7.3. The central $\mathcal{H}_\infty$ filter designed with γ has a guaranteed $\mathcal{H}_\infty$ performance $\tilde{\gamma}$, i.e., the actual $\mathcal{H}_\infty$ filtering performance $\bar{\gamma}$ satisfies

$$\bar{\gamma} \leq \tilde{\gamma} < \gamma$$

where

$$\tilde{\gamma} = \frac{2}{1 + \gamma^{-2}}$$

Proof:

Using Equations 7.1 through 7.3 and Equations 7.50 and 7.51, the filtering error system is given by

$$\dot{e} = (A - PH^T H)e + (B - PH^T D)w$$

$$\tilde{s} = He$$

By applying Lemma 7.1, the $\mathcal{H}_\infty$ norm of the above error system can be assessed by the existence of a matrix $\bar{P} > 0$ such that

$$(A - PH^T H)\bar{P} + \bar{P}(A - PH^T H)^T + \tilde{\gamma}^{-2}\bar{P}H^T H\bar{P} + BB^T + PH^T HP \leq 0$$

Let $\bar{P} = P/\alpha$ with $0 < \alpha \leq 1$ and substitute Equation 7.58 into the above matrix inequality. We obtain that

$$(\alpha + \alpha^{-1}\tilde{\gamma}^{-2} - 1 - \gamma^{-2})PH^T HP + (1 - \alpha)BB^T \leq 0$$

which can be satisfied if

$$\tilde{\gamma}^{-2} \leq \alpha(1 + \gamma^{-2} - \alpha)$$

Minimizing $\tilde{\gamma}$ with respect to α gives the optimal value of α as

$$\alpha^* = \frac{1 + \gamma^{-2}}{2} < 1$$

which leads to

$$\tilde{\gamma}^{-1} = \frac{1 + \gamma^{-2}}{2}$$

Since $1 + \gamma^{-2} \geq 2\gamma^{-1}$, $\tilde{\gamma} \leq \gamma$. Further, $\tilde{\gamma} = \gamma$ if and only if $\gamma = 1$. ∎

REMARK 7.3 The Kalman filter is a special case of the central $\mathcal{H}_\infty$ filter with $\gamma \to \infty$. Hence,

$$\tilde{\gamma} \leq 2$$

when $H = L$. This indicates a remarkable robustness of the Kalman filter which is designed without imposing any $\mathcal{H}_\infty$ specification. ∎

REMARK 7.4 Suppose the desired $\mathcal{H}_\infty$ performance γ_0 satisfying $1 < \gamma_0 < 2$. Then, the central $\mathcal{H}_\infty$ filter designed with

$$\gamma = \sqrt{\frac{\gamma_0}{2 - \gamma_0}} > \gamma_0$$

achieves the desired $\mathcal{H}_\infty$ filtering performance γ_0. ∎

7.4.2 Characterization of All Linear Time-Invariant $\mathcal{H}_\infty$ Filters

Similarly to the finite horizon case, we characterize all the linear stationary $\mathcal{H}_\infty$ filters that achieve the desired suboptimal noise attenuation level, γ.

THEOREM 7.4
Consider the system 7.1 through 7.3 with A, B, H, and D being constant matrices and satisfying Assumptions 7.1 and 7.2. If there exists a stabilizing solution $P = P^T \geq 0$ to the ARE 7.49, then all linear causal $\mathcal{H}_\infty$ filters are parameterized by

$$\dot{\hat{x}}(t) = A\hat{x}(t) + K[z(t) - H\hat{x}(t)] - \gamma^{-2}PL^T\eta(t), \quad \hat{x}(0) = 0 \tag{7.59}$$

$$\hat{s}(t) = L\hat{x}(t) - \eta(t) \tag{7.60}$$

$$\eta(t) = \mathcal{U}[z(t) - H\hat{x}(t)] \tag{7.61}$$

where $\mathcal{U}$ is any proper stable transfer function matrix $U(s)$ satisfying

$$\|U(s)D\|_\infty < \gamma$$

The block diagram of the above parameterization is also shown in Figure 7.4 with K given by Equation 7.52 and P being the stabilizing solution of Equation 7.49.

7.5 Relationship with the Kalman Filter

Similarly to the case of finite-horizon $\mathcal{H}_\infty$ filtering, as $\gamma \to \infty$, the result of Theorem 7.3 reduces to the well-known stationary Kalman filtering result for the system 7.1 through 7.3 under the assumption that the noise signal $w(t)$ is a zero-mean white process with an identity power spectrum density matrix.

Assume that the disturbance signal, w, of Equation 7.1 is a zero-mean Gaussian white noise with identity power spectrum density matrix and that the $\mathcal{H}_\infty$ filter of Equations 7.50 and 7.51 exists. By applying the $\mathcal{H}_\infty$ filter, the filtering error dynamic is given by

$$\dot{e} = (A - KH)e + (B - KD)w \tag{7.62}$$

$$\tilde{s} = Le \tag{7.63}$$

Then, the filtering error variance is given by

$$M = E(\tilde{s}\tilde{s}^T) = L\bar{P}L^T$$

where $\bar{P}$ satisfies

$$(A - KH)\bar{P} + \bar{P}(A - KH)^T + (B - KD)(B - KD)^T = 0 \tag{7.64}$$

By subtracting Equation 7.64 from Equation 7.53 we have

$$(A - KH)(P - \bar{P}) + (P - \bar{P})(A - KH)^T + \gamma^{-2}PL^TLP = 0 \tag{7.65}$$

Since $A - KH$ is asymptotically stable, it is easy to see that $P - \bar{P} \geq 0$, i.e., $P \geq \bar{P}$. Therefore, the filtering error variance is bounded by

$$M \leq LPL^T$$

It is clear that the $\mathcal{H}_\infty$ filter is suboptimal in terms of the filtering error variance when the system under consideration is driven by a white noise. However, the $\mathcal{H}_\infty$ filter has a predictable upper bound for its error variance.

Also, in view of Equation 7.53, the $\mathcal{H}_\infty$ filter can be interpreted as the Kalman filter for the system of Equations 7.1 through 7.3 with an additional white noise input, i.e.,

$$\dot{x} = Ax + Bw + \gamma^{-1}PL^T\tilde{w}$$

$$z = Hx + Dw$$

$$s = Lx$$

where $\tilde{w}$ is a white noise with identity power spectrum matrix and is independent of w.

7.6 Convergence Analysis

Recall that under Assumption 7.3, the DRE associated with the finite horizon $\mathcal{H}_\infty$ filter is given by

$$\dot{P} = AP + PA^T + P\left(\gamma^{-2}L^TL - H^TH\right)P + BB^T, \quad P(0) = P_{x_0} \tag{7.66}$$

and the ARE associated with the stationary $\mathcal{H}_\infty$ filter is

$$0 = AP + PA^T + P\left(\gamma^{-2}L^TL - H^TH\right)P + BB^T \tag{7.67}$$

The question to be answered here is whether the solution 7.66 will be bounded and converge to the nonnegative definite stabilizing solution of 7.67.

For the case of zero initial condition of Equation 7.66, we have the following result (Bolzern *et al.* 1997):

THEOREM 7.5

If (A, H) is detectable and the ARE 7.67 admits a nonnegative definite stabilizing solution P_s, then the solution $P(t)$ of the DRE 7.66 with $P(0) = 0$ exists over $[0, \infty)$ and converges to P_s as $t \to \infty$.

However, this nice convergence property does not carry over for the case of $P(0) \neq 0$ in general. To see this, we consider Example 7.3.

Example 7.3

Let $A = -1$, $H = 1$, $B = \left[\sqrt{3/2} \quad 0\right]$, $D = \left[0 \quad 1\right]$, $L = \sqrt{5}$ and $\gamma = 2$. Then, Equation 7.66 becomes

$$\dot{P} = -2P + \frac{1}{4}P^2 + \frac{3}{2} = \frac{1}{4}\left(P - 4 - \sqrt{10}\right)\left(P - 4 + \sqrt{10}\right)$$

i.e.,

$$\left(\frac{1}{P - 4 - \sqrt{10}} - \frac{1}{P - 4 + \sqrt{10}}\right) \mathrm{d}P = 0.5\sqrt{10}\,\mathrm{d}t.$$

Denote

$$P_a = 4 + \sqrt{10}, \quad P_s = 4 - \sqrt{10}$$

Then, by solving the above differential equation we have

$$\ln \frac{P(t) - P_a}{P(t) - P_s} = 0.5\sqrt{10}t + c_1$$

or yet

$$\frac{P(t) - P_a}{P(t) - P_s} = ce^{0.5\sqrt{10}t}$$

With the initial condition $P(0)$, $c = (P(0) - P_a)/(P(0) - P_s)$. Hence,

$$P(t) = P_s + \frac{2\sqrt{10}\,[P(0) - P_s]}{P(0) - P_s + [P_a - P(0)]\,e^{0.5\sqrt{10}t}}$$

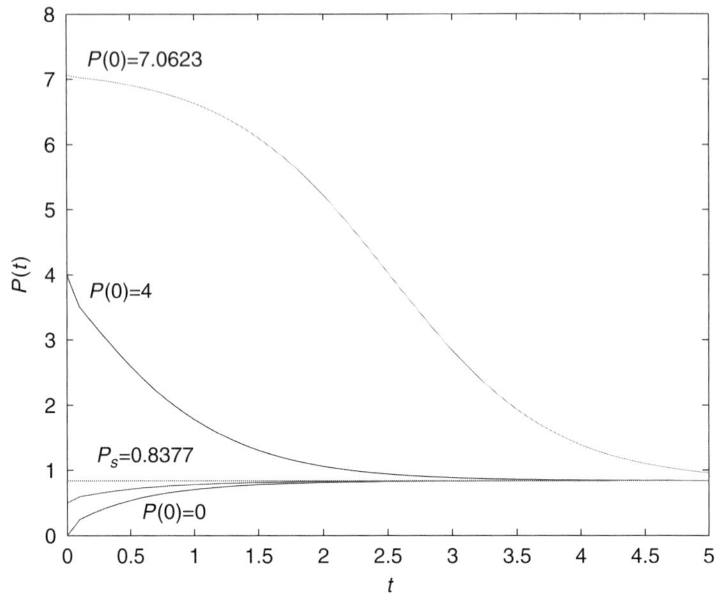

FIGURE 7.9
Convergence of $P(t)$.

Note that both P_a and P_s are solutions of the ARE 7.67 with P_s being the stabilizing one.

Clearly, if $P(0) < P_a$, $P(t) \rightarrow P_s = 4 - \sqrt{10}$ when $t \rightarrow \infty$. The convergence can be seen from Figure 7.9, where $P(0)$ is equal to 0, 0.5, 4, and $4 + \sqrt{10} - 0.1 = 7.0623$, respectively. However, when $P(0) > P_a$, there always exists a finite escape time

$$\tau = \frac{2}{\sqrt{10}} \ln \left(1 + \frac{2\sqrt{10}}{P(0) - P_a} \right)$$

For example, when $P(0) = 4 + \sqrt{10} + 0.1 = 7.2623$, $P(t)$ is shown in Figure 7.10. When $P(0) = P_a$, $P(t) \equiv P_a$. □

Hence, the following issue arises: *Given γ such that the ARE 7.67 admits a nonnegative definite stabilizing solution P_s, for what values of P_{x0} does the solution $P(\cdot)$ of Equation 7.66 remains bounded and converges to P_s?*

First, for a given $\gamma > 0$, it can be easily argued that if the ARE 7.67 admits a nonnegative definite stabilizing solution P_s, then for any P_{x0} satisfying $0 \leq P_{x0} \leq P_s$, the solution $P(t)$ of the DRE 7.66 will remain bounded over $[0, T]$ for any $T > 0$ and converges to P_s. This is because the existence of the stabilizing solution P_s to the ARE 7.67 implies that $P(t) \equiv P_s$ when $P(0) = P_s$. Thus, there exists a filter such that $J(P_s, T) < \gamma$. But $J(P_{x0}, T) \leq J(P_s, T)$ for any $T > 0$. That is to say, there always exists a filter such that $J(P_{x0}, T) < \gamma$.

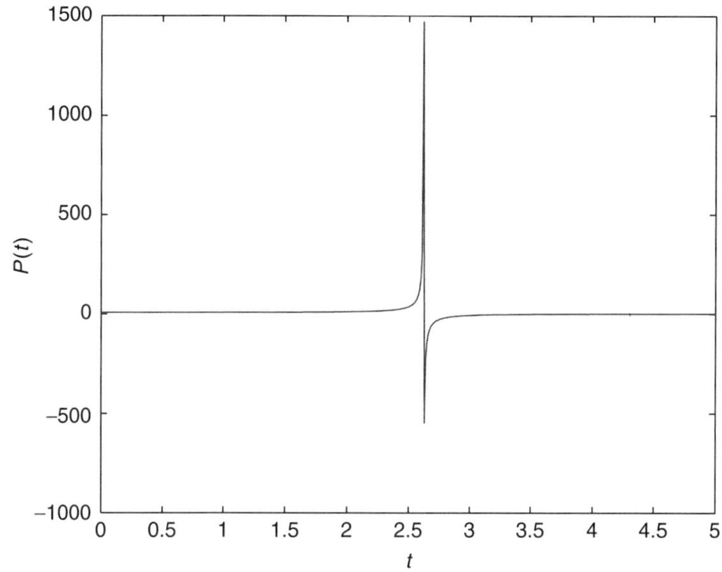

FIGURE 7.10
Finite time escape of $P(t)$.

Hence, for $T>0$ by considering Theorem 7.1 the DRE 7.66 must have a bounded solution $P(t)$ for $t \in [0, T]$. Further, $P_0(t) \le P(t) \le P_s$ (see Problem 6), where $P_0(t)$ is the solution of Equation 7.66 corresponding to $P(0) = P_{x0} = 0$. In view of Theorem 7.5, $P_0(t) \to P_s$. Hence, $P(t) \to P_s$. Note that for Example 7.3, we have verified that when $P_{x0} \le P_s = 4 - \sqrt{10} < P_a$, $P(t)$ is bounded and $P(t) \to P_s$. However, from Example 7.3, we also know that P_{x0} can be greater than P_s for the boundedness of the solution of Equation 7.66 and the convergence. A less-conservative estimate of the upper bound of P_{x0} is provided next.

Since P_s is the stabilizing solution of Equation 7.67, the matrix

$$A_s = A + P_s(\gamma^{-2}L^TL - H^TH)$$

is asymptotically stable. For a symmetric matrix M, decompose M as

$$M = M_+ + M_-$$

where M_+ (M_-) is nonnegative (nonpositive) definite with its nonzero eigenvalues coinciding with the positive (negative) eigenvalues of M. Using these notations, define the Lyapunov equation

$$A_s^TY + YA_s + [\gamma^{-2}L^TL - H^TH]_+ = 0 \tag{7.68}$$

The above Lyapunov equation is guaranteed to have a solution $Y = Y^T \ge 0$. A sufficient condition for the boundedness and convergence of the solution of Equation 7.66 is given below (Bolzern *et al.* 1997).

THEOREM 7.6
Let (A, B) be controllable and the ARE 7.67 admits a nonnegative definite stabilizing solution P_s. Then, if $0 \leq P_{x0} < P_s + (Y + \rho I)^{-1}$, where ρ is a sufficiently small positive constant, the solution of Equation 7.66 converges to P_s as $t \to \infty$.

For Example 7.3,

$$A_s = A + P_s(\gamma^{-2}L^T L - H^T H) = -\frac{\sqrt{10}}{4}$$

and

$$[\gamma^{-2}L^T L - H^T H]_+ = \frac{1}{4}$$

It follows from Equation 7.68 that $Y = 1/2\sqrt{10}$. Take $\rho = 0$ and apply Theorem 7.6. We know that the solution of Equation 7.66 will be bounded over $[0, \infty)$ and converges to P_s if

$$0 \leq P_{x0} < P_s + Y^{-1} = 4 + \sqrt{10}$$

which is also known to be a necessary condition from Example 7.3.

COROLLARY 7.1
From Theorem 7.6, if the ARE 7.67 admits a nonnegative definite stabilizing solution P_s and $\gamma^{-2}L^T L - H^T H < 0$, then for any $P_{x0} \geq 0$, the convergence of $P(t)$ to P_s is guaranteed.

Proof:
In fact, if $\gamma^{-2}L^T L - H^T H < 0$, the Lyapunov Equation 7.68 gives the solution $Y = 0$. Thus, by Theorem 7.6, P_{x0} can be any nonnegative definite matrix which guarantees the convergence of $P(t)$.

Observe that when $\gamma^{-2}L^T L - H^T H < 0$, Equation 7.66 is in fact a Kalman filter type of DRE and Corollary 7.1 is consistent with the convergence result of the Kalman filter. ∎

7.7 $\mathcal{H}_\infty$ Filtering for a Special Class of Signal Models

In Theorem 7.3, we provide a complete solution for suboptimal stationary $\mathcal{H}_\infty$ filtering of the general signal model 7.1 through 7.3. It should be noted that to obtain the optimal $\mathcal{H}_\infty$ filter iteration on γ is required. In the following, we consider the $\mathcal{H}_\infty$ filtering for a class of linear signal models and show that the optimal noise attenuation level can be obtained without going through iteration (Petersen 1989).

Consider the signal model of Equations 7.1 through 7.3 with $D = I$, i.e.,

$$\dot{x}(t) = Ax(t) + Bw(t) \tag{7.69}$$

$$z(t) = Hx(t) + w(t) \tag{7.70}$$

$$s(t) = Lx(t) \tag{7.71}$$

We note that when $D = I$, Assumption 7.2 (b) is reduced to that the matrix $A - BH$ has no purely imaginary eigenvalues (see Problem 7).

First, we consider the case when the matrix $A - BH$ is asymptotically stable. By substituting Equation 7.70 into Equation 7.69 we obtain

$$\dot{x}(t) = Ax(t) + B[z(t) - Hx(t)]$$

i.e.,

$$\dot{x}(t) = (A - BH)x(t) + Bz(t) \tag{7.72}$$

Since $A - BH$ is asymptotically stable, with the zero initial condition the above is an exact state observer. By setting $\hat{s} = Lx$, we have that $\hat{s} = s$, i.e., the filtering error is zero. In another words, the optimal noise attenuation level is

$$\gamma_{\text{inf}} = 0$$

Hence, in the sequel we assume that the matrix $A - BH$ has at least one unstable eigenvalue. We have the following result:

THEOREM 7.7

Consider the system of Equations 7.69 through 7.71 satisfying Assumption 7.2. Let $\gamma > 0$ be a given constant. Then, the following statements are equivalent:

(a) There exists a linear causal filter such that the filtering error dynamic is asymptotically stable and

$$\|G_{\tilde{s}w}(s)\|_\infty < \gamma$$

(b) There exists a stabilizing solution $P = P^T \geq 0$ to the ARE:

$$(A - BH)P + P(A - BH)^T - PH^T HP + \gamma^{-2}PL^T LP = 0 \tag{7.73}$$

Moreover, if any of the above conditions holds, a suitable $\mathcal{H}_\infty$ filter is given by Equations 7.50 and 7.51 with the filter gain given by

$$K = PH^T + B \tag{7.74}$$

Proof:

(b) $\rightarrow$ *(a)*:

First, the ARE 7.73 can be rewritten as

$$(A - KH)P + P(A - KH)^T + PH^T HP + \gamma^{-2}PL^T LP = 0 \tag{7.75}$$

where K is given by Equation 7.74. Since P is the stabilizing solution of Equation 7.73, it is easy to see that $A - KH$ is asymptotically stable. In contrast, the filtering error transfer function matrix is

$$G_{\tilde{s}w}(s) = -L(sI - A + KH)^{-1}PH^T$$

By Lemma 7.1, we conclude that $\|G_{\tilde{s}w}(s)\|_\infty < \gamma$.

$(a) \rightarrow (b)$:

First, it is known from Theorem 7.3 that if $\mathcal{H}_\infty$ filters exist, then there will be a central filter in the form of Equations 7.50 and 7.51. Hence, there exists K in Equations 7.50 and 7.51 such that $\|G_{\tilde{s}w}(s)\|_\infty < \gamma$, where

$$G_{\tilde{s}w}(s) = L(sI - A + KH)^{-1}(B - K)$$

Then, by Lemma 7.1 there exists a matrix $\bar{P} = \bar{P}^T > 0$ satisfying

$$(A - KH)\bar{P} + \bar{P}(A - KH)^T + \gamma^{-2}\bar{P}L^TL\bar{P} + (B - K)(B - K)^T < 0$$

which by completing the square leads to

$$\begin{aligned}&(A - BH)\bar{P} + \bar{P}(A - BH)^T - \bar{P}H^TH\bar{P} + \gamma^{-2}\bar{P}L^TL\bar{P}\\&\quad + [K - (\bar{P}H^T + B)][K - (\bar{P}H^T + B)]^T < 0\end{aligned}$$

i.e.,

$$(A - BH)\bar{P} + \bar{P}(A - BH)^T - \bar{P}H^TH\bar{P} + \gamma^{-2}\bar{P}L^TL\bar{P} < 0$$

This implies by Lemma 7.1 that there exists a stabilizing solution $P = P \geq 0$ to the ARE 7.73. This ends the proof. ∎

Next, we shall determine the optimal noise attenuation level γ_{inf}. Since there exists at least one unstable eigenvalue of the matrix $A - BH$ and $A - BH$ has no purely imaginary eigenvalues, by applying a state-space transformation we can have that

$$A - BH = \begin{bmatrix} A_1 & 0 \\ A_{21} & A_2 \end{bmatrix}$$

$$H = [H_1 \;\; H_2]$$

$$L = [L_1 \;\; L_2]$$

where the matrix A_1 is asymptotically stable and the matrix A_2 is antistable, i.e., $-A_2$ is asymptotically stable. Note that with the detectability of (A, H), (A_2, H_2) is observable.

Next, define the following Lyapunov equations:

$$-A_2^TS - SA_2 + H_2^TH_2 = 0 \tag{7.76}$$

$$-A_2^TT - TA_2 + L_2^TL_2 = 0 \tag{7.77}$$

Since (A_2, H_2) is observable and $-A_2$ is asymptotically stable, there exist solutions $S = S^T > 0$ and $T = T^T \geq 0$ to Equations 7.76 and 7.77, respectively.

THEOREM 7.8
Consider the system 7.69 through 7.71 satisfying Assumption 7.2 and suppose the matrix $A - BH$ has at least one unstable eigenvalue. Then, for a given $\gamma > 0$, there exists a linear causal filter such that the filtering error dynamic is asymptotically stable and $\|G_{\tilde{s}w}(s)\|_\infty < \gamma$ if and only if $S > \gamma^{-2}T$, where $S > 0$ and $T \geq 0$ satisfy Equations 7.76 and 7.77, respectively.

Proof:
Suppose that for given γ, $S > \gamma^{-2}T$. Define $X = S - \gamma^{-2}T$. It follows from Equations 7.76 and 7.77 that X satisfies

$$A_2^T X + XA_2 - H_2^T H_2 + \gamma^{-2} L_2^T L_2 = 0$$

which implies that

$$A_2 P_2 + P_2 A_2^T - P_2 H_2^T H_2 P_2 + \gamma^{-2} P_2 L_2^T L_2 P_2 = 0 \tag{7.78}$$

where $P_2 = X^{-1} > 0$. Also, note that

$$P_2^{-1}\left(A_2 - P_2 H_2^T H_2 + \gamma^{-2} P_2 L^T L\right) P_2 = -A_2^T$$

Since $-A_2$ is asymptotically stable, we conclude that $A_2 - P_2 H_2^T H_2 + \gamma^{-2} P_2 L^T L$ is asymptotically stable. Now, define

$$P = \begin{bmatrix} 0 & 0 \\ 0 & P_2 \end{bmatrix}$$

It is routine to check that P is a solution of the ARE 7.73 and

$$\begin{aligned} &A - BH - PH^T H + \gamma^{-2} PL^T L \\ &= \begin{bmatrix} A_1 & 0 \\ A_{21} - P_2 H_2^T H_1 + \gamma^{-2} P_2 L_2^T L_1 & A_2 - P_2 H_2^T H_2 + \gamma^{-2} P_2 L_2^T L_2 \end{bmatrix} \end{aligned}$$

Since the matrices A_1 and $A_2 - P_2 H_2^T H_2 + \gamma^{-2} P_2 L_2^T L_2$ are both asymptotically stable, then P is the stabilizing solution of Equation 7.73. Therefore, by Theorem 7.4 that there exists a causal linear $\mathcal{H}_\infty$ filter. ∎

The proof of necessity is left for an exercise (see Problem 8).

The above theorem implies that the optimal noise attenuation of the $\mathcal{H}_\infty$ filtering is

$$\gamma_{\inf} = \sqrt{\lambda_{\max}(TS^{-1})}$$

where $\lambda_{\max}(\cdot)$ denotes the maximum eigenvalue. Recall that when the matrix $A - BH$ is stable, $\gamma_{\inf} = 0$.

7.8 Stationary $\mathcal{H}_\infty$ Filter—Linear Matrix Inequality Approach

In Section 7.4, we have discussed the Riccati equation approach to the stationary $\mathcal{H}_\infty$ filter design. There are some inherent restrictions. First, they cannot be applied directly to general singular (the matrix $R = DD^T$ is singular) $\mathcal{H}_\infty$ filtering problem. Secondly, it is required that the transfer function matrix from the noise input to the measurement output has no invariant zeros on the imaginary axis. This section provides an alternate to the Riccati equation approach.

Consider the system of Equations 7.1 through 7.3. In view of the results in Section 7.4, we shall concentrate on the filter of the form

$$\dot{\hat{x}}(t) = A\hat{x}(t) + K[z(t) - H\hat{x}(t)], \quad \hat{x}(0) = 0 \tag{7.79}$$

$$\hat{s}(t) = L\hat{x}(t) \tag{7.80}$$

By defining $e = x - \hat{x}$, the error dynamic of the filter 7.79 and 7.80 is given by

$$\dot{e}(t) = (A - KH)\tilde{x}(t) + (B - KD)w(t), \quad \tilde{x}(0) = 0 \tag{7.81}$$

$$s(t) - \hat{s}(t) = Le(t) \tag{7.82}$$

By Lemma 7.1, we know that for a given $\gamma > 0$, there exists a linear causal filter such that the filtering error dynamics is asymptotically stable and $\|G_{\tilde{s}w}(s)\|_\infty < \gamma$, where $G_{\tilde{s}w}(s)$ denotes the transfer function from w to $\tilde{s}$, if and only if there exists $P = P^T > 0$ such that

$$(A-KH)P+P(A-KH)^T+\gamma^{-2}PL^TLP+(B-KD)(B-KD)^T < 0 \tag{7.83}$$

Now, set $X = P^{-1}$ and $Y = P^{-1}K \in R^{n\times m}$. Then, by multiplying the above from the left and the right by X, Equation 7.83 leads to

$$A^TX + XA - YH - H^TY^T + \gamma^{-2}L^TL + (XB - YD)(XB - YD)^T < 0$$

By applying the Schur complement, the above is equivalent to

$$\Upsilon(X, Y) = \begin{bmatrix} A^TX + XA - YH - H^TY^T + \gamma^{-2}L^TL & XB - YD \\ B^TX^T - D^TY^T & -I \end{bmatrix} < 0 \tag{7.84}$$

The important point to note is that $\Upsilon(X, Y)$ is a jointly linear matrix function of X and Y. Thus, the $\mathcal{H}_\infty$ filter design problem is reduced to a standard convex optimization problem involving a linear matrix inequality. This convex problem can be numerically solved efficiently by interior-point methods. A standard MATLAB package called MATLAB LMI-Toolbox (Gahinet *et al.* 1995) is available to solve the above problem.

When $X > 0$ and Y are obtained, the filter gain is given by

$$K = X^{-1}Y \tag{7.85}$$

We point out that in the LMI approach, Assumptions 7.1 and 7.2(b) are no longer needed, and we are able to address the singular $\mathcal{H}_\infty$ filtering problem.

Example 7.4

For Example 7.1, by applying the LMI approach, we can obtain the optimal γ of 0.2084. For the optimal $\mathcal{H}_\infty$ filtering, the corresponding filter gain is

$$K = \begin{bmatrix} -3.9628 \times 10^6 \\ -2.6778 \times 10^6 \end{bmatrix}$$

which, as expected, is very high. However, by slightly increasing the γ to 0.22 and solving the LMI, we obtain the filter gain of

$$K = \begin{bmatrix} -68.8288 \\ -44.8581 \end{bmatrix}$$

which is much lower than the optimal $\mathcal{H}_\infty$ filter gain. □

Problems

1. Consider the problem of estimating an unknown constant parameter. The problem is modeled as

$$\begin{aligned} \dot{x} &= 0 \\ z &= x + v \end{aligned}$$

where the measurement noise is energy-bounded but otherwise with unknown characteristics. Find a finite horizon $\mathcal{H}_\infty$ filter for estimating the parameter (state). Comment on the effects of the initial condition P_{x_0} of the DRE on the filter.

2. Consider the scalar system

$$\begin{aligned} \dot{x} &= ax + w \\ z &= hx + v \\ s &= x \end{aligned}$$

with

$$a = -1, \qquad h = 1$$

(a) Given $\gamma = 1/\sqrt{2}$, show that a finite horizon $\mathcal{H}_\infty$ filter over $[0, T]$ for any $T > 0$ exists if and only if the weighting for the initial estimate satisfying $0 < P_{x0} < 4$. Obtain an $\mathcal{H}_\infty$ filter when $0 < P_{x0} < 4$.

(b) Show that the optimal stationary $\mathcal{H}_\infty$ filtering performance is $\gamma_{\inf} = 1/\sqrt{2}$.

3. Let the system be described by

$$\dot{x} = \begin{bmatrix} 0 & 1 \\ -\omega_n^2 & -2\alpha \end{bmatrix} x + \begin{bmatrix} 0 \\ 1 \end{bmatrix} w$$

$$z = \begin{bmatrix} 1 & 0 \end{bmatrix} x + v$$

$$s = \begin{bmatrix} 1 & 0 \end{bmatrix} x$$

(a) Write a MATLAB program to compute the optimal stationary $\mathcal{H}_\infty$ filter. Assume that $\omega_n = 1$ and $\alpha = 0.1$.

(b) Verify that the Kalman filter provides an $\mathcal{H}_\infty$ performance of $\|G_e(s)\|_\infty \leq 2$, where $G_e(s)$ is the filtering error transfer function.

4. Consider the problem of estimating motion described by

$$\dot{x} = \begin{bmatrix} 0 & 1 \\ 0 & 0 \end{bmatrix} x + \begin{bmatrix} 0 \\ 1 \end{bmatrix} w$$

where w is a disturbance input with bounded energy. Let the measurement equation be given by

$$z = \begin{bmatrix} 1 & 0 \end{bmatrix} x + v$$

with v the measurement noise which is also energy-bounded. Let $s = [0 \;\; 1]x$ be the signal to be estimated. Given $\gamma = 2$ and $P_{x0} = 0$, find a finite horizon $\mathcal{H}_\infty$ filter and a stationary $\mathcal{H}_\infty$ filter.

5. Consider Problem 2. Let $\gamma = 1$. Find condition on P_{x0} such that the solution of the DRE associated with the finite horizon $\mathcal{H}_\infty$ filtering exists and converges to the stabilizing solution of the ARE associated with the stationary $\mathcal{H}_\infty$ filtering. Obtain a stationary $\mathcal{H}_\infty$ filter.

6. Suppose for a given γ, the DRE 7.66 has bounded solutions $P_1(t)$ and $P_2(t)$ for $P(0) = M_1$ and $P(0) = M_2$, respectively, where $0 \leq M_1 \leq M_2$. Show that $P_1(t) \leq P_2(t)$.

7. Show that when $D = I$, Assumption 7.2(b) is equivalent to that the matrix $A - BH$ has no purely imaginary eigenvalues.

8. Prove the necessity part of Theorem 7.8.

9. Consider the system

$$\begin{aligned}\dot{x} &= \begin{bmatrix} -1 & 0 \\ 2 & 1 \end{bmatrix} x + \begin{bmatrix} 0 \\ 1 \end{bmatrix} w \\ z &= \begin{bmatrix} 1 & 0 \end{bmatrix} x + w \\ s &= \begin{bmatrix} 1 & 0 \end{bmatrix} x\end{aligned}$$

Find the optimal $\mathcal{H}_\infty$ filtering performance γ_{inf}.

10. Consider the $\mathcal{H}_\infty$ filtering problem for the system without measurement error:

$$\begin{aligned}\dot{x}(t) &= Ax(t) + Bw(t) \\ z(t) &= Hx(t) \\ s(t) &= Lx(t)\end{aligned}$$

Assume that A is stable and $L^T L \leq H^T H$. Show that for any given $\gamma > 0$, there exists an $\mathcal{H}_\infty$ stationary filter that achieves the $\mathcal{H}_\infty$ filtering performance γ.

8

$\mathcal{H}_\infty$ *Filtering of Discrete-Time Systems*

In this chapter, we study the $\mathcal{H}_\infty$ filtering of discrete-time systems. The Riccati equation approach is first applied to derive $\mathcal{H}_\infty$ *a priori* and *a posteriori* filters in both the finite horizon and stationary cases (Yaesh and Shaked 1991; Takaba and Katayama 1996). We also present a polynomial approach to the $\mathcal{H}_\infty$ filtering problem whose solution relies on J-spectral factorization (Zhang *et al.* 2000; Colaneri and Ferrante 2002). An application of $\mathcal{H}_\infty$ filtering to channel equalization is given to illustrate the design of an $\mathcal{H}_\infty$ filter.

8.1 Discrete-Time $\mathcal{H}_\infty$ Filtering Problem

Consider the following dynamical system:

$$x_{k+1} = Ax_k + Gw_k \tag{8.1}$$

$$z_k = Hx_k + v_k \tag{8.2}$$

$$s_k = Lx_k \tag{8.3}$$

where $x_k \in R^n$ is the state vector, $z_k \in R^p$ the measurement, and $s_k \in R^r$ the signal to be estimated. The process noise $w_k \in R^l$ and the measurement noise $v_k \in R^p$ are assumed to be deterministic and with bounded energy. The initial state x_0 of the system is unknown with estimate $\bar{x}_0$.

We introduce the following notation for a discrete-time signal $\{z_0, z_1, \ldots, z_N\}$:

$$\|z\|_{2,[k_1,k_2]}^2 = \sum_{i=k_1}^{k_2} z_i^T z_i$$

The signal $\{z_0, z_1, \ldots, z_N\}$ is said to be energy-bounded or from $\ell_2[0,\ N]$ if $\|z\|_{2,[0,N]} < \infty$.

Depending on where there exists a (unity) delay in obtaining the measurement, the filtering problem can be classified as *a priori* filtering or *a posteriori* filtering. Given a prescribed scalar $\gamma > 0$, the $\mathcal{H}_\infty$ *a priori* filtering is concerned with *finding an estimate* $\hat{s}_k$ *of* s_k *based on the measurements* $\{z_0, z_1, \ldots, z_{k-1}\}$ *such that for any nonzero* (x_0, w, v) *with* x_0 *bounded,* $w \in \ell_2[0, N-1]$ *and*

$v \in \ell_2[0, N-1]$,

$$J_{\text{pri}} = \|s - \hat{s}\|^2_{2,[1,N]} - \gamma^2\{(x_0 - \bar{x}_0)^T P_{x_0}^{-1}(x_0 - \bar{x}_0) + \|w\|^2_{2,[0,N-1]} + \|v\|^2_{2,[0,N-1]}\} < 0 \tag{8.4}$$

where P_{x_0} is a positive definite matrix reflecting the uncertainty of the initial state estimate relative to the combined energy of the process and measurement noises.

For the $\mathcal{H}_\infty$ *a posteriori* filtering, the measurement at time k is available for estimating the signal s_k. That is, there is no delay in obtaining the measurement data. In this connection, given a prescribed scalar $\gamma > 0$, the $\mathcal{H}_\infty$ *a posteriori* filtering is stated as *finding an estimate $\hat{s}_k$ of s_k based on the measurements $\{z_1, \ldots, z_k\}$, such that for any nonzero (x_0, w, v) with x_0 bounded, $w \in \ell_2[0, N-1]$ and $v \in \ell_2[1, N]$,*

$$J_{\text{post}} = \|s - \hat{s}\|^2_{2,[1,N]} - \gamma^2 \left\{(x_0 - \bar{x}_0)^T P_{x_0}^{-1} (x_0 - \bar{x}_0) + \|w\|^2_{2,[0,N-1]} + \|v\|^2_{2,[1,N]}\right\} < 0 \tag{8.5}$$

In the remainder of this section, we shall present several versions of bounded real lemma which play a crucial role in deriving an $\mathcal{H}_\infty$ filter.

Consider the discrete-time system of the form

$$x_{k+1} = Ax_k + Gw_k \tag{8.6}$$

$$y_k = Cx_k \tag{8.7}$$

where w_k is the disturbance input which is from $\ell_2[0, N]$ and y_k the output. We introduce the following cost functions associated with the system:

$$J = \|y\|^2_{2,[0,N-1]} - \gamma^2\left(x_0^T P_{x_0}^{-1} x_0 + \|w\|^2_{2,[0,N-1]}\right) \tag{8.8}$$

and

$$J' = \|y\|^2_{2,[1,N]} - \gamma^2\left(x_0^T P_{x_0}^{-1} x_0 + \|w\|^2_{2,[0,N-1]}\right) \tag{8.9}$$

where P_{x_0} is a positive definite matrix.

The bounded real lemma for the system 8.6 and 8.7 is given in Theorem 8.1.

THEOREM 8.1

Consider the system 8.6 and 8.7 and the performance measures 8.8 and 8.9. Then, for a given positive scalar γ, we have the following results:

(a) $J < 0$ for any $0 \neq (x_0, w)$ with $w \in \ell_2[0, N-1]$, if and only if there exists a solution P_k to the differential Riccati equation (DRE):

$$P_{k+1} = AP_kA^T + \gamma^{-2}AP_kC^T\left(I - \gamma^{-2}CP_kC^T\right)^{-1}CP_kA^T + GG^T, \quad P_0 = P_{x_0} \tag{8.10}$$

such that $I - \gamma^{-2} C P_k C^T > 0$*. Moreover, the solution* P_k*, if exists, is positive definite.*

(b) $J' < 0$ *for any* $0 \neq (x_0, w)$ *with* $w \in \ell_2[0, N-1]$ *if and only if there exists a solution* $\bar{P}_k$ *to the DRE*

$$\bar{P}_{k+1} = A\bar{P}_k A^T + \gamma^{-2} A\bar{P}_k C^T \left(I - \gamma^{-2} C\bar{P}_k C^T\right)^{-1} C\bar{P}_k A^T + GG^T \tag{8.11}$$

where $\bar{P}_0 = (P_{x_0}^{-1} + \gamma^{-2} C^T C)^{-1}$*, such that* $I - \gamma^{-2} C\bar{P}_k C^T > 0$*. Moreover, the solution* $\bar{P}_k$*, if it exists, is positive definite.*

Proof:

For simplicity, we assume that A is invertible and set $\gamma = 1$. Note that, for the case of $\gamma \neq 1$ we may just scale the matrix C, i.e., replace C by $\gamma^{-1} C$.

(a) Sufficiency. Since $P_k > 0$, we first apply the matrix inversion lemma to Equation 8.10 to obtain

$$P_{k+1} - GG^T = A(P_k^{-1} - C^T C)^{-1} A^T \tag{8.12}$$

Since A is invertible, taking inversion on both sides of Equation 8.12 leads to

$$A^T (P_{k+1} - GG^T)^{-1} A = P_k^{-1} - C^T C$$

Denote $Q_k = P_k^{-1}$ and apply the matrix inversion lemma again. Then, we obtain the following DRE:

$$A^T Q_{k+1} A - Q_k + A^T Q_{k+1} G \left(I - G^T Q_{k+1} G\right)^{-1} B^T Q_{k+1} A + C^T C = 0 \tag{8.13}$$

where $Q_0 = P_{x_0}^{-1}$. Further, since $I - CP_k C^T > 0$, it can be known from Equation 8.12 that $P_{k+1} - GG^T > 0$ or equivalently $I - G^T Q_{k+1} G > 0$.

Denote $J_k = x_{k+1}^T Q_{k+1} x_{k+1} - x_k^T Q_k x_k$. Then, it follows from Equation 8.6 that

$$\begin{aligned} J_k = {} & x_k^T (A^T Q_{k+1} A - Q_k) x_k + x_k^T A^T Q_{k+1} G w_k + w_k^T G^T Q_{k+1} A x_k \\ & + w_k^T G^T Q_{k+1} G w_k \end{aligned}$$

By utilizing Equations 8.7 and 8.13 and completing the squares, it can be obtained that

$$J_k + y_k^T y_k - w_k^T w_k = -\eta_k^T (I - G^T Q_{k+1} G) \eta_k$$

where $\eta_k = \left(I - G^T Q_{k+1} G\right)^{-1} G^T Q_{k+1} A x_k - w_k$.

By taking summation of the above from $k = 0$ to $N - 1$, we have

$$\|y\|_{2,[0,N-1]}^3 - \|w\|_{2,[0,N-1]}^2 - x_0^T P_{x_0}^{-1} x_0 = -x_N^T Q_N x_N - \sum_{k=0}^{N-1} \eta_k^T (I - G^T Q_{k+1} G) \eta_k$$

That is, $J < 0$ for any nonzero (x_0, w) since $I - G^T Q_k G > 0$ for any $k \in [1, N]$ and $Q_N > 0$. The proof of necessity is left as an exercise.

To prove (b), we note that since $y_0 = Cx_0$ and w_N does not affect y_N (the worst case is $w_N = 0$), the cost function J' with $\gamma = 1$ can be rewritten as

$$J' = \|y\|_{2,[0,N]}^2 - \left(x_0^T \bar{P}_{x_0}^{-1} x_0 + \|w\|_{2,[0,N]}^2\right)$$

where $\bar{P}_{x_0}^{-1} = P_{x_0}^{-1} + C^T C$. The rest of the proof then follows from the proof of (a).

In the case when A is singular, we can replace A by $A + \epsilon I$ so that $A + \epsilon I$ is nonsingular. The above result still holds when taking the limit of $\epsilon \to 0$. Furthermore, $P_k > 0$ will be satisfied.

Consider now $x_0 = 0$ and N approaches ∞. We assume that the system 8.6 is asymptotically stable. In this case, that $J < 0$ for any nonzero $w \in \ell_2[0, \infty)$ is equivalent to $\|G(z)\|_\infty < \gamma$, where $G(z) = C(zI - A)^{-1}G$ is the transfer function matrix of the system and $\|G(z)\|_\infty$ refers to the infinity norm of the stable transfer function $G(z)$, i.e., $\|G(z)\|_\infty = \sup_{0 \leq \omega \leq 2\pi} \sigma_{\max}[G(e^{j\omega})]$ with $\sigma_{\max}(\cdot)$ being the maximum singular value of a matrix.

Then, we have the following result: ∎

THEOREM 8.2

The following statements are equivalent:

(a) *The matrix A is asymptotically stable and $\|G(z)\|_\infty < \gamma$.*

(b) *There exists $\bar{P} = \bar{P}^T > 0$ satisfying*

$$A\bar{P}A^T - \bar{P} + \gamma^{-2} A\bar{P}C^T \left(I - \gamma^{-2} C\bar{P}C^T\right)^{-1} C\bar{P}A^T + GG^T < 0 \quad (8.14)$$

such that $I - \gamma^{-2} C\bar{P}C^T > 0$.

(c) *There exists a stabilizing solution $P = P^T \geq 0$ to the ARE.*

$$APA^T - P + \gamma^{-2} APC^T \left(I - \gamma^{-2} CPC^T\right)^{-1} CPA^T + GG^T = 0 \quad (8.15)$$

such that $I - \gamma^{-2} CPC^T > 0$. Further, if (A, G) is reachable, P is positive definite.

8.2 $\mathcal{H}_\infty$ *a Priori* Filter

In this section, we shall present a solution to the $\mathcal{H}_\infty$ *a priori* filtering for the system 8.1 through 8.3.

8.2.1 Finite Horizon Case

We consider an observer-based filter of the form

$$\hat{x}_{k+1} = A\hat{x}_k + K_k\left(z_k - H\hat{x}_k\right), \quad \hat{x}_0 = \bar{x}_0 \tag{8.16}$$

$$\hat{s}_k = L\hat{x}_k \tag{8.17}$$

It should be noted that there is no loss of generality in assuming the above observer-based filter.

The following result gives a solution to the $\mathcal{H}_\infty$ *a priori* filtering.

THEOREM 8.3

Consider the system 8.1 through 8.3. For a given $\gamma > 0$, there exists an $\mathcal{H}_\infty$ a priori filter over $[0, N]$ if and only if there exists a solution $P_k = P_k^T > 0$, $k = 0, 1, \ldots, N$, to the DRE

$$\begin{aligned} P_{k+1} = {} & AP_kA^T + \gamma^{-2}P_{k+1}L^T\left(I + \gamma^{-2}LP_{k+1}L^T\right)^{-1}LP_{k+1} \\ & - AP_kH^T\left(I + HP_kH^T\right)^{-1}HP_kA^T + GG^T \end{aligned} \tag{8.18}$$

where $P_0 = P_{x_0} > 0$. If the condition is satisfied, a suitable $\mathcal{H}_\infty$ a priori filter can be given by Equations 8.16 and 8.17 with the filter gain of

$$K_k = AP_kH^T\left(I + HP_kH^T\right)^{-1} \tag{8.19}$$

Proof:

Denote $e_k = x_k - \hat{x}_k$ and $\tilde{s}_k = s_k - \hat{s}_k$. Then, it follows from Equations 8.1 through 8.3 and Equations 8.16 and 8.17 that the filtering error dynamic is given by

$$e_{k+1} = (A - K_kH)e_k + [G \quad -K_k]\begin{bmatrix} w_k \\ v_k \end{bmatrix}, \quad e_0 = x_0 - \bar{x}_0 \tag{8.20}$$

$$\tilde{s}_k = Le_k \tag{8.21}$$

By applying Theorem 8.1(b), there exists an $\mathcal{H}_\infty$ *a priori* filter that achieves

$$J_{\text{pri}} = \|\tilde{s}\|^2_{[2,[1,N]} - \gamma^2\left\{e_0^TP_{x_0}^{-1}e_0 + \|w\|^2_{2,[0,N-1]} + \|v\|^2_{[2,[0,N-1]}\right\} < 0$$

for any nonzero (x_0, w, v) if and only if there exists a solution $\bar{P}_k = \bar{P}_k^T > 0$ to the DRE:

$$\begin{aligned} \bar{P}_{k+1} = {} & (A - K_kH)\bar{P}_k(A - K_kH)^T + \gamma^{-2}(A - K_kH)\bar{P}_kL^T\left(I - \gamma^{-2}L\bar{P}_kL^T\right)^{-1} \\ & \times L\bar{P}_k(A - K_kH)^T + GG^T + K_kK_k^T, \quad \bar{P}_0 = (P_{x_0}^{-1} + \gamma^{-2}L^TL)^{-1} \end{aligned}$$

such that $I - \gamma^{-2}L\bar{P}_kL^T > 0$, or yet

$$\bar{P}_{k+1} = (A - K_kH)\left(\bar{P}_k^{-1} - \gamma^{-2}L^TL\right)^{-1}(A - K_kH)^T + GG^T + K_kK_k^T$$

Denote $P_k^{-1} = \bar{P}_k^{-1} - \gamma^{-2} L^T L > 0$. The above implies that

$$\left(P_{k+1}^{-1} + \gamma^{-2} L^T L\right)^{-1} = (A - K_k H) P_k (A - K_k H)^T + GG^T + K_k K_k^T \quad (8.22)$$

where $P_0 = P_{x_0}$.

By completing the squares and applying the matrix inversion lemma, Equation 8.22 leads to

$$\begin{aligned} P_{k+1} = {} & AP_k A^T + \gamma^{-2} P_{k+1} L^T \left(I + \gamma^{-2} L P_{k+1} L^T\right)^{-1} L P_{k+1} \\ & - AP_k H^T R_H^{-1} H P_k A^T + GG^T \\ & + \left(K_k - AP_k H^T R_H^{-1}\right) R_H \left(K_k - AP_k H^T R_H^{-1}\right)^T \end{aligned} \quad (8.23)$$

where $R_H = I + HP_k H^T$.

Hence, the sufficiency is obvious by setting K_k as in Equation 8.19. In this situation, Equation 8.23 reduces to Equation 8.18. The necessity follows from the monotonicity of DRE. That is, if for some given K_k, there exists $P_{1,k} = P_{1,k}^T > 0$ satisfying Equation 8.23, then there exists a solution $P_k = P_k^T > 0$ to Equation 8.18 and $P_k \leq P_{1,k}$. ∎

We now discuss the computation of Equation 8.18. To this end, we show that the existence of P_k to Equation 8.18 is equivalent to the existence of M_k, satisfying $M_k^{-1} - \gamma^{-2} L^T L > 0$, to the following DRE:

$$M_{k+1} = AM_k A^T - AM_k \bar{H}^T \left(\bar{R} + \bar{H} M_k \bar{H}^T\right)^{-1} \bar{H} M_k A^T + GG^T \quad (8.24)$$

where $M_0 = \left(P_{x_0}^{-1} + \gamma^{-2} L^T L\right)^{-1}$ and

$$\bar{H} = \begin{bmatrix} H \\ L \end{bmatrix}, \qquad \bar{R} = \begin{bmatrix} I & 0 \\ 0 & -\gamma^2 I \end{bmatrix}$$

In fact, by the matrix inversion lemma, Equation 8.18 can be rewritten as

$$\left(P_{k+1}^{-1} + \gamma^{-2} L^T L\right)^{-1} = A\left(P_k^{-1} + H^T H\right)^{-1} A^T + GG^T$$

Define $M_k^{-1} = P_k^{-1} + \gamma^{-2} L^T L$. Then, we have

$$M_{k+1} = A\left(M_k^{-1} + H^T H - \gamma^{-2} L^T L\right)^{-1} A^T + GG^T$$

i.e.,

$$M_{k+1} = A\left(M_k^{-1} + \bar{H}^T \bar{R}^{-1} \bar{H}\right)^{-1} A^T + GG^T$$

Thus, Equation 8.24 follows by utilizing the matrix inversion again.

Observe that the condition $P_k > 0$ is equivalent to that of $I > \gamma^{-2} L M_k L^T > 0$. Thus, we can compute M_k via the recursion 8.24. If $I - \gamma^{-2} L M_k L^T > 0$, then $P_k = (M_k^{-1} - \gamma^{-2} L^T L)^{-1}$ and the $\mathcal{H}_\infty$ filter gain of 8.16 and 8.17 is calculated as in Equation 8.19.

Example 8.1

Consider the scalar system

$$x_{k+1} = ax_k + \sqrt{q}w_k \tag{8.25}$$

$$z_k = hx_k + v_k \tag{8.26}$$

$$s_k - x_k \tag{8.27}$$

where a, q, and h are constants. The DRE 8.18 becomes

$$\left(P_{k+1}^{-1} + \gamma^{-2}\right)^{-1} = a^2 P_k - \frac{a^2 h^2 P_k^2}{1 + h^2 P_k} + q$$

i.e.,

$$P_{k+1} = \left(\frac{1 + h^2 P_k}{(a^2 + qh^2)P_k + q} - \gamma^{-2}\right)^{-1} \tag{8.28}$$

where $P_0 = P_{x_0}$ is the weighting factor of the initial estimate. From Equation 8.19, the filter gain is given by

$$K_k = \frac{ahP_k}{1 + h^2 P_k}$$

and the $\mathcal{H}_\infty$ *a priori* filter is

$$\hat{x}_{k+1} = a\hat{x}_k + K_k(z_k - h\hat{x}_k)$$

i.e.,

$$\hat{x}_{k+1} = \frac{a}{1 + h^2 P_k}\hat{x}_k + \frac{ahP_k}{1 + h^2 P_k} z_k$$

To study the effect of the measurement noise, we modify Equation 8.26 as

$$z_k = hx_k + \sqrt{r}v_k$$

i.e.,

$$\frac{z_k}{\sqrt{r}} = \frac{h}{\sqrt{r}} x_k + v_k$$

Then, Equation 8.28 becomes

$$P_{k+1} = \left(\frac{r + h^2 P_k}{(a^2 r + qh^2)P_k + rq} - \gamma^{-2}\right)^{-1} \tag{8.29}$$

and the $\mathcal{H}_\infty$ *a priori* filter is then

$$\hat{x}_{k+1} = \frac{ar}{r + h^2 P_k}\hat{x}_k + \frac{ahP_k}{r + h^2 P_k} z_k \tag{8.30}$$

When $\gamma \to \infty$, Equation 8.29 reduces to

$$P_{k+1} = \frac{a^2 r P_k}{r + h^2 P_k} + q$$

Hence, the $\mathcal{H}_\infty$ *a priori* filter reduces to the Kalman *a priori* filter as in Example 2.5.

a. Perfect measurements

In this case, $r = 0$. Then Equation 8.29 becomes

$$P_{k+1} = (q^{-1} - \gamma^{-2})^{-1}$$

Obviously, $\gamma > \sqrt{q}$ is required. That is, the optimal $\mathcal{H}_\infty$ filter performance is $\gamma_{\text{opt}} = \sqrt{q}$. Also, Equation 8.30 yields the filter

$$\hat{x}_{k+1} = \frac{a}{h} z_k$$

which is the same as the Kalman *a priori* filter. It becomes clear that the Kalman *a priori* filter also achieves the optimal $\mathcal{H}_\infty$ performance.

b. Completely unreliable measurements

Let $r \to \infty$, implying that the measurement is unreliable. In this situation, Equation 8.29 and the filter Equation 8.30 become

$$P_{k+1} = \left(\frac{1}{a^2 P_k + q} - \gamma^{-2} \right)^{-1} \tag{8.31}$$

and

$$\hat{x}_{k+1} = a\hat{x}_k$$

respectively.

For the case $a = 0$, i.e., we consider the system

$$x_{k+1} = \sqrt{q} w_k$$

In the case,

$$P_k = \left(q^{-1} - \gamma^{-2}\right)^{-1}$$

which is positive if and only if $\gamma > \sqrt{q}$. This can be explained as follows: With $\hat{x}_k = 0$, then

$$J_{\text{pri}} = q\|w\|^2_{2,[0,N-1]} - \gamma^2 \left(P_{x_0}^{-1} e_0^2 + \|w\|^2_{2,[0,N-1]} + \|v\|^2_{2,[0,N-1]} \right)$$

which is clearly less than zero if $\gamma > \sqrt{q}$.

When $\gamma \to \infty$, Equation 8.31 reduces to

$$P_{k+1} = a^2 P_k + q$$

which is the same as that of Kalman filter and converges if and only if $|a| < 1$. $\square$

Example 8.2

Consider the system

$$x_{k+1} = \begin{bmatrix} -0.2 & -0.5 \\ 1.5 & 1 \end{bmatrix} x_k + \begin{bmatrix} 0 \\ 1 \end{bmatrix} w_k$$
$$z_k = \begin{bmatrix} -2 & 1 \end{bmatrix} x_k + v_k$$
$$s_k = \begin{bmatrix} 1 & 1 \\ 0 & 1 \end{bmatrix} x_k$$

Let $\gamma = 2.5$, $P_{x_0} = I$, and $N = 100$. Then,

$$M_0 = \left(P_{x_0}^{-1} + \gamma^{-2} L^T L\right)^{-1} = \begin{bmatrix} 0.8767 & -0.1063 \\ -0.1063 & 0.7705 \end{bmatrix}$$

By the Riccati recursion 8.24, we can obtain M_k which satisfies $M_k^{-1} - \gamma^{-2} L^T L > 0$. Then, $P_k = (M_k^{-1} - \gamma^{-2} L^T L)^{-1}$ and the filter gain is computed by Equation 8.19. The filter gain K_k versus k is shown in Figure 8.1. We observe that the filter gain converges within three steps.

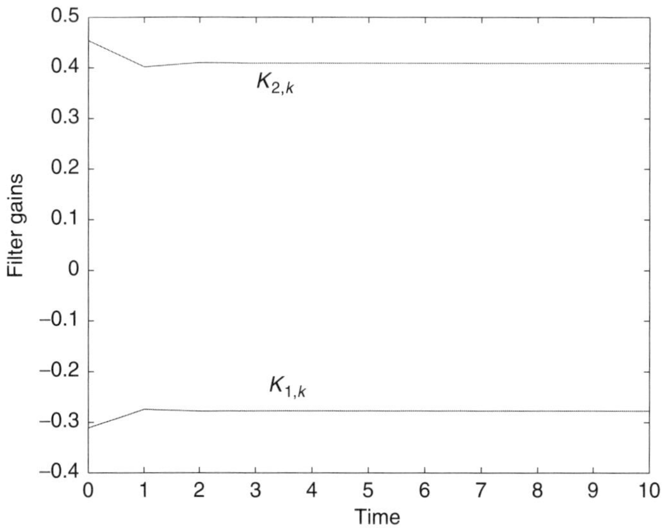

FIGURE 8.1
$\mathcal{H}_\infty$ *a priori* filter gains.

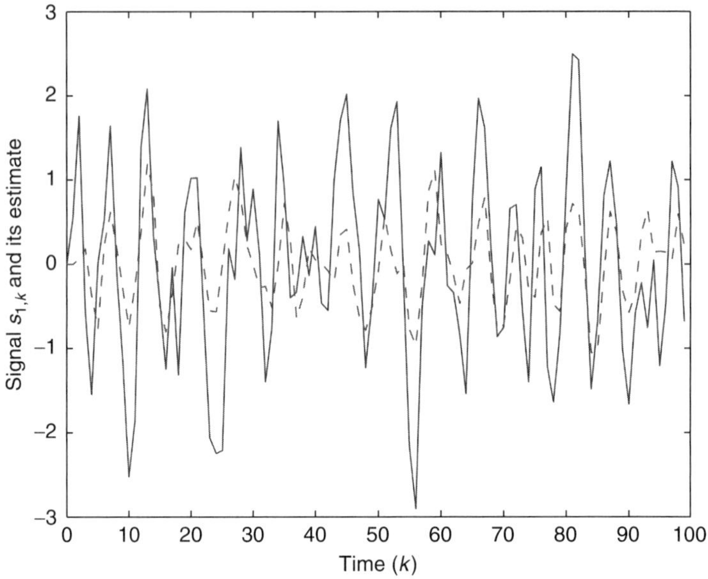

FIGURE 8.2
Comparison between the input signal $s_{1,k}$ and its estimate $\hat{s}_{1,k}$.

Given that w_k and v_k are Gaussian white noises with zero means and variances of 1 and 0.01, respectively, the tracking performance of the $\mathcal{H}_\infty$ *a priori* filter is shown in Figures 8.2 and 8.3, respectively, where the solid line denotes the actual signal $s_{1,k}$ or $s_{2,k}$, where $s_k = [s_{1,k}\ \ s_{2,k}]^T$, and the dashed dot line denotes their estimates. □

For the purpose of making a comparison with the $\mathcal{H}_\infty$ *a posteriori* filtering in the next section, we now consider a modified cost function

$$J'_{\text{pri}} = \|\tilde{s}\|^2_{2,[0,N]} - \gamma^2\left\{e_0^T P_{x_0}^{-1} e_0 + \|w\|^2_{2,[0,N-1]} + \|v\|^2_{2,[0,N-1]}\right\} \tag{8.32}$$

in which we have included $\tilde{s}_0$ in the cost function, i.e., we have taken into consideration the estimation error at $k = 0$.

Note that $\tilde{s}_0 = Le_0$. Hence, the cost function 8.32 can be rewritten as

$$J'_{\text{pri}} = \|\tilde{s}\|^2_{2,[1,N]} - \gamma^2\left\{e_0^T \bar{P}_{x_0}^{-1} e_0 + \|w\|^2_{2,[0,N-1]} + \|v\|^2_{2,[0,N-1]}\right\} \tag{8.33}$$

where $\bar{P}_{x_0}^{-1} = P_{x_0}^{-1} - \gamma^{-2}L^T L$ is required to be positive definite, otherwise J'_{pri} cannot be negative for all e_0.

By applying Theorem 8.1(a) and a similar proof of Theorem 8.3, we have the following result.

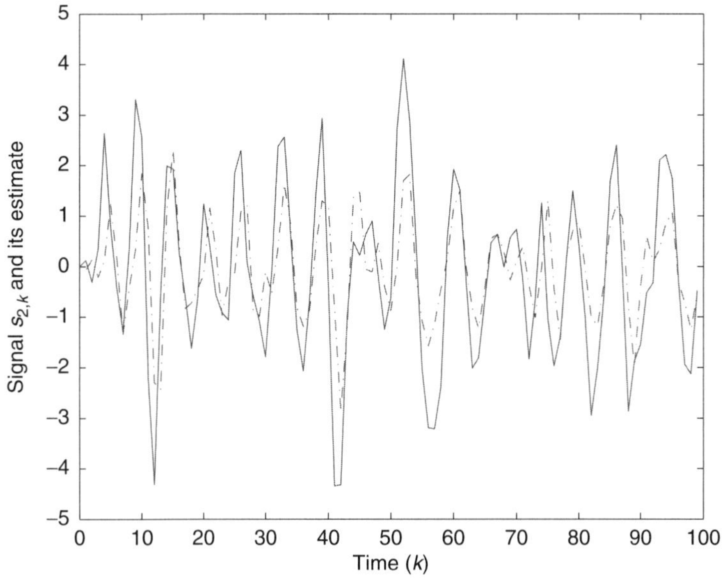

FIGURE 8.3
Comparison between the input signal $s_{2,k}$ and its estimate $\hat{s}_{2,k}$.

THEOREM 8.4
Consider the system 8.1 through 8.3. For a given $\gamma > 0$, there exists an $\mathcal{H}_\infty$ a priori filter over $[0, N]$ such that $J'_{\text{pri}} < 0$ for any nonzero (x_0, w, v) if and only if there exists a solution $M_k > 0$ to Equation 8.24 with the initial condition $M_0 = P_{x_0}$ such that $M_k^{-1} - \gamma^{-2} L^T L > 0$ (or equivalently, $I - \gamma^{-2} L M_k L^T > 0$).

If the condition is satisfied, a suitable $\mathcal{H}_\infty$ a priori filter that achieves $J'_{\text{pri}} < 0$ for any nonzero (x_0, w, v) can be given by Equations 8.16 and 8.17 with the filter gain of

$$K_k = AP_k H^T \left(I + HP_k H^T\right)^{-1} \tag{8.34}$$

where $P_k = \left(M_k^{-1} - \gamma^{-2} L^T L\right)^{-1} > 0.$

It should be noted that the above result is also applicable to linear time-varying systems.

8.2.2 Stationary Case

We now consider the stationary case. In this situation, the initial state is known and the filter 8.16 and 8.17 will be time-invariant, i.e., $K_k = K$ will be a constant matrix.

We make the following assumption concerning the system 8.1 through 8.3.

Assumption 8.1

(i) The pair (H, A) is detectable.

(ii) The pair (A, G) has no unreachable mode on the unit circle.

It should be noted that Assumption 8.1 is necessary for the existence of $\mathcal{H}_\infty$ *a priori* filter. It can be seen from Equations 8.20 and 8.21 that the transfer function matrix from the noise input $[w \quad v]^T$ to $\tilde{s}$ is given by

$$\mathcal{E}(z) = L\left[zI - (A - KH)\right]^{-1}\left[G \quad -K\right] \tag{8.35}$$

For a given scalar $\gamma > 0$, the stationary $\mathcal{H}_\infty$ *a priori* filter design problem is to determine the gain K of the filter 8.16 and 8.17 so that the matrix $A - KH$ is asymptotically stable (with all of its eigenvalues inside the unit disk) and $\|\mathcal{E}(z)\|_\infty < \gamma$.

THEOREM 8.5

Consider the system 8.1 through 8.3 satisfying Assumption 8.1. Then, for a given scalar $\gamma > 0$, the stationary $\mathcal{H}_\infty$ a priori filtering problem is solvable if and only if there exists a stabilizing solution $M = M^T \geq 0$, satisfying $I - \gamma^{-2}LML^T > 0$, to the following ARE:

$$AMA^T - M - AM\bar{H}^T\left(\bar{H}M\bar{H}^T + \bar{R}\right)^{-1}\bar{H}MA^T + GG^T = 0 \tag{8.36}$$

where

$$\bar{H} = \begin{bmatrix} H \\ L \end{bmatrix}, \qquad R = \begin{bmatrix} I & 0 \\ 0 & -\gamma^2 I \end{bmatrix}$$

If the above condition is satisfied, a suitable $\mathcal{H}_\infty$ a priori filter is given by

$$\hat{x}_{k+1} = A\hat{x}_k + K(z_k - H\hat{x}_k), \quad \hat{x}_0 = x_0 \tag{8.37}$$

$$\hat{s}_k = L\hat{x}_k \tag{8.38}$$

with the filter gain of

$$K = APH^T\left(I + HPH^T\right)^{-1} \tag{8.39}$$

where

$$P = M + \gamma^{-2}ML^T\left(I - \gamma^{-2}LML^T\right)^{-1}LM$$

Proof:

From Theorem 8.2, the matrix $A - KH$ is asymptotically stable and $\|\mathcal{E}(z)\|_\infty < \gamma$ if and only if there exists a solution $\bar{M} = \bar{M}^T \geq 0$ to the ARE:

$$\begin{aligned}(A - KH)\bar{M}(A - KH)^T - \bar{M} + \gamma^{-2}(A - KH)\bar{M}L^T\left(I - \gamma^{-2}L\bar{M}L^T\right)^{-1} \\ \times L\bar{M}(A - KH)^T + GG^T + KK^T = 0\end{aligned} \tag{8.40}$$

such that $I - \gamma^{-2}L\bar{M}L^T > 0$ and the matrix

$$\bar{A} = A - KH + \gamma^{-2}(A - KH)\bar{M}L^T\left(I - \gamma^{-2}L\bar{M}L^T\right)^{-1}L$$

is stable.

Similar to the proof of Theorem 8.4, by completing the square, Equation 8.40 is equivalent to (de Souza and Xie 1992)

$$\begin{aligned} &AMA^T - \bar{M} + \gamma^{-2}A\bar{M}L^TU_1^{-1}L\bar{M}A^T - U_3U_2^{-1}U_3^T \\ &\quad + \left(K - U_3U_2^{-1}\right)U_2\left(K - U_3U_2^{-1}\right)^T + GG^T = 0 \end{aligned} \tag{8.41}$$

where

$$\begin{aligned} U_1 &= I - \gamma^{-2}L\bar{M}L^T > 0 \\ U_2 &= I + H\bar{M}H^T + \gamma^{-2}H\bar{M}L^TU_1^{-1}L\bar{M}H^T \\ U_3 &= A\bar{M}H^T + \gamma^{-2}A\bar{M}L^TU_1^{-1}L\bar{M}H^T \end{aligned}$$

By some standard matrix manipulations (left as an exercise), Equation 8.41 can be rewritten as

$$\begin{aligned} &A\bar{M}A^T - \bar{M} - A\bar{M}\bar{H}^T\left(\bar{H}\bar{M}\bar{H}^T + \bar{R}\right)^{-1}\bar{H}\bar{M}A^T \\ &\quad + GG^T + \left(K - U_3U_2^{-1}\right)U_2\left(K - U_3U_2^{-1}\right)^T = 0 \end{aligned} \tag{8.42}$$

and $\bar{A}$ is equivalent to

$$\bar{A} = A - A\bar{M}\bar{H}^T\left(\bar{H}\bar{M}\bar{H}^T + \bar{R}\right)^{-1}\bar{H}$$

By the monotonicity of ARE, it is easy to know that there exists a solution $\bar{M} = \bar{M}^T \geq 0$ to the above ARE such that $\bar{A}$ is stable and $U_1 > 0$ if and only if there exists a stabilizing solution $M = M^T \geq 0$ to the ARE 8.36 such that $I - \gamma^{-2}LML^T > 0$. Furthermore, setting $K = U_3U_2^{-1}$, Equation 8.42 reduces to Equation 8.36 and $\bar{M} = M$. It is easy to see that $K = U_3U_2^{-1}$ is the same as that of Equation 8.39. ∎

Example 8.3

Consider Example 8.2. Let $\gamma = 2.5$. Solving the ARE 8.36 yields the stabilizing solution

$$M = \begin{bmatrix} 0.1607 & -0.3416 \\ -0.3416 & 1.8088 \end{bmatrix}$$

which satisfies $I - \gamma^{-2}LML^T > 0$. The filter gain is then computed by Equation 8.39 as

$$K = \begin{bmatrix} -0.2773 \\ 0.4092 \end{bmatrix}$$

which is the same as the converged gain of Example 8.2. That is, the finite horizon $\mathcal{H}_\infty$ *a priori* filter converges to that of the stationary case.

It can be checked that the minimum achievable γ by the stationary $\mathcal{H}_\infty$ *a priori* filtering is 1.82. □

One should, however, be cautious that unlike the Kalman filter which converges to its stationary filter under very mild conditions, the convergence of the finite horizon $\mathcal{H}_\infty$ filter relies on a proper choice of the initial condition of the DRE 8.24, i.e., on the choice of P_{x_0}. Note that we need to ensure the feasibility condition $M_k^{-1} - \gamma^{-2}L^T L > 0, \forall k \in [1, N]$. For Example 8.2, for the same $\gamma = 2.5$, if $P_{x_0} = \left[\begin{smallmatrix} 2 & 0 \\ 0 & 2 \end{smallmatrix}\right]$, it can be verified that the feasibility condition is violated for $k = 1$.

Suppose that A is invertible and the ARE 8.36 has a stabilizing solution $M > 0$ which is feasible, i.e., $I - \gamma^{-2}LML^T > 0$. Then, the DRE 8.24 has a feasible solution $M_k > 0$ and $M_k \to M$ if $0 < M_0 \le M$.

8.3 $\mathcal{H}_\infty$ *a Posteriori* Filter

8.3.1 Finite Horizon Case

We now study the $\mathcal{H}_\infty$ *a posteriori* filtering problem for the system 8.1 through 8.3. In this situation, the filter is of the form

$$\hat{x}_{k+1} = A\hat{x}_k + K_k(z_{k+1} - HA\hat{x}_k), \quad \hat{x}_0 = \bar{x}_0 \tag{8.43}$$

$$\hat{s}_k = L\hat{x}_k \tag{8.44}$$

where the filter gain K_k is to be determined such that Equation 8.5 is satisfied.

THEOREM 8.6

Consider the system 8.1 through 8.3. For a given $\gamma > 0$, there exists an $\mathcal{H}_\infty$ a posteriori filter over $[0, N]$ if and only if there exists a solution $\bar{P}_k = \bar{P}_k^T > 0$ to the DRE

$$M_{k+1} = A\bar{P}_k A^T + GG^T, \quad \bar{P}_0 = P_{x_0} \tag{8.45}$$

$$\bar{P}_{k+1}^{-1} = M_{k+1}^{-1} + H^T H - \gamma^{-2}L^T L \tag{8.46}$$

If the above condition holds, then a suitable $\mathcal{H}_\infty$ a posteriori filter is given by Equations 8.43 and 8.44 with the filter gain of

$$K_k = M_{k+1}H^T\left(I + HM_{k+1}H^T\right)^{-1} \tag{8.47}$$

Proof:
For the simplicity of presentation, we assume that A is invertible. First, it follows from Equations 8.1 through 8.3 and Equations 8.43 and 8.44 that the filtering error dynamic is given by

$$e_{k+1} = (I - K_k H)Ae_k + [(I - K_k H)G \quad -K_k] \begin{bmatrix} w_k \\ v_{k+1} \end{bmatrix} \tag{8.48}$$

$$\tilde{s}_k = Le_k \tag{8.49}$$

Denote $\bar{w}_k = \left[\begin{smallmatrix} w_k \\ v_{k+1} \end{smallmatrix}\right]$. Then, the cost function (Equation 8.5) can be rewritten as

$$J_{\text{post}} = \|\tilde{s}\|^2_{2,[1,N]} - \gamma^2 \left\{ e_0^T P_{x_0}^{-1} e_0 + \|\bar{w}\|^2_{2,[0,N-1]} \right\}$$

Hence, by applying Theorem 8.1(b), there exists an $\mathcal{H}_\infty$ *a posteriori* filter if and only if there exists a solution $\hat{P}_k = \hat{P}_k^T > 0$ to the DRE

$$\begin{aligned} \hat{P}_{k+1} = {} & (I - K_k H)A\hat{P}_k A^T (I - K_k H)^T \\ & + \gamma^{-2}(I - K_k H)A\hat{P}_k L^T \left(I - \gamma^{-2} L\hat{P}_k L^T\right)^{-1} L\hat{P}_k A^T (I - K_k H)^T \\ & + (I - K_k H)GG^T (I - K_k H)^T + K_k K_k^T, \quad \hat{P}_0 = \left(P_{x_0}^{-1} + \gamma^{-2} L^T L\right)^{-1} \end{aligned}$$

such that $I - \gamma^{-2} L\hat{P}_k L^T > 0$. Denote $\bar{P}_k^{-1} = \hat{P}_k^{-1} - \gamma^{-2} L^T L > 0$. By applying the matrix inversion lemma, we get

$$\left(\bar{P}_{k+1}^{-1} + \gamma^{-2} L^T L\right)^{-1} = (I - K_k H) \left(A\bar{P}_k A^T + GG^T\right) (I - K_k H)^T + K_k K_k^T$$

which, by completing the squares, leads to

$$\begin{aligned} \left(\bar{P}_{k+1}^{-1} + \gamma^{-2} L^T L\right)^{-1} = {} & M_{k+1} - M_{k+1} H^T R_H^{-1} H M_{k+1}^T \\ & + \left(K_k - M_{k+1} H^T R_H^{-1}\right) R_H \left(K_k - M_{k+1} H^T R_H^{-1}\right)^T \end{aligned} \tag{8.50}$$

where $R_H = I + HM_{k+1}H^T$ and M_k is as in Equation 8.45 and is positive definite.

Again, by applying the matrix inversion lemma to the right-hand side of Equation 8.50, we have

$$\begin{aligned} \left(\bar{P}_{k+1}^{-1} + \gamma^{-2} L^T L\right)^{-1} = {} & \left(M_{k+1}^{-1} + H^T H\right)^{-1} \\ & + \left(K_k - M_{k+1} H^T R_H^{-1}\right) R_H \left(K_k - M_{k+1} H^T R_H^{-1}\right)^T \end{aligned} \tag{8.51}$$

Hence, the sufficiency is obvious by setting K_k as in Equation 8.47. The necessity follows from the monotonicity of DRE. That is, if for some given

K_k, there exists $\bar{P}_{1,k} = \bar{P}_{1,k}^T > 0$ satisfying Equation 8.51, then there exists a solution $\bar{P}_k = \bar{P}_k^T > 0$ to Equations 8.45 and 8.46 and $\bar{P}_k \leq \bar{P}_{1,k}$. ∎

REMARK 8.1 Observe that

$$\left(M_k^{-1} + H^T H - \gamma^{-2} L^T L\right)^{-1} = \left(M_k^{-1} + \bar{H}^T \bar{R}^{-1} \bar{H}\right)^{-1}$$
$$= M_k - M_k \bar{H}^T \left(\bar{R} + \bar{H} M_k \bar{H}^T\right)^{-1} \bar{H} M_k$$

where

$$\bar{H} = \begin{bmatrix} H \\ L \end{bmatrix}, \qquad \bar{R} = \begin{bmatrix} I & 0 \\ 0 & -\gamma^2 I \end{bmatrix} \tag{8.52}$$

Hence, the DREs 8.45 and 8.46 can be replaced by a single one:

$$M_{k+1} = A M_k A^T - A M_k \bar{H}^T \left(\bar{R} + \bar{H} M_k \bar{H}^T\right)^{-1} \bar{H} M_k A^T + G G^T \tag{8.53}$$

where $M_0 = \left(P_{x_0}^{-1} + \gamma^{-2} L^T L - H^T H\right)^{-1}$. ∎

Example 8.4

Consider now the $\mathcal{H}_\infty$ *a posteriori* filtering of Example 8.2. With the same $\gamma = 2.5$ and $P_{x_0} = I$, we know that the $\mathcal{H}_\infty$ *a posteriori* filtering problem is solvable as the $\mathcal{H}_\infty$ *a priori* filtering has been solved in Example 8.2.

Setting $\bar{P}_0 = P_{x_0}$ and solving Equations 8.45 and 8.46 recursively, we can obtain M_k and $\bar{P}_k$ and verify that $\bar{P}_k > 0$ for $k = 1, 2, \ldots, N$. The filter gains are then computed using Equation 8.47 and are shown in Figure 8.4. The gain converges to

$$K = \begin{bmatrix} -0.1376 \\ 0.5172 \end{bmatrix}$$

Given that w_k and v_k are Gaussian white noises with zero means and variances of 1 and 0.01, respectively, the comparison between the signal s_k and its $\mathcal{H}_\infty$ *a posteriori* filtering estimate is shown in Figures 8.5 and 8.6, respectively. Clearly, it shows much better tracking performance than the $\mathcal{H}_\infty$ *a priori* filter.

The minimum γ achievable by the $\mathcal{H}_\infty$ *a posteriori* filtering is 1.23 which is in comparison with the minimum γ of 2.40 by the $\mathcal{H}_\infty$ *a priori* filtering.

By reducing the uncertainty in the initial estimate, that is, by decreasing P_{x_0}, the minimum achievable γ can be improved. Indeed, by choosing $P_{x_0} = 0.01I$, the minimum achievable γ by the $\mathcal{H}_\infty$ *a posteriori* filtering is 1.07. □

We now modify the cost function J_{post} by taking into consideration the initial estimation error as in Equation 8.32. That is, we consider

$$J'_{\text{post}} = \|\tilde{s}\|_{2,[0,N]}^2 - \gamma^2 \left\{ e_0^T P_{x_0}^{-1} e_0 + \|w\|_{2,[0,N-1]}^2 + \|v\|_{2,[0,N]}^2 \right\} \tag{8.54}$$

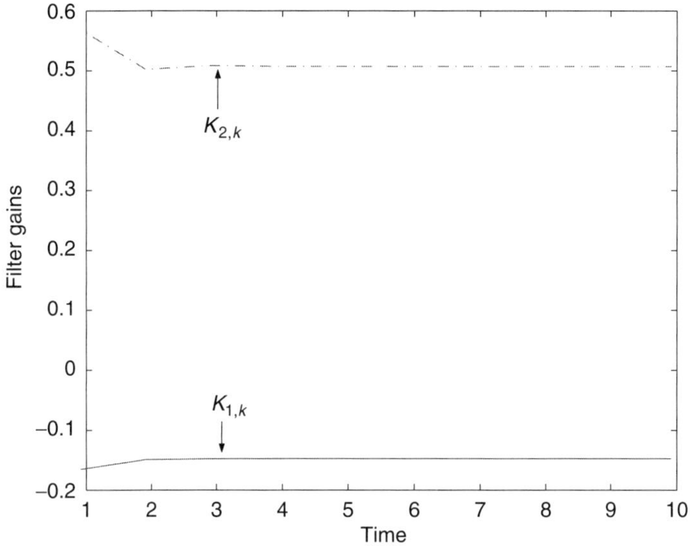

FIGURE 8.4
$\mathcal{H}_\infty$ *a posteriori* filter gains.

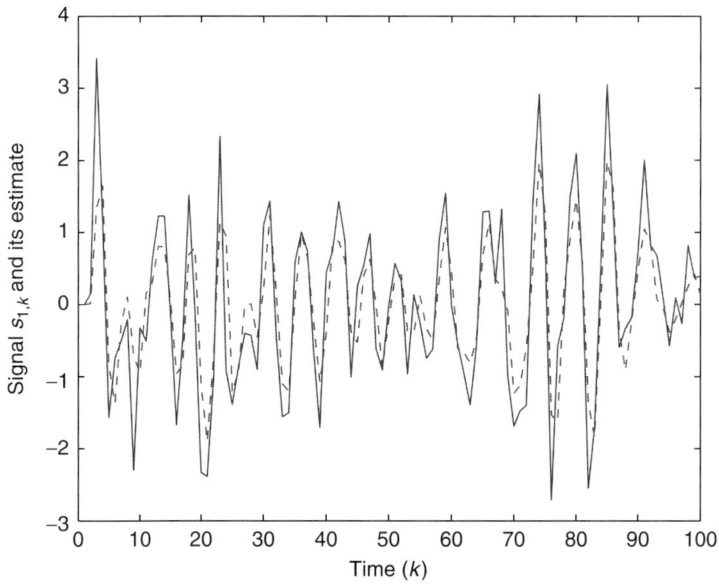

FIGURE 8.5
Signal $s_{1,k}$ and its $\mathcal{H}_\infty$ *a posteriori* filtering estimate.

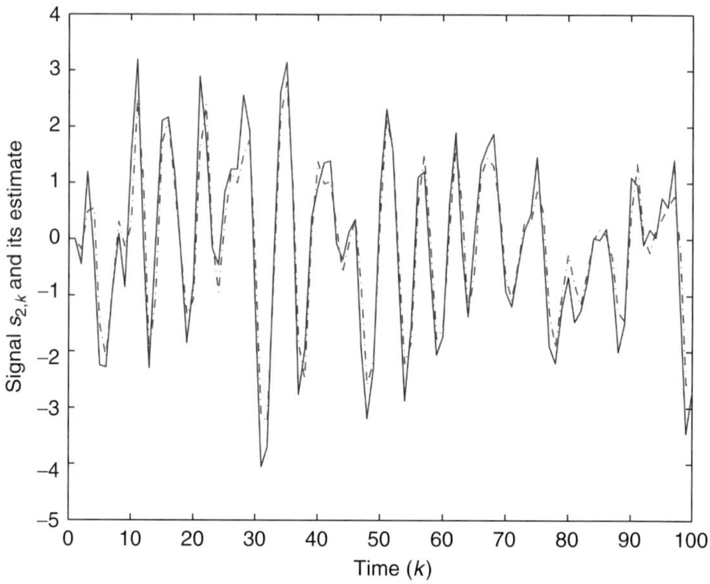

FIGURE 8.6
Signal $s_{2,k}$ and its $\mathcal{H}_\infty$ *a posteriori* filtering estimate.

Note that $\tilde{s}_0 = Le_0$, whereas the initial measurement error can be given as

$$v_0 = Hx_0 - H\bar{x}_0 = He_0$$

Thus, Equation 8.54 is given by

$$J'_{\text{post}} = \|\tilde{s}\|^2_{2,[1,N]} - \gamma^2 \left\{ e_0^T \bar{P}_{x_0}^{-1} e_0 + \|w\|^2_{2,[0,N-1]} + \|v\|^2_{2,[1,N]} \right\} \tag{8.55}$$

where $\bar{P}_{x_0}^{-1} = P_{x_0}^{-1} + H^T H - \gamma^2 L^T L$. By applying a similar proof as in Theorem 8.6, we have the following:

THEOREM 8.7
Consider the system 8.1 through 8.3. For a given $\gamma > 0$, there exists an $\mathcal{H}_\infty$ a posteriori filter over $[0, N]$ such that $J'_{\text{post}} < 0$ for all (x_0, w, v) if and only if there exists a solution M_k to the DRE (8.24) with $M_0 = P_{x_0}$ such that

$$M_k^{-1} + H^T H - \gamma^{-2} L^T L > 0 \tag{8.56}$$

If the above condition is satisfied, then a suitable $\mathcal{H}_\infty$ a posteriori filter is given by Equations 8.43 and 8.44 with the filter gain of Equation 8.47.

REMARK 8.2 Observe that both the $\mathcal{H}_\infty$ *a priori* and *a posteriori* filtering require the existence of a solution M_k to Equation 8.24. Note from

Theorem 8.4 that the former requires $M_k^{-1} - \gamma^{-2} L^T L > 0$, which implies Equation 8.56 of the latter in Theorem 8.7. Thus, it is concluded that if the $\mathcal{H}_\infty$ *a priori* filtering is solvable, then the $\mathcal{H}_\infty$ *a posteriori* filtering is solvable for the same $\gamma > 0$. The converse is not true in general. This is because the latter has made use of the measurements y_i, $1 \le i \le k$ when estimating s_k, whereas the former only uses the measurements y_i, $1 \le i \le k-1$. ∎

REMARK 8.3 When γ approaches ∞, the $\mathcal{H}_\infty$ *a priori* filter and *a posteriori* filter reduce to their Kalman filter counterparts. In this case, $\bar{P}_k$ and P_k are respectively the *a posteriori* and *a priori* Kalman filtering error covariances. From Theorem 8.6, when $\gamma \to \infty$, $P_k = M_k$, whereas $\bar{P}_k = (M_k^{-1} + H^T H)^{-1} \le M_k = P_k$. P_k is the prediction error covariance, whereas $\bar{P}_k$ is the error covariance of the updated estimate. ∎

8.3.2 Stationary Case

We now consider the stationary $\mathcal{H}_\infty$ *a posteriori* filtering. In this situation, the filter 8.16 and 8.17 under consideration will be time-invariant. It can be seen from Equations 8.48 and 8.49 that the transfer function from the noise $[w^T \;\; v^T]^T$ to the filtering error $\tilde{s}$ is given by

$$\mathcal{E}'(z) = L\,[zI - (I - KH)A]^{-1}\,[(I - KH)G \quad -Kz] \tag{8.57}$$

The stationary $\mathcal{H}_\infty$ *a posteriori* filtering problem is to design the filter gain K such that the matrix $(I - KH)A$ is asymptotically stable and

$$\|\mathcal{E}'(z)\|_\infty < \gamma$$

For simplicity, we replace Assumption 8.1(ii) by the following assumption.

Assumption 8.2 *The pair (A, G) is reachable.*

THEOREM 8.8

Consider the system 8.1 through 8.3 satisfying Assumptions 8.1(i) and 8.2. Then, for a given scalar $\gamma > 0$, the stationary $\mathcal{H}_\infty$ a posteriori filtering problem is solvable if and only if there exists a stabilizing solution $M = M^T > 0$ to the ARE

$$AMA^T - M - AM\bar{H}^T\left(\bar{R} + \bar{H}M\bar{H}^T\right)^{-1}\bar{H}MA^T + GG^T = 0 \tag{8.58}$$

such that $M^{-1} + H^T H - \gamma^{-2} L^T L > 0$, where $\bar{H}$ and $\bar{R}$ are as given in Equation 8.52.

If the above condition is satisfied, a suitable $\mathcal{H}_\infty$ a posteriori filter is given by Equations 8.43 and 8.44 with the filter gain of

$$K = MH^T(I + HMH^T)^{-1} \tag{8.59}$$

Proof:
Noting the fact that the $\mathcal{H}_\infty$-norm of $\mathcal{E}'(z)$ is the supreme of the square root of the maximum eigenvalue of $\mathcal{E}'(e^{j\omega})\mathcal{E}'^T(e^{-j\omega})$ over $\omega \in [0, 2\pi]$, we set

$$\mathcal{E}'(z) = L\,[zI - (I - KH)A]^{-1}\,[(I - KH)G \quad -K]$$

In view of Theorem 8.2, the $\mathcal{H}_\infty$ *a posteriori* filtering problem is solvable if and only if there exists a matrix $\hat{P} = \hat{P}^T \geq 0$ to the matrix equation

$$\begin{aligned}&(I - KH)A\hat{P}A^T(I - KH)^T - \hat{P}\\ &\quad + \gamma^{-2}(I - KH)A\hat{P}L^T\left(I - \gamma^{-2}L\hat{P}L^T\right)^{-1}L\hat{P}A^T(I - KH)^T\\ &\quad + (I - KH)GG^T(I - KH)^T + KK^T = 0\end{aligned}$$

such that $I - \gamma^{-2}L\hat{P}L^T > 0$ and the matrix

$$\bar{A} = (I - KH)A + \gamma^{-2}(I - KH)A\hat{P}L^T\left(I - \gamma^{-2}L\hat{P}L^T\right)^{-1}L$$

is asymptotically stable.

Note that the reachability of (A, G) implies that of $((I - KH)A, [(I - KH)G - K])$. Hence, in view of Theorem 8.2, $\hat{P} > 0$.

Next, following a similar proof of Theorem 8.6, it can be shown that the above is equivalent to the existence of $\bar{P} = \bar{P}^T > 0$ such that

$$M = A\bar{P}A^T + GG^T, \qquad \bar{P}^{-1} = M^{-1} + H^TH - \gamma^{-2}L^TL \tag{8.60}$$

Note that $M > 0$ since $\bar{P} > 0$ and (A, G) is reachable. By using the notations in Equation 8.52 and the matrix inversion lemma,

$$\begin{aligned}\left(M^{-1} + H^TH - \gamma^{-2}L^TL\right)^{-1} &= \left(M^{-1} + \bar{H}^T\bar{R}^{-1}\bar{H}\right)^{-1}\\ &= M - M\bar{H}^T\left(\bar{R} + \bar{H}M\bar{H}^T\right)^{-1}\bar{H}M\end{aligned}$$

Hence, Equation 8.60 becomes

$$AMA^T - M - AM\bar{H}^T\left(\bar{R} + \bar{H}M\bar{H}^T\right)^{-1}\bar{H}MA^T + GG^T = 0 \tag{8.61}$$

We now show that $\hat{A} = A - AM\bar{H}^T\left(\bar{R} + \bar{H}M\bar{H}^T\right)^{-1}\bar{H}$ is similar to the matrix

$$\bar{A} = (I - KH)A + \gamma^{-2}(I - KH)A\hat{P}L^T\left(I - \gamma^{-2}L\hat{P}L^T\right)^{-1}L$$

Note that, with K as given in Equation 8.59,

$$I - KH = I - MH^T(I + HMH^T)^{-1}H = \left(M^{-1} + H^TH\right)^{-1}M^{-1}$$

Recall from the proof of Theorem 8.6 that $M^{-1} + H^TH = \bar{P}^{-1} + \gamma^{-2}L^TL = \hat{P}^{-1}$. Hence,

$$\begin{aligned}\bar{A} &= (I - KH)A\left[I + \gamma^{-2}\hat{P}L^T\left(I - \gamma^{-2}L\hat{P}L^T\right)^{-1}L\right]\\ &= (I - KH)A\left(\hat{P}^{-1} - \gamma^{-2}L^TL\right)^{-1}\hat{P}^{-1}\\ &= \hat{P}M^{-1}A\bar{P}\hat{P}^{-1}\\ &= \hat{P}M^{-1}A\bar{P}M^{-1}(M\hat{P}^{-1})\end{aligned}$$

The above implies that $\bar{A}$ is similar to the matrix

$$
\begin{aligned}
A\bar{P}M^{-1} &= A\left(M^{-1} + H^T H - \gamma^{-2} L^T L\right)^{-1} M^{-1} \\
&= A - AM\bar{H}^T\left(\bar{R} + \bar{H}M\bar{H}^T\right)^{-1}\bar{H} = \hat{A}
\end{aligned}
$$

Hence, $\bar{A}$ is stable if and only if $\hat{A}$ is stable. ∎

REMARK 8.4 It is worth pointing out that in Theorem 8.8, Assumption 8.2 can be replaced by Assumption 8.1(ii). In this situation, the solution M of the ARE 8.58 is only required to be positive semidefinite. ∎

Example 8.5

Consider Example 8.2. Let $\gamma = 1.2$. Solving the ARE 8.58 yields the stabilizing solution

$$
M = \begin{bmatrix} 2.0240 & -4.5289 \\ -4.5289 & 11.2719 \end{bmatrix}
$$

which satisfies $M > 0$ and $M^{-1} + H^T H - \gamma^{-2} L^T L > 0$. The filter gain is then computed by Equation 8.59 as

$$
K = \begin{bmatrix} -0.2229 \\ 0.5283 \end{bmatrix}
$$

It can be obtained that the minimum achievable γ by the stationary $\mathcal{H}_\infty$ *a posteriori* filtering is 1.07, which is the same as that of the finite horizon $\mathcal{H}_\infty$ *a posteriori* filtering when P_{x_0} is small. □

REMARK 8.5 Let (A, G) and (H, A) be respectively controllable and observable and $H = L$. Then, the following hold (Bolzern *et al.* 2001):

(i) For the case when $H = L$ and A is unstable, then the infimum achievable γ by an $\mathcal{H}_\infty$ *a posteriori* filter is 1. That is, $\gamma_{\text{inf}} = 1$. If A is stable, then

$$
\gamma_{\text{inf}} = \frac{\mu}{\sqrt{1 + \mu^2}} \tag{8.62}
$$

where $\mu = \|H(zI - A)^{-1}G\|_\infty$.

(ii) Assume that $\gamma > 1$. Then, the actual $\mathcal{H}_\infty$ *a posteriori* filtering performance $\tilde{\gamma}$ achieved by the central $\mathcal{H}_\infty$ filter satisfies

$$
\tilde{\gamma} \leq \frac{2}{1 + \gamma^{-2}}
$$

The Kalman filter is a special case of the central $\mathcal{H}_\infty$ filter with $\gamma \to \infty$. Hence,

$$
\tilde{\gamma} \leq 2
$$

(iii) Suppose the desired $\mathcal{H}_\infty$ performance γ_0 satisfying $1 < \gamma_0 < 2$. Then, the central $\mathcal{H}_\infty$ filter designed with

$$\gamma = \sqrt{\frac{\gamma_0}{2-\gamma_0}} > \gamma_0$$

achieves the desired $\mathcal{H}_\infty$ filtering performance γ_0. ∎

8.4 Polynomial Approach to $\mathcal{H}_\infty$ Estimation

In this section, we study the stationary $\mathcal{H}_\infty$ estimation using a polynomial approach. We note that a polynomial approach is often preferred in signal processing and communications due to its efficient computation.

We recall the linear time-invariant system under consideration:

$$x_{k+1} = Ax_k + B\eta_k \tag{8.63}$$

$$z_k = Hx_k + D\eta_k \tag{8.64}$$

$$s_k = Lx_k \tag{8.65}$$

where $z_k \in R^p$ is the measurement output, $s_k \in R^r$ is the signal to be estimated, and $\eta_k \in R^l$ is the input noise. Note that the above modelcontains the model 8.1 through 8.3 as a special case since the latter can be obtained by setting $\eta_k = [w_k^T \quad v_k^T]^T$, $B = [G \quad 0]$, and $D = [0 \quad I]$.

Define the transfer functions from the noise input η to z and s, respectively, as

$$T_1(z) = H(zI - A)^{-1}B + D \tag{8.66}$$

$$T_2(z) = L(zI - A)^{-1}B \tag{8.67}$$

The stationary $\mathcal{H}_\infty$ filtering problem is concerned with the design of a stable causal filter $\mathcal{F}(z)$, such that the filtering error dynamic $\mathcal{F}(z)T_1(z) - T_2(z)$ is asymptotically stable and with its $\mathcal{H}_\infty$ norm less than a prescribed value γ.

We introduce the following standard assumption.

Assumption 8.3 *$(H,\ A)$ is detectable and $(A,\ G)$ has no unreachable mode on the unit circle.*

The above assumption is necessary. The detectability of $(H,\ A)$ ensures the existence of a filter such that the filtering error dynamic is asymptotically stable.

Denote

$$T(z) = \begin{bmatrix} T_1(z) & 0 \\ T_2(z) & I_r \end{bmatrix} \tag{8.68}$$

and

$$J_{i,j} = \begin{bmatrix} I_i & 0 \\ 0 & -\gamma^2 I_j \end{bmatrix} \tag{8.69}$$

where I_i is the identity matrix of dimension $i \times i$.

Then, we have the following result.

THEOREM 8.9

Consider the system 8.63 through 8.65. The $\mathcal{H}_\infty$ filtering problem has a solution if and only if there exists a square transfer function matrix

$$\Omega(z) = \begin{bmatrix} \Omega_{11}(z) & \Omega_{12}(z) \\ \Omega_{21}(z) & \Omega_{22}(z) \end{bmatrix}$$

such that

(i) $\Omega^{-1}(z), \Omega^{-1}(z)T(z)$ *and* $\Theta(z) = \Omega_{22}(z) - \Omega_{21}(z)\Omega_{11}^{-1}(z)\Omega_{12}(z)$ *are stable.*

(ii)
$$\Omega(z)J_{p,r}\Omega^*(z) = T(z)J_{l,r}T^*(z) \tag{8.70}$$

If the above conditions are met, a class of filters that solve the $\mathcal{H}_\infty$ filtering problem is given by

$$\mathcal{F}(z) = [\Omega_{21}(z) - \Omega_{22}(z)U(z)]\,[\Omega_{11}(z) - \Omega_{12}(z)U(z)]^{-1} \tag{8.71}$$

where $U(z)$ is any stable transfer function matrix satisfying $\|U(z)\|_\infty < \gamma$.

Proof:

We first prove the sufficiency. From condition (ii), the (1,1)-block in Equation 8.70 gives

$$T_1(z)T_1^*(z) = \Omega_{11}(z)\Omega_{11}^*(z) - \gamma^2\Omega_{12}(z)\Omega_{12}^*(z)$$

i.e.,

$$\gamma^{-2}I - \Omega_{11}^{-1}(z)\Omega_{12}(z)\Omega_{12}^*(z)\Omega_{11}^{-*}(z) = \gamma^{-2}\Omega_{11}^{-1}(z)T_1(z)T_1^*(z)\Omega_{11}^{-*}(z) \geq 0$$

for all z on the unit circle, where $\Omega_{11}^{-*}(z) = \left[\Omega_{11}^{-1}(z)\right]^*$. Further, it can be easily checked that $\Omega_{11}^{-1}(z)\Omega_{12}(z) = -\left[\Omega(z)^{-1}\right]_{12}\Theta$ which is stable. Hence, $\|\Omega_{11}^{-1}(z)\Omega_{12}(z)\|_\infty \leq \gamma^{-1}$. In view of $\|U(z)\|_\infty < \gamma$, using the small gain theorem we know that

$$\Gamma(z) = \left(I - U(z)\Omega_{11}^{-1}(z)\Omega_{12}(z)\right)^{-1}$$

is stable. Next, we show that

$$[\mathcal{F}(z) \quad -I] = -\Theta(z)\Gamma(z)\,[U(z) \quad I]\,\Omega^{-1}(z) \tag{8.72}$$

To this end, we note that

$$[\mathcal{F}(z) \quad -I]\,\Omega(z) = [\mathcal{F}(z)\Omega_{11}(z) - \Omega_{21}(z) \quad \mathcal{F}(z)\Omega_{12}(z) - \Omega_{22}(z)] \tag{8.73}$$

By taking into account Equation 8.71,

$$\begin{aligned}
&\mathcal{F}(z)\Omega_{11}(z) - \Omega_{21}(z)\\
&\quad = [\Omega_{21}(z) - \Omega_{22}(z)U(z)]\,[\Omega_{11}(z) - \Omega_{12}(z)U(z)]^{-1}\,\Omega_{11}(z) - \Omega_{21}(z)\\
&\quad = [\Omega_{21}(z) - \Omega_{22}(z)U(z)]\,\left[I - \Omega_{11}^{-1}(z)\Omega_{12}(z)U(z)\right]^{-1} - \Omega_{21}(z)\\
&\quad = -\Theta(z)U(z)\left[I - \Omega_{11}^{-1}(z)\Omega_{12}(z)U(z)\right]^{-1}\\
&\quad = -\Theta(z)\left[I - U(z)\Omega_{11}^{-1}(z)\Omega_{12}(z)\right]^{-1}U(z)\\
&\quad = -\Theta(z)\Gamma(z)U(z)
\end{aligned}$$

Similarly, we can show that

$$\mathcal{F}(z)\Omega_{12}(z) - \Omega_{22}(z) = -\Theta(z)\Gamma(z)$$

Hence, in the light of Equation 8.73, the above implies Equation 8.72.

It is clear from Equation 8.72 that $\mathcal{F}(z)$ and $\mathcal{F}(z)T_1(z) - T_2(z) = [\mathcal{F}(z) \quad -I]T(z)$ are both stable. Next, we show that the filtering error transfer function matrix has an $\mathcal{H}_\infty$ norm of less than γ.

$$\begin{aligned}
&[\mathcal{F}(z)T_1(z) - T_2(z)]\,[\mathcal{F}(z)T_1(z) - T_2(z)]^* - \gamma^2 I\\
&\quad = [\mathcal{F}(z) \quad -I]\,T(z)J_{m,r}T^*(z)\,[\mathcal{F}(z) \quad -I]^*\\
&\quad = [\mathcal{F}(z) \quad -I]\,\Omega(z)J_{p,r}\Omega^*(z)\,[\mathcal{F}(z) \quad -I]^*\\
&\quad = \Theta(z)\Gamma(z)\left[U(z)U^*(z) - \gamma^2 I\right]\Gamma^*(z)\Theta^*(z)
\end{aligned}$$

which implies that $\|\mathcal{F}(z)T_1(z) - T_2(z)\|_\infty < \gamma$ if and only if $\|U(z)\|_\infty < \gamma$.

As seen from Theorem 8.9, the polynomial approach to the $\mathcal{H}_\infty$ filtering relies on the J-spectral factorization as in Equation 8.70. An ARE-based method will be given next (Zhang *et al.* 2001). ∎

8.5 J-Spectral Factorization

First, define the ARE:

$$APA^T - P - (AP\bar{H}^T + [BD^T \quad 0])Q_w^{-1}(AP\bar{H}^T + [BD^T \quad 0])^T + BB^T = 0 \tag{8.74}$$

where $Q_w = R + \bar{H}P\bar{H}^T$ and

$$R = \begin{bmatrix} DD^T & 0 \\ 0 & -\gamma^2 I_r \end{bmatrix}, \qquad \bar{H} = \begin{bmatrix} H \\ L \end{bmatrix}$$

If the ARE 8.74 has a nonnegative definite stabilizing solution P, and the matrices Q_w and $J_{p,r}$ have the same inertia (same numbers of positive, negative eigenvalues), then there exists a nonsingular constant matrix W such that

$$Q_w = WJ_{p,r}W^T$$

In fact, W can be computed in terms of P as

$$W = \begin{bmatrix} (DD^T + HPH^T)^{1/2} & 0 \\ LPH^T(DD^T + HPH^T)^{-1/2} & V^{1/2} \end{bmatrix} \tag{8.75}$$

where

$$V = I - \gamma^{-2}L\left[P - PH^T\left(DD^T + HPH^T\right)^{-1}HP\right]L^T$$

Further, a J-spectral factor $\Omega(z)$ of Equation 8.70 with $\Omega^{-1}(z)$ and $\Omega_{11}(z)$ being stable can be computed as

$$\Omega(z) = \bar{H}(zI - A)^{-1}KW + W \tag{8.76}$$

where

$$K = AP\bar{H}^TQ_w^{-1}$$

Example 8.6

Consider again the system in Example 8.2

$$\begin{aligned} x_{k+1} &= \begin{bmatrix} -0.2 & -0.5 \\ 1.5 & 1 \end{bmatrix} x_k + \begin{bmatrix} 0 \\ 1 \end{bmatrix} w_k \\ z_k &= \begin{bmatrix} -2 & 1 \end{bmatrix} x_k + v_k \\ s_k &= \begin{bmatrix} 1 & 1 \\ 0 & 1 \end{bmatrix} x_k \end{aligned}$$

Let $\gamma = 1.2$. By solving the ARE 8.74, we obtain

$$P = \begin{bmatrix} 0.6006 & -1.3243 \\ -1.3243 & 4.0415 \end{bmatrix}$$

It follows from Equation 8.75 that

$$W = \begin{bmatrix} 0.2802 & 0 & 0 \\ -0.3269 & 0.7052 & -0.2523 \\ -0.5251 & -0.2523 & 0.7545 \end{bmatrix}$$

Thus, a J-spectral factor can be computed using Equation 8.76 as

$$\Omega(z) = \frac{1}{z^2 - 0.8z + 0.55}\begin{bmatrix} 3.5694z^2 - 0.4513z + 0.1541 & -1.7999z + 0.3092 & -1.5783z + 0.3692 \\ 1.1667z^2 - 0.9160z & 0.7052z^2 - 1.1086z + 0.8860 & -0.2523z^2 - 0.2256z + 0.2963 \\ 1.8743z^2 - 0.6864z & -0.2523z^2 - 0.7610z + 0.2963 & 0.7545z^2 - 1.4147z + 0.8281 \end{bmatrix}$$

Note that $\Omega(z)$ is stable and so is

$$\Omega_{11}^{-1}(z) = \frac{z^2 - 0.8z + 0.55}{3.5694z^2 - 0.4513z + 0.1541}$$

A class of filters that achieve the $\mathcal{H}_\infty$ performance $\gamma = 1.2$ can be parameterized by Equation 8.71. In particular, the central $\mathcal{H}_\infty$ filter (setting $U(z) \equiv 0$) is given by

$$\mathcal{F}(z) = \Omega_{21}(z)\Omega_{11}^{-1}(z) = \begin{bmatrix} \frac{1.1667z^2 - 0.9160z}{3.5694z^2 - 0.4513z + 0.1541} \\ \frac{1.8743z^2 - 0.6864z}{3.5694z^2 - 0.4513z + 0.1541} \end{bmatrix}$$

The singular value plots of the filtering error transfer function matrix $\mathcal{F}(z)T_1(z) - T_2(z)$ with sampling period of 0.01 are shown in Figure 8.7.

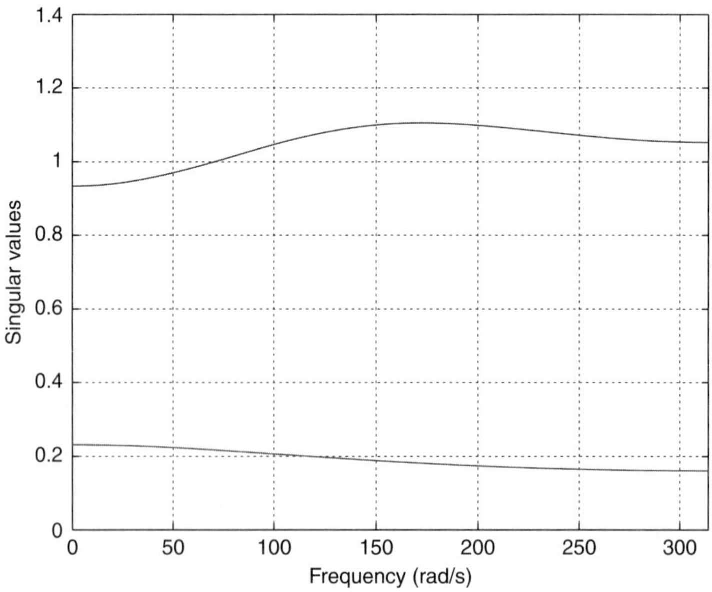

FIGURE 8.7
The singular value plots of the filtering error transfer function matrix.

It can be known that the actual $\mathcal{H}_\infty$ norm of the filtering error transfer function matrix is 1.055, i.e.,

$$\|\mathcal{F}(z)T_1(z) - T_2(z)\|_\infty = 1.055$$

which is below the prescribed performance requirement. $\square$

Example 8.7

Consider the unstable system

$$\begin{aligned} x_{k+1} &= \begin{bmatrix} 5 & 0.5 & 0 \\ 0 & 2 & 1 \\ 0 & 0 & 3 \end{bmatrix} x_k + \begin{bmatrix} 0 & 0 \\ 0 & 1 \\ 1 & 0 \end{bmatrix} w_k \\ z_k &= \begin{bmatrix} 1 & 2 & 0 \\ 1 & 0 & 0 \end{bmatrix} x_k + v_k \\ s_k &= \begin{bmatrix} 1 & 2 & 0 \\ 1 & 0 & 0 \end{bmatrix} x_k \end{aligned}$$

Let $\gamma = 1.1$. By solving the ARE 8.74, we obtain

$$P = \begin{bmatrix} 124.4740 & -44.0057 & -58.3527 \\ -44.0057 & 84.4010 & 183.6707 \\ -58.3527 & 183.6707 & 452.0973 \end{bmatrix}$$

It follows from Equation 8.75 that

$$W = \begin{bmatrix} 16.8927 & 1.3014 & 0 & 0 \\ 1.3014 & 11.1257 & 0 & 0 \\ 16.8329 & 1.3084 & 0.4202 & -0.0010 \\ 1.3084 & 11.0350 & -0.0010 & 0.4247 \end{bmatrix}$$

Then, $\Omega^{-1}(z)$ and $\Omega_{11}^{-1}(z)$ can be obtained as

$$\Omega^{-1}(z) = \left[\begin{matrix} \frac{0.0597z^3-1.6049z^2+2.4807z-0.4590}{z^3-0.9362z^2+0.2830z-0.0271} & \frac{-0.0070z^3-0.1870z^2+0.0856z-0.0357}{z^3-0.9362z^2+0.2830z-0.0271} \\ \frac{-0.0070z^3+0.0467z^2-0.1367z-0.0354}{z^3-0.9362z^2+0.2830z-0.0271} & \frac{0.0907z^3-2.4099z^2+1.7180z-0.3017}{z^3-0.9362z^2+0.2830z-0.0271} \\ \frac{-2.3710z^3+1.9989z^2-0.3187z-0.0001}{z^3-0.9362z^2+0.2830z-0.0271} & \frac{-0.0082z^3+0.0176z^2-0.0171z-0.0001}{z^3-0.9362z^2+0.2830z-0.0271} \\ \frac{-0.0082z^3+0.0412z^2-0.0708z}{z^3-0.9362z^2+0.2830z-0.0271} & \frac{-2.3350z^3+1.6928z^2-0.3010z}{z^3-0.9362z^2+0.2830z-0.0271} \end{matrix}\right.$$

$$\left.\begin{matrix} \frac{1.2801z^2-2.0365z+0.3780}{z^3-0.9362z^2+0.2830z-0.0271} & \frac{0.1602z^2-0.0721z+0.0296}{z^3-0.9362z^2+0.2830z-0.0271} \\ \frac{-0.0332z^2+0.1114z+0.0294}{z^3-0.9362z^2+0.2830z-0.0271} & \frac{1.9217z^2-1.3988z+0.2474}{z^3-0.9362z^2+0.2830z-0.0271} \\ \frac{2.3800z^3-2.0457z^2+0.3823z-0.0113}{z^3-0.9362z^2+0.2830z-0.0271} & \frac{0.0058z^3-0.0136z^2+0.0139z+0.0001}{z^3-0.9362z^2+0.2830z-0.0271} \\ \frac{0.0058z^3-0.0331z^2+0.0582z}{z^3-0.9362z^2+0.2830z-0.0271} & \frac{2.3540z^3-1.7962z^2+0.3688z-0.0115}{z^3-0.9362z^2+0.2830z-0.0271} \end{matrix}\right]$$

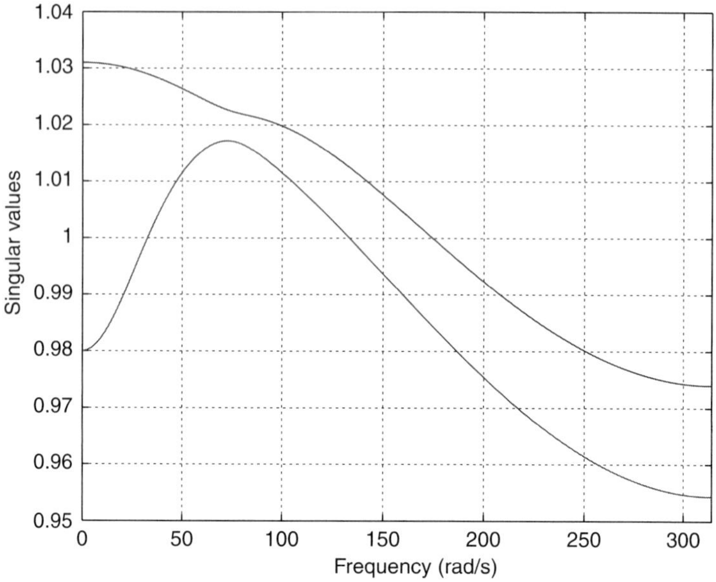

FIGURE 8.8
The singular value plots of the filtering error transfer function matrix.

$$\Omega_{11}^{-1}(z) = \begin{bmatrix} \frac{0.0597z^3-0.3142z^2+0.3897z-0.0146}{z^3-0.6860z^2+0.0467z-0.0008} & \frac{-0.0070z^3-0.0289z^2+0.0098z-0.0009}{z^3-0.6860z^2+0.0467z-0.0008} \\ \frac{-0.0070z^3+0.0139z^2-0.0291z-0.0011}{z^3-0.6860z^2+0.0467z-0.0008} & \frac{0.0907z^3-0.4817z^2+0.2774z-0.0092}{z^3-0.6860z^2+0.0467z-0.0008} \end{bmatrix}$$

Both $\Omega^{-1}(z)$ and $\Omega_{11}^{-1}(z)$ are verified to be stable. Then, the central $\mathcal{H}_\infty$ filter is given by

$$\mathcal{F}(z) = \Omega_{21}(z)\Omega_{11}^{-1}(z) = \begin{bmatrix} \frac{0.9964z^3-0.6679z^2+0.0241z}{z^3-0.6867z^2+0.0476z-0.0009} & \frac{0.0011z^3-0.0016z^2+0.0014z}{z^3-0.6867z^2+0.0476z-0.0009} \\ \frac{0.0011z^3-0.0035z^2+0.0054z}{z^3-0.6867z^2+0.0476z-0.0009} & \frac{0.9917z^3-0.6432z^2+0.0225}{z^3-0.6867z^2+0.0476z-0.0009} \end{bmatrix}$$

The singular value plots of the filtering error transfer function matrix $\mathcal{F}(z)$ $T_1(z) - T_2(z)$ with sampling period of 0.01 are shown in Figure 8.8.

From the figure, it can be known that the actual $\mathcal{H}_\infty$ norm of the filtering error dynamics is 1.0310 which is below the prescribed value of 1.1. □

8.6 Applications in Channel Equalization

Channel equalization is illustrated in Figure 8.9, where $\mathcal{H}$ denotes the channel model and $\mathcal{F}$ is the filter to be designed to reconstruct the input signal s_{k-d} at the time instant k, with d the allowable delay. More specifically, given the measurements $\{z_1, z_2, \ldots, z_k\}$, the equalization problem is to find the filter $\mathcal{F}$

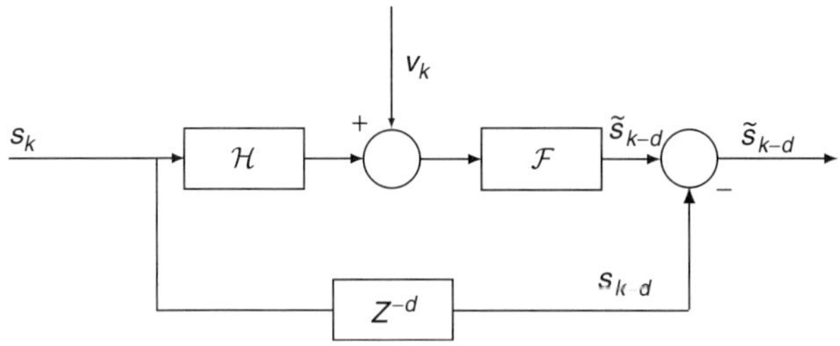

FIGURE 8.9
Equalization filtering problem.

so as to minimize $s_{k-d} - \hat{s}_{k-d}$ in a certain sense. This problem is also termed as deconvolution.

We consider a linear time-invariant channel model described by an nth order finite impulse response (FIR) filter of the form

$$\bar{z}_k = h_0 s_k + h_1 s_{k-1} + h_2 s_{k-2} + \cdots + h_n s_{k-n} \tag{8.77}$$

and the equalization problem with one step delay ($d=1$). By taking into consideration the measurement noise,

$$z_k = \bar{z}_k + v_k$$

Then, a state space representation of the model and the measurement can be given by

$$x_{k+1} = Ax_k + Bw_k \tag{8.78}$$

$$z_k = Hx_k + Dw_k \tag{8.79}$$

$$s_{k-1} = Lx_k \tag{8.80}$$

where $x_k = [s_{k-n} \;\; s_{k-n+1} \cdots s_{k-1}]^T$, $w_k = [s_k \;\; v_k]^T$, and

$$A = \begin{bmatrix} 0 & 1 & 0 & \cdots & 0 \\ 0 & 0 & 1 & \cdots & 0 \\ \vdots & \vdots & \vdots & \ddots & \vdots \\ 0 & 0 & 0 & \cdots & 1 \\ 0 & 0 & 0 & \cdots & 0 \end{bmatrix}, \qquad B = \begin{bmatrix} 0 & 0 \\ 0 & 0 \\ \vdots & \vdots \\ 0 & 0 \\ 1 & 0 \end{bmatrix}$$

$$H = \begin{bmatrix} h_n & h_{n-1} & \cdots & h_2 & h_1 \end{bmatrix}, \qquad D = [h_0 \;\; 1]$$

$$L = \begin{bmatrix} 0 & 0 & \cdots & 0 & 1 \end{bmatrix}$$

Using the polynomial approach in the last sections, an $\mathcal{H}_\infty$ equalization filter can be designed.

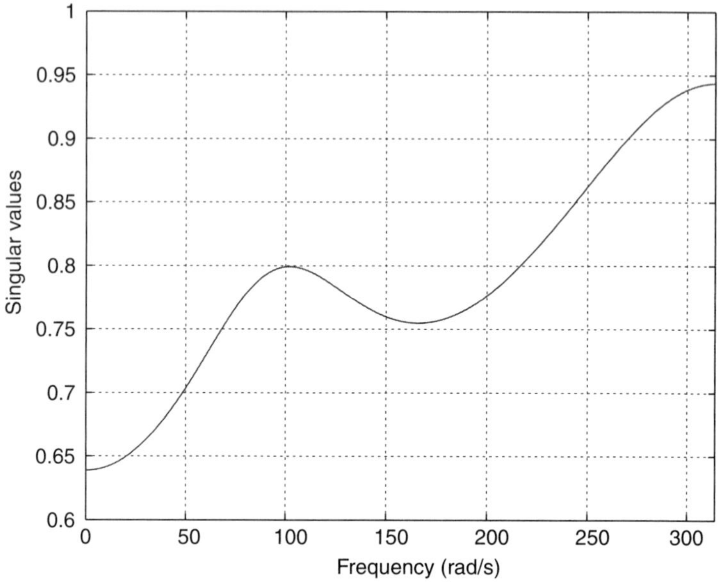

FIGURE 8.10
The singular value plots of the filtering error transfer function matrix.

Example 8.8

Assume a third-order FIR channel model with

$$h_0 = 1, \quad h_1 = 0.5, \quad h_2 = -0.5, \quad h_3 = 0.5$$

and $d = 1$.

Let the prescribed $\mathcal{H}_\infty$ performance $\gamma = 0.95$. Then, following the approach of the last section, the central $\mathcal{H}_\infty$ filter can be designed which is given by

$$\mathcal{F}(z) = \frac{0.3520z^2(z + 0.4960)}{z^3 - 0.3498z^2 + 0.0887z + 0.0873}$$

The singular value plot of the filtering error transfer function is shown in Figure 8.10.

The actual $\mathcal{H}_\infty$ performance achieved is 0.9451. We note that the channel model is nonminimum phase and there exists measurement noise. Thus, perfect reconstruction is not possible.

Let the input signal s_k be a uniformly distributed signal from $[0, 1]$ and the measurement noise be a white noise with variance of 0.01. The initial states of the system and filter are zeros. The comparison between the input signal and its estimate is given in Figure 8.11.

Now, let us take a look at minimum-phase channel model:

$$h_0 = 1, \quad h_1 = -0.5, \quad h_2 = 0.5, \quad h_3 = 0.3$$

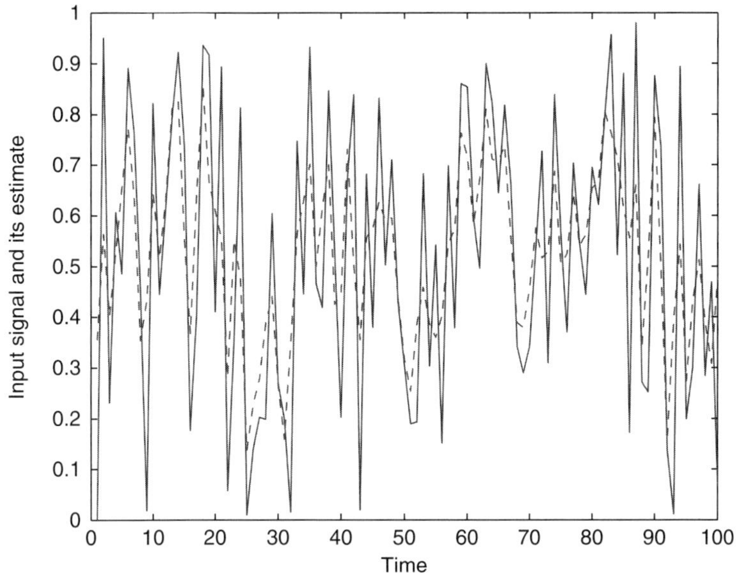

FIGURE 8.11
Comparison between the input signal s and its estimate $\hat{s}$.

where the FIR model has zeros at $0.4334 \pm 0.7938i$ and -0.3668. Let us assume that $D = [h_0 \;\; 0.001]$, i.e., the measurement is almost perfect. With $\gamma = 0.01$, we can obtain the filter

$$\mathcal{F}(z) = \frac{z^{-1} + 0.001z^{-2} + 0.001z^{-3}}{1 - 0.5z^{-1} - 0.5z^{-2} + 0.3z^{-3}}$$

which results in

$$\mathcal{H}(z)\mathcal{F}(z) = z^{-1} + 0.001z^{-2} + 0.001z^{-3}$$

By neglecting the measurement error, the tracking error with respect to the input is given by

$$\hat{s}_{k-1} - s_{k-1} = 0.001(s_{k-2} + s_{k-3})$$

which is negligible. This can be seen from the singular value plot of the filtering error transfer function shown in Figure 8.12 as well.

Again, let the input signal s_k be a uniformly distributed signal between 0 and 1 and the measurement noise be a white noise with variance of 0.01. The initial states of the system and filter are zeros. The comparison between the input signal and its estimate is given in Figure 8.13, where both the signals almost overlap with each other. ∎

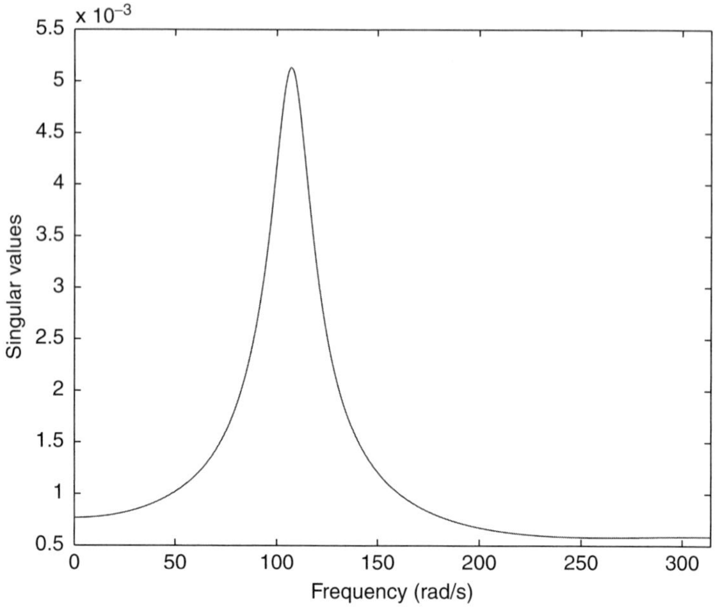

FIGURE 8.12
The singular value plots of the filtering error transfer function matrix.

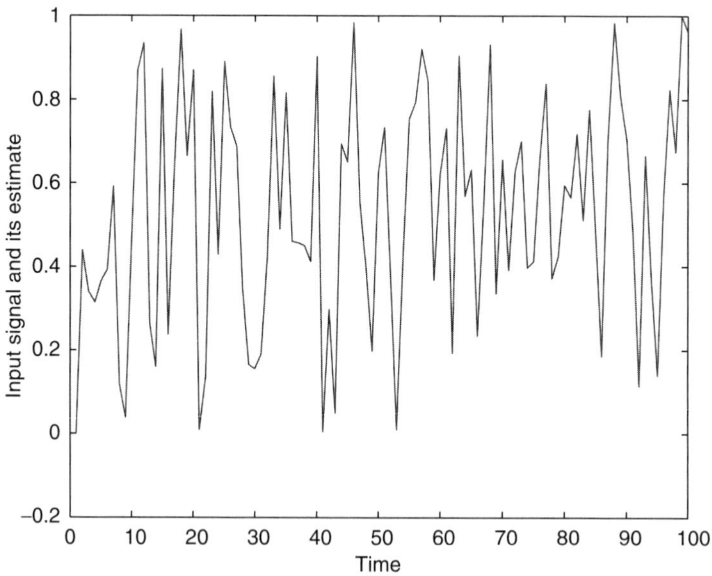

FIGURE 8.13
Comparison between the input signal s and its estimate $\hat{s}$.

Problems

1. Consider the problem of estimating a constant parameter. The problem can be formulated as an estimation problem for the system

$$\begin{aligned} x_{k+1} &= x_k \\ z_k &= x_k + v_k \\ s_k &= x_k \end{aligned}$$

(a) Show that for any horizon $N > 0$ and any initial state weighting P_{x_0}, there exists an $\mathcal{H}_\infty$ *a priori* filter if and only if $\gamma > 1$.

(b) For a given $\gamma > 1$, compare the central $\mathcal{H}_\infty$ filter with the Kalman filter. Assume that the covariance of the initial state estimate for the Kalman filter is P_{x_0}.

(c) Obtain an $\mathcal{H}_\infty$ filter for $\gamma = \sqrt{2}$.

2. In the least mean square (LMS) estimation, we consider the problem of estimating the unknown vector w based on an observed sequence $\{z_k\}$, where

$$z_k = \phi_k^T w + v_k$$

with $\phi_k = [\phi_{k1}\ \phi_{k2} \dots \phi_{kn}]$ being known. Let $\hat{w}_k$ be an estimate of w based on $\{z_0, z_1, \dots, z_k\}$. The objective is to minimize the prediction error:

$$\tilde{s}_i = \phi_k^T w - \phi_k^T \hat{w}_{k-1}$$

Given an initial estimate $\hat{w}_0$, formulate an $\mathcal{H}_\infty$ estimation for the above problem and show that the achievable $\mathcal{H}_\infty$ performance cannot be less than 1.

3. Consider the system

$$\begin{aligned} x_{k+1} &= \begin{bmatrix} 0 & 1 \\ -1 & -0.2 \end{bmatrix} x_k + \begin{bmatrix} 1 \\ 0.5 \end{bmatrix} w_k \\ z_k &= \begin{bmatrix} 1 & 0 \end{bmatrix} x_k + v_k \end{aligned}$$

The signal to be estimated is $s_k = \begin{bmatrix} 1 & 0 \end{bmatrix} x_k$. The weighting matrix for the initial estimate is $P_{x_0} = I$. Use MATLAB to compute the following:

(a) For $N = 20$ and $\gamma = 2$, derive an $\mathcal{H}_\infty$ *a priori* filter.

(b) Find the optimal stationary $\mathcal{H}_\infty$ *a priori* filtering performance.

(c) Obtain the singular value plot of the optimal $\mathcal{H}_\infty$ filtering error transfer function.

4. Consider the $\mathcal{H}_\infty$ *a posteriori* filtering for the system in Problem 8.1. Compare that the optimal achievable $\mathcal{H}_\infty$ *a posteriori* filtering performance that of the $\mathcal{H}_\infty$ *a priori* filtering. Explain the result.

5. Repeat Problem 3 by considering the stationary $\mathcal{H}_\infty$ *a posteriori* filtering. For the given $\gamma = 1.1$, verify the result in Remark 8.5.

6. Prove Equation 8.62 in Remark 8.5.

7. Consider the system

$$\begin{aligned} x_{k+1} &= \begin{bmatrix} 1 & 1 \\ 0 & 1 \end{bmatrix} x_k + \begin{bmatrix} 1 \\ 1 \end{bmatrix} w_k \\ z_k &= \begin{bmatrix} 1 & 0 \end{bmatrix} x_k + v_k \\ s_k &= \begin{bmatrix} 1 & 0 \end{bmatrix} x_k \end{aligned}$$

Use Matlab, find the optimal $\mathcal{H}_\infty$ *a priori* and *a posteriori* filtering performances. Also, verify that the Kalman *a posteriori* filter and *a priori* filter provide $\mathcal{H}_\infty$ filtering performances of $\gamma = 2$ and $\gamma = 3.0810$, respectively.

8. For a given $\gamma > 0$, suppose the DRE 8.53 has feasible solutions $M_k > 0$ and $M_k' > 0$ for the given initial condition $M_0 > 0$ and $M_0' > 0$, respectively. Show if $M_0' > M_0$, then $M_k' \geq M_k$ for any positive integer k.

9. Consider again the Problem 7. Given $\gamma = 2$, obtain an $\mathcal{H}_\infty$ filter via the spectral factorization approach.

10. Consider Example 8.8. Suppose the allowable equalization delay is $d = 2$. Formulate the equalization problem as an $\mathcal{H}_\infty$ filtering problem via state augmentation. Use MATLAB to verify that a better $\mathcal{H}_\infty$ filtering can be achieved than the case of $d = 1$.

Part III

Optimal Stochastic Control

9

Stochastic Control for State Variable Systems

In Part I we showed how to estimate the state vector of a stochastic system given the noise-corrupted measurements $z(t)$ of the state. We shall now show how to use this state estimate for the optimal control of stochastic systems. Since nature is inherently noisy and seldom gives us information in the form in which we need it, the techniques we shall discuss here are usually required for optimal control of the actual physical systems. The deterministic optimal control theory is generally useful only in some ideal noise-free situations where the entire state can be exactly measured.

Although our presentation stands on its own, an introduction to the optimal control of deterministic systems providing full state information will make the next two chapters more meaningful.

9.1 Dynamic Programming Approach

Dynamic programming is based on Bellman's *principle of optimality* (Bellman 1957, Bellman and Dreyfus 1962, Bellman and Kalaba 1965):

> An optimal policy has the property that no matter what the previous decisions (i.e., controls) have been, the remaining decisions must constitute an optimal policy with regard to the state resulting from those previous decisions. (9.1)

We shall see that the principle of optimality serves to limit the number of potentially optimal control strategies that must be investigated. It also implies that optimal control strategies must be determined by working *backward* from the final stage; the optimal control problem is inherently a *backward-in-time problem*, in direct contrast to the optimal filtering problem which is a forward-in-time problem.

Dynamic programming can easily be used to find optimal controls for nonlinear stochastic systems. In this section we shall assume full state information; that is, that the entire state x_k is exactly measurable at time k. Thus, we only consider the effects of process noise, not measurement noise. First, we cover discrete, and then continuous, systems.

9.1.1 Discrete-Time Systems

Consider the nonlinear stochastic plant

$$x_{k+1} = f^k(x_k, u_k) + Gw_k, \quad k \geq i \tag{9.2}$$

where w_k is a white noise sequence. Superscript k indicates time dependence. The initial state x_i is a random variable. To make the plant exhibit a desired behavior over a time interval $[i, N]$, we could try to select the control input u_k to minimize the *performance index*

$$J_i(x_i) = \phi(N, x_N) + \sum_{k=i}^{N-1} L^k(x_k, u_k) \tag{9.3}$$

where ϕ and L^k are deterministic functions. However, since x_k is a random variable due to the random process noise w_k and initial state x_i, so also is the cost $J_i(x_i)$. To define a control problem that does not depend on the particular x_i or sequence w_k, let us select u_k to minimize not $J_i(x_i)$, but its expected value, the *expected cost*

$$j_i = E[J_i(x_i)] \tag{9.4}$$

We shall assume that x_k is exactly measurable at time k.

Example 9.1 Some Useful Performance Indices

To clarify the problem formulation, it is worthwhile to discuss some common performance indices (Equation 9.3) that we can select for the given system (Equation 9.2).

a. Minimum time problems

Suppose we want to find the control u_k to drive the system from the given initial state x_i to a desired final state $x \in R^n$ in minimum time. Then we could select the performance index

$$J = N - i = \sum_{k=i}^{N-1} 1 \tag{9.5}$$

and specify the *boundary condition*

$$x_N = x \tag{9.6}$$

In this case $\phi = N - i$ and $L = 0$, or equivalently $\phi = 0$ and $L = 1$.

b. Minimum fuel problems

To find the scalar control u_k to drive the system from x_i to a desired final state x at a fixed time N using minimum fuel, we could use

$$J = \sum_{k=i}^{N-1} |u_k| \tag{9.7}$$

since the fuel burned is proportional to the magnitude of the control vector. Then $\phi = 0$ and $L^k = |u_k|$. The boundary condition $x_N = x$ would again apply.

c. Minimum energy problems

Suppose we want to find u_k to minimize the expected energy of the final state and all intermediate states, and also of the control used to achieve this. Let the final time N be fixed. Then we could use

$$J = \frac{1}{2} s x_N^T x_N + \frac{1}{2} \sum_{k=i}^{N-1} \left(q x_k^T x_k + r u_k^T u_k \right) \tag{9.8}$$

where q, r, and s are scalar weighting factors. Then $\phi = \frac{1}{2} s x_N^T x_N$ and $L = \frac{1}{2}(q x_k^T x_k + r u_k^T u_k)$ are quadratic functions. □

Minimizing the expected energy corresponds in some sense to keeping the state and the control close to zero. If it is more important to us that the intermediate state be small, then we should choose q *large* to weight it heavily in J, which we are trying to minimize. If it is more important that the control energy be small, then we should select a large value of r. If we are more interested in a small final state, then s should be large. For more generality, we could select weighting *matrices* Q, R, S instead of scalars.

At this point, several things should be clearly understood. First, the system dynamics (Equation 9.2) are *given* by the physics of the problem, while the performance index (Equation 9.3) is what we *choose* to achieve the desired system response. Second, to achieve different control objectives, different types of performance indices J may be selected. Finally, the optimal control problem is characterized by *compromises and tradeoffs*, with different weighting factors in J resulting in different balances between conformability with performance objectives and magnitude of the required optimal controls.

Returning to the problem at hand, using Bayes' rule we may write Equation 9.4 as

$$j_i = E_{x_i}\{E[J_i(x_i)/x_i]\} \tag{9.9}$$

where E_{x_i} is the expected value with respect to x_i. We shall find it very useful at times to consider the initial state x_i fixed. Let us therefore define

$$\bar{J}_i(x_i) = E[J_i(x_i)/x_i] \tag{9.10}$$

the expected cost with fixed initial state, or *conditional expected cost*. Then

$$j_i = E_{x_i}[\bar{J}_i(x_i)] \tag{9.11}$$

It can be shown (Åström 1970) that, if asterisks denote optimal values,

$$j_i^* = \min_{u_k}(j_i) = E_{x_i}\{\min_{u_k}[\bar{J}_i(x_i)]\} = E_{x_i}[\bar{J}_i^*(x_i)] \tag{9.12}$$

that is, minimization with respect to the control sequence $u_k, i \leq k \leq N-1$ and expectation with respect to x_i commute. Thus, to minimize j_i we can equivalently minimize $\bar{J}_i(x_i)$ for each x_i (cf. Equation 1.8). This is the approach we shall take.

To see what the principle of optimality (9.1) looks like for this stochastic system, we proceed as follows (see Fel'dbaum 1965 and Åström 1970). Suppose we have computed the optimal conditional expected cost $\bar{J}^*_{k+1}(x_{k+1})$ and the associated optimal control sequence $u^*_{k+1}, u^*_{k+2}, \ldots, u^*_{N-1}$ for each admissible value of x_{k+1}; that is, the optimal control problem on the interval $[k+1, N]$ has been solved. To solve the optimal control problem on $[k, N]$, first decompose $J_k(x_k)$ into two parts as

$$J_k(x_k) = L^k(x_k, u_k) + J_{k+1}(x_{k+1}) \tag{9.13}$$

We assume throughout that x_k is fixed (i.e., known by some measurement process) at time k. Suppose first that the transition

$$\Delta x_k = x_{k+1} - x_k \tag{9.14}$$

is deterministic (i.e., fixed). Then so is x_{k+1}, since x_k is exactly known. Taking expected values of Equation 9.13 conditioned on Δx_k fixed, or equivalently on x_{k+1} fixed, yields

$$E[J_k(x_k)/x_{k+1}] = E[L^k(x_k, u_k)/x_{k+1}] + E[J_{k+1}(x_{k+1})/x_{k+1}] \tag{9.15}$$

The quantity $L^k(x_k, u_k)$ is a deterministic function of the control u_k which we are trying to determine, since x_k is given. Therefore,

$$E[J_k(x_k)/x_{k+1}] = L^k(x_k, u_k) + \bar{J}_{k+1}(x_{k+1}) \tag{9.16}$$

According to the optimality principle (9.1), the only costs that need be considered at time k in determining u^*_k are those *the last portions of which are optimal.* This means that we may restrict our attention to the *admissible costs*

$$E[J_k(x_k)/x_{k+1}] = L^k(x_k, u_k) + \bar{J}^*_{k+1}(x_{k+1}) \tag{9.17}$$

We now realize that in fact x_{k+1} is not deterministic. The transition Δx_k depends on the process noise w_k, so it is a random variable. According to Equation 9.14, so is x_{k+1}. Using Bayes' rule and denoting expectation with respect to x_{k+1} as $E_{x_{k+1}}$, we obtain

$$\bar{J}_k(x_k) = L^k(x_k, u_k) + E_{x_{k+1}}[\bar{J}^*_{k+1}(x_{k+1})] \tag{9.18}$$

since $L^k(x_k, u_k)$ is deterministic. Referring to the principle of optimality again, we see that at time k the optimal conditional cost is

$$\bar{J}^*_k(x_k) = \min_{u_k}\{L^k(x_k, u_k) + E_{x_{k+1}}[\bar{J}^*_{k+1}(x_{k+1})]\} \tag{9.19a}$$

and the optimal control u^*_k is the value of u_k achieving the minimum.

This equation is the principle of optimality for the discrete nonlinear stochastic control problem with complete state information. Optimal controls found using this equation have the form of a *state-variable feedback.*

Because of the one-to-one relation (9.14) between Δx_k and x_{k+1}, we can replace the expectation in Equation 9.19a by $E_{\Delta x_k}$, which is sometimes more convenient. Note also that since $L^k(x_k, u_k)$ is deterministic, Equation 9.19a can be written as

$$\bar{J}_k^*(x_k) = \min_{u_k} E_{x_{k+1}}[L^k(x_k, u_k) + \bar{J}_{k+1}^*(x_{k+1})] \tag{9.19b}$$

Examining Equations 9.3 and 9.10, it is evident that the boundary condition for Equation 9.19 is

$$\bar{J}_N^*(x_N) = E[\phi(N, x_N)/x_N] = \phi(N, x_N) \tag{9.20}$$

if we assume that x_N is known at time $k = N$.

In terms of $\bar{J}_i^*(x_i)$, the minimal expected cost is

$$j_i^* = E_{x_i}[\bar{J}_i^*(x_i)] \tag{9.21}$$

Example 9.2 Dynamic Programming for a Scalar Discrete Stochastic System

To illustrate the use of Equation 9.19, consider the stochastic plant

$$x_{k+1} = x_k + u_k + w_k \tag{9.22}$$

where process noise w_k is white with the PDF shown in Figure 9.1. The performance index selected is

$$J_0(x_0) = (x_N - 1)^2 + \frac{1}{2}\sum_{k=0}^{N-1} u_k^2 \tag{9.23}$$

For simplicity, let $N = 2$. The *admissible controls* are

$$u_k = -0.5,\ 0,\ 0.5 \tag{9.24}$$

and the *admissible state values* are half-integers:

$$x_k = 0,\ \pm 0.5,\ \pm 1, \ldots \tag{9.25}$$

It is desired to find the control input u_k to minimize $j_0 = E[J_0(x_0)]$, the expected cost. This will tend to keep the final state x_N near 1, but we would also like to ensure that the *final state constraint*

$$0 \le x_N \le 1.5 \tag{9.26}$$

holds. That is, the final state must belong to the *target set*

$$x_N \in (0,\ 0.5,\ 1,\ 1.5) \tag{9.27}$$

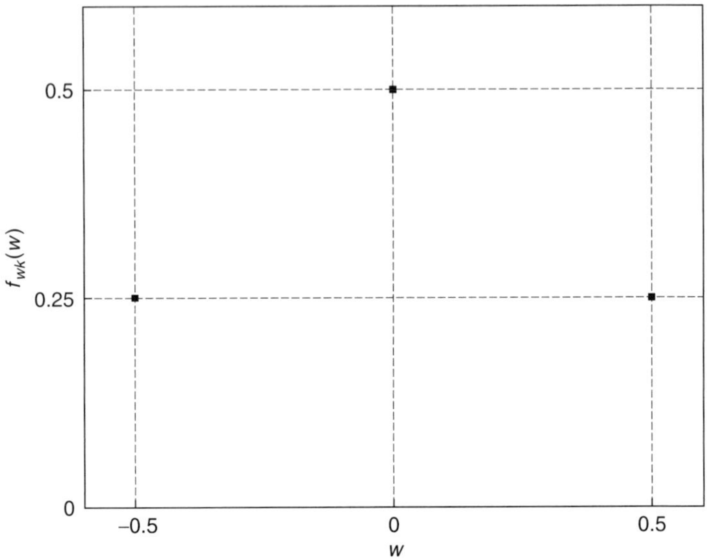

FIGURE 9.1
Process noise probability density function.

TABLE 9.1
Final Costs

x_2	$\bar{J}_2(x_2) = (x_2 - 1)^2 = \bar{J}_2^*(x_2)$
1.5	0.25
1.0	0.0
0.5	0.25
0.0	1.0

To solve this optimal control problem using Equation 9.19, we can construct a decision grid in tabular form, showing the decision process at each stage separately.

First, set $k = N = 2$ to determine the final cost $\bar{J}_N(x_N)$. Since at $k = N$ no decision needs to be made, these costs are also equal to $\bar{J}_N^*(x_N)$. This step is simple, since we must only compute the deterministic quantity

$$\bar{J}_2^*(x_2) = (x_2 - 1)^2 \tag{9.28}$$

for each admissible value of x_2 in Equation 9.27. The result is shown in Table 9.1.

Now decrement to $k = N - 1 = 1$ to find u_1^*. For each admissible value of state x_1, we must apply each admissible value of the control as u_1 and compute the costs

$$J_1(x_1) = \frac{u_1^2}{2} + \bar{J}_2^*(x_2) \tag{9.29}$$

for each resulting x_2. (Note: This is actually $J_1(x_1)/x_2$.) We also need to compute the probability $P(x_2)$ of each value of x_2 arising when u_1 is applied to x_1. Then Equation 9.18 can be used to compute the admissible expected costs

$$\bar{J}_1(x_1) = E_{x_2}[J_1(x_1)] = \sum_{x_2} J_1(x_1)P(x_2) \tag{9.30}$$

Finally, we must use Equation 9.19, comparing the $\bar{J}_1(x_1)$ for each value of u_1 to determine the optimal expected cost

$$\bar{J}_1^*(x_1) = \min_{u_1}[\bar{J}_1(x_1)] \tag{9.31}$$

and to select u_1^*.

See Table 9.2. To accomplish all of this, suppose first that $x_1 = 1.5$. Then $u_1 = 0.5$ is clearly a bad choice since it will result in

$$x_2 = x_1 + u_1 + w_1 = 2 + w_1 \tag{9.32}$$

which does not satisfy Equation 9.26 if $w_1 = 0$ or 0.5. Therefore, control $u_1 = 0.5$ is *inadmissible* if $x_1 = 1.5$. Let us next examine the control $u_1 = 0$.

If $x_1 = 1.5$ and $u_1 = 0$, then

$$x_2 = x_1 + u_1 + w_1 = 1.5 + w_1 \tag{9.33}$$

so that, using Figure 9.1,

$$x_2 = \begin{cases} 2.0 & \text{with probability } 0.25 \\ 1.5 & \text{with probability } 0.5 \\ 1.0 & \text{with probability } 0.25 \end{cases} \tag{9.34}$$

This information is recorded in Table 9.2. Since $x_2 = 2$ is not admissible, and since it occurs with probability 0.25 if we select $u_1 = 0$, this control is inadmissible if $x_1 = 1.5$. We symbolize this by a line in the $J_1(x_1)$ column, which simply means that further computations using $u_1 = 0$ are not required, since we cannot apply that control if $x_1 = 1.5$.

Now apply $u_1 = -0.5$ when $x_1 = 1.5$. The result is

$$x_2 = x_1 + u_1 + w_1 = 1.0 + w_1 \tag{9.35}$$

so that, using PDF of w_1, the possible values of x_2 and their probabilities $P(x_2)$ are

$$x_2 = \begin{cases} 1.5 & \text{with probability } 0.25 \\ 1.0 & \text{with probability } 0.5 \\ 0.5 & \text{with probability } 0.25 \end{cases} \tag{9.36}$$

This is recorded in the table. Since all values of x_2 are in the target set, $u_1 = -0.5$ is an admissible control, so we can proceed to complete the information required in the Table 9.2.

TABLE 9.2
First Stage Optimization

x_1	u_1	x_2	$P(x_2)$	$J_1(x_1) =$ $u_1^2/2 + \bar{J}_2^*(x_2)$	$\bar{J}_1(x_1)$	u_1^*	$\bar{J}_1^*(x_1)$
2.0	−0.5	2.0	0.25	–			
		1.5	0.5				
		1.0	0.25				
1.5	0.0	2.0	0.25	–			
		1.5	0.5				
		1.0	0.25				
	−0.5	1.5	0.25	0.375	0.25	−0.5	0.25
		1.0	0.5	0.125			
		0.5	0.25	0.375			
1.0	0.0	1.5	0.25	0.25	0.125	0.0	0.125
		1.0	0.5	0.0			
		0.5	0.25	0.25			
	−0.5	1.0	0.25	0.125	0.5		
		0.5	0.5	0.375			
		0.0	0.25	1.125			
0.5	0.5	1.5	0.25	0.375	0.25	0.5	0.25
		1.0	0.5	0.125			
		0.5	0.25	0.375			
	0.0	1.0	0.25	0.0	0.375		
		0.5	0.5	0.25			
		0.0	0.25	1.0			
0.0	0.5	1.0	0.25	0.125	0.5	0.5	0.5
		0.5	0.5	0.375			
		0.0	0.25	1.125			
−0.5	0.5	0.5	0.25	–			
		0.0	0.5				
		−0.5	0.25				

The cost (Equation 9.29) is given by

$$J_1(x_1 = 1.5) = \frac{u_1^2}{2} + \bar{J}_2^*(x_2) = 0.125 + \bar{J}_2^*(x_2) = \begin{cases} 0.375 & \text{if } x_2 = 1.5 \\ 0.125 & \text{if } x_2 = 1.0 \\ 0.375 & \text{if } x_2 = 0.5 \end{cases} \tag{9.37}$$

where $\bar{J}_2^*(x_2)$ is found from Table 9.1. Now, the expected cost (Equation 9.30) is found so as to remove the dependence on the particular value of w_1:

$$\begin{aligned} \bar{J}_1(x_1 = 1.5) &= E_{x_2}[J_1(x_1)] \\ &= 0.375 \times 0.25 + 0.125 \times 0.5 + 0.375 \times 0.25 = 0.25 \end{aligned} \tag{9.38}$$

At this point, $\bar{J}_1(x_1)$ has been determined for $x_1 = 1.5$ for all admissible values of u_1 (i.e., only one value: $u_1 = -0.5$). Then, according to Equation 9.31, $u_1^* = -0.5$ and $\bar{J}_1^*(x_1 = 1.5) = 0.25$.

We now assume $x_1 = 1.0$ and determine $x_2, P(x_2), J_1(x_1 = 1.0)$, and $\bar{J}_1(x_1 = 1.0)$ for values of u_1 equal to 0 and -0.5 (clearly $u_1 = 0.5$ is inadmissible, since if the noise $w_1 = 0.5$, it results in $x_2 = x_1 + u_1 + w_1 = 2$). We obtain (see Table 9.2)

$$\bar{J}_1(x_1 = 1) = \begin{cases} 0.125 & \text{if } u_1 = 0 \\ 0.5 & \text{if } u_1 = -0.5 \end{cases} \tag{9.39}$$

Performing the minimization (Equation 9.31), we select

$$u_1^* = 0 \quad \bar{J}_1^*(x_1 = 1) = 0.125 \tag{9.40}$$

Similarly, we fill in the table for $x_1 = 0.5$ and $x_1 = 0$ using only control values which ensure that Equation 9.26 is satisfied. We have also filled in the table for $x_1 = 2$ and $x_1 = -0.5$. Note, however, that $x_1 = 2$ and $x_1 = -0.5$ are not admissible states, since no control drives them to a final state satisfying Equation 9.26 with certainty.

Now, decrement k to zero to find the optimal costs $\bar{J}_0^*(x_0)$ and u_0^* for each possible initial state x_0. In exactly the manner just described, Table 9.3 is completed. The required values of $\bar{J}_1^*(x_1)$ are taken from Table 9.2. Any combination of the state x_0 and control u_0 which results in $x_1 > 1.5$ or $x_1 < 0$ is not allowed since according to Table 9.2 these are inadmissible values of x_1.

TABLE 9.3
Zeroth Stage Optimization

x_0	u_0	x_1	$P(x_1)$	$J_0(x_0) = u_0^2/2 + \bar{J}_1^*(x_1)$	$\bar{J}_0(x_0)$	u_0^*	$\bar{J}_0^*(x_0)$
1.5	−0.5	1.5	0.25	0.375	0.3125	−0.5	0.3125
		1.0	0.5	0.25			
		0.5	0.25	0.375			
1.0	0.0	1.5	0.25	0.25	0.1875	0.0	0.1875
		1.0	0.5	0.125			
		0.5	0.25	0.25			
	−0.5	1.0	0.25	0.25	0.40625		
		0.5	0.5	0.375			
		0.0	0.25	0.625			
0.5	0.5	1.5	0.25	0.375	0.3125		
		1.0	0.5	0.25			
		0.5	0.25	0.375			
	0.0	1.0	0.25	0.125	0.28125	0.0	0.28125
		0.5	0.5	0.25			
		0.0	0.25	0.5			
0.0	0.5	1.0	0.25	0.25	0.40625	0.5	0.40625
		0.5	0.5	0.375			
		0.0	0.25	0.625			

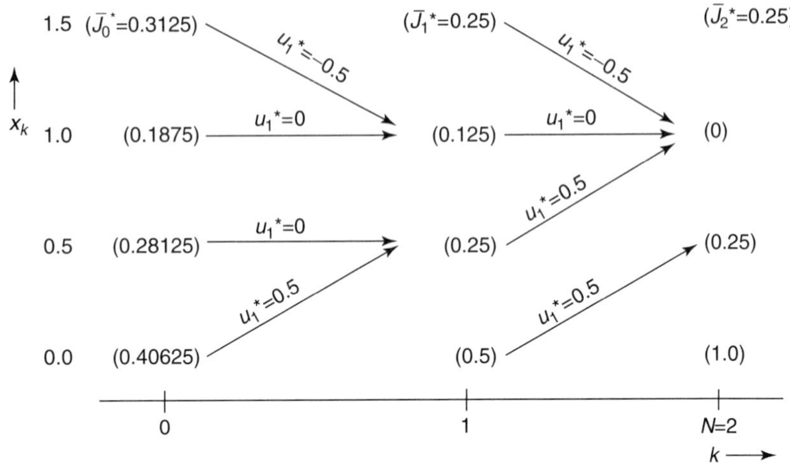

FIGURE 9.2
State feedback optimal control law.

The tables we have constructed contain the optimal feedback control law. To display it more clearly, a graph as in Figure 9.2 can be constructed, which shows for each state value the optimal control and *cost-to-go*. The arrows indicate $x_{k+1} = x_k + u_k$ for the given state and control; they ignore the effect of the noise. Such a graph could be called a "stochastic field of extremals" (Bryson and Ho 1975) since it has the form of a vector field. Changing the weightings in the performance index (Equation 9.23) will change the form of the graph (see problems at the end of the chapter).

The figure indicates the optimal conditional cost $\bar{J}_0^*(x_0)$ for each x_0. To compute the optimal expected cost j_0^*, we require some statistics for the initial state x_0. Let us suppose it is known that

$$x_0 = \begin{cases} 0 & \text{with probability } 0.5 \\ 0.5 & \text{with probability } 0.25 \\ 1.0 & \text{with probability } 0.125 \\ 1.5 & \text{with probability } 0.125 \end{cases} \tag{9.41}$$

Then, performing the expectation with respect to x_0 yields

$$j_0^* = E_{x_0}[\bar{J}_0^*(x_0)] = \sum_{x_0} \bar{J}^*(x_0)P(x_0) = 0.3359 \tag{9.42}$$

□

We have not yet discussed a very important aspect of optimal control for stochastic systems. We know how to find the optimal control minimizing an expected cost using Equation 9.19, and the Example 9.2 demonstrated how easy it is to ensure that the final state x_N belongs to a specified target set.

However, once the optimal control has been constructed, if we are given an initial state x_i the actual value of the final state is still not known! Given an initial state x_i, it is impossible to say with certainty what value x_N will take on if the optimal control u_k^* is applied, since w_k is unknown. Unlike the deterministic control problem, the optimal control sequence and state trajectory cannot be exactly predicted given x_i. This is because the extremal field is now stochastic, depending as it does on the values of the process noise w_k.

Given the transition probabilities that were computed to determine u_k^* (i.e., $P(x_2)$ in Table 9.2 and $P(x_1)$ in Table 9.3), we can determine the transition PDF $f(x_N/x_i)$ using the Chapman–Kolmogorov equation. Let $i = 0$ for notational simplicity. Since the process 9.2 is Markov, we have $f(x_2/x_1, x_0) = f(x_2/x_1)$ and so on, so that

$$f(x_N/x_0) = \sum_{x_1, x_2, \ldots, x_{N-1}} f(x_N/x_{N-1}) \cdots f(x_2/x_1) f(x_1/x_0) \tag{9.43}$$

Let us demonstrate using Example 9.1.

Example 9.1 (continued)

Suppose that it is required to determine the possible values of the final state x_2 and their probabilities if $x_0 = 0.5$ and the optimal control scheme of Figure 9.2 is used.

Equation 9.43 becomes

$$f(x_2/x_0 = 0.5) = \sum_{x1} f(x_2/x_1) f(x_1/x_0 = 0.5) \tag{9.44}$$

The PDFs $f(x_2/x_1)$ and $f(x_1/x_0)$ are found using Tables 9.2 and 9.3, respectively. Note that what we called $P(x_2)$ and $P(x_1)$ in those tables are actually the *conditional* probabilities $P(x_2/x_1)$ and $P(x_1/x_0)$. These are called transition probabilities.

If $x_0 = 0.5$ and $u_0^* = 0$ is applied, then according to Table 9.3, x_1 takes on values of

$$x_1 = \begin{cases} 1.0 & \text{with probability } 0.25 \\ 0.5 & \text{with probability } 0.5 \\ 0 & \text{with probability } 0.25 \end{cases} \tag{9.45}$$

The conditional PDF $f(x_1/x_0 = 0.5)$ is therefore as shown in Figure 9.3. Note that due to Equation 9.22, $u_0^* = 0$.

$$f(x_1/x_0 = 0.5) = f_{w_k}(w - 0.5) \tag{9.46}$$

Equation 9.44 can now be written as

$$f(x_2/x_0 = 0.5) = \frac{1}{4} f(x_2/x_1 = 1) + \frac{1}{2} f(x_2/x_1 = 0.5) + \frac{1}{4} f(x_2/x_1 = 0) \tag{9.47}$$

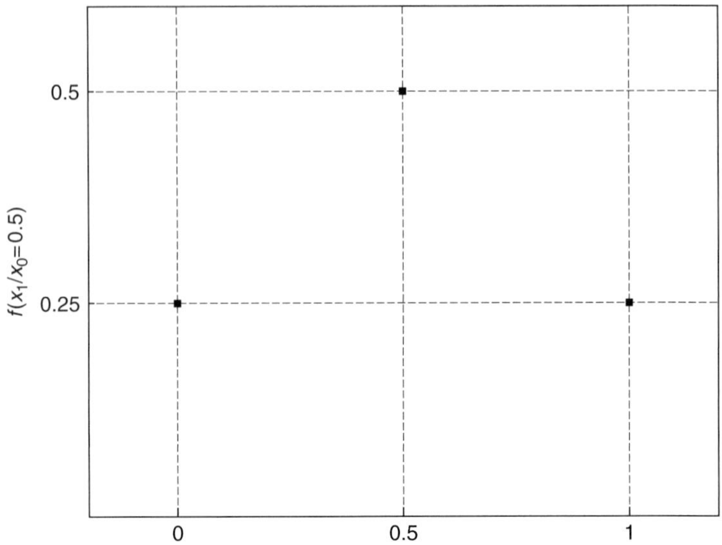

FIGURE 9.3
Transition PDF for $x_0 = 0.5$.

Next, we determine the transition PDFs $f(x_2/x_1)$, assuming u_1^* is applied, from Table 9.2. The results are

$$f(x_2/x_1 = 1) = \begin{cases} \frac{1}{4}, & x_2 = 1.5 \\ \frac{1}{2}, & x_2 = 1.0 \\ \frac{1}{4}, & x_2 = 0.5 \end{cases} \tag{9.48}$$

and so on. Using these values in Equation 9.47 yields

$$f(x_2/x_0 = 0.5) = \begin{cases} \frac{3}{16}, & x_2 = 1.5 \\ \frac{7}{16}, & x_2 = 1.0 \\ \frac{5}{16}, & x_2 = 0.5 \\ \frac{1}{16}, & x_2 = 0 \end{cases} \tag{9.49}$$

Thus, for example, the probability that $x_2 = 1$ given that $x_0 = 0.5$ and the optimal control used is

$$P(x_2 = 1/x_0 = 0.5) = \frac{7}{16} \tag{9.50}$$

□

We shall now discuss an alternative to the performance index (Equation 9.3). Suppose it is desired to maximize the probability that x_N takes on a value in a given target set S (which might contain a single value r_N, the desired or "reference" final state). Then we could define the performance index

$$J_i(x_i) = P(x_N \in S/x_i) \tag{9.51}$$

Let us determine the form of the optimality principle in this case. Using the Chapman–Kolmogorov equation, decompose $J_k(x_k)$ into two parts as

$$\begin{aligned} J_k(x_k) &= \sum_{x_{k+1}} P(x_N \in S/x_{k+1})P(x_{k+1}/x_k,\, u_k) \\ &= \sum_{x_{k+1}} J_{k+1}(x_{k+1})P(x_{k+1}/x_k, u_k) \end{aligned} \tag{9.52}$$

where $P(x_{k+1}/x_k, u_k)$ is the transition probability from x_k to x_{k+1} if the control u_k is applied.

According to Equation 9.1, we only need to consider costs the last portions of which are optimal:

$$J_k(x_k) = \sum_{x_{k+1}} P(x_{k+1}/x_k, u_k)J^*_{k+1}(x_{k+1}) \tag{9.53}$$

Maximizing over these admissible costs yields

$$J^*_k(x_k) = \max_{u_k} \sum_{x_{k+1}} P(x_{k+1}/x_k, u_k)J^*_{k+1}(x_{k+1}) \tag{9.54}$$

This is the principle of optimality for the plant (Equation 9.2) with performance index given by Equation 9.51. The optimal control u^*_k is the value of u_k for which the maximum is attained.

We shall not present an example illustrating Equation 9.54, but its use is very similar to that of Equation 9.19. See Elbert (1984). ▯

9.1.2 Continuous-Time Systems

The discrete stochastic principle of optimality (Equation 9.19) can be applied to continuous systems if they are first discretized and quantized. Note that the process noise values must also be quantized. This approach to the optimal control of stochastic systems is very suitable for programming on a digital computer, although the *curse of dimensionality* (Bellman and Kalaba 1965) is a limitation on the size of the systems that can be dealt with and the accuracy that can be achieved.

We shall now derive a partial differential equation that is satisfied by the optimal conditional expected cost in a continuous stochastic control problem. The steps are similar to those leading up to Equation 9.19. See Fel'dbaum (1965). Consider the nonlinear stochastic plant

$$\dot{x} = a(x, u, t) + w \tag{9.55}$$

where $w(t) \sim N(0, \sigma^2)$ is white noise and $x(t_0)$ is random. For simplicity we consider the scalar case $x \in R$. Let us associate with Equation 9.55 the performance index

$$J[x(t_0), t_0] = \phi[x(T), T] + \int_{t_0}^{T} L(x, u, t)\mathrm{d}t, \tag{9.56}$$

where ϕ and L are deterministic functions.

We want to find the optimal control that minimizes the expected cost

$$j(t_0) = E\{J[x(t_0), t_0]\} \tag{9.57}$$

and ensures that the final state satisfies

$$\psi[x(T), T] = 0 \tag{9.58}$$

for a given deterministic function ψ. Final time T is fixed. The entire state $x(t)$ is assumed to be known by measurement at time t.

Using Bayes' rule we have

$$j(t_0) = E_{x(t_0)}\{\bar{J}[x(t_0), t_0]\} \tag{9.59}$$

where the *conditional expected cost* assuming $x(t_0)$ is fixed is defined as

$$\bar{J}[x(t_0), t_0] = E\{J[x(t_0), t_0]/x(t_0)\} \tag{9.60}$$

It can be shown (see Equation 1.8) that

$$j^*(t_0) = E_{x(t_0)}\{\bar{J}^*[x(t_0), t_0]\} \tag{9.61}$$

so that the optimal control problem can be solved by minimizing the conditional cost. To solve this modified problem, we shall find a partial differential equation that can be solved for $\bar{J}^*(x, t)$.

Before we do this, the conditional PDF of the increment Δx given x and t must be determined, since it will soon be needed. From the state equation, for small Δt we may write

$$\Delta x = a\Delta t + w\Delta t \tag{9.62}$$

with the process $w\Delta t \sim N(0, \sigma^2 \Delta t)$. Since $w\Delta t = \Delta x - a\Delta t$ we have (function of a random variable)

$$f(\Delta x/x, t) = \frac{1}{\sqrt{2\pi\sigma^2\Delta t}} e^{-(\Delta x - a\Delta t)^2/2\sigma^2\Delta t} \tag{9.63}$$

so that Δx with x, t fixed is $N(a\Delta t, \sigma^2\Delta t)$. The PDF in Equation 9.63 is conditional since $a(x, u, t)$ depends on x and t (and, to be complete, on $u(t)$). Since Δx depends only on the current state and not previous states, the process $x(t)$ is Markov.

Now, let us focus our attention on a fixed current state and time (x, t). Let Δx be a deterministic (i.e., fixed) increment in x occurring over a small time Δt. Then Equation 9.56 can be decomposed into two parts as

$$J(x, t) = \int_t^{t+\Delta t} L(x, u, \tau)\mathrm{d}\tau + J(x + \Delta x, t + \Delta t) \tag{9.64}$$

Taking expected values conditioned on Δx being fixed yields

$$E[J(x, t)/\Delta x] = E\left[\int_t^{t+\Delta t} L(x, u, \tau)\mathrm{d}\tau/\Delta x\right] + E[J(x + \Delta x, t + \Delta t)/\Delta x] \tag{9.65}$$

but the integral is a deterministic function of the control $u(\tau), t \le \tau \le t + \Delta t$, since $x(t)$ is fixed. Therefore,

$$E[J(x,t)/\Delta x] = \int_t^{t+\Delta t} L(x,u,\tau)\mathrm{d}\tau + \bar{J}(x+\Delta x, t+\Delta t) \tag{9.66}$$

(Note: x and Δx fixed implies $x + \Delta x$ fixed.)

According to Bellman's principle (Equation 9.1), the only costs we must consider at time t in determining the optimal policy are those of the form

$$E[J(x,t)/\Delta x] = \int_t^{t+\Delta t} L(x,u,\tau)\mathrm{d}\tau + \bar{J}^*(x+\Delta x, t+\Delta t) \tag{9.67}$$

Now we consider the fact that Δx is not fixed; it depends on the process noise and so it is a random variable. Its conditional PDF is given by Equation 9.63. Therefore, using Bayes' rule and denoting expectation with respect to Δx as E_{Δ_x}, we obtain

$$\begin{aligned}\bar{J}(x,t) &= E_{\Delta x}\left[\int_t^{t+\Delta t} L(x,u,\tau)\mathrm{d}\tau + \bar{J}^*(x+\Delta x, t+\Delta t)\right] \\ &= \int_t^{t+\Delta t} L(x,u,\tau)\mathrm{d}\tau + E_{\Delta x}[\bar{J}^*(x+\Delta x, t+\Delta t)]\end{aligned} \tag{9.68}$$

Principle 9.1 therefore yields

$$\begin{aligned}\bar{J}^*(x,t) = \min_{\substack{u(\tau)\\ t\le\tau\le t+\Delta t}} &\left[\int_t^{t+\Delta t} L(x,u,\tau)\mathrm{d}\tau \right. \\ &\left. + \int \bar{J}^*(x+\Delta x, t+\Delta t) f(\Delta x/x,t)\, d(\Delta x)\right]\end{aligned} \tag{9.69}$$

with $f(\Delta x/x,t)$ given by Equation 9.63. This is the continuous stochastic principle of optimality assuming complete state information.

To find a method for determining analytic expressions for the optimal conditional cost $\bar{J}^*(x,t)$ and control $u^*(t)$, let us derive a partial differential equation satisfied by $\bar{J}^*(x,t)$. Take a first-order approximation to the first integral in Equation 9.69, and perform a Taylor series expansion of $\bar{J}^*(x+\Delta x, t+\Delta t)$ about (x,t). Then (recall $x \in R$)

$$\begin{aligned}\bar{J}^*(x,t) = \min_{\substack{u(\tau)\\ t\le\tau\le t+\Delta t}} &\left[L(x,u,\tau)\Delta t + \int \bar{J}^*(x,t) f(\Delta x/x,t)\mathrm{d}(\Delta x) \right. \\ &+ \int \bar{J}_x^* \Delta x f(\Delta x/x,t)\mathrm{d}(\Delta x) + \int \bar{J}_t^* \Delta t f(\Delta x/x,t)\mathrm{d}(\Delta x) \\ &+ \frac{1}{2}\int \bar{J}_{xx}^* (\Delta x)^2 f(\Delta x/x,t)\mathrm{d}(\Delta x) + \frac{1}{2}\int \bar{J}_{tt}^* (\Delta t)^2 f(\Delta x/x,t)\mathrm{d}(\Delta x) \\ &\left. + \int \bar{J}_{xt}^* \Delta x \Delta t f(\Delta x/x,t)\mathrm{d}(\Delta x) + \cdots\right]\end{aligned} \tag{9.70}$$

Now observe that $\Delta t, \bar{J}^*$, and the partials $\bar{J}_x^*, \bar{J}_t^*, \bar{J}_{xx}^*, \bar{J}_{tt}^*, \bar{J}_{xt}^*$ can be removed from the integrals, and that the remaining mathematical expectations are either the area under the PDF (i.e., equal to 1) or the moments of $\Delta x \sim N(a\Delta t, \sigma^2 \Delta t)$. Therefore,

$$\bar{J}^*(x,t) = \min_{\substack{u(\tau)\\ t\leq\tau\leq t+\Delta t}} \Big[L(x,u,\tau)\Delta t + \bar{J}^*(x,t) + \bar{J}_x^* a\Delta t + \bar{J}_t^*\Delta t + \frac{1}{2}\bar{J}_{xx}^*\sigma^2\Delta t + \frac{1}{2}\bar{J}_{tt}^*(\Delta t)^2 + \bar{J}_{xt}^*\Delta t(a\Delta t) + \cdots \Big] \tag{9.71}$$

(To first order approximation, $E(\Delta x^2) = \sigma^2\Delta t$.) Removing from under the minimization the quantities $\bar{J}^*(x,t)$ and $\bar{J}_t^*\Delta t$, which are independent of $u(\tau)$ for $t \leq \tau \leq t + \Delta t$, and neglecting second-order terms in Δt yields

$$0 = \bar{J}_t^*\Delta t + \min_{\substack{u(\tau)\\ t\leq\tau\leq t+\Delta t}} \left[L(x,u,\tau)\Delta t + \bar{J}_x^* a\Delta t + \frac{1}{2}\bar{J}_{xx}^*\sigma^2\Delta t \right] \tag{9.72}$$

Finally, we divide by Δt and let it become small to get

$$-\frac{\partial \bar{J}^*}{\partial t} = \min_{u(t)} \left[L + \frac{\partial \bar{J}^*}{\partial x} a + \frac{1}{2}\frac{\partial^2 \bar{J}^*}{\partial x^2}\sigma^2 \right] \tag{9.73}$$

This is the stochastic *Hamilton–Jacobi–Bellman* (HJB) *equation* for the case of completely known state $x(t)$. In the vector case

$$\dot{x} = a(x,u,t) + Gw \tag{9.74}$$

with $x \in R^n$ and white process noise $w(t) \sim (0, Q')$, it generalizes to

$$-\frac{\partial \bar{J}^*}{\partial t} = \min_{u(t)} \left[L + \left(\frac{\partial \bar{J}^*}{\partial x}\right)^T a + \frac{1}{2}\text{trace}\left(\frac{\partial^2 \bar{J}^*}{\partial x^2} GQ'G^T\right) \right] \tag{9.75}$$

The stochastic HJB equation is solved backward for the optimal conditional cost with a boundary condition of

$$\bar{J}^*[x(T),T] = \phi[x(T),T] \text{ on the hypersurface } \psi[x(T),T] = 0 \tag{9.76}$$

Then $j^*(t)$ is given using Equation 9.61 for general t.

Once $u(t)$ has been eliminated, the stochastic HJB equation becomes a *Kolmogorov equation*. In principle, it can be solved for analytic expressions for the optimal conditional cost and control, yielding a state feedback control law. In practice, the HJB equation has been solved for only a few cases. Fortunately, these cases are quite important, as the next example shows.

Example 9.3 HJB Derivation of Stochastic Linear Quadratic Regular

Let the linear stochastic plant

$$\dot{x} = Ax + Bu + Gw \tag{9.77}$$

with white process noise $w(t) \sim N(0, Q')$ and $x(t_0)$ random have the associated *quadratic performance index*

$$J[x(t_0), t_0] = \frac{1}{2}x^T(T)S(T)x(T) + \frac{1}{2}\int_{t_0}^{T}(x^T Qx + u^T Ru)\mathrm{d}t \tag{9.78}$$

with the final state $x(T)$ free. It is desired to minimize the expected cost

$$j(t_0) = E\{J[x(t_0), t_0]\} \tag{9.79}$$

if the final time T is fixed. The control $u(t)$ can depend on $x(t)$, which is assumed exactly known at time t. Assume $Q \geqq 0, R > 0$.

a. Solution of HJB equation

The stochastic HJB equation for the optimal conditional cost is

$$-\bar{J}_t^* = \min_{u(t)}\left[\frac{1}{2}x^T Qx + \frac{1}{2}u^T Ru + (\bar{J}_x^*)^T(Ax + Bu) + \frac{1}{2}\text{trace}(\bar{J}_{xx}^* GQ'G^T)\right] \tag{9.80}$$

Since the input is not constrained, the minimization can be performed by differentiating, so that

$$0 = Ru + B^T\bar{J}_x^* \tag{9.81}$$

and the optimal control is given in terms of $\bar{J}_x^*$ by

$$u = -R^{-1}B^T\frac{\partial\bar{J}^*}{\partial x} \tag{9.82}$$

Using Equation 9.82 to eliminate $u(t)$ in Equation 9.80 yields the Kolmogorov equation

$$-\bar{J}_t^* = \frac{1}{2}x^T Qx - \frac{1}{2}(\bar{J}_x^*)^T BR^{-1}B^T\bar{J}_x^* + (\bar{J}_x^*)^T Ax + \frac{1}{2}\text{trace}(\bar{J}_{xx}^* GQ'G^T) \tag{9.83}$$

Let us initially make the naive assumption that $\bar{J}^*(x, t)$ has the same form as it does for the deterministic case (Bryson and Ho 1975, Lewis 1986). That is, assume

$$\bar{J}^*(x, t) = \frac{1}{2}x^T Sx \tag{9.84}$$

for some as yet unknown deterministic symmetric matrix function $S(t)$. Then

$$\bar{J}_x^* = Sx, \quad \bar{J}_t^* = \frac{1}{2}x^T\dot{S}x, \quad \bar{J}_{xx}^* = S \tag{9.85}$$

Using these in Equation 9.83 results in

$$-\frac{1}{2}x^T\dot{S}x = \frac{1}{2}x^T Qx + x^T SAx - \frac{1}{2}x^T SBR^{-1}B^T Sx + \frac{1}{2}\text{trace}(SGQ'G^T) \tag{9.86}$$

$$0 = \frac{1}{2}x^T(\dot{S} + Q + A^T S + SA - SBR^{-1}B^T S)x + \frac{1}{2}\text{trace}(SGQ'G^T) \tag{9.87}$$

where we have replaced $x^T SAx$ by the symmetric form $\frac{1}{2}x^T(A^T S + SA)x$, which has the same value (Appendix).

Unfortunately, Equation 9.87 has no solution for $S(t)$ which is independent of $x(t)$. Evidently the assumption (Equation 9.84) is not valid. $\bar{J}^*(x,t)$ must contain a term whose partial derivative yields a term like $\frac{1}{2}$ trace $(SGQ'G^T)$. Let us therefore assume that

$$\bar{J}^*(x,t) = \frac{1}{2}\left[x^T Sx + \text{trace}\int_t^T SGQ'G^T \mathrm{d}\tau\right] \tag{9.88}$$

Then $\bar{J}_x^*$ and $\bar{J}_{xx}^*$ are as in Equation 9.85, but now

$$\bar{J}_t^* = \frac{1}{2}[x^T \dot{S}x - \text{trace}(SGQ'G^T)] \tag{9.89}$$

Using these new partials in the HJB Equation 9.83 results in

$$0 = \frac{1}{2}x^T(\dot{S} + A^T S + SA - SBR^{-1}B^T S + Q)x \tag{9.90}$$

since the trace terms cancel. Now we can select $S(t)$ as the solution of

$$-\dot{S} = A^T S + SA - SBR^{-1}B^T S + Q \tag{9.91}$$

with boundary condition (see Equation 9.76) of

$$S(T) \text{ given} \tag{9.92}$$

This is the same Riccati equation appearing in the deterministic LQ regulator! The optimal control can be computed as a state variable feedback, since Equations 9.82 and 9.85 yield

$$U = -R^{-1}B^T Sx \tag{9.93}$$

The optimal conditional cost (Equation 9.88) on $[t, T]$ is given in terms of the state $x(t)$ at time t. To eliminate the need for knowing $x(t)$, we can take expectations with respect to $x(t)$ to get the optimal expected cost

$$\begin{aligned} j^*(t) &= \frac{1}{2}E_x\left[x^T Sx + \text{trace}\int_t^T SGQ'G^T \mathrm{d}\tau\right] \\ &= \frac{1}{2}\left[S(t)X(t) + \text{trace}\int_t^T SGQ'G^T \mathrm{d}\tau\right] \end{aligned} \tag{9.94}$$

We have used trace $(AB) = \text{trace}(BA)$ and the fact that $j^*(t)$ is a scalar. We have also defined

$$X(t) = E[x(t)x^T(t)] \tag{9.95}$$

the *mean-square value of* $x(t)$ (i.e., correlation of $x(t)$). The optimal expected cost over $[t_0, T]$ is thus

$$j^*(t_0) = \frac{1}{2}\left[S(t_0)X(t_0) + \text{trace}\int_t^T SGQ'G^T \mathrm{d}\tau\right] \tag{9.96}$$

The effect of the process noise $w(t)$ is now clear: the solution to the stochastic optimal control problem, Equations 9.91 and 9.93, is exactly the same as in the deterministic case. The only difference is that the term

$$\frac{1}{2}\text{trace}\int_t^T SGQ'G^T \mathrm{d}\tau \tag{9.97}$$

must be added to the optimal cost. This represents the increase in cost due to the disturbing process noise with covariance Q'.

b. Average closed-loop behavior (Bryson and Ho 1975)

In the deterministic case, as $(Q - R)$ tends to infinity, the LQ regulator will drive the state to zero as $(T - t_0)$ becomes large. In the stochastic case, this is not true. Owing to the influence of the process noise $w(t)$, the state is continually disturbed from zero and at steady state the behavior of the state $x(t)$ is determined by a balance between the disturbing effect of $w(t)$ and the regulating effect of the optimal control. The net result is that the mean-square value $X(t)$ of the state does not go to zero at steady state.

To find the mean-square state $X(t)$, write the closed-loop system under the influence of the optimal control (Equation 9.93) as

$$\dot{x} = (A - BK)x + Gw \tag{9.98}$$

where $w(t) \sim (0, Q')$ and

$$K(t) = R^{-1}B^T S(t) \tag{9.99}$$

is the Kalman (control) gain. This is a stochastic system, and so from Example 3.1 the mean-square state is given by the Lyapunov equation

$$\dot{X} = (A - BK)X + X(A - BK)^T + GQ'G^T \tag{9.100}$$

with initial condition

$$X(t_0) = E[x(t_0)x^T(t_0)] = P_0 + \bar{x}_0\bar{x}_0^T \tag{9.101}$$

with $x_0 \sim (\bar{x}_0, P_0)$.

Note that the Riccati Equation (9.91) is solved backward in time. Since it is independent of $x(t)$ it can be solved off-time. Then $K(t)$ is found using Equation 9.99. Next, $X(t)$ can be determined using Equation 9.100 by integrating *forward* in time given the initial value $X(t_0)$. Finally, the optimal expected cost is given by Equation 9.96. We can therefore determine both the expected cost and the mean-square value of the state *before* we actually apply the optimal control to the plant. Thus the performance of the control can be evaluated before we build the controller.

We can also determine the mean-square value of the control before it is applied, for by Equations 9.93 and 9.99

$$E(uu^T) = E(Kxx^TK^T) = KX(t)K^T \tag{9.102}$$

At steady state, the mean-square value of $x(t)$ is found by solving the algebraic Lyapunov equation

$$0 = (A - BK)X + X(A - BK)^T + GQ'G^T \tag{9.103}$$

c. Non-Gaussian process noise

The regulator (Equations 9.91 and 9.93) is the optimal solution of the control problem Equations 9.77 through 9.79 if $w(t)$ is not normal, then it provides the best control $u(t)$ which is a linear function of the state $x(t)$. See Kwakernaak and Sivan (1972) and Åstrom (1970). ▯

Example 9.4 Optimal Feedback Control and Simulation of a Scalar Stochastic System

To show how to apply the results of the previous example and how to simulate stochastic control systems on a computer, let us consider the scalar plant

$$\dot{x} = ax + bu + gw \tag{9.104}$$

with white Gaussian process noise $w(t) \sim N(0, q')$ and $x(0) \sim N(\bar{x}_0, p_0)$. The performance index is

$$J = \frac{1}{2}s(T)x^2(T) + \frac{1}{2}\int_0^T (qx^2 + ru^2)\mathrm{d}t \tag{9.105}$$

It is desired to find a control to minimize the expected cost

$$j(0) = E(J) \tag{9.106}$$

and to find the resulting mean-square value of the state. The state $x(t)$ is assumed exactly measurable at each time t.

The LQ regulator is

$$u(t) = -K(t)x(t) \tag{9.107}$$

with Kalman gain

$$K(t) = s(t)b/r \tag{9.108}$$

where

$$-\dot{s} = 2as - b^2s^2/r + q, \quad t \le T \tag{9.109}$$

with $s(T)$ given. This could easily be solved by separation of variables as in Example 3.3.

The mean-square state, $X(t) = E[x^2(t)]$, resulting when this control is applied is given by the Lyapunov equation

$$\dot{X} = 2(a - bK)X + g^2 q', \quad t \geq 0 \tag{9.110}$$

with $X(0) = p_0 + \bar{x}_0 \bar{x}_0^T$. The optimal expected cost on $[0, T]$ is

$$j^*(0) = \frac{1}{2}\left[s(0)X(0) + g^2 q' \int_0^T s(\tau)d\tau \right] \tag{9.111}$$

Implementation of the stochastic LQ regulator is very easy. First the Riccati equation 9.82 is integrated backward to find $s(t)$, and the gain $K(t)$ is found and stored. We actually implement

$$\dot{s} = 2as - b^2 s^2/r + q, \quad t \leq 0 \tag{9.112}$$

with $s(0)$ given, and then reverse the resulting function $s(t)$ in time. (Note if $\tau = T - t$ then $\mathrm{d}/\mathrm{d}\tau = -\mathrm{d}/\mathrm{d}t$.) ▯

The Lyapunov equation 9.83 must be integrated forward in time. Since the simulation using control (Equation 9.80) on the plant also runs forward in time, we can integrate Equations 9.77 and 9.83 simultaneously. Implement the plant dynamics (Equation 9.77) and also Equation 9.83 in time update function, where the vector in the argument X is integrated forward by the Runge–Kutta routine (ODE23 in our case). To do this, assign the components of argument X as $X(1) = x(t)$ the state, and $X(2) = X(t)$ the mean-square value of the state. This subroutine is also shown in Figure 9.4.

A word on generating the normal process $w(t) \sim N(0, q')$ is in order. MATLAB's routine RANDN( ) generates a random variable z that is uniformly distributed on $[0, 1]$. To convert this to a normally distributed RV, we use the line of code in Figure 9.4 that defines a function of z that is normal with mean of zero and variance of q'.

The results of a computer simulation using this software are shown in Figure 9.5. The final state weighting was $s(T) = 0$, and the control weighting was $r = 1$. Note that as state weighting q increases, the steady-state values of the mean and mean-square of the state $x(t)$ both decrease. The Runge–Kutta step size was 10 ms.

9.2 Continuous-Time Linear Quadratic Gaussian Problem

In this section we discuss the linear stochastic plant

$$\dot{x} = Ax + Bu + Gw \tag{9.113}$$

```
% Simulation of stochastic control system
% To run the program pass the process noise covariance q=1,10,100,e.g.
f6p1d4(10)
function f6p1d4(q1)
    global u i gnoise q;
    q=q1;
    b=1; r=1; i=1; s=0; x1=[10 10]; % initial values
    gnoise=randn(102,1); % white gaussian noise
    options = odeset('RelTol',1e-6,'AbsTol',[1e-61e-6]);
    for t=0:0.1:10
        [td s]=ode23(@sfunc,[t t+0.1],s);
        s=s(end);
        % Optimal Feedback gains & control input
        k=b*s/r; u(1)=-k*x1(end,1); u(2)=k;
        [td x]=ode23(@xfunc,[t t+0.1],x1(end,:),options);
        x1(i,:)=x(end,:); t1(i)=td(end);
        i=i+1;
    end
    % Plotting
    plot(t1,x1(:,1),'k-',t1,x1(:,2),'k-.'); grid on;
    legend('x(t)','X(t)'); title(strcat(' q = ',num2str(q)));
    xlabel('Time (secs)'); axis([0 10 -5 12]);
return

% Function for Ricatti equation solution
function ds=sfunc(t,s)
    global q;
    a=0.05; b=1; r=1;
    ds(1)=2*a*s(1)-b^2*(s(1)^2)/r+q;
return

% Function for plant dynamics and Lyapunov equation
function dx=xfunc(t,x)
    global u i gnoise;
    a=0.05; b=1; qp=5; g=1;
    dx(1)=a*x(1)+b*u(1)+g*qp*gnoise(i,1);
    dx(2)=2*(a-b*u(2))*x(2)+(g^2)*qp;
    dx=dx';
return
```

FIGURE 9.4
Subroutines for use with CTLQR for stochastic LQ regulator.

with white process noise $w(t) \sim N(0, Q')$ and $x(t_0) \sim N(\bar{x}_0, P_0)$ a random variable. The associated performance index is the quadratic form

$$J[x(t_0), t_0] = \frac{1}{2}x^T(T)S(T)x(T) + \frac{1}{2}\int_{t_0}^{T} (x^T Qx + u^T Ru)dt \tag{9.114}$$

with symmetric weighting matrices $S(T) \geq 0, Q \geq 0, R > 0$. The plant and weighting matrices can be functions of time. It is desired to determine the control $u^*(t)$ on $[t_0, T]$ which minimizes the expected cost

$$j(t_0) = E\{J[x(t_0), t_0]\} \tag{9.115}$$

with $x(T)$ free and T fixed.

This is called the *linear quadratic Gaussian* (LQG) *control problem.*

Two cases must be considered. In the case of *complete state information*, the state vector $x(t)$ is exactly known at time t, so $u(t)$ can be expressed as a

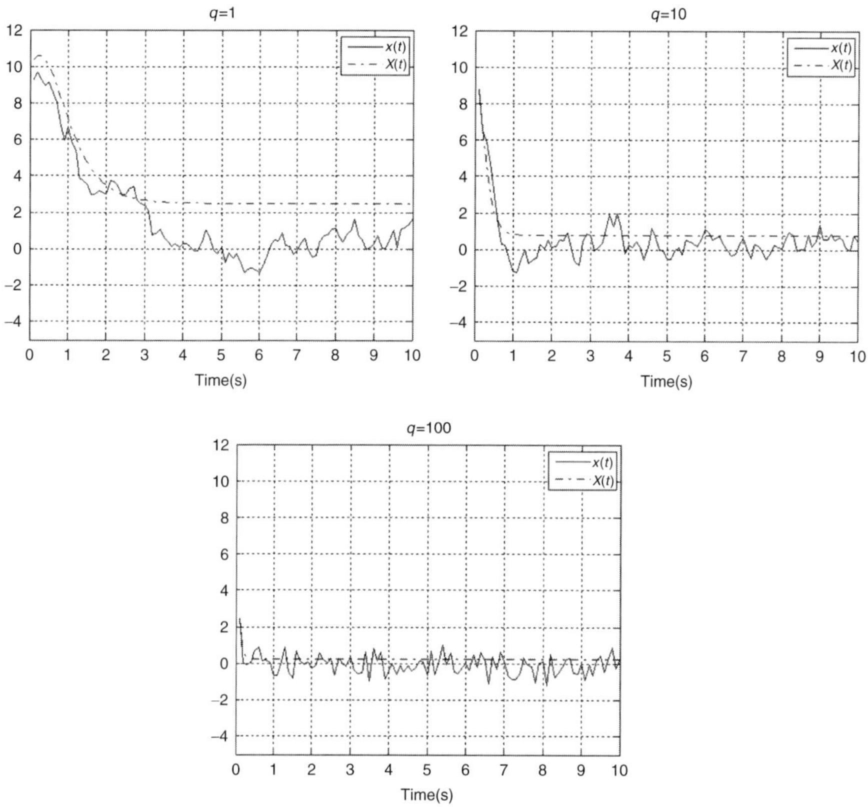

FIGURE 9.5
State and mean-square value of state for stochastic LQ regulator.

function of $x(t)$; that is, as a state variable feedback. In the case of *incomplete state information*, the state is measured in the presence of Gaussian noise to produce the observation, or measurement, vector

$$Z = Hx + v \tag{9.116}$$

with measurement noise $v(t) \sim N(0, R')$. Now the control $u(t)$ can only depend on the *estimate* $\hat{x}(t)$ of the state given the measurements.

The results we are about to derive represent the optimal control if the noise(s) are Gaussian. In the situation where the noise statistics are not Gaussian, they yield the best *linear* control law (Åström 1970, Kwakernaak and Sivan 1972).

9.2.1 Complete State Information

Here we rederive the results of Example 9.3 by a different approach. Suppose that the state $x(t)$ of Equation 9.113 is exactly measurable at time t. Then admissible feedback control laws are of the form

$$u(t) = K[x(t), t] \tag{9.117}$$

where K is a general nonlinear function. Admissible *linear* feedback control laws have the form

$$u(t) = K(t)x(t) \tag{9.118}$$

We shall see that the optimal control is the linear state variable feedback (Equation 9.118) if all noises are normal.

As a preliminary step we would like to find an expression for $(\mathrm{d}/\mathrm{d}t)E(x^T Sx)$, where $S(t)$ is a symmetric function of time and $x(t)$ described by Equation 9.113 is stochastic with a PDF of $f_x(x)$. Equation 5.44 provides an expression for $(\mathrm{d}/\mathrm{d}t)E\{\phi[x(t)]\}$, where $\phi[x(t)]$ is an arbitrary twice-differentiable scalar function of $x(t)$. The quadratic form $x^T(t)S(t)x(t)$ is of the form $\phi[x(t), t]$. Carrying out a derivation like the one leading to Equation 5.44 for the case where ϕ is an explicit function of x *and* of t, we obtain instead of Equation 5.44:

$$\begin{aligned}\frac{\mathrm{d}}{\mathrm{d}t}E[\phi(x,t)] &= \frac{\mathrm{d}}{\mathrm{d}t}\int \phi(x,t) f_x(x)\mathrm{d}x \\ &= \int \left[\frac{\partial\phi(x,t)}{\partial t} f_x(x) + \phi(x,t)\frac{\partial f_x(x)}{\partial t}\right]\mathrm{d}x \\ &= E\left[\frac{\partial\phi}{\partial t}\right] + \int \phi(x,t)\frac{\partial f_x(x)}{\partial t}\mathrm{d}x \end{aligned} \tag{9.119}$$

Continuing the derivation, the final result is clearly (note that $a(x, t) = Ax + Bu$)

$$\frac{\mathrm{d}}{\mathrm{d}t}E[\phi(x,t)] = E(\phi_t) + E[\phi_x^T(Ax + Bu)] + \frac{1}{2}\mathrm{trace}E(GQ'G^T\phi_{xx}) \tag{9.120}$$

Substituting $\phi(x, t) = x^T Sx$, we obtain

$$\begin{aligned}\frac{\mathrm{d}}{\mathrm{d}t}E(x^T Sx) &= E(x^T\dot{S}x) + 2E[x^T S(Ax + Bu)] + \mathrm{trace}E(GQ'G^T S) \\ &= E(x^T\dot{S}x) + E[x^T S(Ax + Bu) + (Ax + Bu)^T Sx] \\ &\quad + \mathrm{trace}(SGQ'G^T)\end{aligned} \tag{9.121}$$

where we have used a symmetric form for the scalar $x^T S(Ax + Bu)$ (Appendix). This holds for any symmetric matrix $S(t)$.

Now, note the identity

$$\frac{1}{2}\int_{t_0}^{T}\frac{\mathrm{d}}{\mathrm{d}t}E(x^T Sx)\mathrm{d}t = \frac{1}{2}E[x^T(T)S(T)x(T)] - \frac{1}{2}E[x^T(t_0)S(t_0)x(t_0)] \tag{9.122}$$

Using these two preliminary results, we can solve the LQG control problem with complete state information. Add zero in the form of the left side of Equation 9.122 minus its right side to Equation 9.114, and then use Equation 9.121

to obtain for the expected cost

$$j(t_0) = \frac{1}{2}E[x^T(t_0)S(t_0)x(t_0)] + \frac{1}{2}E\int_{t_0}^{T}[x^T(\dot{S} + A^TS + SA + Q)x + x^TSBu + u^TB^TSx + u^TRu]\mathrm{d}t + \frac{1}{2}\mathrm{trace}\int_{t_0}^{T} SGQ'G^T\mathrm{d}t \quad (9.123)$$

Suppose that $S(t)$ is selected to satisfy the Riccati equation

$$-\dot{S} = A^TS + SA - SBR^{-1}B^TS + Q \quad (9.124)$$

for $t \le T$. Then Equation 9.123 becomes

$$j(t_0) = \frac{1}{2}E[x^T(t_0)S(t_0)x(t_0)] + \frac{1}{2}E\int_{t_0}^{T}[x^T\dot{S}BR^{-1}B^TSx + x^TSBu + u^TB^TSx + u^TRu]\mathrm{d}t + \frac{1}{2}\mathrm{trace}\int_{t_0}^{T} SGQ'G^T\mathrm{d}t \quad (9.125)$$

and the first integrand is now the perfect square of a weighted norm (McReynolds 1966), so that

$$j(t_0) = \frac{1}{2}E[x^T(t_0)S(t_0)x(t_0)] + \frac{1}{2}E\int_{t_0}^{T}\|R^{-1}B^TSx + u\|_R^2\mathrm{d}t + \frac{1}{2}\mathrm{trace}\int_{t_0}^{t} SGQ'G^T\mathrm{d}t \quad (9.126)$$

Note that the Riccati equation arises quite naturally in completing the square in the first integrand. Since only the second term depends on $u(t)$, the minimum value of $j(t_0)$ is

$$j^*(t_0) = \frac{1}{2}E[x^T(t_0)S(t_0)x(t_0)] + \frac{1}{2}\mathrm{trace}\int_{t_0}^{T} SGQ'G^T\mathrm{d}t \quad (9.127)$$

and it is achieved for the optimal control

$$u^*(t) = -R^{-1}B^TS(t)x(t) \quad (9.128)$$

where $S(t)$ satisfies Equation 9.124. The boundary condition for the Riccati equation is $S(T)$, the final weighting matrix in $J(t_0)$, since then $j^*(T) = j(T)$.

We have just discovered that the solution to the stochastic LQG control problem with complete state information is *identical to the deterministic* LQ *regulator* (Bryson and Ho 1975, Kwakernaak and Sivan 1972, Lewis 1986), except that a term is added to $j^*(t_0)$ to account for the cost increase due to the process noise with covariance Q'. This is the same result obtained by dynamic programming in Example 9.3. For computational purposes, note that the trace and integration operators commute.

In terms of the Kalman gain

$$K(t) = R^{-1}B^T S(t) \tag{9.129}$$

the optimal control can be written as

$$u(t) = -K(t)x(t) \tag{9.130}$$

which is a linear time-varying state variable feedback.

It would be useful to know the answer to another question. Suppose that the control input is selected as the suboptimal feedback (Equation 9.130), where $K(t)$ is any given state feedback gain, not necessarily the optimal gain (Equation 9.129). Then what is the resulting value of the expected cost $j(t_0)$? The motivation for our interest is that it is often convenient to use a suboptimal feedback of simplified form; for example, the constant steady-state feedback

$$K_\infty = R^{-1}B^T S^-(\infty) \tag{9.131}$$

where $S^-(\infty)$ is the limiting Riccati solution. We might want to determine the resulting $j(t_0)$ to see if the closed-loop plant behavior using this simplified control law is acceptable.

To answer this question, use (Equation 9.130) in (Equation 9.123) to obtain

$$\begin{aligned} j(t_0) &= \frac{1}{2}E[x^T(t_0)S(t_0)x(t_0)] \\ &+ \frac{1}{2}E\int_{t_0}^{T} x^T[\dot{S} + (A-BK)^T S + S(A-BK) + Q + K^T RK]x\,\mathrm{dt} \\ &+ \frac{1}{2}\mathrm{trace}\int_{t_0}^{T} SGQ'G^T\mathrm{dt} \end{aligned} \tag{9.132}$$

Now select $S(t)$ to satisfy the Lyapunov equation

$$-\dot{S} = (A-BK)^T S + S(A-BK) + Q + K^T RK \tag{9.133}$$

for $t \geq T$, which is expressed in terms of the known plant matrix of the closed-loop system

$$\dot{x} = (A-BK)x + Gw \tag{9.134}$$

Then the expected cost resulting from any feedback control of the form $u = -Kx$ is equal to

$$j(t_0) = \frac{1}{2}E[x^T(t_0)S(t_0)x(t_0)] + \frac{1}{2}\mathrm{trace}\int_{t_0}^{T} SGQ'G^T\mathrm{dt} \tag{9.135}$$

where $S(t)$ satisfies Equation 9.133. The boundary condition for the Lyapunov equation is $S(T)$.

If $K(t)$ is selected as the optimal feedback gain (Equation 9.129), then Equation 9.133 becomes the Joseph-stabilized Riccati equation.

To determine how successful the control is at regulating the state in the presence of process noise, we can examine the mean-square value of the state

$$X(t) = E[x(t)x^T(t)] \tag{9.136}$$

Since Equation 9.134 is a stochastic system (Example 3.1), we have

$$\dot{X} = (A - BK)X + X(A - BK)^T + GQ'G^T \tag{9.137}$$

for $t \geq t_0$. $X(t_0) = P_0 + \bar{x}_0\bar{x}_0^T$ is given. If $X(t)$ is small, then the regulating action of the control is satisfactory.

It is worth comparing the two Lyapunov equations (Equations 9.133 and 9.137). In terms of the mean-square value of the initial state, the optimal expected cost can be written as

$$j(t_0) = \frac{1}{2}\text{trace}\left[s(t_0)X(t_0) + \int_{t_0}^{T} SGQ'G^T \,\mathrm{d}t\right] \tag{9.138}$$

According to Equation 9.130, the mean-square value of the control is given by

$$E[u(t)u^T(t)] = K(t)X(t)K^T(t) \tag{9.139}$$

It is useful in predicting the control effort required by the LQG regulator. An illustration of the software simulation of the LQG regulator is provided by Example 9.4.

9.2.2 Incomplete State Information and the Separation Principle

Now let us assume that the entire state is not available. Instead, there are available measurements $z(t)$ given by Equation 9.116. Then the optimal estimate of the state for *any* input $u(t)$ is given by the Kalman filter equations from Table 3.1:

$$\dot{P} = AP + PA^T + GQ'G^T - PH^T(R')^{-1}HP, \quad t \geq t_0, \quad P(t_0) = P_0 \tag{9.140}$$

$$L = PH^T(R')^{-1} \tag{9.141}$$

$$\dot{\hat{x}} = A\hat{x} + Bu + L(z - H\hat{x}), \quad \hat{x}(0) = \bar{x}_0 \tag{9.142}$$

where $P(t)$ is the error covariance, $L(t)$ the Kalman filter gain, and $\hat{x}(t)$ the state estimate. Recall that $\hat{x}(t)$ and $P(t)$ are also the conditional mean and covariance of $x(t)$ given the control input and measurements prior to time t (Section 1.1).

The admissible controls are now functions of the *estimate* $\hat{x}(t)$, and not the unknown state $x(t)$. To find the optimal control under these circumstances, we return to Equation 9.126.

Only the second term in Equation 9.142 depends on the input $u(t)$. Defining the set of data available at time t as

$$Z_t = \{z(\tau)\,|\,t_0 \leq \tau \leq t\} \tag{9.143}$$

we can write this term as

$$\frac{1}{2}E\int_{t_0}^{T} \|R^{-1}B^TSx + u\|_R^2 \mathrm{d}t = \frac{1}{2}E_{Z_t}\int_{t_0}^{T} E[(u + Kx)^T R(u + Kx)/Z_t]\mathrm{d}t \tag{9.144}$$

with the Kalman control gain defined in Equation 9.129 and E_{Z_t} denoting expectation over Z_t.

To proceed, we require a preliminary result (Åström 1970). Suppose that an arbitrary random variable x has mean $\bar{x}$ and covariance P_x. Then we can write, for a symmetric deterministic $R \geq 0$,

$$E(x^T Rx) = E(x - \bar{x})^T R(x - \bar{x}) + \bar{x}^T R\bar{x} = \bar{x}^T R\bar{x} + \mathrm{trace}E[R(x - \bar{x})(x - \bar{x})^T]$$

or

$$E(x^T Rx) = \bar{x}^T R\bar{x} + \mathrm{trace}(RP_x) \tag{9.145}$$

This expression also holds if $\bar{x}$ and P_x are a conditional mean and covariance, and E is a conditional expectation. Using this identity, we can write Equation 9.144 as

$$\begin{aligned}\frac{1}{2}E\int_{t_0}^{T} \|R^{-1}B^TSx + u\|_R^2 \mathrm{d}t &= \frac{1}{2}E_{Z_t}\int_{t_0}^{T} (u + K\hat{x})^T R(u + K\hat{x})\mathrm{d}t \\ &\quad + \frac{1}{2}\mathrm{trace}\int_{t_0}^{T} K^T RKP\mathrm{d}t\end{aligned} \tag{9.146}$$

Since minimization with respect to $u(t)$ and the operator E_{Z_t} commute, the optimal control is

$$u(t) = -K(t)\hat{x}(t) \tag{9.147}$$

Using the minimum value of Equation 9.146 in Equation 9.126 yields the optimal expected cost

$$\begin{aligned}j^*(t_0) = \frac{1}{2}E[x^T(t_0)S(t_0)x(t_0)] + \frac{1}{2}\mathrm{trace}\int_{t_0}^{T} SGQ'G^T\mathrm{d}t \\ + \frac{1}{2}\mathrm{trace}\int_{t_0}^{T} K^T RKP\mathrm{d}t\end{aligned} \tag{9.148}$$

(For computational purposes, trace and integration may be interchanged.)

This is quite an important result known as the *separation principle*. (It is also called the *certainty equivalence principle*; however, in adaptive control theory this term means something a little bit different.) It can be stated as follows. The solution to the *stochastic* LQG control problem with incomplete state information is to use the optimal *deterministic* LQ regulator, except that the control is a feedback (Equation 9.147) of the optimal state *estimate* which is constructed using the Kalman filter in Table 3.1. The resulting optimal expected cost is given by Equation 9.148, where the first term comes from the deterministic LQ regulator, the second term represents the increase in cost due to the disturbing influence on $x(t)$ of the process noise, and the last term is the increase in cost due to measurement uncertainties.

The importance of the separation principle is that the LQG regulator design procedure can be accomplished in two separate stages: the Kalman filter design and the control feedback design. This means that all results derived separately for the deterministic optimal control problem and the optimal estimation problem are still valid. For instance, if the process and measurement noises are correlated, we need only use the modified filter gain in Section 3.9. If the noise statistics are not Gaussian, then the separation principle yields the best *linear* regulator in terms of the best *linear* estimate of the state.

The form of the LQG regulator is shown in Figure 9.6, where a reference or command input $r(t)$ has been included and $s(t)$ is an intermediate signal to be used later. The main regulator equations are

$$\dot{\hat{x}} = (A - LH)\hat{x} + Bu + Lz \tag{9.149}$$

$$u = -K\hat{x} + r \tag{9.150}$$

which describe a *dynamic output feedback*. That is, the plant output $z(t)$ and the control input $u(t)$ are processed by a dynamical system (Equation 9.152), whose output $K(t)\hat{x}(t)$ is returned as the control. The function of the added dynamics in such a scheme is to reconstruct information about the state that is not directly measurable at the output. In our present application, the state of the dynamic output feedback system is actually an optimal estimate for the state of the plant. This output feedback should be contrasted with the state feedback we used in the case of complete state information.

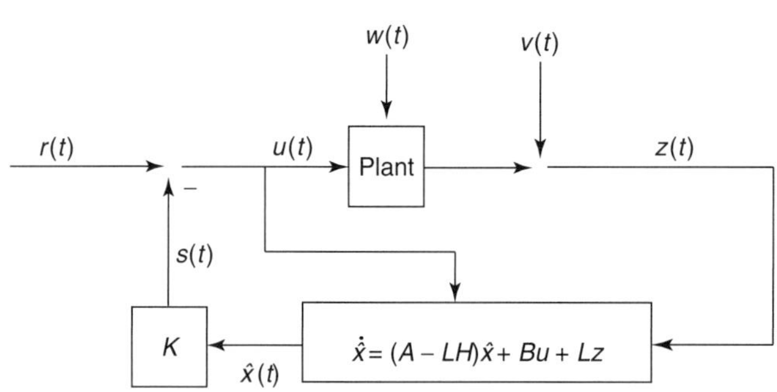

FIGURE 9.6
LQG regulator with incomplete state information.

Since we are considering a stochastic plant, we would like some feel for how small the regulator keeps the state. Thus, it would be useful to determine the mean-square value of the state $X(t) = E[x(t)x^T(t)]$ under the influence in our LQG regulator. See Bryson and Ho (1975).

The dynamics of the estimation error

$$\tilde{x}(t) = x(t) - \hat{x}(t) \tag{9.151}$$

are given by (Section 3.4)

$$\dot{\tilde{x}} = (A - LH)\tilde{x} + Gw - Lv \tag{9.152}$$

Note that $\tilde{x}(t)$ is independent of the control, which is one of the reasons the separation principle works. (The other is our ability to express Equation 9.146 in terms of $\hat{x}(t)$.) Since Equation 9.152 is a stochastic system, we can use Example 3.1 to see that the error covariance satisfies

$$\dot{P} = (A - LH)P + P(A - LH)^T + GQ'G^T + LR'L^T, \quad t \geq t_0 \tag{9.153}$$

which is a Joseph-stabilized version of Equation 9.140.

Now we turn to Equation 9.142. Using the control (Equation 9.147) and introducing the residual

$$\tilde{z}(t) = H\tilde{x}(t) + v(t) \tag{9.154}$$

we can write the estimate dynamics as

$$\dot{\hat{x}} = (A - BK)\hat{x} + L\tilde{z} \tag{9.155}$$

Since $\tilde{z}(t)$ is a white noise process with mean zero and covariance R' which is independent of $\hat{x}$ (Section 3.4), the mean-square value of the *estimate*

$$\hat{X}(t) \triangleq E[\hat{x}(t)\hat{x}^T(t)] \tag{9.156}$$

propagates according to

$$\dot{\hat{X}} = (A - BK)\hat{X} + \hat{X}(A - BK)^T + LR'L^T \tag{9.157}$$

What we are interested in determining here is $X(t)$. However,

$$X(t) = E[xx^T] = E[(\hat{x} + \tilde{x})(\hat{x} + \tilde{x})^T]$$

so that

$$X(t) = \hat{X}(t) + P(t) \tag{9.158}$$

where we have used the fact that $E(\tilde{x}) = 0$. Therefore, the mean-square value of the state $X(t)$ is the sum of the mean-square value of the estimate $\hat{X}(t)$ and the error covariance $P(t)$, which are found using Equations 9.157 and 9.153.

Since the initial mean $\bar{x}_0$ and covariance P_0 of the state are known, we have

$$X(t_0) = P_0 + \bar{x}_0\bar{x}_0^T \tag{9.159}$$

Therefore, by Equation 9.158,

$$\hat{X}(t_0) = X(t_0) - P_0 = x_0x_0^T \tag{9.160}$$

is the initial condition for Equation 9.157.

In terms of the mean-square value of the initial state, the optimal expected cost can be written as

$$j(t_0) = \frac{1}{2}\text{trace } S(t_0)X(t_0) + \frac{1}{2}\text{trace} \int_{t_0}^{T} SGQ'G^T \text{d}t + \frac{1}{2}\text{trace} \int_{t_0}^{T} K^T RKP\text{d}t \tag{9.161}$$

The mean-square value of the control is useful to know so that we can predict the control effort required. According to Equation 9.147, it is given by

$$E[u(t)u^T(t)] = K(t)\hat{X}(t)K^T(t) \tag{9.162}$$

The software simulation of the LQG regulator with incomplete state information is a straightforward extension of the procedure in Example 9.4.

9.3 Discrete-Time Linear Quadratic Gaussian Problem

Our concern here is the discrete linear stochastic plant

$$x_{k+1} = Ax_k + Bu_k + Gw_k, \quad k \geqslant i \tag{9.163}$$

with white process noise $w_k \sim N(0, Q')$ and initial state $x_i \sim N(\bar{x}_i, P_i)$. The performance index over the time interval of interest $[i, N]$ is

$$J_i(x_i) = \frac{1}{2}x_N^T S_N x_N + \frac{1}{2}\sum_{k=i}^{N-1}\left(x_k^T Qx_k + u_k^T Ru_k\right) \tag{9.164}$$

with $S_N \geq 0, Q \geq 0, R > 0$. If the plant and weighting matrices are time-varying, the subscript k can be added to our results.

We want to find the control u_k^* on the time interval $[i, N]$ which minimizes the expected cost

$$j_i = E[J_i(x_i)] \tag{9.165}$$

with x_N free and final time N fixed. This is the *discrete* LQG problem.

The results for complete and incomplete state information are presented in turn. They represent the optimal controller if the noise is Gaussian, and for arbitrary noise statistics they give the best linear control law.

We do not derive the results. Derivations would be along the lines of the previous section, though somewhat more involved due to the more complicated form of the curvature matrix $(B^T S_{k+1} B + R)$ and residual covariance $(HP_k H^T + R')$ for discrete-time systems.

9.3.1 Complete State Information

If the state x_k is completely measurable at time k, then the LQG regulator is the same as the deterministic LQ regulator (Bryson and Ho 1975, Kwakernaak and Sivan 1972, Lewis 1986). Thus the optimal control is

$$u_k = -K_k x_k \tag{9.166}$$

where the Kalman control gain is

$$K_k = (B^T S_{k+1} B + R)^{-1} B^T S_{k+1} A \tag{9.167}$$

and S_k satisfies the Riccati equation

$$S_k = A^T [S_{k+1} - S_{k+1} B (B^T S_{k+1} B + R)^{-1} B^T S_{k+1}] A + Q, \quad k < N \tag{9.168}$$

with boundary condition of S_N.

The optimal expected cost is

$$j_i^* = \frac{1}{2} \text{trace} \left[S_i X_i + \sum_{k=i}^{N-1} S_{k+1} G Q' G^T \right] \tag{9.169}$$

where the mean-square state

$$X_k \triangleq E\left[x_k x_k^T\right] \tag{9.170}$$

has initial value of

$$X_i = P_i + \bar{x}_i \bar{x}_i^T \tag{9.171}$$

The second term in j_i^* is the cost increase due to the disturbing effect of the process noise.

A derivation of these results could be carried out using the discrete stochastic optimality principle (Equation 9.19). Alternatively, we could use an approach like that in Section 9.2.

To see how effectively this LQG controller regulates the state, we can examine the mean-square value of the state. To find an equation for X_k, write the closed-loop system

$$x_{k+1} = (A - BK) x_k + G w_k \tag{9.172}$$

Now note that x_k and w_k are orthogonal, post-multiply this equation by its transpose and take expectations to get

$$X_{k+1} = (A - BK)X_k(A - BK)^T + GQ'G^T \tag{9.173}$$

for $k > i$. The initial condition is given by Equation 9.171.

Using Equation 9.165, the mean-square value of the optimal control is

$$E\big(u_k u_k^T\big) = K_k X_k K_k^T \tag{9.174}$$

9.3.2 Incomplete State Information

If the state is not completely measurable at time k, but if noisy measurements are available of the form

$$z_k = Hx_k + v_k \tag{9.175}$$

with white measurement noise $v_k \sim N(0, R')$, then a dynamic output feedback must be used to control the plant.

Carrying out a derivation like the one in Section 9.2 shows that the LQG regulator with incomplete state information is composed of a state estimator like the one in Table 2.1, followed by a deterministic LQ regulator that feeds back the estimate $\hat{x}_k$ and not x_k. Thus the separation principle also holds for discrete systems. We gather the equations into one place here for convenience.

If $\hat{x}_k, P_k$ are the *a posteriori* estimate and error covariance, and $\hat{x}_k^-, P_k^-$ are the *a priori* quantities, then the optimal stochastic LQG regulator is given by

Dynamic Output Feedback

$$u_k = -K_k\hat{x}_k \tag{9.176}$$

$$\hat{x}_{k+1}^- = A\hat{x}_k + Bu_k, \quad \hat{x}_i = \bar{x}_i \tag{9.177}$$

$$\hat{x}_k = \hat{x}_k^- + L_k(z_k - H\hat{x}_k^-) \tag{9.178}$$

Computation of Gains

$$K_k = (B^T S_{k+1} B + R)^{-1} B^T S_{k+1} A \tag{9.179}$$

$$L_k = P_k^- H^T (HP_k^- H^T + R')^{-1} = P_k H^T (R')^{-1} \tag{9.180}$$

For $k < N$

$$S_k = A^T S_{k+1} A - K_k^T (B^T S_{k+1} B + R) K_k + Q, \quad S_N \text{ given} \tag{9.181}$$

For $k \geq i$

$$P_{k+1}^- = AP_k A^T + GQ'G^T, \quad P_i \text{ given} \tag{9.182}$$

$$P_k = P_k^- - L_k\big(HP_k^- H^T + R'\big)L_k^T \tag{9.183}$$

The control gain is K_k and the filter gain is L_k. (We have written the Riccati equations in this form because it will be convenient later.)

The optimal expected cost using this LQG regulator is

$$j_i^* = \frac{1}{2}\text{trace}\left[S_iX_i + \sum_{k=i}^{N-1} S_{k+1}GQ'G^T + \sum_{k=i}^{N-1} S_{k+1}BK_kP_kA^T\right] \tag{9.184}$$

where the third term is the cost increase due to measurement uncertainties.

If the plant, covariance, and cost-weighting matrices depend on time, the subscript k should be appended everywhere. To determine the success of the LQG regulator at keeping the state small in the presence of process noise and to see how much control effort will be required, we can proceed as follows (Bryson and Ho 1975).

Define the *a priori* and *a posteriori* estimation errors as

$$\tilde{x}_k^- = x_k - \hat{x}_k^- \tag{9.185}$$

$$\tilde{x}_k = x_k - \hat{x}_k \tag{9.186}$$

Then the mean-square value of the state is

$$X_k = P_k^- + \hat{X}_k^- \tag{9.187}$$

(We used Equation 9.185, where the estimate $\hat{x}_k^-$ and the estimation error $\tilde{x}_k^-$ are orthogonal.) P_k^- is given by Equation 9.182 but the mean-square value of the *a priori* estimate

$$\hat{X}_k^- \triangleq E[\hat{x}_k^-(\hat{x}_k^-)^T] \tag{9.188}$$

remains to be determined.

To find $\hat{X}_k^-$, use Equations 9.175 through 9.178 to write the dynamics of the *a priori* estimate (cf. Table 2.3) as

$$\begin{aligned}\hat{x}_{k+1}^- &= (A - BK_k)\hat{x}_k = (A - BK_k)[\hat{x}_k^- + L_k(z_k - H\hat{x}_k^-)]\\ &= (A - BK_k)[\hat{x}_k^- + L_kH\tilde{x}_k^- + L_kv_k]\end{aligned} \tag{9.189}$$

Now note that $\hat{x}_k^-, \tilde{x}_k^-$, and v_k are pairwise orthogonal; multiply Equation 9.189 on the right by its transpose, and take expectations to obtain

$$\hat{X}_{k+1}^- = (A - BK_k)\big(\hat{X}_k^- + L_kHP_k^-H^TL_k^T + L_kR'L_k^T\big)(A - BK_k)^T \tag{9.190}$$

Using Equation 9.183, there results the forward recursion

$$\hat{X}_{k+1}^- = (A - BK_k)(\hat{X}_k^- + P_k^- - P_k)(A - BK_k)^T \tag{9.191}$$

for the mean-square value of the *a priori* estimate.

We know the initial mean and covariance of the state, $\bar{x}_i$ and P_i. From Equations 9.171 and 9.187

$$\hat{X}_i^- = X_i - P_i^- = P_i - P_i^- + \bar{x}_i\bar{x}_i^T \tag{9.192}$$

Using this in Equation 9.191 yields

$$\hat{X}_{i+1}^{-} = (A - BK_i)\bar{x}_i\bar{x}_i^T \cdot (A - BK_i)^T \tag{9.193}$$

which is used to start (Equation 9.191).

Using $\hat{X}_k^-$ in Equation 9.187 we can determine X_k, which allows us to tell if the regulating action of the controller will be satisfactory. To find the mean-square value of the control input, write

$$u_k = -K_k\hat{x}_k = -K_k(\hat{x}_k^- + L_kH\tilde{x}_k^- + L_kv_k) \tag{9.194}$$

so that

$$E\big(u_ku_k^T\big) = K_k(\hat{X}_k^- + P_k^- - P_k)K_k^T \tag{9.195}$$

It is of some interest to compare the differencing between *a priori* and *a posteriori* error covariances in these equations with the Rauch–Tung–Striebel smoother in Section 2.8. For additional insight on the LQG regulator, see Åström (1970), Bryson and Ho (1975), and Kwakernaak and Sivan (1972).

Problems

Section 9.1

1. For the system in Example 9.2, select the control sequence to maximize the probability that $x_3 = 0$ for any initial state x_0.

2. Effect of Control Weighting on Field of Extremals. Redo Example 9.2 if

$$J_0 = (x_N - 1)^2 + \frac{1}{2}\sum_{k=0}^{N-1} ru_k^2$$

for $r = \frac{1}{4}$, 4, and 8 and compare the resulting fields of extremals.

3. Stochastic Bilinear System. Consider the scalar bilinear system

$$x_{k+1} = x_ku_k + u_k^2 + w_k$$

with cost index

$$J_0 = x_N^2 + \sum_{k=0}^{N-1} |x_ku_k|$$

where $N = 2$. The control is constrained to take on values of $u_k = \pm 1$, and the state x_k to take on values of $-1, 0, 1$, or 2. The process noise w_k has PDF

$$f_{w_k}(w) = \begin{cases} 1 & \text{with probability } \frac{1}{2} \\ 0 & \text{with probability } \frac{1}{4} \\ -1 & \text{with probability } \frac{1}{4} \end{cases}$$

(a) Find an optimal state feedback control law minimizing $E(J_0)$.

(b) Suppose the control of (a) is used. Find the conditional PDF of x_2 given that $x_0 = 1$.

(c) Suppose x_0 takes on each of the allowed state values with probability 1/2. Find the expected cost using the control in (a).

(d) Now find a control law to maximize the probability that x_2 is either 1 or 2.

4. Consider Newton's system

$$\ddot{x} = u + w$$

with process noise $w(t) \sim N(0, q')$ and $x(0) \sim N(\bar{x}_0, p_0)$. Select

$$J(0) = \frac{1}{2}x^2(T) + \frac{1}{2}\int_{t_0}^{T} ru^2 \mathrm{d}t$$

where $T = 2$.

(a) Write the HJB equation, eliminating $u(t)$.

(b) Assume that

$$\bar{J}^*(x, t) = \frac{1}{2}s_1(t)x^2(t) + s_2 x(t)\dot{x}(t) + \frac{1}{2}s_3(t)\dot{x}^2(t) + \frac{1}{2}\int_t^T q' s_3(\tau)\mathrm{d}\tau$$

for some functions s_1, s_2, s_3. Use the HJB equation to find coupled scalar equations for these s_i. Find boundary conditions for the equations. Express the optimal control as a linear state feedback in terms of the $s_i(t)$.

Section 9.2

5. The plant is

$$\dot{x} = u + w$$
$$z = x + v$$

with $w \sim (0, q')$, $v \sim (0, r')$, and performance index

$$J = \frac{1}{2}s(T)x^2 + \frac{1}{2}\int_0^T (qx^2 + ru^2)\mathrm{d}t$$

Determine analytic expressions for the optimal feedback gain and optimal cost for the cases of

(a) Complete state information.

(b) Incomplete state information. For this case, determine also the transfer function at steady from the output $z(t)$ to the control $u(t)$. Draw a block diagram of the optimal steady-state regulator.

(c) Write MATLAB code to simulate the controller in (b).

6. Polynomial Formulation of Optimal Regulator. Examine Figure 9.6. Assume a single-input single-output system.

(a) Find transfer functions from $z(t)$ to intermediate signal $s(t)$ and from $u(t)$ to $s(t)$.

(b) Show that the regulator can be put into the form in Figure 9.7, where $T(s)$ is the Kalman filter characteristic polynomial (i.e., find $R(s), S(s)$, and $T(s)$). This is called a *two-degree of freedom* regulator.

7. Scalar Plant. To the plant in Example 9.4, add measurements of

$$z = hx + v$$

with $v(t) \sim N(0, r')$ white and uncorrelated with $w(t)$ and $x(0)$.

(a) Write down the complete set of equations for the LQG regulator.

(b) Write MATLAB code to implement the regulator using continuous time LQR.

(c) Find analytic solutions for $S(t)$ and the error covariance $P(t)$.

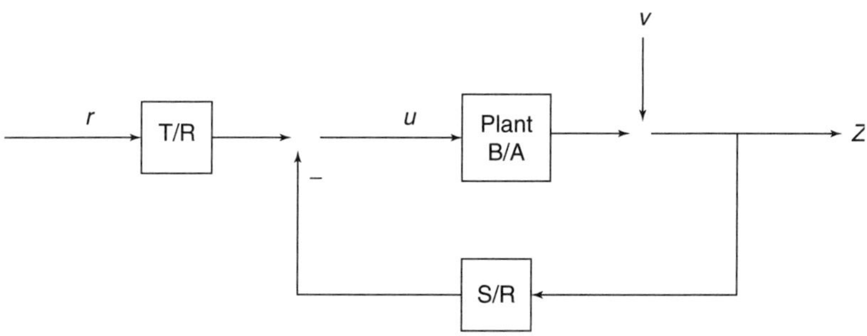

FIGURE 9.7
Two-degree of freedom regulator.

8. Suboptimal Scalar LQG Regulator. Let

$$\dot{x} = ax + bu + gw$$
$$z = hx + v$$

with $w(t) \sim N(0, q')$ and $v(t) \sim N(0, r')$ white and uncorrelated with each other and with $x(0) \sim N(\tilde{x}_0, p_0)$. Suppose

$$J(0) = \frac{1}{2}\int_{t_0}^{T} (qx^2 + ru^2)\mathrm{d}t$$

Let $x \in R$

(a) Find steady-state feedback gain K_∞.

(b) Find steady-state Kalman gain L_∞. Find transfer function from z to $\hat{x}$ of the Wiener filter

$$\dot{\hat{x}} = (a - L_\infty h)\hat{x} + bu + L_\infty z$$

Now we will combine the results of (a) and (b) to get a suboptimal regulator. Accordingly, suppose that gains K_∞, L_∞ are used for all time.

(c) Find transfer function of the steady-state regulator (Equation 9.149), (Equation 9.150) from z to u. Draw a block diagram of the closed-loop system.

(d) Find the associated suboptimal error covariance $p(t)$.

(e) Find the resulting suboptimal mean-square state $X(t)$.

9. Steady-State LQG Regulator for Newton's System. Let

$$\ddot{y} = u + w$$
$$z = y + v$$

with $w(t) \sim N(0, \sigma^2), v(t) \sim N(0, \rho^2)$ white and uncorrelated. Select the *infinite horizon* cost

$$J = \frac{1}{2}\int_{0}^{\infty} [q(y^2 + \dot{y}^2) + ru^2]\mathrm{d}t$$

(a) Determine the optimal feedback gain K if the state is $x = [y \quad \dot{y}]^T$.

(b) Sketch the poles of the closed-loop plant $(A - BK)$ as q/r varies from 0 to ∞. Find the closed-loop damping ratio and natural frequency.

(c) In terms of the signal-to-noise ratio $\Lambda = \sigma/\rho$ determine the steady-state Kalman gain.

(d) Find the transfer function $H(s) = \hat{X}(s)/Z(s)$ of the Wiener filter.

(e) Now suppose the Wiener filter is used to estimate the state for all t. Find the transfer function from the data z to the control u for the resulting suboptimal regulator.

(f) The *observer polynomial* is $|sI - A + LH|$. Plot the observer poles. Find the observer damping ratio and natural frequency (in terms of Λ).

(g) Find the suboptimal error covariance and mean-square state analytically.

(h) Write MATLAB code to implement the *optimal* time-varying regulator using continuous-time LQR.

10. The plant is described by

$$\dot{x} = u + v$$

and the state is measured according to

$$z = x + v$$

where noise $v(t)$ has spectral density

$$\Phi_z(\omega) = \frac{2\sigma^2}{\omega^2 + 1}$$

It is desired to find a control $u(t)$ to minimize

$$J = \frac{1}{2}x^2(T) + \frac{1}{2}\int_0^T (qx^2 + u^2)\mathrm{d}t$$

(a) Find the optimal steady-state feedback and observer gains, and the steady-state regulator transfer function from $z(t)$ to $u(t)$.

(b) Plot root loci of steady-state closed-loop plant and Wiener filter poles versus σ^2 and q.

(c) If the steady-state regulator is used for all t, find the resulting mean-square state $X(t)$.

Section 9.3

11. Derive the discrete LQG regulator equations in Section 9.3 for the case of:

(a) Complete state information

(b) Incomplete state information

12. Feedback of *A priori* Estimate. Derive the discrete LQG regulator equations if a feedback of the form

$$u_k = -K_k\hat{x}_k^-$$

is used, where $\hat{x}_k^-$ is a one-step-ahead prediction of x_k. This feedback might be useful in a sampled system where there is a delay in implementing the control input.

13. Scalar Discrete LQG Regulator. Let

$$\begin{aligned} x_{k+1} &= x_k + u_k \\ z_k &= x_k + v_k \end{aligned}$$

with $v_k \sim N(0, r')$ white and uncorrelated with $x_0 \sim (\bar{x}_0, p_0)$. Select

$$J = \frac{1}{2}x_N^2 + \frac{1}{2}\sum_{k=0}^{N-1} ru_k^2$$

(a) Write the LQG regulator equations.

(b) Solve for the observer and feedback gains.

14. Steady-State LQG Regulator. Let

$$\begin{aligned} x_{k+1} &= ax_k + bu_k + w_k \\ z_k &= hx_k + v_k \end{aligned}$$

with $w_k \sim N(0, q')$ and $v_k \sim N(0, r')$ white and uncorrelated. Select the infinite horizon cost

$$J = \frac{1}{2}\sum_{k=0}^{\infty}(qx_k^2 + ru_k^2)$$

For simplicity, let $x_k \in R$.

(a) Find the optimal feedback gain K_∞.

(b) Find the steady-state Kalman gain L_∞. Find the transfer function from z_k to $\hat{x}_k$ of the Wiener filter.

(c) Suppose that the Wiener filter is used to estimate x_k for all k. Find the transfer function from z_k to u_k of the resulting suboptimal regulator.

10

Stochastic Control for Polynomial Systems

In this chapter, we discuss the optimal control of stochastic polynomial systems. Only discrete systems are covered, so that the results will be useful in digital control schemes. In the problems, it is shown that the results also apply to continuous systems. We treat single-input single-output systems, but the extension to multivariable systems is not too difficult (Borison 1979, Koivo 1980).

10.1 Polynomial Representation of Stochastic Systems

Here, discrete linear stochastic systems will be represented as

$$A(z^{-1})y_k = z^{-d}B(z^{-1})u_k + C(z^{-1})w_k \tag{10.1}$$

where $u_k \in R$, $y_k \in R$ are the control input and measured output, and $w_k \in R$ is a zero-mean white noise process with variance of q. The polynomials

$$A(z^{-1}) = 1 + a_1 z^{-1} + \cdots + a_n z^{-n} \tag{10.2}$$

$$B(z^{-1}) = b_0 + b_1 z^{-1} + \cdots + b_m z^{-m}, \qquad b_0 \neq 0 \tag{10.3}$$

$$C(z^{-1}) = 1 + c_1 z^{-1} + \cdots + c_n z^{-n} \tag{10.4}$$

are expressed in terms of the delay operator z^{-1} (i.e., $z^{-1}\ y_k = y_{k-1}$). It is assumed that $C(z^{-1})$ is stable. The control delay is d.

Let us motivate the formulation 10.1 using two points of view. Suppose y_k is a stochastic process with known spectral density $\Phi_y\ (e^{j\omega})$. Then (Section 1.5) we can perform the factorization

$$\Phi_y(e^{j\omega}) = H(e^{j\omega})qH(e^{-j\omega}) \tag{10.5}$$

where $H(e^{j\omega})$ is stable, with all zeros inside or on the unit circle. We shall assume $H(e^{j\omega})$ is of minimum phase, with relative degree of zero and monic numerator and denominator. Represent $H(z)$ as

$$H(z) = \frac{z^n + c_1 z^{n-1} + \cdots + c_n}{z^n + a_1 z^{n-1} + \cdots + a_n} \tag{10.6}$$

(i.e., q has been selected to make $c_0 = 1$). Now, it is seen that Equation 10.1 with $u_k = 0$ represents the process y_k. Next, consider the single-input single-output state representation

$$x_{k+1} = Fx_k + Gu_k + Je_k \tag{10.7a}$$

$$y_k = Hx_k + v_k \tag{10.7b}$$

where F, G, J, and H are matrices, process noise e_k is $(0, Q')$, measurement noise v_k is $(0, R')$, and the noises e_k and v_k are independent. In terms of transfer functions, we have

$$Y(z) = H(zI - F)^{-1}GU(z) + H(zI - F)^{-1}JE(z) + V(z) \tag{10.8}$$

Represent the coefficient of $E(z)$ by $H_E(Z)$ so that the noise term in Equation 10.8 is $H_E(z)E(z) + V(z)$. Now, perform a spectral factorization to determine $H(z)$ (minimum phase) and q in

$$H(e^{j\omega})qH(e^{-j\omega}) = H_E(e^{j\omega})Q'H_E(e^{-j\omega}) + R' \tag{10.9}$$

Defining

$$H(z) = \frac{C_1(z)}{A_1(z)} \tag{10.10}$$

with both $C_1(z)$ and $A_1(z)$ monic, and

$$A_2(z) = |zI - F| \tag{10.11}$$

$$B_2(z) = H[\text{adj}(zI - F)]G \tag{10.12}$$

Equation 10.8 can be written as

$$Y(z) = \frac{B_2(z)}{A_2(z)}U(z) + \frac{C_1(z)}{A_1(z)}W(z) \tag{10.13}$$

where $w_k \sim (0, q)$ is white noise whose effect is equivalent to the combined effects of e_k and v_k.

If $n_1 = \deg[A_1(z)] = \deg[C_1(z)]$, $n_2 = \deg[A_2(z)]$, and $m_2 = \deg[B_2(z)]$, then it is evident that Equation 10.1 represents Equation 10.7 if

$$A(z^{-1}) = z^{-(n1+n2)}A_1(z)A_2(z) \tag{10.14}$$

$$B(z^{-1}) = z^{-(n1+m2)}A_1(z)B_2(z) \tag{10.15}$$

$$C(z^{-1}) = z^{-(n1+n2)}C_1(z)A_2(z) \tag{10.16}$$

and the control delay is

$$d = n_2 - m_2 \tag{10.17}$$

10.2 Optimal Prediction

The treatment of the control problem is deferred to the next section; at this point, we merely want to find an optimal d-step-ahead prediction for the output of Equation 10.1 given any control sequence. That is, it is desired to predict y_{k+d} in terms of quantities that occur at times k or before. We shall follow Åström (1970).

In this section, we shall assume $A(z^{-1})$ is stable. To solve this problem, first perform a long division of $A(z^{-1})$ into $C(z^{-1})$ to define quotient $F(z^{-1})$ and remainder $z^{-d}G(z^{-1})$:

$$\begin{array}{r l} & F(z^{-1}) \\ A(z^{-1}) & \overline{\big)\, C(z^{-1})} \\ & \vdots \\ & \overline{z^{-d}G(z^{-1})} \end{array} \tag{10.18}$$

The division is carried out in d steps, that is, until z^{-d} can be factored out of the remainder to yield $G(z^{-1})$. Then

$$AF + z^{-d}G = C \tag{10.19}$$

This is a *Diophantine equation* for $F(z^{-1})$ and $G(z^{-1})$. The solution to Equation 10.19 is not unique, but the particular solution $F(z^{-1})$ and $G(z^{-1})$ determined by long division has the form

$$F(z^{-1}) = 1 + f_1 z^{-1} + \cdots + f_{d-1} z^{-(d-1)} \tag{10.20}$$

$$G(z^{-1}) = g_0 + g_1 z^{-1} + \cdots + g_{n-1} z^{-(n-1)} \tag{10.21}$$

The degree $(d-1)$ of $F(z^{-1})$ will be particularly important.

Multiply Equation 10.1 by $F(z^{-1})$ to get

$$AFy_{k+d} = BFu_k + CFw_{k+d} \tag{10.22}$$

then the Diophantine equation gives

$$Cy_{k+d} = z^{-d}Gy_{k+d} + BFu_k + CFw_{k+d}$$

or

$$y_{k+d} = \frac{G}{C}y_k + \frac{BF}{C}u_k + Fw_{k+d} \tag{10.23}$$

Denote the optimal prediction of y_{k+d} in terms of $y_k, y_{k-1}, \ldots, u_k, u_{k-1}, \ldots$ as $\hat{y}_{k+d/k}$. The d-step prediction error is

$$\tilde{y}_{k+d/k} \triangleq y_{k+d} - \hat{y}_{k+d/k} \tag{10.24}$$

Let us determine $\hat{y}_{k+d/k}$ to minimize the mean-square error,

$$j_k = E(\tilde{y}^2_{k+d/k}) = E[(y_{k+d} - \hat{y}_{k+d/k})^2] \tag{10.25}$$

Using Equation 10.23, this becomes

$$j_k = E\left[\left(\frac{G}{C}y_k + \frac{BF}{C}u_k - \hat{y}_{k+d/k}\right) + Fw_{k+d}\right]^2 \tag{10.26}$$

The term in parentheses depends only on the input and output at times k and before, since $\hat{y}_{k+d/k}$ is only allowed to be expressed in terms of these values. Therefore, the term in parentheses depends only on the noise values $w_k, w_{k-1}, \ldots$. The remaining term looks like

$$F(z^{-1})w_{k+d} = w_{k+d} + f_1 w_{k+d-1} + \cdots + f_{d-1}w_{k+1} \tag{10.27}$$

since the degree of $F(z^{-1})$ is $(d-1)$, so it only depends on noise values subsequent to and including time $(k+1)$. Since the noise sequence is zero-mean and white, we concluded that the term in parentheses is orthogonal to $f(z^{-1})w_{k+d}$. Therefore, since y_k, u_k, and $\hat{y}_{k+d/k}$ are deterministic, the performance index separates into a deterministic part and a stochastic part:

$$j_k = \left(\frac{G}{C}y_k + \frac{BF}{C}u_k - \hat{y}_{k+d/k}\right)^2 + E(Fw_{k+d})^2 \tag{10.28}$$

It is now evident that the minimum value of j_k occurs if we define the d-step prediction of the output as

$$\hat{y}_{k+d/k} = \frac{G}{C}y_k + \frac{BF}{C}u_k \tag{10.29}$$

Then the prediction error is

$$\tilde{y}_{k+d/k} = Fw_{k+d} \tag{10.30}$$

Thus, the optimal prediction is the output of a linear deterministic system whose inputs are u_k and y_k, and the error is a moving average of the noise sequence. See Figure 10.1. Since we assumed $C(z^{-1})$ is stable, the predictor is a stable system.

The output can be decomposed into two parts as

$$y_{k+d} = \hat{y}_{k+d/k} + \tilde{y}_{k+d/k} \tag{10.31}$$

These parts are orthogonal since the prediction depends only on values of the noise at times k and before, and the error depends only on noise values subsequent to time k. This decomposition is written explicitly as Equation 10.23. Equation 10.23 is called the *predictive formulation* of the plant 10.1. We shall see that the solution to stochastic control problems depends on using this representation for the plant dynamics. In the present application, the predictive

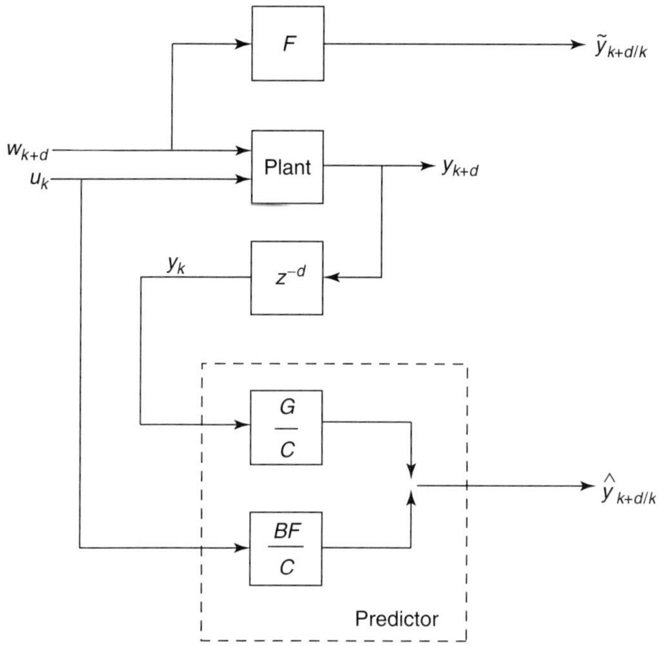

FIGURE 10.1
The d-step ahead predictor.

formulation allowed us to decompose the performance index J into two parts so we could solve the optimal prediction problem.

The minimum mean-square error is given by

$$\begin{aligned} j_k^* &= E\big[F(z^{-1})w_{k+d}\big]^2 \\ &= E(w_{k+d} + f_1 w_{k+d-1} + \cdots + f_{d-1}w_{k+1})^2 \\ &= E\big(w_{k+d}^2\big) + f_1^2 E\big(w_{k+d-1}^2\big) + \cdots + f_{d-1}^2 E\big(w_{k+1}^2\big) \end{aligned}$$

or

$$j_k^* = q\big(1 + f_1^2 + \cdots + f_{d-1}^2\big) \tag{10.32}$$

The cross-terms of the square drop out due to the whiteness of the noise. Since $E(\tilde{y}_{k+d/k}) = 0$, Equation 10.32 is the minimum error *variance*, hence Equation 10.29 is a *minimum variance* estimate.

The results we have just derived are the equivalent for polynomial systems of the discrete Kalman filter in Chapter 2. The Kalman filter requires the solution of a Riccati equation, while the polynomial system predictor requires the solution of a Diophantine equation.

Example 10.1 Minimum Variance Predictor

If the discretized plant dynamics are described by the difference equation

$$y_k + a_1 y_{k-1} + a_2 y_{k-2} = b_0 u_{k-d} + w_k + c_1 w_{k-1} \tag{10.33}$$

with white noise $w_k \sim (0, q)$, then Equation 10.1 becomes

$$(1 + a_1 z^{-1} + a_2 z^{-2}) y_k = z^{-d} b_0 u_k + (1 + c_1 z^{-1}) w_k \tag{10.34}$$

a. 1-step predictor

Let the control delay be

$$d = 1 \tag{10.35}$$

Then, dividing $A(z^{-1})$ into $C(z^{-1})$ yields

$$\begin{array}{r l} & \quad 1 \\ 1 + a_1 z^{-1} + a_2 z^{-2} & \overline{)1 + c_1 z^{-1}} \\ & \underline{1 + a_1 z^{-1} + a_2 z^{-2}} \\ & z^{-1}[(c_1 - a_1) - a_2 z^{-1}] \end{array} \tag{10.36}$$

so that

$$F(z^{-1}) = 1 \tag{10.37}$$

$$G(z^{-1}) = (c_1 - a_1) - a_2 z^{-1} \tag{10.38}$$

Note that, for $d = 1$, there always results $F(z^{-1}) = 1$ and

$$G(z^{-1}) = [C(z^{-1}) - A(z^{-1})]z \tag{10.39}$$

The optimal 1-step prediction satisfies Equation 10.29, or

$$C(z^{-1})\hat{y}_{k+1/k} = z[C(z^{-1}) - A(z^{-1})]y_k + B(z^{-1})u_k \tag{10.40}$$

so that $\hat{y}_{k+1/k}$ is given by the difference equation

$$\hat{y}_{k+1/k} = -c_1 \hat{y}_{k/k-1} + (c_1 - a_1) y_k - a_2 y_{k-1} + b_0 u_k, \quad k > 0 \tag{10.41}$$

The initial conditions can be taken as

$$\hat{y}_{0/-1} = 0, \qquad y_{-1} = 0 \tag{10.42}$$

Since $C(z^{-1})$ is stable, the prediction error will eventually reach its optimal value regardless of the initial conditions.

The minimum error variance (Equation 10.32) is

$$j_k^* = q \tag{10.43}$$

b. 2-step predictor

Let

$$d = 2 \tag{10.44}$$

so that we now need to predict ahead two steps. Carrying out the long division of $A(z^{-1})$ and $C(z^{-1})$ out steps yields

$$\begin{array}{r|l} & 1+(c_1-a_1)z^{-1} \\ \cline{2-2} 1+a_1z^{-1}+a_2z^{-2} & 1+c_1z^{-1} \\ & 1+a_1z^{-1}+a_2z^{-2} \\ \cline{2-2} & (c_1-a_1)z^{-1}-a_2z^{-2} \\ & (c_1-a_1)z^{-1}+a_1(c_1-a_1)z^{-2}+a_2(c_1-a_1)z^{-3} \\ \cline{2-2} & z^{-2}\{[-a_2-a_1(c_1-a_1)]-a_2(c_1-a_1)z^{-1}\} \end{array} \tag{10.45}$$

so that

$$F(z^{-1}) = 1 + (c_1 - a_1)z^{-1} \tag{10.46}$$

$$G(z^{-1}) = -[a_2 + a_1(c_1 - a_1)] - a_2(c_1 - a_1)z^{-1} \tag{10.47}$$

The predictor recursion (Equation 10.29) is now

$$\begin{aligned} \hat{y}_{k+2/k} &= -c_1\hat{y}_{k+1/k-1} - [a_2 + a_1(c_1 - a_1)]y_k \\ &\quad - a_2(c_1 - a_1)y_{k-1} + b_0u_k + b_0(c_1 - a_1)u_{k-1} \end{aligned} \tag{10.48}$$

and the optimal 2-step error variance is

$$y_k^* = q + (c_1 - a_1)^2 q \tag{10.49}$$

Note that as we attempt to predict further into the future, the error variance increases. Note also that $G(z^{-1})$ generally has degree of $n-1$, while the degree $d-1$ of $F(z^{-1})$ increases with the delay. Thus, the d-step-ahead predictor requires $n-1$ previous values of the output and $d-1$ previous values of the input. □

10.3 Minimum Variance Control

Now, let us determine one type of optimal controller for the polynomial system

$$A(z^{-1})y_k = z^{-d}B(z^{-1})u_k + C(z^{-1})w_k \tag{10.50}$$

with white noise $w_k \sim N(0, q)$, and polynomials given by Equations 10.2 through 10.4. Our result will be the following: the control u_k will be selected to zero the d-step prediction $\hat{y}_{k+d/k}$ of the output. To derive this result, we shall use the predictive formulation 10.23 of the plant, again following Åström (1970).

Select as a performance index the mean-square value of the output

$$j_k = E(y_{k+d}^2) \tag{10.51}$$

and use predictive formulation 10.23 to say that

$$j_k = E\left[\left(\frac{G}{C}y_k + \frac{BF}{C}u_k\right) + Fw_{k+d}\right]^2 \tag{10.52}$$

We want to select u_k to minimize j_k. Admissible controls u_k depend on previous values of the control, and on $y_k, y_{k-1}, \ldots$. Since the term in parentheses in Equation 10.52 depends only on $w_k, w_{k-1}, \ldots$, and $F(z^{-1})w_{k+d}$ depends only on $w_{k+1}, w_{k+2}, \ldots, w_{k+d}$, we have the decomposition

$$j_k = E\left(\frac{G}{C}y_k + \frac{BF}{C}u_k\right)^2 + E(Fw_{k+d})^2 \tag{10.53}$$

Thus, it is clear that the minimum value of j_k is attained if the control is selected to satisfy

$$u_k = -\frac{G}{BF}y_k \tag{10.54}$$

Then, the optimal cost is

$$\begin{aligned} j_k^* &= E\left[F(z^{-1})w_{k+d}\right]^2 \\ &= q\left(1 + f_1^2 + \cdots + f_{d-1}^2\right) \end{aligned} \tag{10.55}$$

exactly as Equation 10.32.

This control scheme is shown in Figure 10.2, where a reference signal s_k has been added. Since the feedback is a function of z^{-1}, it contains some dynamics. Therefore, it is a dynamic output feedback, exactly like the LQG regulator in Chapter 9, which contained a state estimator and a deterministic LQ regulator. It can also be described as a *one-degree-of-freedom regulator*, since there is no feedforward term so that only the plant poles, and not the zeros, can be moved.

By examining Equation 10.30, it is seen that the optimal control u_k given by Equation 10.54 simply zeros the d-step prediction $\hat{y}_{k+d/k}$ of the output. The resulting mean-square value (Equation 10.55) of the output is exactly equal to the prediction error variance! It is therefore apparent that the dynamic output feedback (Equation 10.54) implicitly contains an estimator for the output. The upshot of all this is that the control law we have just derived could be interpreted as a *separation principle* for polynomial systems (Åström 1970).

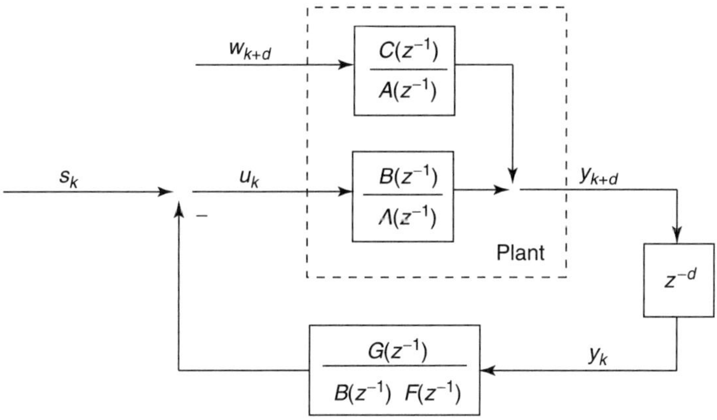

FIGURE 10.2
Minimum variance controller.

That is, to derive the minimum variance control, we could first find the optimal prediction of the output and *then* select the control input to zero the prediction.

We must now discuss the closed-loop behavior under the influence of our control law. Using Equation 10.54 in Equation 10.50, the transfer characteristics from the noise w_k to the output y_k are seen to be described by

$$\left(A + z^{-d}\frac{BG}{BF}\right)y_k = Cw_k \tag{10.56}$$

or

$$y_k = \frac{BFC}{B(AF + z^{-d}G)}w_k \tag{10.57}$$

According to Equation 10.19,

$$y_k = \frac{BFC}{BC}w_k \tag{10.58}$$

$C(z^{-1})$ is assumed stable. If $B(z^{-1})$ is stable, that is, if the plant $B(z^{-1})/A(z^{-1})$ is minimum phase, then we can cancel to obtain

$$y_k = F(z^{-1})w_k \tag{10.59}$$

which is the *regulator error*. (This agrees with the result obtained by using Equation 10.54 in Equation 10.23.) Therefore, the minimum variance regulation error is simply equal to the prediction error (Equation 10.30).

According to Equation 10.58, if w_k has a mean of zero, then so does y_k in the closed-loop system (as long as reference input s_k is zero) (see Section 1.5). Thus, j_k in Equation 10.51 is the *variance* of the output, which justifies calling our control law a *minimum variance* controller.

Now, let us turn to the closed-loop transfer relation between the reference input s_k and y_{k+d}. From Figure 10.2,

$$y_{k+d} = \frac{B/A}{1 + GBz^{-d}/BAF} s_k$$

or

$$y_{k+d} = \frac{B^2 F}{B(AF + z^{-d}G)} s_k \tag{10.60}$$

Taking into account the Diophantine equation gives

$$y_{k+d} = \frac{B^2 F}{BC} s_k \tag{10.61}$$

The closed-loop characteristic polynomial is therefore

$$\Delta^{\text{cl}}(z) = B(z^{-1})C(z^{-1})z^{m+n} \tag{10.62}$$

(multiplication by z^{m+n} removes terms in z^{-1}). If the plant $B(z^{-1})/A(z^{-1})$ is minimum phase, then the closed-loop system is stable (since $C(z^{-1})$ is stable by assumption). Then we can cancel to obtain

$$y_{k+d} = \frac{B(z^{-1})F(z^{-1})}{C(z^{-1})} s_k \tag{10.63}$$

so that the closed-loop poles are the zeros of the noise polynomial $C(z^{-1})$.

Example 10.2 Minimum Variance Controller

We can easily find the minimum variance controllers for the two cases in Example 10.1, for the work has already been done.

a. 1-step controller

If the control delay is $d = 1$, then the 1-step predicted output can be set to zero using Equation 10.54, which becomes

$$B(z^{-1})u_k = -z[C(z^{-1}) - A(z^{-1})]y_k \tag{10.64}$$

or, in our case

$$u_k = \frac{-(c_1 - a_1)}{b_0} y_k + \frac{a_2}{b_0} y_{k-1} \tag{10.65}$$

With this control, the regulation error is given by Equation 10.59, or

$$y_k = w_k \tag{10.66}$$

and the regulation error variance is

$$j_k^* = q \tag{10.67}$$

b. 2-step controller

If $d = 2$, then the 2-step predicted output can be zeroed using the control defined by the recursion

$$b_0\left[1 + (c_1 - a_1)z^{-1}\right]u_k = \left\{[a_2 + a_1(c_1 - a_1)] + a_2(c_1 - a_1)z^{-1}\right\}y_k \tag{10.68}$$

or

$$u_k = -(c_1 - a_1)u_{k-1} + \frac{a_2 + a_1(c_1 - a_1)}{b_0}y_k + \frac{a_2(c_1 - a_1)}{b_0}y_{k-1} \tag{10.69}$$

The regulation error is now

$$y_k = w_k + (c_1 - a_1)w_{k-1} \tag{10.70}$$

with a variance of

$$j_k^* = q + q(c_1 - a_1)^2 \tag{10.71}$$

Note that the regulation error variance increases with the control delay. □

10.4 Polynomial Linear Quadratic Gaussian Regulator

The minimum variance controller was found in the previous section by minimizing the cost index (Equation 10.51). In this section, it is desired to find the control minimizing a more general cost index.

The linear plant is described by

$$Ay_k = z^{-d}Bu_k + Cw_k \tag{10.72}$$

with white noise $w_k \sim N(0, q)$, and

$$\begin{aligned} A(z^{-1}) &= 1 + a_1z^{-1} + \cdots + a_nz^{-n} \\ B(z^{-1}) &= b_0 + b_1z^{-1} + \cdots + b_mz^{-m}, \qquad b_0 \neq 0 \\ C(z^{-1}) &= 1 + c_1z^{-1} + \cdots + c_nz^{-n} \end{aligned} \tag{10.73}$$

Let us select the quadratic performance index

$$J_k = (Py_{k+d} - Qs_k)^2 + (Ru_k)^2 \tag{10.74}$$

with weighting polynomials

$$\begin{aligned} P(z^{-1}) &= 1 + p_1 z^{-1} + \cdots + p_{n_P} z^{-n_P} \\ Q(z^{-1}) &= q_0 + q_1 z^{-1} + \cdots + q_{n_Q} z^{-n_Q} \\ R(z^{-1}) &= r_0 + r_1 z^{-1} + \cdots + r_{n_R} z^{-n_R} \end{aligned} \tag{10.75}$$

Signal s_k is a given reference signal.

We have chosen this cost index for two reasons: it results in a very nice solution, and it can be selected to specify a wide range of designs that result in desirable closed-loop behavior for different applications. Let us discuss some examples.

To keep the output small without using too much control energy, we might simply select $P = 1$, $Q = 0$, and $R = r_0$, so that

$$J_k = y_{k+d}^2 + r_0^2 u_k^2 \tag{10.76}$$

(In this case the reference input is not needed.) If we are more concerned about keeping the *changes* small in the control input, we could select $P = 1$, $Q = 0$, and $R = r_0(1 - z^{-1})$, so that

$$J_k = y_{k+d}^2 + r_0^2(u_k - u_{k-1})^2 \tag{10.77}$$

By choosing $P = Q = 1$ and $R = r_0$, we get the cost index

$$J_k = (y_{k+d} - s_k)^2 + r_0^2 u_k^2 \tag{10.78}$$

which is a polynomial tracker; the controlled output will follow a (delayed) reference signal s_k. We can also design a regulator that makes the plant Bz^{-d}/A behave like any desired model, for if R is chosen as zero, then the optimal controller will attempt to ensure that

$$Py_{k+d} = Qs_k \tag{10.79}$$

This results in *model-following behavior*, where the plant output tracks the output of a desired model $Q(z^{-1})z^{-d}/P(z^{-1})$ which has input s_k. To see this, write Equation 10.79 as

$$y_k = \frac{Q(z^{-1})z^{-d}}{P(z^{-1})} s_k \tag{10.80}$$

Therefore, $Q(z^{-1})$ and $P(z^{-1})$ in Equation 10.74 need only be selected to be the specified model numerator and denominator dynamics. Finally, if $P = 1$, $Q = 0$, and $R = 0$, then $E(J_k)$ reduces to (Equation 10.51).

The *polynomial linear quadratic Gaussian (LQG) problem* is to determine a control sequence that minimizes the expected cost

$$j_k = E(J_k) \tag{10.81}$$

Admissible controls u_k depend on u_{k-1}, $u_{k-2}, \ldots$, y_k, $y_{k-1}, \ldots$. To solve this problem, we shall follow Clarke and Gawthrop (1975).

We cannot immediately differentiate the expected cost j_k to determine the optimal control u_k, since y_{k+d} depends on u_k. This dependence must first be made explicit. To do this, we shall find an optimal j-step-ahead predictor for values y_{k+j} of the output for $j \leq d$.

According to Equation 10.31, y_{k+d} can be decomposed into two orthogonal parts as

$$y_{k+d} = \hat{y}_{k+d/k} + \tilde{y}_{k+d/k} \tag{10.82}$$

where

$$\hat{y}_{k+d/k} = \frac{G_d}{C} y_k + \frac{BF_d}{C} u_k \tag{10.83}$$

is the optimal prediction of y_{k+d} given the values $y_k, y_{k-1}, \ldots, u_k, u_{k-1}, \ldots$, and

$$\tilde{y}_{k+d/k} = F_d w_{k+d} \tag{10.84}$$

is the prediction error. The polynomials $F_d(z^{-1})$ and $G_d(z^{-1})$ are defined by the Diophantine equation

$$AF_d + z^{-d} G_d = C \tag{10.85}$$

which amounts to a d-step long division of $A(z^{-1})$ into $C(z^{-1})$.

We would like to find an optimal prediction for each y_{k+j}, $j \leq d$, in terms of quantities that occur at times k and before, since the values of the output appearing in Equation 10.74 are y_{k+d}, $y_{k+d-1}, \ldots$, y_{k+d-n_p}, with n_p the degree of $P(z^{-1})$. If $\hat{y}_{k+j/k}$ denotes the optimal prediction of y_{k+j} in terms of quantities that occurs at times k and before, and $\tilde{y}_{k+j/k}$ is the associated prediction error, then

$$y_{k+j} = \hat{y}_{k+j/k} + \tilde{y}_{k+j/k}, \quad j \leq d \tag{10.86}$$

with $\hat{y}_{k+j/k}$ and $\tilde{y}_{k+j/k}$ orthogonal.

For $j \leq 0$, it is clear that

$$\hat{y}_{k+j/k} = y_{k+j}, \quad j \leq 0 \tag{10.87}$$

since, then y_{k+j} itself is part of the known data. For $j = d$, we know that $\hat{y}_{k+j/k}$ and $\tilde{y}_{k+j/k}$ are given by Equations 10.83 and 10.84. However, it is *not* true that we can generalize Equation 10.83 to say, for $j < d$,

$$\hat{y}_{k+j/k} = \frac{G_d}{C} y_{k-d+j} + \frac{BF_d}{C} u_{k-d+j} \tag{10.88}$$

The reason is that this estimate disregards the available data $y_{k-d+j+1}, \ldots,$ y_k, so it is not optimal. To construct a minimum variance estimate for y_{k+j}, $0 < j \leq d$, in terms of $y_k, y_{k-1}, \ldots, u_k, u_{k-1}, \ldots,$ we proceed as follows. Perform a j-step long division of $A(z^{-1})$ into $C(z^{-1})$ to define $F_j(z^{-1})$ and $G_j(z^{-1})$ by

$$AF_j + z^{-j}G_j = C \tag{10.89}$$

Then, $F_j(z^{-1})$ has the form

$$F_j(z^{-1}) = 1 + f_1 z^{-1} + \cdots + f_{j-1} z^{-(j-1)} \tag{10.90}$$

The polynomials $F_j(z^{-1})$ and $G_j(z^{-1})$ must be determined for all j such that $d - n_p \leq j \leq d$. If we perform a single d-step long division of $A(z^{-1})$ into $C(z^{-1})$, the $F_j(z^{-1})$ is the quotient and $z^{-j}G_j(z^{-1})$ is the remainder at the jth step. See Example 10.3.

Multiply the plant 10.72 by F_j to get

$$AF_j y_{k+j} = BF_j u_{k-d+j} + F_j C w_{k+j} \tag{10.91}$$

and use the Diophantine equation 10.89 to see that

$$Cy_{k+j} = z^{-j}G_j y_{k+j} + BF_j u_{k-d+j} + F_j C w_{k+j}$$

or

$$y_{k+j} = \left(\frac{G_j}{C} y_k + \frac{BF_j}{C} u_{k-d+j} \right) + F_j w_{k+j} \tag{10.92}$$

Since the term in parentheses depends only on $w_k, w_{k-1}, \ldots,$ and $F_j(z^{-1})w_{k+j}$ depends only on $w_{k+1}, w_{k+2}, \ldots,$ these two terms are orthogonal. Therefore, an argument like the one in Section 10.2 shows that Equation 10.86 holds, where the optimal j-step prediction is

$$\hat{y}_{k+j/k} = \frac{G_j}{C} y_k + \frac{BF_j}{C} u_{k-d+j}, \quad 0 < j \leq d \tag{10.93}$$

and the prediction error is

$$\tilde{y}_{k+j/k} = F_j w_{k+j} \tag{10.94}$$

The error variance is equal to

$$E(\tilde{y}_{k+j/k}^2) = q(1 + f_1^2 + \cdots + f_{j-1}^2) \tag{10.95}$$

where the f_i are the coefficients of $F_j(z^{-1})$.

Note that the prediction 10.93 uses *all* the available outputs, and that it reduces to Equation 10.83 if $j = d$. At this point, we have succeeded in deriving a predictive formulation 10.86 for y_{k+j} for all $j \leq d$. The estimate and the error in Equation 10.86 are orthogonal, and are given for $0 \leq j \leq d$ by Equations 10.93 and 10.94. We may include Equation 10.87 in 10.93 by defining

$$F_j = 0, \qquad G_j = z^j C, \quad \text{for } j \leq 0 \tag{10.96}$$

for then Equations 10.86, 10.93, and 10.94 hold for all $j \leq d$.

What is now required is to use the orthogonal decomposition 10.86 for y_{k+j} in the cost index 10.81. Then

$$j_k = E\{[P(\hat{y}_{k+d/k} + \tilde{y}_{k+d/k}) - Qs_k]^2 + (Ru_k)^2\} \tag{10.97}$$

Since each error $\tilde{y}_{k+j/k}$ depends only on w_{k+1} and subsequent noise values, $P(z^{-1})\tilde{y}_{k+d/k}$ is orthogonal to all the other terms. Furthermore, these other terms are all deterministic. Therefore,

$$j_k = (P\hat{y}_{k+d/k} - Qs_k)^2 + (Ru_k)^2 + E(P\tilde{y}_{k+d/k})^2 \tag{10.98}$$

The importance of the predictive formulation 10.86, which is explicitly stated in Equation 10.92, is now clear; it has allowed us to decompose the cost index into a stochastic portion independent of u_k, and a deterministic portion that can be minimized by differentiation. Note that $\hat{y}_{k+d/k}$ depends on u_k, and that $\hat{y}_{k+j/k}$ for $j < d$ does not depend on u_k, but only on earlier values of the input. Note further that

$$\frac{\partial \hat{y}_{k+d/k}}{\partial u_k} = \frac{B(0)F_d(0)}{C(0)} = b_0 \tag{10.99}$$

Furthermore,

$$\begin{aligned}\frac{\partial}{\partial u_k}(Ru_k)^2 &= \frac{\partial}{\partial u_k}(r_0 u_k + r_1 u_{k-1} + \cdots + r_{n_R} u_{k-n_R})^2 \\ &= 2r_0 Ru_k\end{aligned} \tag{10.100}$$

Therefore, differentiating j_k yields

$$\frac{\partial j_k}{\partial u_k} = 2b_0(P\hat{y}_{k+d/k} - Qs_k) + 2r_0 Ru_k = 0 \tag{10.101}$$

The optimal LQG control law is therefore defined by the recursion

$$P\hat{y}_{k+d/k} + \frac{r_0}{b_0}Ru_k - Qs_k = 0 \tag{10.102}$$

for u_k in terms of $u_{k-1}, u_{k-2}, \ldots, y_k, y_{k-1}, \ldots, s_k, s_{k-1}, \ldots$. The polynomial LQG control problem is now solved. We can find a more explicit recursion for

u_k as follows. Use the definition of $P(z^{-1})$ to write Equation 10.102 as

$$\sum_{j=0}^{n_p} p_j \hat{y}_{k+d-j/k} + \frac{r_0}{b_0} R u_k - Q s_k = 0$$

with $p_0 = 1$. Now, by Equation 10.93, this is

$$\sum_{j=0}^{n_p} p_j G_{d-j} y_k + \sum_{j=0}^{n_p} p_j B F_{d-j} u_{k-j} + \frac{r_0}{b_0} C R u_k - C Q s_k = 0 \tag{10.103}$$

Defining new (composite) polynomials

$$G'(z^{-1}) = \sum_{j=0}^{n_p} p_j G_{d-j}(z^{-1}) \tag{10.104}$$

$$F'(z^{-1}) = \sum_{j=0}^{n_p} p_j B(z^{-1}) F_{d-j}(z^{-1}) z^{-j} + \frac{r_0}{b_0} C(z^{-1}) R(z^{-1}) \tag{10.105}$$

this becomes

$$F' u_k = -G' y_k + C Q s_k \tag{10.106}$$

It is now clear that what we have on our hands is a two-degree-of-freedom regulator (see problems in Chapter 9). See Figure 10.3. Since there are feedback and feedforward components, the plant poles *and* zeros can now be moved (e.g., see Equation 10.80).

If we select the weighting polynomials $P(z^{-1}) = 1$, $Q(z^{-1}) = 0$, and $R(z^{-1}) = 0$, then

$$j_k = E\left(y_{k+d}^2\right) \tag{10.107}$$

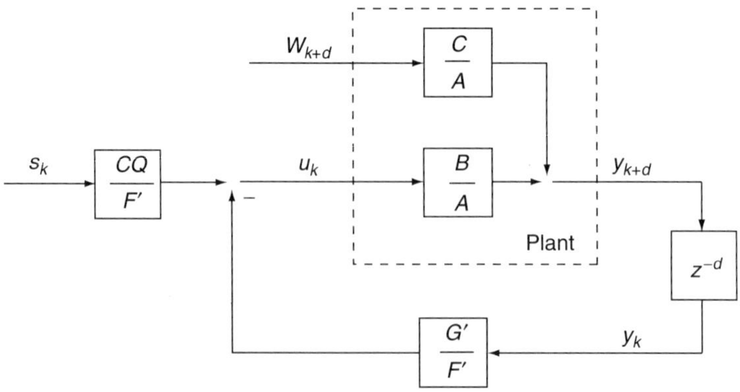

FIGURE 10.3
Polynomial LQG controller.

In this case, $G' = G_d$, $F' = BF_d$, and the LQG control law (Equation 10.106) reduces to

$$BF_d u_k = -G_d y_k \tag{10.108}$$

This is exactly the minimum variance controller (Equation 10.54).

The polynomial LQG controller depends on the solution to the Diophantine equation(s) 10.89. Its dynamics contain an implicit estimator for the portion of the internal state not appearing directly in the output. It should be compared to the state variable LQG controller in Section 6.3, whose structure is considerably more complex! According to Figure 10.3, the closed-loop transfer function from s_k to y_{k+d} is given by

$$H^{\text{cl}}(z) = \frac{CQ}{F'} \cdot \frac{BF'}{AF' + BG'z^{-d}} \tag{10.109}$$

Thus, the closed-loop characteristic polynomial is

$$\Delta^{\text{cl}}(z) = (AF' + BG'z^{-d})z^h \tag{10.110}$$

where h is the highest power of z^{-1} in $H^{\text{cl}}(z)$ after canceling $F'(z^{-1})$. To show how simple the design of a polynomial LQG controller really is, let us consider an example.

Example 10.3 Polynomial LQG Controller

For purpose of illustration, we shall consider a discrete plant with simple and unrealistic numbers. Suppose that the discretized plant has a transfer function description of

$$Y(z) = \frac{z - 0.5}{z^3 + 1.5z^2 - z}U(z) + \frac{z^3 + 0.25z^2}{z^3 + 1.5z^2 - z}W(z) \tag{10.111}$$

with $w_k \sim N(0, q)$ a white noise process. Then the polynomial description is

$$(1 + 1.5z^{-1} - z^{-2})y_k = z^{-2}(1 - 0.5z^{-1})u_k + (1 + 0.25z^{-1})w_k \tag{10.112}$$

which is

$$Ay_k = z^{-d}Bu_k + Cw_k \tag{10.113}$$

with a control delay of

$$d = 2 \tag{10.114}$$

The plant is unstable with poles at

$$z = 0.5, -2 \tag{10.115}$$

Let the performance index be

$$J_k = [(y_{k+2} - y_{k+1}) - (s_k - s_{k-1})]^2 + [u_k - u_{k-1}]^2 \tag{10.116}$$

or

$$J_k = [(1 - z^{-1})y_{k+2} - (1 - z^{-1})s_k]^2 + [(1 - z^{-1})u_k]^2 \tag{10.117}$$

which is

$$J_k = (Py_{k+2} - Qs_k)^2 + (Ru_k)^2 \tag{10.118}$$

where s_k is a given reference input. This cost function will try to keep the changes small in input u_k. It will also try to make changes in the output following changes in the (delayed) reference input. Thus, we are not concerned about matching the DC level in y_k to the DC level in the reference signal.

a. LQG tracker

We call this a tracker since changes in the output will follow changes in s_k.

Carrying out a long division of $A(z^{-1})$ into $C(z^{-1})$ to $d = 2$ steps yields

$$\begin{array}{r|l}
 & 1 + 1.25z^{-1} \\
1 + 1.5z^{-1} - z^{-2} & 1 + 0.25z^{-1} \\
 & \underline{1 + 1.5z^{-1} - z^{-2}} \\
 & -1.25z^{-1} + z^{-2} \\
 & \underline{-1.25z^{-1} - 1.875z^{-2} + 1.25z^{-3}} \\
 & \qquad 2.875z^{-2} - 1.25z^{-3}
\end{array}$$

whence we can immediately read off

$$\begin{aligned}
&F_1(z^{-1}) = 1, \qquad G_1(z^{-1}) = -1.25 + z^{-1} \\
&F_2(z^{-1}) = 1 - 1.25z^{-1}, \qquad G_2(z^{-1}) = 2.875 - 1.25z^{-1}
\end{aligned} \tag{10.119}$$

The composite polynomials in Equations 10.104 and 10.105 are thus

$$\begin{aligned}
G'(z^{-1}) &= p_0G_2 + p_1G_1 \\
&= (2.875 - 1.25z^{-1}) - (-1.25 + z^{-1}) \\
&= 4.125 - 2.25z^{-1}
\end{aligned} \tag{10.120}$$

$$\begin{aligned}
F'(z^{-1}) &= B(p_0F_2 + p_1F_1z^{-1}) + \frac{r_0}{b_0}CR \\
&= (1 - 0.5z^{-1})(1 - 1.25z^{-1} - z^{-1}) + (1 + 0.25z^{-1})(1 - z^{-1}) \\
&= 2 - 3.5z^{-1} + 0.875z^{-2}
\end{aligned} \tag{10.121}$$

Also,

$$CQ = (1 + 0.25z^{-1})(1 - z^{-1}) = 1 - 0.75z^{-1} - 0.25z^{-2} \tag{10.122}$$

The LQG regulator is given by Equation 10.106:

$$\begin{aligned}(2 - 3.5z^{-1} + 0.875z^{-2})u_k &= -(4.125 - 2.25z^{-1})y_k \\ &\quad + (1 - 0.75z^{-1} - 0.25z^{-2})s_k\end{aligned} \tag{10.123}$$

so that the optimal control can be determined by the recursion

$$u_k = \tfrac{1}{2}(3.5u_{k-1} - 0.875u_{k-2} - 4.125y_k + 2.25y_{k-1} + s_k - 0.75s_{k-1} - 0.25s_{k-2}) \tag{10.124}$$

b. Closed-loop system

According to Equation 10.110 the closed-loop characteristic polynomial is

$$\Delta^{cl}(z) = (AF' + BG'z^{-d})z^h \tag{10.125}$$

where h is the power of z required to clear negative powers of z. Performing the required operations,

$$\Delta^{cl}(z) = (2 - 0.5z^{-1} - 2.25z^{-2} + 0.5z^{-3} + 0.25z^{-4})z^h \tag{10.126}$$

and it is now evident that $h = 4$, so that

$$\Delta^{cl}(z) = 2z^4 - 0.5z^3 - 2.25z^2 + 0.5z + 0.25 \tag{10.127}$$

The closed-loop poles are at

$$z = -0.25, 0.5, -1, 1 \tag{10.128}$$

and the closed-loop system is marginally stable even though the original plant was unstable. □

Problems

Section 10.1

1. A process y_k is defined by

$$\begin{aligned} x_{k+1} &= \tfrac{1}{2}x_k + u_{k-1} + e_{k+1} \\ y_k &= x_k + v_k \end{aligned}$$

with measurement noise $v_k \sim N(0,1)$ white and process noise e_k having a spectral density of

$$\Phi_E(\omega) = \frac{(1-a^2)\sigma^2}{1+a^2-2a\cos\omega}$$

where $a = 3/4$. Find the optimal d-step predictors for x_k and for y_k as a function of σ^2. Find both prediction error variances.

2. Prediction of Sampled Signals. A Markov process $x(t)$ with spectral density

$$\phi_X(\omega) = \frac{2}{\omega^2+1}$$

is measured at discrete intervals kT according to

$$y_k = x(kT) + v_k$$

with $v_k \sim N(0,1)$ white. Find the optimal 1-step and 2-step predictors for y_k as a function of the sampling period T. Find the prediction error variances.

Section 10.2

3.

(a) For the plant in Problem 1. find the control input minimizing the variance of y_k. Compute the closed-loop transfer function and poles, and find the regulation error variance.

(b) Repeat for the control input minimizing the variance of x_k, assuming that x_k is exactly measurable.

Section 10.3

4. For the plant in Problem 1, find the control minimizing the expected value of:

(a) $J_k = (y_{k+d} - y_{k+d-1})2$ (this control will keep small the *changes* in the output).

(b) $J_k = y_{k+d}^2 + ru_k^2$ (a weighted-input minimum variance control).

(c) $J_k = y_{k+d}^2 + r(u_k - u_{k-1})^2$

5. Model Following Controller

(a) Find a control to make the plant

$$y_k = \frac{z-\frac{1}{4}}{(z-\frac{1}{2})^2}u_k + \frac{z^2}{(z-\frac{1}{2})^2}w_k$$

behave like the model

$$y_k = \frac{(z-\frac{1}{2})}{(z-\frac{1}{4})^2}u_k$$

(b) Find the first five terms of the impulse response of

(i) The original plant.

(ii) The model.

(iii) The plant plus controller.

6. Model Following with Control Weighting. Solve Problem 2, including a term like ru_k^2 in the performance index. As r becomes small the control gives better model following behavior. Plot the closed-loop poles as a function of the control weighting r.

7. A system

$$x_{k+1} = 1.1x_k + u_{k-1}$$

is driven by a noisy actuator described by

$$u_{k+1} = 0.9u_k + v_k + e_k$$

where v_k is the input command and e_k is noise with spectral density

$$\Phi_E(\omega) = \frac{\frac{3}{4}}{\frac{5}{4} - \cos\omega}$$

Find the regulator to generate input v_k from measurements described by

$$y_k = x_k + n_k$$

where noise n_k has

$$\Phi_N(\omega) = \frac{\frac{15}{16}}{\frac{17}{16} - \frac{1}{2}\cos\omega}$$

if we want to minimize

$$J_k = (y_{k+d} - s_k)^2 + \frac{u_k^2}{10}$$

where d is the control delay and s_k is a reference input. Sketch your regulator.

8. Regulator Error. Find the regulation error for the optimal controller 10.106. Find the regulation error variance if $w_k \sim (0, q)$ is white.

9. Tracker for Newton's System. A plant

$$\dot{d} = v$$
$$\dot{v} = u + w$$

with $w(t) \sim N(0, \sigma^2)$ a white process noise, has discrete position measurements

$$y_k = d(kT) + n_k$$

with white measurement noise $n_k \sim N(0, 1)$. The cost index is

$$J_k = (y_{k+d} - s_k)^2 + (ru_k)^2$$

with s_k a reference input, and d the control delay.

(a) Let $T=0.2$ ms and $r=0.1$. Find the optimal regulator. Sketch the regulation error variance as a function of signal-to-noise ratio σ^2.

(b) Let $T=0.2$ ms and $\sigma^2=0.1$. Find the optimal regulator. Sketch the regulation error variance as a function of control weighting r.

(c) Let $\sigma^2=0.1$ and $r=0.1$. Find the optimal regulator. Sketch the regulation error variance as a function of sampling period T.

10. Polynomial Control for Continuous Systems. To explore the possibility of extending our results to the design of continuous controllers, consider the following. It is desired to compute a controller for the system of Problem 6 of the form

$$u = -H_1(s)y + H_2(s)z \tag{10.129}$$

where $H_1(s)$ and $H_2(s)$ are feedback and feedforward regulators, respectively, and $z(t)$ is a reference input. We want to use the cost index

$$J = [y(t) - z(t)]^2 + [ru(t)]^2 \tag{10.130}$$

Assume position measurements of the form $y = d + n$, with noise $n \sim N(0,1)$.

(a) Carry out the design if there is no process noise so that $\sigma^2=0$. That is, find H_1 and H_2.

(b) Repeat if $\sigma^2 \neq 0$.

(c) Repeat if velocity measurements are taken so that $y = v + n$.

Appendix

Review of Matrix Algebra

We present here a brief review of some concepts which are assumed as background for the text. Gantmacher (1977) and Brogan (1974) are good references.

A.1 Basic Definitions and Facts

The *determinant* of a square $n \times n$ matrix is symbolized as $|A|$. If A and B are both square, then

$$|A| = |A^T| \tag{A.1}$$

$$|AB| = |A| \cdot |B| \tag{A.2}$$

where superscript T represents transpose. If $A \in C^{m\times n}$ and $B \in C^{n\times m}$ (where n can equal m), then trace(AB) = trace(BA), where the trace of a matrix is the sum of the diagonal elements. Also

$$|I_m + AB| = |I_n + BA| \tag{A.3}$$

with I_n the $n \times n$ identity matrix. (C represents the complex numbers.)

For any matrices A and B,

$$(AB)^T = B^T A^T \tag{A.4}$$

and if A and B are nonsingular, then

$$(AB)^{-1} = B^{-1}A^{-1} \tag{A.5}$$

The *Kronecker product* of two matrices $A = [a_{ij}] \in C^{m\times n}$ and $B = [b_{ij}] \in C^{p\times q}$ is

$$A \otimes B = [a_{ij}B] \in C^{mp\times nq} \tag{A.6}$$

(It is sometimes defined as $A \otimes B = [Ab_{ij}]$.) If $A = [a_1 \quad a_2 \quad \dots \quad a_n]$, where a_i are the columns of A, the *stacking operator* is defined by

$$s(A) = \begin{bmatrix} a_1 \\ a_2 \\ \vdots \\ a_n \end{bmatrix} \tag{A.7}$$

It converts $A \in C^{m \times n}$ into a vector $s(A) \in C^{mn}$. An identity that is often useful is

$$s(ABD) = (D^T \otimes A)s(B) \tag{A.8}$$

If $A \in C^{m \times m}$ and $B \in C^{p \times p}$, then

$$|A \otimes B| = |A|^p \cdot |B|^m \tag{A.9}$$

See Brewer (1978) for other results.

If λ_i is an eigenvalue for A with eigenvector v_i, then $1/\lambda_i$ is an eigenvalue for A^{-1} with the same eigenvector, for

$$Av_i = \lambda_i v_i \tag{A.10}$$

implies that

$$\lambda_i^{-1} v_i = A^{-1} v_i \tag{A.11}$$

If λ_i is an eigenvalue of A with eigenvector v_i, and μ_j is an eigenvalue of B with eigenvector w_j, then $\lambda_i \mu_j$ is eigenvalue of $A \otimes B$ with eigenvector $v_i \otimes w_j$ (Brewer 1978).

A.2 Partitioned Matrices

If

$$D = \begin{bmatrix} A_{11} & 0 & 0 \\ 0 & A_{22} & 0 \\ 0 & 0 & A_{33} \end{bmatrix} \tag{A.12}$$

where A_{ij} are matrices, then we write $D = \text{diag}(A_{11}, A_{22}, A_{33})$ and call D *block diagonal*. If the A_{ii} are square, then

$$|D| = |A_{11}| \cdot |A_{22}| \cdot |A_{33}| \tag{A.13}$$

and if $|D| \neq 0$, then

$$D^{-1} = \text{diag}\left(A_{11}^{-1}, A_{22}^{-1}, A_{33}^{-1}\right) \tag{A.14}$$

If

$$D = \begin{bmatrix} A_{11} & A_{12} & A_{13} \\ 0 & A_{22} & A_{23} \\ 0 & 0 & A_{33} \end{bmatrix} \tag{A.15}$$

where A_{ij} are matrices, then D is *upper block triangular* and Equation A.13 still holds. *Lower block triangular* matrices have the form of the transpose of Equation A.15 and still obey Equation A.15.

If

$$A = \begin{bmatrix} A_{11} & A_{12} \\ A_{21} & A_{22} \end{bmatrix} \tag{A.16}$$

we define the *Schur complement of* A_{22} as

$$D_{22} = A_{22} - A_{21}A_{11}^{-1}A_{12} \tag{A.17}$$

and the *Schur complement of* A_{11} as

$$D_{11} = A_{11} - A_{12}A_{22}^{-1}A_{21} \tag{A.18}$$

The inverse of A can be written as

$$A^{-1} = \begin{bmatrix} A_{11}^{-1} + A_{11}^{-1}A_{12}D_{22}^{-1}A_{21}A_{11}^{-1} & -A_{11}^{-1}A_{12}D_{22}^{-1} \\ -D_{22}^{-1}A_{21}A_{11}^{-1} & D_{22}^{-1} \end{bmatrix} \tag{A.19}$$

$$A^{-1} = \begin{bmatrix} D_{11}^{-1} & -D_{11}^{-1}A_{12}A_{22}^{-1} \\ -A_{22}^{-1}A_{21}D_{11}^{-1} & A_{22}^{-1} + A_{22}^{-1}A_{21}D_{11}^{-1}A_{12}A_{22}^{-1} \end{bmatrix} \tag{A.20}$$

or

$$A^{-1} = \begin{bmatrix} D_{11}^{-1} & -A_{11}^{-1}A_{12}D_{22}^{-1} \\ -A_{22}^{-1}A_{21}D_{11}^{-1} & D_{22}^{-1} \end{bmatrix} \tag{A.21}$$

depending of course on whether $|A_{11}| \neq 0$, $|A_{22}| \neq 0$, or both. These can be verified by checking that $AA^{-1} = A^{-1}\ A = I$. By comparing these various forms, we obtain the *well-known matrix inversion lemma*

$$\left(A_{11}^{-1} + A_{12}A_{22}A_{21}\right)^{-1} = A_{11} - A_{11}A_{12}\left(A_{21}A_{11}A_{12} + A_{22}^{-1}\right)^{-1}A_{21}A_{11} \tag{A.22}$$

The Schur complement arises naturally in the solution of linear simultaneous equations. If

$$\begin{bmatrix} A_{11} & A_{12} \\ A_{21} & A_{22} \end{bmatrix} \begin{bmatrix} X \\ Y \end{bmatrix} = \begin{bmatrix} 0 \\ Z \end{bmatrix} \tag{A.23}$$

then from the first equation

$$X = -A_{11}^{-1}A_{12}Y$$

and using this in the second equation yields

$$(A_{22} - A_{21}A_{11}^{-1}A_{12})Y = Z \tag{A.24}$$

If A is given by Equation A.16, then

$$|A| = |A_{11}| \cdot |A_{22} - A_{21}A_{11}^{-1}A_{12}| = |A_{22}| \cdot |A_{11} - A_{12}A_{22}^{-1}A_{21}| \tag{A.25}$$

Therefore, the determinant of A is the product of the determinant of A_{11} (respectively, A_{22}) and the determinant of the Schur complement of A_{22} (respectively, A_{11}).

A.3 Quadratic Forms and Definiteness

If $x \in R^n$ is a vector, then the square of the Euclidean norm is

$$\|x\|^2 = x^T x \tag{A.26}$$

If S is any nonsingular transformation, the vector Sx has a norm squared of $(Sx)^T Sx = x^T S^T Sx$. Letting $P = S^T S$, we write

$$\|x\|_P^2 = x^T Px \tag{A.27}$$

as the norm squared of Sx. We call $\|x\|_P$ the *norm of x with respect to P*.

We call

$$x^T Qx \tag{A.28}$$

a *quadratic form*. We shall assume Q is real.

Every real square matrix Q can be decomposed into a symmetric part Q_s (i.e., $Q_s^T = Q_s$) and an antisymmetric part Q_a (i.e., $Q_a^T = -Q_a$):

$$Q = Q_s + Q_a \tag{A.29}$$

where

$$Q_s = \frac{Q + Q^T}{2} \tag{A.30}$$

$$Q_a = \frac{Q - Q^T}{2} \tag{A.31}$$

If the quadratic form $x^T\ Ax$ has A antisymmetric, then it must be equal to zero since $x^T Ax$ is a scalar so that $x^T Ax = (x^T Ax)^T = -x^T A^T x$. For a general real square Q, then

$$x^T Qx = x^T (Q_s + Q_a)x = x^T Q_s x = x^T (Q + Q^T)x/2 \tag{A.32}$$

We can therefore assume without loss of generality that Q in Equation A.28 is symmetric. Let us do so.

We say Q is

Positive definite $(Q > 0)$ if $x^T Qx > 0$ for all nonzero x.

Positive semidefinite $(Q \geq 0)$ if $x^T Qx \geq 0$ for all nonzero x.

Negative semidefinite $(Q \leq 0)$ if $x^T Qx \leq 0$ for all nonzero x.

Negative definite $(Q < 0)$ if $x^T Qx < 0$ for all nonzero x.

Indefinite if $x^T Qx > 0$ for some x, $x^T Qx < 0$ for other x.

We can test for definiteness independently of the vectors x. If λ_i are the eigenvalues of Q, then

$$\begin{aligned} Q > 0 &\quad \text{if all } \lambda_i > 0 \\ Q \geq 0 &\quad \text{if all } \lambda_i \geq 0 \\ Q \leq 0 &\quad \text{if all } \lambda_i \leq 0 \\ Q < 0 &\quad \text{if all } \lambda_i < 0 \end{aligned} \tag{A.33}$$

Another test is provided as follows. Let $Q = [q_{ij}] \in R^{n \times n}$. The leading minors of Q are

$$m_1 = q_{11}, \qquad m_2 = \begin{vmatrix} q_{11} & q_{12} \\ q_{21} & q_{22} \end{vmatrix}, \qquad m_3 = \begin{vmatrix} q_{11} & q_{12} & q_{13} \\ q_{21} & q_{22} & q_{23} \\ q_{31} & q_{32} & q_{33} \end{vmatrix}, \ldots, m_n = |Q| \tag{A.34}$$

In terms of the leading minors, we have

$$\begin{aligned} &Q > 0 \text{ if } m_i > 0, \quad \text{all } i \\ &Q < 0 \text{ if } \begin{cases} m_i < 0, & \text{all odd } i \\ m_i > 0, & \text{all even } i \end{cases} \end{aligned} \tag{A.35}$$

Any positive semidefinite matrix Q can be factored into *square roots* either as

$$Q = \sqrt{Q}\sqrt{Q}^T \tag{A.36}$$

or as

$$Q = \sqrt{Q}^T\sqrt{Q} \tag{A.37}$$

The ("left" and "right") square roots in Equations A.36 and A.37 are not in general the same. Indeed, Q may have several roots since each of these factorizations is not even unique. If $Q > 0$, then all square roots are nonsingular.

If $P > 0$, then Equation A.27 is a norm. If $P \geq 0$, it is called a *seminorm* since $x^T Px$ may be zero even if x is not.

A.4 Matrix Calculus

Let $x = [x_1 \quad x_2 \quad \ldots \quad x_n]^T \in C^n$ be a vector, $s \in C$ a scalar, and $f(x) \in C^m$ an m vector function of x. The differential in x is

$$\mathrm{d}x = \begin{bmatrix} \mathrm{d}x_1 \\ \mathrm{d}x_2 \\ \vdots \\ \mathrm{d}x_m \end{bmatrix} \tag{A.38}$$

and the derivative of x with respect to s (which could be time) is

$$\frac{\mathrm{d}x}{\mathrm{d}s} = \begin{bmatrix} \mathrm{d}x_1/\mathrm{d}s \\ \mathrm{d}x_2/\mathrm{d}s \\ \vdots \\ \mathrm{d}x_n/\mathrm{d}s \end{bmatrix} \tag{A.39}$$

If s is a function of x, then the *gradient* of s with respect to x is the *column* vector

$$s_x \triangleq \frac{\partial s}{\partial x} = \begin{bmatrix} \partial s/\partial x_1 \\ \partial s/\partial x_2 \\ \vdots \\ \partial s/\partial x_n \end{bmatrix} \tag{A.40}$$

(The gradient is defined as a row vector in some references.) Then the total differential in s is

$$\mathrm{d}s = \left(\frac{\partial s}{\partial x}\right)^T \mathrm{d}x = \sum_{i=1}^{n} \frac{\partial s}{\partial x_i} \mathrm{d}x_i \tag{A.41}$$

If s is a function of two vectors x and y, then

$$\mathrm{d}s = \left(\frac{\partial s}{\partial x}\right)^T \mathrm{d}x + \left(\frac{\partial s}{\partial y}\right)^T \mathrm{d}y \tag{A.42}$$

The *Hessian* of s with respect to x is the second derivative

$$s_{xx} \triangleq \frac{\partial^2 s}{\partial x^2} = \left[\frac{\partial^2 s}{\partial x_i \partial x_j}\right] \tag{A.43}$$

which is a symmetric $n \times n$ matrix. In terms of the gradient and the Hessian, the *Taylor series expansion* of $s(x)$ about x_0 is

$$s(x) = s(x_0) + \left(\frac{\partial s}{\partial x}\right)^T (x - x_0) + \frac{1}{2}(x - x_0)^T \frac{\partial^2 s}{\partial x^2}(x - x_0) + 0(3) \tag{A.44}$$

where $0(3)$ represents terms of order 3 and derivatives are evaluated at x_0.

The *Jacobian* of f with respect to x is the $m \times n$ matrix

$$f_x \triangleq \frac{\partial f}{\partial x} = \begin{bmatrix} \frac{\partial f}{\partial x_1} & \frac{\partial f}{\partial x_2} & \cdots & \frac{\partial f}{\partial x_n} \end{bmatrix} \tag{A.45}$$

so that the total differential of f is

$$\mathrm{d}f = \frac{\partial f}{\partial x}\,\mathrm{d}x = \sum_{i=1}^{n} \frac{\partial f}{\partial x_i}\,\mathrm{d}x_i \tag{A.46}$$

We shall use the shorthand notation

$$\frac{\partial f^T}{\partial x} \triangleq \left(\frac{\partial f}{\partial x}\right)^T \in C^{n\times m} \tag{A.47}$$

If y is a vector, and A, B, D, and Q are matrices, all with dimensions so that the following expressions make sense, then we have the following results:

$$\frac{\mathrm{d}}{\mathrm{d}t}(A^{-1}) = -A^{-1}\dot{A}A^{-1} \tag{A.48}$$

Some useful gradients are

$$\frac{\partial}{\partial x}(y^T x) = \frac{\partial}{\partial x}(x^T y) = y \tag{A.49}$$

$$\frac{\partial}{\partial x}(y^T Ax) = \frac{\partial}{\partial x}(x^T A^T y) = A^T y \tag{A.50}$$

$$\frac{\partial}{\partial x}[y^T f(x)] = \frac{\partial}{\partial x}[f^T(x) y] = f_x^T y \tag{A.51}$$

$$\frac{\partial}{\partial x}(x^T Ax) = Ax + A^T x \tag{A.52}$$

and if Q is symmetric, then

$$\frac{\partial}{\partial x}(x^T Qx) = 2Qx \tag{A.53}$$

$$\frac{\partial}{\partial x}(x-y)^T Q(x-y) = 2Q(x-y) \tag{A.54}$$

The chain rule for two vector functions becomes

$$\frac{\partial}{\partial x}(f^T y) = f_x^T y + y_x^T f \tag{A.55}$$

Some useful Hessians are

$$\frac{\partial^2 x^T Ax}{\partial x^2} = A + A^T \tag{A.56}$$

and if Q is symmetric

$$\frac{\partial^2 x^T Q x}{\partial x^2} = 2Q \tag{A.57}$$

$$\frac{\partial^2}{\partial x^2}(x-y)^T Q(x-y) = 2Q \tag{A.58}$$

Some useful Jacobians are

$$\frac{\partial}{\partial x}(Ax) = A \tag{A.59}$$

(contrast this with Equation A.49), and the chain rule (with s a scalar)

$$\frac{\partial}{\partial x}(sf) = \frac{\partial}{\partial x}(fs) = sf_x + fs_x^T \tag{A.60}$$

(contrast this with Equation A.55).

If $A = [a_{ij}] \in R^{n\times n}$ is a matrix and s a scalar, then the derivative of s with respect to A is defined as

$$\frac{\partial s}{\partial A} = \left[\frac{\partial s}{\partial a_{ij}}\right] \in R^{n\times n} \tag{A.61}$$

Some useful derivatives involving the trace and determinant are

$$\frac{\partial}{\partial A}\text{trace}(A) = I \tag{A.62}$$

$$\frac{\partial}{\partial A}\text{trace}(BAD) = B^T D^T \tag{A.63}$$

$$\frac{\partial}{\partial A}\text{trace}(ABA^T) = 2AB, \quad \text{if } B = B^T \tag{A.64}$$

$$\frac{\partial}{\partial A}|BAD| = |BAD|A^{-T} \tag{A.65}$$

where $A^{-T} \triangleq (A^{-1})^T$.

References

Anderson, B. D. O., "An Algebraic Solution to the Spectral Factorization Problem," *IEEE Trans. Autom. Control*, **AC-12**(4), 410–414, 1967.

Anderson, B. D. O., "Exponential Data Weighting in the Kalman-Bucy Filter," *Inf. Sci.*, **5**, 217–230, 1973.

Anderson, B. D. O. and J. B. Moore, *Optimal Filtering*, Englewood Cliffs, NJ: Prentice-Hall, 1979.

Åström, K. J., *Introduction to Stochastic Control Theory*, New York: Academic Press, 1970.

Åström, K. J., "Self-Tuning Regulators-Design Principles and Applications," *Applications of Adaptive Control*, K. S. Narendra and R. V. Monopoli, eds., New York: Academic Press, 1980, pp. 1–68.

Bartle, R. G., *The Elements of Real Analysis*, New York: Wiley, 2nd Edition, 1976.

Bellman, R. E., *Dynamic Programming*, Princeton, NJ: Princeton University Press, 1957.

Bellman, R. E. and S. E. Dreyfus, *Applied Dynamic Programming*, Princeton, NJ: Princeton University Press, 1962.

Bellman, R. E. and R. E. Kalaba, *Dynamic Programming and Modern Control Theory*, New York: Academic Press, 1965.

Benedict, T. R. and G. W. Bordner, "Synthesis of an Optimal Set of Radar Track-While-Scan Smoothing Equations," *IRE Trans. Autom. Control*, **7**(4), 27–32, 1962.

Bennet, A., *Inverse Modeling of the Ocean and Atmosphere*, Cambridge: Cambridge University Press, 2002.

Bierman, G. J., *Factorization Methods for Discrete Sequential Estimation*, New York: Academic Press, 1977.

Bierman, G. J., "Efficient Time Propagation of U-D Covariance Factors," *IEEE Trans. Autom. Control*, **AC-26**(4), 890–894, 1981.

Blakelock, J. H., *Automatic Control of Aircraft and Missiles*, New York: Wiley, 1965.

Bode, H. W. and C. E. Shannon, "A Simplified Derivation of Linear Least Square Smoothing and Prediction Theory," *Proc. IRE*, **38**, 417–425, 1950.

Bolzern, P., P. Colaneri, and G. de Nicolao, "Finite Escape and Convergence Properties of Guaranteed-Cost Robust Filters," *Automatica*, **33**(1), 31–47, 1997.

Bolzern, P., P. Colaneri, and G. de Nicolao, "H_∞-Differential Riccati Equations: Convergence, Properties and Finite Escape Phenomena," *IEEE Trans. Autom. Control*, **AC-42**(1), 113–118, 1997.

Bolzern, P., P. Colaneri, G. de Nicolao, and U. Shaked,"Guaranteed H_∞ Robustness Bounds for Wiener Filtering and Prediction," *Int. J. Robust Nonlinear Contr.*, **12**, 41–56, 2001.

Bolzern, P. and M. Maroni, "New Conditions for the Convergence of H_∞ Filters and Predictors," *IEEE Trans. Autom. Control*, **AC-44**(8), 1564–1568, 1999.

Borison, U., "Self-Tuning Regulators for a Class of Multivariable Systems." *Automatica*, **15**, 209–215, 1979.

Brewer, J. W., "Kronecker Products and Matrix Calculus in System Theory," *IEEE Trans. Circuits Syst.*, **CAS-25**(9), 772–781, 1978.

Brogan, W. L., *Modern Control Theory*, New York: Quantum, 1974.

Bryson, Jr., A. E. and Y. C. Ho, *Applied Optimal Control*, New York: Hemisphere, 1975.

Bryson, Jr., A. E. and D. E. Johansen, "Linear Filtering for Time-Varying System Using Measurements Containing Colored Noise," *IEEE Trans. Autom. Control*, **AC-10**(1), 4–10, 1965.

Bucy, R. S., "Optimal Filtering for Correlated Noise," *J. Math. Anal. Appl.*, **2**, 1–8, 1967.

Businger, P. and G. H. Golub, "Linear Least Squares Solution by Householder Transformations," *Numer. Math.*, **7**, 269–276, 1965.

Casti, J., *Dynamical Systems and Their Applications: Linear Theory*, New York: Academic, 1977.

Clarke, D. W. and P. J. Gawthrop, "Self-Tuning Controller," *Proc. IEE*, **122**(9), 929–934, 1975.

Colaneri, P. and A. Ferrante, "A J-Spectral Factorization Approach for H_∞ Estimation Problems in Discrete Time," *IEEE Trans. Autom. Control*, **AC-47**(12), 2108–2113, 2002.

D'Appolito, J. A., "The Evaluation of Kalman Filter Designs for Multisensor Integrated Navigation Systems," The Analytic Sciences Corp., AFAL-TR-70–271, (AD 881286), January 1971.

D'Appolito, J. A. and C. E. Hutchinson, "Low Sensitivity Filters for State Estimation in the Presence of Large Parameter Uncertainties," *IEEE Trans. Autom. Control*, **14**(3), 310–312, 1969.

Darmon, C. A., "A Ship Tracking System Using a Kalman-Schmidt Filter," *Practical Aspects of Kalman Filtering Implementation*, AGARD-LS-82, Chapter 6, March 1976.

Davis, M. C., "Factoring the Spectral Matrix," *IEEE Trans. Autom. Control*, **AC-8**(4), 296–305, 1963.

de Souza, C. E. and A. Trofino,"An LMI Approach to the Design of Robust H_2 Filters," in *Recent Advances on Linear Matrix Inequality Methods in Control,* L. El Ghaoui and S. Niculescu (Eds.), SIAM, 1999.

de Souza, C. E. and L. Xie, "On the Discrete-Time Bounded Real Lemma with Application in the Characterization of Static State Feedback H_∞ Controllers," *Systems Control Lett.*, **18**, 61–71, 1992.

de Souza, C. E. and L. Xie, "Robust H_∞ Filtering," in *Advances in Control and Dynamic Systems: Digital Signal Processing Techniques and

Applications, C. T. Leondes (Ed.), pp. 323–377, New York: Academic Press, 1994.

Deyst, J. J., "A Derivation of the Optimum Continuous Linear Estimator for Systems with Correlated Measurement Noise," *AIAA J.*, **7**(11), 2116–2119, 1969.

Dyer, P. and S. R. McReynolds, "Extension of Square Root Filtering to Include Process Noise," *J. Optim. Theory Appl.*, **3**(6), 444 (1969).

Elbert, T. F., *Estimation and Control of Systems*, New York: Van Nostrand, 1984.

Fel'dbaum, A. A., *Optimal Control Systems*, New York: Academic Press, 1965.

Fenwick, J., P. Newman, and J. Leonard, "Cooperative Concurrent Mapping and Localization," in Proceedings of the IEEE International Conference on Robotics and Automation, pp. 1810–1817, Washington, DC, May, 2002.

Franklin, G. F. and J. D. Powell, *Digital Control of Dynamic Systems*, Reading, MA: Addison-Wesley, 1980.

Fraser, D. C., "A New Technique for the Optimal Smoothing of Data," Ph.D. Thesis, Massachusetts Institute of Technology, January 1967.

Friedland, B., "Treatment of Bias in Recursive Filtering," *IEEE Trans. Autom. Control*, **AC-14**(4), 359–367, 1969.

Fu, M., "Interpolation Approach to H_∞ Estimation and Its Interconnections to Loop Transfer Recovery," *Systems Control Lett.*, **17**, 29–36, 1991.

Fu, M., C. E. de Souza, and Z. Luo, "Finite Horizon Robust Kalman Filter Design," *IEEE Trans. Signal Processing,* **49**(9), 2103–2112, 2001.

Fu, M., C. E. de Souza, and L. Xie, "$H\infty$ Estimation for Uncertain Systems," *Int. J. Robust Nonlinear Control*, **2**, 87–105, 1992.

Gahinet, P. and P. Apkarian, "A Linear Matrix Inequality Approach to $\mathcal{H}_\infty$Control," *Int. J. Robust Nonlinear Control,* **4**, 421–448, 1994.

Gahinet, P., A. Nemirovski, A. J. Laub, and M. Chilali, *LMI Control Toolbox.* Natick, MA: The Math Works, Inc., 1995.

Gantmacher, F. R., *The Theory of Matrices*, New York: Chelsea, 1977.

Gauss, K. G., *Theory of Motion of Heavenly Bodies*, New York: Dover, 1963.

Gawthrop, P. J., "Some Interpretations of the Self-Tuning Controller," *Proc. Inst. Elec. Eng.*, **124**(10), 889–894, 1977.

Gelb, A., ed., *Applied Optimal Estimation*, Cambridge, MA: MIT Press, 1974.

Geromel, J. C., J. Bernussou, G. Garcia, and M. C. de Oliveira, "H_2 and $\mathcal{H}_\infty$ Robust Filtering for Discrete-Time Linear Systems," *SIAM J. Control Optimization,* **38**(5), 1353–1368, 2000.

Geromel, J. C., M. de Oliveira, and J. Bernussou, "Robust Filtering of Discrete-Time Linear Systems with Parameter Dependent Lyapunov Functions," *SIAM J. Control Optimization*, **41**(3), 700–711, 2002.

Green, M. and D. J. N. Limebeer, *Linear Robust Control,* Englewood Cliffs, NJ: Prentice-Hall, 1995.

Huddle, J. R., "On Suboptimal Linear Filter Design," *IEEE Trans. Autom. Control*, **12**(3), 317–318, 1967.

Huddle, J. R. and D. A. Wismer, "Degradation of Linear Filter Performance Due to Modeling Error," *IEEE Trans. Autom. Control*, 421–423, 1968.

IMSL, *Library Contents Document*, 8th Edition, IMSL, Houston, TX, 1980.

Jain, B. N., "Guaranteed Error Estimation in Uncertain Systems," *IEEE Trans. Autom. Control,* **20**, 230–232, 1975.

Jazwinski, A. H., *Stochastic Processes and Filtering Theory*, New York: Academic Press, 1970.

Kailath, T., "An Innovations Approach to Least-Squares Estimation, Part I: Linear Filtering in Additive White Noise," *IEEE Trans. Autom. Control*, **AC-13**(6), 646–655, 1968.

Kailath T., "A View of Three Decades of Linear Filtering Theory," *IEEE Trans. Inf. Theory*, **20**(2), 146–181, 1974.

Kailath, T., *Lectures on Wiener and Kalman Filtering*, CISM Courses and Lectures, No. 140, New York: Springer, 1981.

Kailath, T., *Linear Systems*, Englewood Cliffs, NJ: Prentice-Hall, 1980.

Kailath, T., A. H. Sayed, and B. Hassibi, *Linear Estimation,* Upper Saddle River, NJ: Prentice-Hall, 1999.

Kalman, R. E., "A New Approach to Linear Filtering and Prediction Problems," *Trans. ASME J. Basic Eng.*, **82**, 34–35, 1960.

Kalman, R. E., "New Methods in Wiener Filtering Theory," in Proceedings of the Symposium on Engineering Applications of Random Function Theory of Probability, New York: Wiley, 1963.

Kalman, R. E. and R. S. Bucy, "New Results in Linear Filtering and Prediction Theory," *Trans. ASME J. Basic Eng.*, **83**, 95–108, 1961.

Kaminski, P. G., A. E. Bryson, and S. F. Schmidt, "Discrete Square Root Filtering: A Survey of Current Techniques," *IEEE Trans. Autom. Control*, **AC-16**(6), 727–736, 1971.

Koivo, H. N., "A Multivariable Self-Tuning Controller," *Automatica*, **16**, 351–366, 1980.

Kolmogorov, A. N., "Sur l'interpolation et Extrapolation des Suites Stationnaires," *C. R. Acad. Sci.*, **208**, 2043, 1939.

Kolmogorov, A., "Interpolation and Extrapolation," *Bull. Acad. Sci.*, U.S.S.R., Ser. Math., **5**, 3–14, 1941.

Kolmogorov, A. N., "Interpolation and Extrapolation of Stationary Random Sequences," translated by W. Doyle and J. Selin, Rept. RM-3090-PR, Santa Monica, CA: RAND Corp., 1962.

Kortüm, W., "Design and Analysis of Low-Order Filters Applied to the Alignment of Inertial Platforms," *Practical Aspects of Kalman Filtering*, AGARD-LS-82, Chapter 7, 1976.

Kumar, V., G. A. S. Pereira, A. K. Das, and M. F. M. Campos, "Decentralized Motion Planning for Multiple Robots Subject to Sensing and Communications Constraints," in Proceedings of the Second Multi Robot Systems Workshop, 2003.

Kwakernaak, H. and R. Sivan, *Linear Optimal Control Systems*, New York: Wiley-Interscience, 1972.

Letov, A. M., "Analytical Controller Design, I, II," *Autom. Remote Control*, **21**, 303–306, 1960.

Lewis, F. L., *Optimal Control*, New York: Wiley, 1986.

Li, H. and M. Fu, "A Linear Matrix Inequality Approach to Robust $\mathcal{H}_\infty$ Filtering," *IEEE Trans. Signal Processing*, **45**, 2338–2350, 1997.

Limebeer, D. J. and U. Shaked, "New Results on H_∞ Filtering: Recent Advances in Mathematical Theory of Systems, Control, Networks and Signal Processing," in Proceedings of the International Symposium MTNS-91, Kobe, Japan, vol. 1, pp. 317–322, June 1991.

Limebeer, D. J. and U. Shaked, "New Results on Infinity Norm Optimal Filtering," in Proceedings of the Workshop on Robust Control System Design Using H_∞ and Related Methods, Cambridge, U.K., 21–22, March 1991.

Luenberger, D. G., *Optimization by Vector Space Methods*, New York: Wiley, 1969.

Luenberger, D. G., *Introduction to Dynamic Systems*, New York: Wiley, 1979.

Mataric, M. J., A. Howard, and G. S. Sukhatme, "Mobile Sensor Network Deployment Using Potential Fields: A Distributed, Scalable Solution to the Area Coverage Problem," in Proceedings of the 6th International Symposium on Distributed Autonomous Robotics Systems, 2002.

Maybeck, P. S., *Stochastic Models, Estimation, and Control*, Vol. 1, Mathematics in Science and Engineering, Series, Vol. 141-1, New York: Academic Press, 1979.

McReynolds, S. R., Ph.D. Thesis, Harvard University, Cambridge, MA, 1966.

Melsa, J. L. and D. L. Cohn, *Decision and Estimation Theory*, New York: McGraw-Hill, 1978.

Morf, M. and T. Kailath, "Square Root Algorithms for Least-Squares Estimation," *IEEE Trans. Autom. Control*, **AC-20**(4), 487–497, 1975.

Nagpal, K. M. and P. P. Khargonekar, "Filtering and Smoothing in an H_∞ Setting," *IEEE Trans. Autom. Control,* **AC-36**(2), 152–166, 1991.

Nash, R. A., J. A. D'Appolito, and K. J. Roy, "Error Analysis of Hybrid Inertial Navigation Systems," AIAA Guidance and Control Conference, Paper No. 72-848, Stanford, CA, August 1972.

Papoulis, A., *Probability, Random Variables, and Stochastic Processes*, New York: McGraw-Hill, 1965.

Petersen, I. R., "Complete Results for a Class of State Feedback Disturbance Attenuation Problems," *IEEE Trans. Autom. Control,* **AC-34**(11), 1196–1199, 1989.

Petersen, I. R. and D. C. McFarlane, "Robust State Estimation for Uncertian Systems," in Proceedings of the 30th IEEE Conference on Decision and Control, Brighton, U.K., 2630–2631, 1991.

Petersen, I. R. and D. C. McFarlane, "Optimal Guaranteed Cost Control and Filtering for Uncertain Linear Systems," *IEEE Trans. Autom. Control,* **39**, 1971–1977, 1994.

Petersen, I. R. and D. C. McFarlane, "Optimal Guaranteed Cost Filtering for Uncertain Discrete-Time Linear Systems," *Int. J. Robust Nonlinear Control,* **6**, 267–280, 1996.

Parkus, H., *Optimal Filtering*, New York: Springer, 1971.

Popa, D. O., M. F. Mysorewala, and F. L. Lewis, "EKF-Based Adaptive Sampling with Mobile Robotic Sensor Nodes," Intelligent Robots and Systems, (IROS 2006). IEEE International Conference, October 2006.

Popa, D. O., A. C. Sanderson, R. J. Komerska, S. S. Mupparapu, D. R. Blidberg, and S. G. Chappel, "Adaptive Sampling Algorithms for Multiple Autonomous Underwater Vehicles," Autonomous Underwater Vehicles, 2004 IEEE/OES, pp. 108–118, 17–18 June.

Potter, J. E., "Matrix Quadratic Solutions," *J. SIAM Appl. Math.*, **14**(3), 496–501, 1966.

Rauch, H. E., F. Tung, and C. T. Striebel, "Maximum Likelihood Estimates of Linear Dynamic Systems," *AIAA J.* **3**(8), 1445–1450, 1965.

Rhodes, I. B., "A Tutorial Introduction to Estimation and Filtering," *IEEE Trans. Autom. Control*, **16**(6), 688–706, 1971.

Sayed, A. H., "A Framework for State-Space Estimation with Uncertain Models," *IEEE Trans. Autom. Control,* **AC-46**(7), 998–1013, 2001.

Schmidt, S. F., "Application of State-Space Methods to Navigation Problems," *Advan. Control Syst.*, **3**, 293–340, 1966.

Schmidt, S. F., "Estimation of State with Acceptable Accuracy Constraints," TR 67-16, Palo Alto: Analytical Mechanics Assoc., CA, 1967.

Schmidt, S. F., "Computational Techniques in Kalman Filtering," *Theory and Applications of Kalman Filtering*, Chapter 3, NATO Advisory Group for Aerospace Research and Development, AGARDograph 139, February 1970.

Schooler, S. F., "Optimal α–β Filters for Systems with Modeling Inaccuracies," *IEEE Trans. Aerospace Electron. Syst.*, **AES-11**(6), 1300–1306, 1975.

Schultz, D. G. and J. L. Melsa, *State Functions and Linear Control Systems*, New York: McGraw-Hill, 1967.

Shaked, U., "A General Transfer Function Approach to Linear Stationary Filtering and Steady-State Optimal Control Problems," *Int. J. Control*, **24**(6), 741–770, 1976.

Shaked, U. and C. E. de Souza, "Robust Minimum Variance Filtering," *IEEE Trans. Autom. Control,* **AC-43**(11), 2474–2483, 1995.

Shaked, U. and C. E. de Souza, "Robust Minimum Variance Filtering," *IEEE Trans. Signal Processing,* **43**, 2474–2483, 1995.

Shaked, U., L. Xie, and Y. C. Soh, "New Approaches to Robust Minimum Variance Filter Design," *IEEE Trans. Signal Processing,* **49**, 2620–2629, 2001.

Simon, K. W. and A. R. Stubberud, "Reduced Order Kalman Filter," *Int. J. Control*, **10**(5), 501–509, 1969.

Sorenson, H. W., "Comparison of Kalman, Bayesian, and Maximum-Likelihood Estimation Techniques," AGARDograph 139, Chapter 6, 1970a.

Sorenson, H. W., "Least-Squares Estimation: From Gauss to Kalman," *IEEE Spectrum*, 63–68, 1970b.

Stear, E. B. and A. R. Stubberud, "Optimal Filtering for Gauss-Markov Noise," *Int. J. Control*, **8**(2), 123–130, 1968.

Strintzis, M. G., "A Solution to the Matrix Factorization Problem," *IEEE Trans. Inform. Theory*, **IT-18**(2), 225–232, 1972.

Swerling, P., "A Proposed Stagewise Differential Correction Procedure for Satellite Tracking and Prediction," Report P-1292, CA: Santa Monica, The Rand Corp., January 1958.

Swerling, P., "Topics in Generalized Least Squares Signal Estimation," *J. SIAM Appl. Math.*, **14**(5), 998–1031, 1966.

Swerling, P., "Modern State Estimation Methods from the Viewpoint of the Method of Least Squares," *IEEE Trans. Autom. Control*, **AC-16**(6), 707–719, 1971.

Takaba, K. and T. Katayama, "Discrete-Time H_∞ Algebraic Riccati Equation and Parametrization of all H_∞ Filters," *Int. J. Control,* **64**(6), 1129–1149, 1996.

Theodor, Y. and U. Shaked, "Robust Discrete-Time Minimum-Variance Filtering," *IEEE Trans. Signal Processing,* **44**, 181–189, 1996.

Tuan, H. D., P. Apkarian, and T. Q. Nguyen, "Robust and Reduced-Order Filtering: New LMI-Based Characterizations and Methods," *IEEE Trans. Signal Processing*, **49**(12), 2975–2984, 2001.

Van Trees, H. L., *Detection, Estimation, and Modulation Theory, Part 1*, New York: Wiley, 1968.

Wagner, W. E., "Re-Entry Filtering, Prediction, and Smoothing," *J. Spacecraft Rockets*, **3**, 1321–1327, 1966.

Wauer, J. C., "Practical Considerations in Implementing Kalman Filters," *Practical Aspects of Kalman Filtering Implementation*, AGARD-LS-82, Chapter 2, March 1976.

Weiss, I. M., "A Survey of Discrete Kalman-Bucy Filtering with Unknown Noise," AIAA Paper No. 70-955, AIAA Guidance, Control, and Flight Mechanics, Conference, August 1970.

Wiener, N., *The Extrapolation, Interpolation, and Smoothing of Stationary Time Series*, New York: Wiley, 1949.

Wiener, N. and P. Masani, "The Prediction Theory of Multivariable Stochastic Processes, I. The Regularity Condition," *Acta Math.*, **98**, 111–150, 1957.

Wiener, N. and P. Masani, The Prediction Theory of Multivariable Stochastic Processes, II. The Linear Predictor. *Acta Math.*, **99**, 93–137, 1958.

Wong, E. and J. B. Thomas, "On the Multidimensional Prediction and Filtering Problem and the Factorization of Spectral Matrices," *J. Franklin Inst.*, **272**(2), 87–99, 1961.

Xie, L., "H_∞ Output Feedback Control of Systems with Parameter Uncertainty," *Int. J. Control,* **63**, 741–750, 1996.

Xie, L., "On Robust H2 Estimation," *Acta Automatica Sinca,* **31**(1), 1–12, 2005.

Xie, L., C. E. de Souza, and M. Fu, "H_∞ Estimation for Discrete-Time Linear Uncertain Systems," *Int. J. Robust Nonlinear Control,* **1**, 111–123, 1991.

Xie, L., L. Lu, D. Zhang, and H. Zhang, "Improved Robust H_2 and H_∞ Filtering for Uncertain Discrete-Time Systems," *Automatica,* **40**, 873–880, 2004.

Xie, L. and Y. C. Soh, "Robust Kalman Filtering for Uncertian Systems," *Systems Control Lett.,* **22**, 123–129, 1994.

Xie, L. and Y. C. Soh, "Guaranteed Cost Control of Uncertain Discrete-Time Systems," *Control Theory Advanced Technology, Japan,* **10**(4), Part 2, 1235–1251, 1995.

Xie, L., Y. C. Soh, and C. E. de Souza, "Robust Kalman Filtering for Uncertain Discrete-Time Systems," *IEEE Trans. Autom. Control,* **39**, 1310–1314, 1994.

Xie, L., Y. C. Soh, and C. Du, "Robust H_2 Estimation and Control," in Proceedings of 7th International Conference on Control, Automation, Robotics and Vision, Singapore, December 2002.

Yaesh, I., Linear Optimal Control and Estimation in the Minmum H_∞-norm Sense, Ph.D. thesis, Tel-Aviv University, 1992.

Yaesh, I. and U. Shaked, "H_∞ Optimal State Estimation of Discrete-Time Linear Systems," in Proceedings of the International Symposium on Mathematical Theory Network System, Kobe, Japan, June 1991.

Yaesh, I. and U. Shaked, "A Transfer Function Approach to the Problems of Discrete-Time Systems H_∞—Optimal Contol and Filtering," *IEEE Trans. Autom. Control,* **AC-36**, 1264–1271, 1991.

Yaesh, I. and U. Shaked, "Game Theory Approach to Optimal Linear State Estimation and Its Relation to H_∞ Norm Estimation," *IEEE Trans. Autom. Control,* **AC-37**, 828–831, 1992.

Youla, D. C. "On the Factorization of Rational Matrices," *IRE Trans. Inform. Theory,* **IT-17**(3), 172–189, 1961.

Yovits, M. C. and J. L. Jackson, "Linear Filter Optimization with Game Theory Considerations," *IRE Natl. Convention Rec.*, Pt. **4**, 193–199, 1955.

Zhang, H., L. Xie, and Y. C. Soh, "H-Infinity Deconvolution Filtering, Prediction and Smoothing: A Krein Space Polynomial Systems Approach," *IEEE Trans. Signal Processing, USA*, **SP-48**(3), 888–892, 2000.

Zhang, H., L. Xie, and Y. C. Soh, "Discrete J-Spectral Factorization," *Systems Control Lett.,* **43**, 275–286, 2001.

Zhou, K., J. C. Doyle, and K. Glover, *Robust and Optimal Control,* Upper Saddle River, NJ: Prentice-Hall, 1995.

Zhu, X., Y. C. Soh, and L. Xie, "Design and Analysis of Discrete-Time Robust Kalman Filters," *Automatica, UK*, **38**(6), 1069–1077, 2002.

Zhu, X., Y. C. Soh, and L. Xie, "Feasibility and Convergence Analysis of Discrete-Time H_∞ A Priori Filters," in Proceedings of the American Control Conference, Chicago, Illinois, pp. 4179–4183, 2000.

Index

A

Ackermann's formula
  Chang-Letov equation, Kalman filter design, 115–118
  discrete-time Kalman filtering, deterministic state observer, 60–64
actual error correlation
  continuous reduced-order filters, 222–223
  Kalman filters, model mismatch, continuous Kalman filter, 222
  reduced-order Kalman filter decoupling, suboptimal filters, weakly coupled systems, 240–244
actual norms, stationary H∞ filters, 371–373
adaptive sampling, extended Kalman filter, 283–304
  algorithms, 296
  closed-form estimation, linear field, 286–290
  closed-form estimation, uncertainty in linear field, 290–295
  combined adaptive sampling technique, 284–286
  measurement update, 296
  mobile robot
    color field sampling, 301–304
    localization, 284
    simultaneous localization and sampling, 297–304
  parameter-linear field, 296–297
  system equations, 295–296
  time update, 295
additive noise, deterministic state observer, 63–64
aircraft longitudinal dynamics
  continuous-time Kalman filter derivation, gust noise, 211
  gust noise, 125–127
algebraic Lyapunov equation
  continuous-time Kalman filter derivation, solutions, 204
  solutions, 140–141
algebraic Riccati equation (ARE)
  continuous-time Kalman filter derivation
    analytic solution, 207
    error dynamics, 179–180
    solutions, 208–209
    spectral factorization, 181–183
  discrete Wiener filter, 118–123
  dscrete-time H∞ filtering, stationary case, 397–400
  error dynamics, statistical steady state, 103–110
  H∞ filtering
    continuous-time systems, 356–357
    convergence analysis, 374–378
    J-spectral factorization, 410–414
  robust Kalman filters
    asymptotically stable filters, 342–343
    convergence analysis and, 326–329
  robust stationary Kalman filtering, 322–326
  scalar variant, 145
  stabilization via, 140
  stationary H∞ filters, 367–373
    a posteriori filtering, 406–408
  steady-state Kalman filter, 122–123
  weighted minimum-variance estimate, 110
alpha-beta-gamma radar tracking, 95
alpha-beta radar tracking
  continuous-discrete Kalman filter, 195–199
  discrete-time Kalman filter, 92–98
angle-of-attack measurements, Kalman and Wiener filtering, 121–123
a posteriori filtering
  discrete-time H∞ filtering, 400–408
    finite horizon case, 400–405

a posteriori filtering (*contd.*)
overview, 388–390
stationary case, 405–408
discrete-time Kalman filter, 71–75
discrete-time linear quadratic Gaussian problem, incomplete information, 456–457
error covariance recursion, 141–142
error systems, 110
Kalman filters, modeling errors, 214–223
linear measurement updates, 269–271
mean-square estimation, 5–6
measurement update approximation, extended Kalman filter, 272–273
optimal smoothing, 136–140
robust Kalman filters, 338–340
apparent divergence, Kalman filter, 223–226
a priori filtering
correlated measurement and process noise, 128–132
discrete-time Kalman filter, 71–75
discrete-time linear quadratic Gaussian problem, incomplete information, 456–457
dscrete-time H∞ filtering, 390–400
finite horizon case, 391–397
stationary case, 397–400
error system, 102
Kalman filters, modeling errors, 214–223
linear mean-square estimation, 15–19
linear measurement updates, 269–271
measurement update approximation, extended Kalman filter, 272–273
optimal smoothing, 136–140
robust finite horizon Kalman filter, 317–321
robust Kalman filtering, linear matrix inequality, 334–335
robust stationary Kalman filter, 321–326
robust stationary Kalman filters, convergence analysis, 329–331
stochastic control, feedback of, 461–462
asymptotically stable filters
H∞ filtering, signal models, 378–381
robust Kalman filtering, 342–343
stationary H∞ filters
a posteriori filtering, 405–408
Riccati equation approach, 365–373
autocorrelation function, Wiener filtering, 34
autonomous vehicles (AVs), extended Kalman filter, combined adaptive sampling, 285–286
average closed-loop behavior, stochastic control, state variable systems, 441–443

B

backward filter updates
continuous-discrete Kalman filter, information, 200–201
optimal smoothing, 134–140
ballistic coefficient, extended Kalman filter, 310–312
batch processing, recursive estimation, 28–30
Bayesian estimation, mean-square estimation, 4
Bayes' rule
hyperstate propagation, discrete systems, 263
mean-square estimation, 6–7
stochastic control, state variable systems
continuous-time systems, 436–443
discrete-time systems, 425–426
Bellman's principle of optimality
dynamic programming, stochastic control, state variable systems, 423–424
stochastic control, state variable systems, continuous-time systems, 437–443
block diagonals, sequential estimation, 55
block triangular filter structure, Kalman filtering, 256
boundary conditions, stochastic control, state variable systems, discrete-time systems, 424
bounded behavior
H∞ filtering, convergence analysis, 376–378
Kalman filter divergence, process noise injection, 230
robust Kalman filters, difference Riccati equation, 343–345

robust stationary Kalman filters, 325–326
Butterworth spectrum, FM demodulation, extended Kalman filter, 275–278

C

causal steady-state filtering
 H∞ filtering, continuous-time systems, finite horizon linear filters, 362–365
 H∞ filtering, signal models, 379–381
 stationary H∞ filters, Riccati equation approach, 365–373
 Wiener-Hopf equation, 41–49
certainty equivalence principle, continuous-time linear quadratic Gaussian problem, 450–453
chain rule, hyperstate propagation, discrete systems, 260–263
Chang-Letov equation
 continuous-time Kalman filter derivation, 184–188
 Kalman filter design, frequency-domain solution, 115–118
channel equalization, H∞ filtering applications, 414–418
Chapman-Kolmogorov equation
 hyperstate propagation, discrete systems, 260–263
 stochastic control, state variable systems, 431–435
Cholesky decomposition, scalar measurement updating, Kalman filter, 253–254
classical estimation theory
 Cramér-Rao bound, 25–28
 maximum-likelihood estimation, 19–25
 mean-square estimation, 3–19
 recursive estimation, 28–34
 Wiener filtering, 34–49
closed-form estimation, linear field
 extended Kalman filter, 286–290
 localization uncertainty, 290–295
closed-loop system
 Lyapunov equation, 142
 stochastic control, polynomial systems, 472–473
color field sampling, mobile robot localization, 301–304
colored measurement noise
 continuous-time Kalman filter derivation, 190–193, 210
 discrete-time Kalman filtering, 130–132
 tachometer vibration, 148
colored noise process, continuous-time Kalman filter derivation, 188–189
colored process noise, 149
 shaping filters, 124–127
combined adaptive sampling, extended Kalman filter, 284–286
completely unreliable measurements
 continuous-time Kalman filter derivation, scalar system, 161–166
 dscrete-time H∞ filtering, finite horizon case, 394–397
 scalar system filtering, 83
complete state information
 continuous-time linear quadratic Gaussian problem, 445–449
 discrete-time linear quadratic Gaussian problem, 454–455
conditional expected cost, stochastic control, state variable systems, 425–426
 continuous-time systems, 436–443
conditional mean estimate
 linear Gaussian measurements, 11–12
 mean-square estimation, 6–7
 nonlinear mean-square estimation, 8–9
conditional probabilities, stochastic control, state variable systems, 433–435
constant parameter estimation
 continuous stochastic systems, 88–89
 H∞ filter estimation, 419
constant unknown condition, Kalman filters, modeling errors
 Newton's system, 255
 Wiener filtering, 218–223
continuous-continuous extended Kalman filter, 275
continuous-discrete Kalman filter
 convergence of, 211–212
 nonlinear estimates, 273–274
 system and measurement models, 193–197
continuous estimation updating, 51

continuous reduced-order filters, model mismatch, 222–223
continuous spectrum-shaping filters, properties of, 126
continuous stochastic systems, discrete-time Kalman filtering, 85–98
continuous-time Kalman filter
  actual error correlation, 222
  constant scalar unknown estimation, 157–159
  discrete Kalman filter derivation, 151–157
  discrete Kalman measurements, 84–101
  discrete measurements, 193–197
  error covariance differential equation, 170–171
  error dynamics and statistical steady state, 177–180
  estimate differential equation, 169–170
  frequency domain results, 180–188
    Chang-Letov design, 184–188
    linear stochastic system spectral densities, 181
    spectral factorization, 181–183
  kinematic modeling, 244–248
  multiple sampling rates, 101
  multiple-system rates, 98–101
  noise correlation and shaping, 188–193
    colored measurement noise, 190–193
    colored process noise, 188–189
    correlated measurement and process noise, 189–190
  optimal impulse response differential equation, 168–169
  optimal smoothing, 197–203
    estimation, 201–202
    information filter, 200–201
    Rauch-Ting-Striebel smoother, 203
  robust filtering for, 341–342
  scalar systems, 160–166
  shaping filter derivation, 167–168
  stochastic system discretization, 85–98
  Wiener-Hopf equation, 166–167
continuous-time systems
  extended Kalman filter applications, 273–283
  H∞ filtering
    convergence analysis, 375–378
    finite horizon linear filter, 357–365
    guaranteed *vs.* actual norms, 371–373
    Kalman filter *vs.*, 373–374
    linear matrix inequality, 382–383
    linear time-invariant filters, 373
    overview, 353–355
    signal model applications, 378–381
    stationary Riccati equation, 365–371
    two-person zero-sum game, 356–357
  hyperstate propagation, nonlinear estimation, 263–265
  stochastic control, polynomial systems, 484
  stochastic control, state variable systems
    dynamic programming, 435–443
    linear quadratic Gaussian problem, 443–453
      complete state information, 445–449
      incomplete state information, separation principle, 449–453
continuous Wiener process, suboptimal gains, 250–253
convergence analysis
  H∞ filtering, 374–378
  robust stationary Kalman filters, 326–331
    ε-switching strategy, 329–331
    feasibility, 326–329
convex optimization, robust Kalman filter modeling, linear matrix inequality, 337–340
correlated measurement
  continuous-time Kalman filter
    actual error correlation, 222
    derivation, 189–190
    measurement noise and, 211
correlated noise
  continuous-time Kalman filter
    derivation, 188–193
    colored measurement noise, 190–193
    colored process noise, 188–189
    correlated measurement and process noise, 189–190
  linear measurements, 52
  measurement and process noise, 127–132, 149

correlation functions, linear estimation, 37–38
cost function
  dscrete-time H∞ filtering, finite horizon case, 396–397
  H∞ filtering, continuous-time systems, 356–357
  mean-square estimation, 4
covariances
  closed-form estimation, extended Kalman filter, 287–290
  finite horizon robust Kalman *a priori* filter, 317–321
  Kalman filter, exponential data weighting, 234–236
  linear stochastic systems, 65–70
  nonlinear updates, 265–266
  radar tracking model, 92
Cramér-Rao bound, basic principles, 25–28
cross-variance, linear Gaussian measurements, 11–12
curse of dimensionality, continuous-time systems, dynamic programming, 435–443

D

damped harmonic oscillator, continuous-time Kalman filter derivation, 164–166
data prewhitening
  causal steady-state filtering, 41–49
  recursive estimation, 33–34
dead-reckoning estimates, mobile robot localization, simultaneous localization and sampling, 300–304
decoupling, reduced-order Kalman filters, 242–248
  Kalman-Schmidt error covariance equations, 248
  kinematic modeling, continuous systems, 244–248
derived measurements
  colored measurement noise, 131–132
  continuous-time Kalman filter derivation, colored measurement noise, 190–193
  reduced-order Kalman filter decoupling, 243–244
detectability parameters, H∞ filter estimation, polynomial approach, 408–410
deterministic asymptotic estimation, discrete-time Kalman filtering, 59–64
deterministic functions, continuous-time systems, dynamic programming, 435–443
deterministic inputs, linear stochastic systems, 64
deterministic linear quadratic regulation, stochastic control, state variable systems, 447–449
deterministic state observer, discrete-time Kalman filtering, 59–64
deterministic variable, Cramér-Rao bound, 25–28
Deyst filter, continuous-time Kalman filter derivation, 192–193
difference Riccati equation (DRE)
  discrete-time H∞ filtering, 388–390
  dscrete-time H∞ filtering, finite horizon case, 391–397
  finite horizon H∞ filters, *a posteriori* filtering, 400–405
  finite horizon robust Kalman *a priori* filter, 318–321
  H∞ filtering
    continuous-time systems, 356–357
    convergence analysis, 374–378
    finite horizon linear filter, 357–365
  robust Kalman filters
    convergence analysis and, 326–329
    proof, 343–345
  robust stationary Kalman filters, 324–326
differential drive robots, simultaneous localization and sampling, 298–301
Diophantine equation, stochastic control, polynomial systems
  closed-loop system, 472–473
  linear quadratic Gaussian control, 476–481
  optimal prediction, 465–469
discrete Markov process, linear stochastic systems, 64–65
discrete polynomial tracking filter, continuous-time Kalman filter derivation, 206
discrete spectrum-shaping filters, properties of, 123

discrete stochastic dynamical equation, linear stochastic systems, 64
discrete systems, hyperstate propagation, nonlinear estimation, 259–263
discrete-time H∞ filtering
  a priori filter, 390–400
    finite horizon case, 391–397
    stationary case, 397–400
  channel equalization, 414–418
  J-spectral factorization, 410–414
  overview, 388–390
  polynomial approach, 408–410
  stationary case, 400–408
discrete-time Kalman filter
  a posteriori error system, 110
  a priori recursive formulation, 74–75
  algebraic Riccati equation, 103–110
  algorithm, 71–73
  alternative measurement update equations, 73–74
  Chang-Letov design procedure, 115–118
  colored measurement noise, 130–132
  colored process noise, 124–127
  continuous-time discrete measurements, 84–98
  continuous-time Kalman filter derivation, 151–157
  correlated measurement and process noise, 127–129
  correlated noise and shaping, 123–124
  deterministic state observer, 59–64
  error dynamics, 101–102
  formulation, 71
  frequency domain results, 112–114
  information filter, 133–136
  innovations representation, 114
  innovations sequence, 102–103
  linear stochastic systems, 64–70
  multiple sampling rates, 98–101
  optimal smoothing, 132–138
  orthogonality principle, 84
  Rauch-Tung-Striebel smoother, 138–140
  recursive estimation, constant scalar unknown, 78–81
  scalar system filtering, 81–84
  ship navigational fixes, 75–78
  spectral factorization, 112–114
  suboptimal gains, 247–253
  time-varying systems, 101, 110–112
  weighted minimum-variance estimate, 110
  Wiener filter derivation, 118–123
discrete-time systems, stochastic control
  dynamic programming, 424–435
  linear quadratic Gaussian problem, stochastic control, 453–457
    complete state information, 454–455
    incomplete state information, 455–457
discretization, radar tracking, 91–92
divergence
  continuous Kalman filter, 254–255
  Kalman filters, 223–226
    process noise injection and prevention of, 227–230
dscrete-time H∞ filtering, *a posteriori* filtering, 400–408
  finite horizon case, 400–405
  stationary case, 405–408
dynamic output feedback
  continuous-time linear quadratic Gaussian problem, 451–453
  discrete-time linear quadratic Gaussian problem, incomplete information, 455–457
dynamic programming, stochastic control, state variable systems, 424–443
  continuous-time systems, 435–443
  discretetime systems, 424–435

E

ε-switching strategy, robust stationary Kalman filters, convergence analysis, 329–331
efficiency condition, Cramér-Rao bound, 27–28
eigenvalues, H∞ filtering, signal models, 379–381
electric circuit analog, maximum-likelihood estimation, 24–25
equality, Cramér-Rao bound, 27–28
equalization filtering, H∞ filtering applications, channel equalization, 414–418
error covariance
  continuous-time Kalman filter derivation, 153–157
    differential equation, 170–172

continuous-time linear quadratic Gaussian problem, 449–453
continuous Wiener process, suboptimal gains, 251–253
Cramér-Rao bound, 26–28
discrete-time Kalman filter, 72–73
finite horizon robust Kalman *a priori* filter, 318–321
H∞ filtering *vs.* Kalman filtering, 373–374
Kalman filters
exponential data weighting, 232–236
modeling errors, 215–223
Wiener filtering, constant unknown condition, 218–223
timing diagram, 70
Kalman-Schmidt filter, 248
linear estimation, 37–38
linear Gaussian measurements, 12, 22–25
linear mean-square estimation, 16–19
linear measurement updates, 269–271
maximum-likelihood estimation, electric circuit analog, 24–25
mean-square estimation, 4–5, 7
optimal smoothing, 134–140
preliminary analysis, Kalman filter simplification, 95–98
recursive estimation, scalar unknown, 79–81
robust Kalman filters
convergence analysis, 329–331
stationary filter, 323–326
uncertainty in, 316–317
scalar linear measurements, 46–49
scalar uknown, independent measurement, 23–25
sequential maximum-likelihood estimation, 32–33
ship navigational fixes, 77–78
stationary H∞ filters, Kalman filters *vs.*, 370–373
stochastic control, polynomial systems, optimal prediction, 467–469
suboptimal gain filtering, 249–253
error covariance update equation, sequential maximum-likelihood estimation, 32–33
error dynamics
continuous-time Kalman filter derivation, 177–180
discrete-time Kalman filters, 101–102
time-variant plant, 110–112
weighted minimum-variance estimate, 110
error system
a posteriori, 110
algebraic Ricatti equation, 108–110
discrete-time Kalman filtering, 60–64
Kalman filters, modeling errors, 217–223
statistical steady state, 101–102
suboptimal gain filtering, 249–253
error transfer function
H∞ filtering
J-spectral factorization, 412–414
performance analysis, 419
stationary H∞ filters, algebraic Riccati equation, 368–373
estimate differential equation, causal steady-state filtering, 169–170
estimate updates, linear measurements, 55
estimation error
continuous-time Kalman filter, Wiener-Hopf equation, 166–167
continuous-time linear quadratic Gaussian problem, 451–453
linear estimation, 37–38
linear Gaussian measurements, 21–25
mean-square estimation, 3–4
estimators, linear Gaussian measurements, 22
Euclidean norm, orthogonality principle, 13–14
Euler's approximation
colored process noise, 127
continuous stochastic systems, discrete-time Kalman filtering, 87
expected cost
stochastic control, polynomial systems, 474–481
stochastic control, state variable systems
continuous-time linear quadratic Gaussian problem, 446–449
discrete-time systems, 424–435
exponential data weighting, Kalman filter, 230–236
covariance expression, 235–236
fictitious process noise, 226–230

exponential data weighting, Kalman filter (*contd.*)
- filter stability margin, 255
- stability margin, 255

extended Kalman filter (EKF)
- adaptive sampling, 295–297
- nonlinear system estimation, 271–283
  - adaptive sampling applications, 283–304
  - continuous-continuous model, 275
  - continuous-discrete model, 273–274
  - FM demodulation, 275–278
  - measurement update approximation, 272–273
  - satellite orbit estimation, 278–283
  - time update approximation, 271–272
- pendulum analysis, 307
- rocket dynamics, 308
- spacecraft reentry trajectory estimation, 308–309
- waste sewage treatment, 312

F

feasibility analysis, robust stationary Kalman filters, convergence analysis and, 326–329

feedback gain
- scalar linear measurements, 48–49
- stochastic control, state variable systems, 448–449

fictitious process noise, Kalman filter, 226–230

finite horizon H∞ filters
- *a posteriori* filtering, 400–405
- characterization, 361–365
- dscrete-time H∞ filtering, 391–397

finite horizon robust Kalman filter
- *a priori* model, 317–321
- convergence analysis, 326–329

finite time horizons, H∞ filtering, continuous-time systems, 354–355
- linear filters, 357–361

first-order statistics
- continuous-time Kalman filter derivation, scalar system, 160–166
- linear mean-square estimation, 16–19
- time update approximation, extended Kalman filter, 271–272

first stage optimization, stochastic system, dynamic programming, 430–435

Fisher information matrix
- Cramér-Rao bound, 25–28
- time-variant plant, 111–112

fixed interval smoothing
- continuous-discrete Kalman filter, 202
- discrete-time Kalman filtering, 132–140

FM demodulation, extended Kalman filter, 275–278

Fokker-Planck equation
- hyperstate propagation, continuous systems, 264–265
- Markov process, 305
- time update estimates, 267–268
- Wiener process, 305

Fourier transform theory
- causal steady-state filtering, 41–49
- Wiener-Hopf equation, infinite-delay steady-state smoothing, 38–41

free fall tracking, extended Kalman filter, 310–312

frequency-domain solution
- causal steady-state filtering, 43–49
- Chang-Letov design procedure, 115–118
- continuous-time Kalman filter, 180–188
  - Chang-Letov design, 184–188
  - linear stochastic system spectral densities, 181
  - spectral factorization, 181–183
- discrete Wiener filter derivation, 118–123
- innovations representation, 114
- spectral factorization result, 112–114

full rank processes, Wiener filtering, 35

G

Gaussian conditional mean and covariance
- linear mean-square estimation, 16–19
- linear stochastic systems, 67–70
- mean-square estimation, 9–11

Gaussian field sampling, extended Kalman filter, 288–290

Gaussian measurements, maximum-likelihood estimation, 21–25

guaranteed norms, stationary H∞ filters, 371–373
gust noise, aircraft longitudinal dynamics, 125–127

H

Hamiltonian matrix, continuous-time Kalman filter derivation, 209
Hamilton-Jacobi-Bellman (HJB) equation
  Newton's system, 458
  stochastic control, state variable systems, continuous-time systems, 438–443
harmonic oscillators, continuous-time Kalman filter derivation
  damped oscillator, 164–166
  position measurements, 205, 206
  stochastic oscillators, 205
Hessian matrices, time update estimates, 267–268
H∞ filtering
  continuous-time systems
    convergence analysis, 375–378
    finite horizon linear filter, 357–365
    guaranteed *vs.* actual norms, 371–373
    Kalman filter *vs.*, 373–374
    linear matrix inequality, 382–383
    linear time-invariant filters, 373
    overview, 353–355
    signal model applications, 378–381
    stationary Riccati equation, 365–371
    two-person zero-sum game, 356–357
  discrete-time systems
    a priori filter, 390–400
    channel equalization, 414–418
    J-spectral factorization, 410–414
    overview, 388–390
    polynomial approach, 408–410
    stationary case, 400–408
hyperstate propagation
  nonlinear system estimation, 259–265
    continuous systems, 263–265
    discrete systems, 259–263
  plant stability/instability, 69–70

I

identity matrix, linear Gaussian measurements, 22–23
inaccurate noise parameters, 52–54
incomplete state information
  continuous-time linear quadratic Gaussian problem, 445–449
  discrete-time linear quadratic Gaussian problem, 455–457
independent measurements
  optimal combinations, 54–55
  scalar unknown, 23–25
infinite-delay steady-state smoothing, Wiener-Hopf equation, 38–41
information formulation, Kalman filter
  continuous-discrete Kalman filter, 200–201
  matrix for, 142
  optimal smoothing, 133–140
innovations sequence
  continuous-time Kalman filter derivation
    error dynamics, 178–180
    spectral factorization, 182–183
  error dynamics, statistical steady state, 102
  frequency domain results, 114
input signals, H∞ filter estimation, channel equalization, 416–418
intermittent measurements, prediction and, 143–144

J

Jacobian matrices
  closed-form linear field estimation, 294–295
  measurement update approximation, extended Kalman filter, 272–273
  time update approximation, extended Kalman filter, 272
Joseph stabilization
  continuous-time linear quadratic Gaussian problem, 449
  optimal smoothing, 135–140
J-spectral factorization, H∞ filter estimation, 410–414

K

Kalman filters
- angle-of-attack measurements, 121–123
- continuous-time Kalman filter
  - constant scalar unknown estimation, 157–159
  - discrete Kalman filter
    - derivation, 151–157
    - measurements, 84–101
  - discrete measurements, 193–197
  - error covariance differential equation, 170–171
  - error dynamics and statistical steady state, 177–180
  - estimate differential equation, 169–170
  - frequency domain results, 180–188
    - Chang-Letov design, 184–188
    - linear stochastic system spectral densities, 181
    - spectral factorization, 181–183
  - multiple sampling rates, 101
  - multiple-system rates, 98–101
  - noise correlation and shaping, 188–193
    - colored measurement noise, 190–193
    - colored process noise, 188–189
    - correlated measurement and process noise, 189–190
  - optimal impulse response differential equation, 168–169
  - optimal smoothing, 197–203
    - estimation, 201–202
    - information filter, 200–201
    - Rauch-Ting-Striebel smoother, 203
  - scalar systems, 160–166
  - shaping filter derivation, 167–168
  - stochastic system discretization, 85–98
  - Wiener-Hopf equation, 166–167
- continuous-time linear quadratic Gaussian problem, 449–453
- decoupling and parallel processing, 236–242
  - block triangular filter structure, 256
  - process noise mismodeling effect, 256–257
  - suboptimal filters, weakly coupled systems, 237–242
- discrete-time filters
  - *a posteriori* error system, 110
  - *a priori* recursive formulation, 74–75
  - algebraic Riccati equation, 103–110
  - algorithm, 71–73
  - alternative measurement update equations, 73–74
  - Chang-Letov design procedure, 115–118
  - colored measurement noise, 130–132
  - colored process noise, 124–127
  - continuous-time discrete measurements, 84–98
  - correlated measurement and process noise, 127–129
  - correlated noise and shaping, 123–124
  - deterministic state observer, 59–64
  - error dynamics, 101–102
  - formulation, 71
  - frequency domain results, 112–114
  - information filter, 133–136
  - innovations representation, 114
  - innovations sequence, 102–103
  - linear stochastic systems, 64–70
  - multiple sampling rates, 98–101
  - optimal smoothing, 132–138
  - orthogonality principle, 84
  - Rauch-Tung-Striebel smoother, 138–140
  - recursive estimation, constant scalar unknown, 78–81
  - scalar system filtering, 81–84
  - ship navigational fixes, 75–78
  - spectral factorization, 112–114
  - time-varying systems, 101, 110–112
  - weighted minimum-variance estimate, 110
  - Wiener filter derivation, 118–123
- divergence, 223–226
- exponential data weighting, 230–236
  - covariance expression, 235–236
  - filter stability margin, 255
- fictitious process noise injection, 226–230
- finite horizon H∞ filters, *a posteriori* filtering, 405

H∞ filtering
comparisons, 373–374
continuous-time systems, finite horizon linear filter, 360–361
stationary filters, 368–373
modeling errors, 213–223
continuous reduced-order filter mismatch, 222–223
continuous-time Kalman filter divergence, 254–255
continuous-time Kalman filter error correlation, 222
Newton's system as constant unknown, 255
reduced-order filter mismatch, 221–222
Wiener process inaccuracies, 218–221
reduced-order filters, 242–248
Kalman-Schmidt error covariance equations, 248
kinematic modeling, continuous systems, 244–248
robust filters
continuous-time systems, 341–349
convergence analysis, 326–331
finite horizon *a priori* filter, 317–321
linear matrix inequality, 331–340
modeling uncertainties, 315–317
stationary *a priori* filter, 321–326
scalar measurement updating, 253–254, 257–258
suboptimal gains, 249–253
continuous Wiener process, 250–253
filtering problems, 257
Kalman gain
continuous stochastic systems, 89
continuous-time Kalman filter derivation, 151–157, 172–173
discrete-time Kalman filter, 72
exponential data weighting, 232–236
steady-state solution, 233–236
Kalman filter divergence, 224–226
process noise injection, 227–230
mean-square estimation, 13
preliminary analysis, Kalman filter simplification, 95–98
reduced-order Kalman filter decoupling, suboptimal filters, weakly coupled systems, 238–244
scalar system filtering, 81–84
Kalman-Schmidt filter
error covariance equations, 248
kinematic modeling, 247–248
Keplerian orbital elements, satellite orbit determination, extended Kalman filter approximation, 280–283
kinematic modeling, continuous filter systems, 244–248
Kronecker delta
continuous stochastic systems, discrete-time Kalman filtering, 86–87
linear stochastic systems, 64
Lyapunov equation, vector formulation, 204

L

lateral aircraft dynamics, continuous-time Kalman filter derivation, Chang-Letov equation, 184–188
law of large numbers, recursive estimation, scalar unknown, 81
least mean square (LMS) estimation, H∞ filters, 419
least-squares solutions
closed-form linear field estimation, 290–295
linear equations, 50
Leibniz's rule, continuous-time Kalman filter, optimal impulse response, 169–170
likelihood function
linear Gaussian measurements, 21–25
maximum-likelihood estimation, 20–25
limiting behavior, discrete-time Kalman filtering, 147
linear data combination, scalar unknown, 54
linear equations, least-squares solution, 50
linear estimation, Wiener filtering, 36–38
linear field
closed-form estimation, extended Kalman filter, 286–290
one-dimensional/two-dimensional sampling, extended Kalman filter, 287–290

linear field (*contd.*)
  parameter-linear field, extended Kalman filter estimation, 296–297
linear filters
  H∞ filtering
    continuous-time systems, 361–365
      finite time horizons, 357–361
    signal models, 379–381
linear functions, unknown factor, 51–52
linear Gaussian measurements
  conditional mean estimate, 11–12
  Cramér-Rao bound, 25–28
  maximum-likelihood estimation, 21–25
  scalar unknown, 23–25
  sequential maximum-likelihood estimation, 31–33
linearized Kalman filter, continuous systems, 273–274
linear matrix inequality (LMI)
  H∞ filtering, 382–383
  robust Kalman filters, 331–340
    norm-bounded uncertainty, 332–335
    polytopic uncertainty, 335–340
linear mean-square estimation, 14–19
  orthogonality principle, 17–19
  quadratic estimation, 19
linear measurements
  continuous-time Kalman filter, shaping filter, 167–168
  correlated noise, 52
  estimate updates, 55
  updates, nonlinear systems, 269–271
  Wiener filtering, 46–49
linear quadratic Guassian (LQG) regulators
  continuous-time linear quadratic Gaussian problem, 443–453
    complete state information, 445–449
    incomplete state information, 445–449
  discrete-time linear quadratic Gaussian problem, 453–457
    complete state information, 454–455
    incomplete state information, 455–457
  scalar discrete regulator, 462
  steady-state regulator, 463–465
  stochastic control
    polynomial systems, 473–481
    state variable systems, 439–443
  suboptimal regulator, 460
linear stochastic systems
  continuous-time Kalman filter
    derivation, 156–157
    spectral densities, 181
  discrete-time Kalman filtering, 64–70
linear time-invariant filters
  H∞ filter estimation, polynomial approach, 408–410
  robust Kalman filter modeling, linear matrix inequality, 332–340
  robust stationary Kalman filtering, 322–326
  stationary H∞ filters, 373
localization uncertainty
  closed-form linear field estimation, 290–295
  mobile robot localization, simultaneous localization and sampling, 297–301
Lyapunov equation
  algebraic Riccati equation, 104–105
  closed-loop system equation, 142
  continuous-discrete Kalman filter, 196–197
  continuous-time Kalman filter derivation, 156–157
    spectral densities, linear stochastic systems, 181
  finite horizon robust Kalman filter, convergence analysis, 328–329
  H∞ filtering
    convergence analysis, 377–378
    signal models, 379–381
  hyperstate propagation, continuous systems, 264–265
  Ricatti equation as, closed-loop system equation, 149
  robust Kalman filter modeling, linear matrix inequality, 332–339
  robust stationary Kalman filter, 322–326
  steady-state covariance, 69–70
  stochastic control, state variable systems
    continuous-time linear quadratic Gaussian problem, 449
    continuous-time systems, 443
  time update estimates, 267–268
  vector equation, 140
  vector formulation, 204

M

Markov process
  causal steady-state filtering, 45–49
  colored process noise, 127
  Fokker-Planck equation, 305
  hyperstate propagation, discrete systems, 261–263
  linear stochastic systems, 64–65
  sampled signal predictions, 482
  stochastic control, state variable systems, continuous-time systems, 436–443
  Wiener filtering, second-order process, 57
  Wiener-Hopf equation, infinite-delay steady-state smoothing, 39–40
matrix inversion lemma, finite horizon H∞ filters, *a posteriori* filtering, 401–405
maximum a posteriori estimation, mean-square estimation, 4
maximum-likelihood estimation (MLE), 19–25
  Cramér-Rao bound, 26–28
  discrete-time Kalman filter, 74–75
  electric circuit analog example, 24–25
  linear Gaussian measurements, 20–25
  nonlinear, 19–20
  recursive estimation, scalar unknown, 81
  sequential MLE, 31–33
means propagation, linear stochastic systems, 65–70
mean-square error
  defined, 3–4
  stochastic control, polynomial systems, optimal prediction, 467–469
mean-square estimation (MSE)
  continuous-time linear quadratic Gaussian problem, 444–453
  criterion, 3–4
  discrete-time linear quadratic Gaussian problem, complete information, 454–455
  Gaussian conditional mean and covariance, 9–11
  hyperstate propagation, discrete systems, 259–263
  Kalman gain, 13
  linear estimation, 14–19
  nonlinear example, 8–9
  orthogonality principle, 13–14
  random variable X, 4–13
  stochastic control, state variable systems, 440–443
mean updates, nonlinear estimation, 265–266
measurement model
  H∞ filtering, continuous-time systems, 353–355
  radar tracking, 89–91
measurement noise
  colored measurement noise, 130–132
  continuous stochastic systems, Newton's laws, sampling motion, 88
  continuous-time Kalman filter derivation, correlated measurement, 211
  delayed process noise, 129
  deterministic state observer, 63–64
  dscrete-time H∞ filtering, finite horizon case, 393–397
  linear Gaussian measurements, 11–12
  linear stochastic systems, 64
  process noise and, 127–132
  Wiener-Hopf equation, infinite-delay steady-state smoothing, 39–40
measurement update equations
  continuous-discrete Kalman filter, 196–197
  discrete-time Kalman filter, 73
  extended Kalman filter/adaptive sampling algorithm, 295–297
  extended Kalman filter approximation, 272–273
    continuous systems, 273–283
    FM demodulation, 277–278
    satellite orbit determination, 282–283
  hyperstate propagation, discrete systems, 261–263
  nonlinear systems, 268–269
  recursive estimation, scalar unknown, 79–81
  ship navigational fixes, 76–78
  uniform noise, 306
Millman's theorem
  continuous-discrete Kalman filter, optimal smoothed esimate, 201–202
  maximum-likelihood estimation, electric circuit analog, 25
  optimal smoothing estimate, 136–140

minimum energy problems, stochastic control, state variable systems, discrete-time systems, 425
minimum fuel problems, stochastic control, state variable systems, discrete-time systems, 424–425
minimum time problems, stochastic control, state variable systems, discrete-time systems, 424
minimum variance control, polynomial systems, 469–471
minimum variance predictor, stochastic control, polynomial systems, optimal prediction, 468–469
mismatch errors, Kalman filters, modeling errors, 214–223
mobile robot localization, extended Kalman filter sampling, 284
  simultaneous localization and sampling, 297–301
model-following behavior
  linear quadratic Guassian (LQG) regulators, stochastic control, polynomial systems, 474–481
  stochastic control
    controller, 482–483
    control weighting, 484
modeling errors
  Kalman filters, 213–223
    continuous reduced-order filter mismatch, 222–223
    continuous-time Kalman filter divergence, 254–255
    continuous-time Kalman filter error correlation, 222
    Newton's system as constant unknown, 255
    reduced-order filter mismatch, 221–222
    Wiener process inaccuracies, 218–221
  Kalman-Schmidt filter, 247–248
  robust Kalman filters, uncertainties, 315–317
Monte Carlo measurements, linear field sampling, 287–290
Moore-Penrose matrix, Kalman filters, model mismatch, reduced-order filters, 221–222
motion estimation
  continuous-time Kalman filter derivation, 174–175
  extended Kalman filter approximation, satellite orbit determination, 278–283
  H∞ filtering, 383–384
multiple sampling rates, Kalman filtering, 98–101
multiplicative noise, estimation theory, 52
multivariable root locus techniques, Chang-Letov equation, 118

N

Newton's system
  Kalman filter modeling, constant unknown condition, 255
  sampling motion
    continuous stochastic systems, discrete-time Kalman filtering, 87–88
    continuous-time Kalman filter derivation, 174–175, 204
      velocity measurements, 207
  steady-state LQG regulator, 460–461
  stochastic control, 458
    polynomial systems, 483–484
noise covariance
  H∞ filtering, signal models, 379–381
  H∞ filtering, continuous-time systems, 354–355
  inaccuracy, 52–53
noise sequence, shaping filter correlation, 124–132
noise-weighting matrix, nonlinear system estimation, 307
nonlinear least-squares estimation, 50
nonlinear maximum-likelihood estimation, 19–20
nonlinear mean-square estimation, example of, 8–9
nonlinear system estimation
  conversion to linear systems, 310
  dynamics measurement, 306
  extended Kalman filter, 271–283
    adaptive sampling applications, 283–304
    continuous-continuous model, 275
    continuous-discrete model, 273–274
    FM demodulation, 275–278

measurement update approximation, 272–273
satellite orbit estimation, 278–283
time update approximation, 271–272
hyperstate update, 259–265
continuous systems, 263–265
discrete systems, 259–263
linear measurement update, 269–271
mean and covariance updates, 265–266
measurement update, 268–269
noise-weighting matrix, 307
time update, 266–268
nonstationary Markov process, linear stochastic systems, 64–65
norm-bounded uncertainty
finite horizon robust Kalman *a priori* filter, 320–321
robust Kalman filter modeling, 316–317
linear matrix inequality, 332–335
stationary H∞ filters, guaranteed *vs.* actual norms, 371–373

O

observability gramian, time-variant plant, 111–112
one-degree-of-freedom regulator, stochastic control, polynomial systems, 469–471
optimal combination, independent measurements, 54–55
optimal error variances
finite horizon robust Kalman filter, 320–321
reduced-order Kalman filter decoupling, suboptimal filters, weakly coupled systems, 239–244
optimal feedback control, stochastic control, state variable systems
polynomial formulation, 459
scalar systems, 442–443
optimal filters
alternative design methods, 55–56
continuous-time Kalman filter derivation, Chang-Letov equation, 187–188
H∞ filtering, continuous-time systems, 354–355
linear estimation, 37–38
robust Kalman filtering
difference Riccati equation, 343–345
linear matrix inequality, 338–340
optimal impulse response, continuous-time Kalman filter, 169–170
optimality principle, dynamic programming, stochastic control, state variable systems, 423–424
continuous-time systems, 435–443
discrete-time systems, 425–435
optimal prediction, stochastic control, polynomial systems, 465–469
optimal smoothing
a priori estimates, 149
continuous-time Kalman filter, 197–203
estimation, 201–202
information filter, 200–201
Newton's system, 212
Rauch-Ting-Striebel smoother, 203
scalar system, 212
discrete-time Kalman filtering, 132–140
estimates, 136–140
information formulation, 133–140
Rauch-Tung-Striebel formulation, 138–140
orthogonality principle
Kalman filter derivation, 84
linear mean-square estimation, 17–19
mean-square estimation, 13–14
stochastic control, polynomial systems, linear quadratic Gaussian control, 476–481
output feedback, continuous-time linear quadratic Gaussian problem, 451–453

P

parallel combination, maximum-likelihood estimation, electric circuit analog, 24–25
parallel Kalman filtering, alpha-beta radar tracking, 92

parallel processing, reduced-order Kalman filter decoupling, 236–244
  suboptimal filters, weakly coupled systems, 237–244
parameter-linear field, extended Kalman filter estimation, 296–297
parametric measured field, extended Kalman filter, combined adaptive sampling, 284–286
partial differential equation, continuous-time systems, dynamic programming, 435–443
pendulum analysis, extended Kalman filter, 307
perfect measurements
  dscrete-time H∞ filtering, finite horizon case, 394–397
  scalar system filtering, 82
performance analysis
  continuous filter systems, kinematic modeling, 246–248
  discrete-time H∞ filtering, 388–390
  H∞ filtering
    continuous-time systems, 354–355
    error transfer function, 419
  Kalman filters, modeling errors, Wiener filtering, constant unknown condition, 220–221
  robust *a priori* and *a posteriori* filters, 340
  stationary H∞ filters, guaranteed *vs.* actual norms, 372–373
performance indices
  linear quadratic Guassian (LQG) regulators, stochastic control, polynomial systems, 473–481
  stochastic control, polynomial systems, 469–471
  stochastic control, state variable systems, discrete-time systems, 424–435
plant matrix, deterministic state observer, 62–64
plant system, Kalman filters, modeling errors, 213–223
polynomial approach
  H∞ filter estimation, 408–410
    channel equalization, 415–418
  stochastic control, state variable systems, 459
polynomial systems, stochastic control
  linear quadratic Gaussian regulator, 473–481
  minimum variance control, 469–473
  optimal prediction, 465–469
  overview, 463–464
polytopic uncertainty, robust Kalman filtering, linear matrix inequality, 335–340
positive definite matrix
  discrete-time H∞ filtering, 387–390
  robust stationary *a priori* Kalman filter, 320–321
predicted error covariance
  Kalman filter divergence, process noise injection, 227–230
  Kalman filters, modeling errors, 217–223
    Wiener filtering, constant unknown condition, 218–223
  reduced-order Kalman filter decoupling, suboptimal filters, weakly coupled systems, 239–244
prediction
  intermittent measurements and, 143–144
  Wiener filtering and, 56
predictive formulation, stochastic control, polynomial systems, optimal prediction, 466–469
predictor-corrector formulation, discrete-time Kalman filter, 71–72
prefiltered data, recursive estimation, 33–34
preliminary analysis, Kalman filter simplification, 95–98
prewhitening of data, scalar measurement updating, Kalman filter, 253–254
probability density function (PDF)
  a priori estimates, 4–13
  conditional mean estimate, 6–7
  continuous-time linear quadratic Gaussian problem, complete state information, 445–449
  Cramér-Rao bound, 27–28
  Gaussian conditional mean and covariance, 9–11
  hyperstate propagation
    continuous systems, 264–265

discrete systems, 259–263
linear stochastic systems, 67–70
maximum-likelihood estimation, 20–25
mean-square estimation, 4
scalar discrete stochastic system, dynamic programming, 427–435
stochastic control, state variable systems, continuous-time systems, 436–443
process noise
algebraic Ricatti equation, 106–107
continuous-time Kalman filter derivation, 189–190, 206
scalar system, 162–166
correlated measurement and, 127–132
delayed noise, measurement correlation with, 129–132
deterministic state observer, 63–64
Kalman filter mismodeling, 256–257
scalar discrete stochastic system, dynamic programming, 427–435
scalar system filtering, absence of, 83–84
stochastic control, state variable systems, non-Gaussian noise, 442
proof of necessity, H∞ filtering, continuous-time systems, finite horizon linear filter, 359–360
proof of sufficiency, H∞ filtering, continuous-time systems, finite horizon linear filter, 358–359
Pythagoras' theorem, orthogonality principle, 13–14

Q

quadratic linear mean-square estimation, 19
quadratic performance index, Hamilton-Jacobi-Bellman (HJB) equation derivation, 439–443
quadratic stability, robust stationary Kalman filter, 321–326

R

radar tracking
alpha-beta tracker, 92–98
discrete-time Kalman filtering, 89–92
extended Kalman filter, 309–310
multiple-rate Kalman filter, 99–101
random variables, nonlinear systems, sum of, 305
Rauch-Tung-Striebel formulation
continuous-discrete Kalman filter, 203
discrete-time linear quadratic Gaussian problem, incomplete information, 457
optimal smoothing, 138–140
RBF coefficients, Gaussian field sampling, extended Kalman filter, 288–290
reachability, time-variant plant, 111–112
recursive estimation
data prewhitening, 33–34
discrete-time Kalman filter, 74–75
dscrete-time H∞ filtering, finite horizon case, 395–397
extended Kalman filter/adaptive sampling algorithm, 295–297
hyperstate propagation, discrete systems, 261–263
line of best fit, 143
scalar unknown, 78–81
sequential measurement processing, 28–30
sequential maximum-likelihood estimation, 31–33
reduced-order Kalman filters
decoupling and, 242–248
Kalman-Schmidt error covariance equations, 248
kinematic modeling, continuous systems, 244–248
model mismatch, 221–222
regulator error, stochastic control, polynomial systems, 471–472, 484
representation theorem
causal steady-state filtering, 42–49
Wiener filtering, 35
residuals
continuous-time Kalman filter derivation, 155–157
covariance, 207

residuals (*contd.*)
linear Gaussian measurements, 21–25
linear mean-square estimation, 15–16
Riccati equation. *See also* algebraic Riccati equation; difference Riccati equation
continuous filter systems, kinematic modeling, 244–248
continuous-time Kalman filter derivation, 153–157
damped harmonic oscillator, 164–166
error covariance, 171–172
periodic solution, 205
scalar system, 160–166
continuous Wiener process, suboptimal gains, 250–253
difference Riccati equation, finite horizon robust Kalman *a priori* filter, 318–321
discrete-time Kalman filtering, deterministic state observer, 61–64
discrete-time linear quadratic Gaussian problem, complete information, 454–455
dscrete-time H∞ filtering, finite horizon case, 395–397
error dynamics, statistical steady state, 102
H∞ filtering, linear matrix inequality, 382–383
H∞ filtering, stationary filters, 365–373
guaranteed and actual norm, 371–373
linear time-invariant filters, 373
innovations sequence, 102
Lyapunov closed-loop system equation, 142, 149
reduced-order Kalman filter decoupling, suboptimal filters, weakly coupled systems, 237–244
robust Kalman filter modeling, linear matrix inequality, 334–335
scalar system filtering, 81–84
square-root formulation, 142–143
stabilization via, 140
stochastic control, state variable systems, 440–443
continuous-time linear quadratic Gaussian problem, 447–449
time-variant plant, 111–112
robust Kalman filters
continuous-time systems, 341–349
convergence analysis, 326–331
finite horizon *a priori* filter, 317–321
linear matrix inequality, 331–340
modeling uncertainties, 315–317
stationary *a priori* filter, 321–326
rocket dynamics, extended Kalman filter, 308
root locus, error systems
algebraic Ricatti equation, 108–110
Chang-Letov equation, 117–118
continuous-time Kalman filter derivation, 187–188
Runge-Kutta integrator
continuous-discrete Kalman filter, 194–197
continuous-time Kalman filter derivation, damped harmonic oscillator, 165–166
reduced-order Kalman filter decoupling, suboptimal filters, weakly coupled systems, 238–244
stochastic control, state variable systems, continuous-time systems, 443
time update approximation, extended Kalman filter, 272

S

sampled signal predictions, stochastic control, 482
sample mean and variance, recursive estimation, 28–30
sampling motion, continuous stochastic systems, discrete-time Kalman filtering, 87–88
satellite orbit determination, optimal smoothing, 132–140
scalar system
continuous-time Kalman filter derivation, 160–166
deterministic observer, Riccati equation design, 61–64
discrete stochastic system, dynamic programming, 427–435
dscrete-time H∞ filtering, finite horizon case, 392–397
H∞ filtering, 382–383
Kalman filtering, 81–84
linear measurements

steady-state filtering, 107–108
Wiener filtering, 46–49
mean propagation, linear stochastic systems, 67–70
measurement updating, Kalman filters, 253–254, 257–258
optimal feedback control, 442–443
robust stationary Kalman filters, 324–326
stochastic control, state variable systems, 459
scalar unknown
continuous-time Kalman filter derivation, 157–158
linear data combination, 54
linear Gaussian measurements, 23–25
recursive estimation, 30
Kalman filter, 78–81
scaling parameters, finite horizon robust Kalman *a priori* filter, 321
Schur complement
Gaussian conditional mean and covariance, 10–11
robust Kalman filtering, 346–349
second-order autoregressive model, robust Kalman filter modeling, 316–317
second-order polynomial tracking Kalman filter, alpha-beta radar tracking, 93
second-order statistics, linear mean-square estimation, 16–19
separation principle
continuous-time linear quadratic Gaussian problem, 450–453
stochastic control, polynomial systems, 469–471
sequential maximum-likelihood estimation
block diagonals, 55
recursive estimation, 31–33
sequential measurement processing, recursive estimation, 28–30
shaping filters
colored process noise, 124–127
continuous-time Kalman filter, 167–168
correlated noise, 188–193
ship navigational fixes, discrete-time Kalman filter, 75–78
signal generating mechanisms
discrete-time H∞ filtering, 387–390
H∞ filtering, continuous-time systems, 353–355
signal models, H∞ filtering, 378–381
simultaneous localization and mapping algorithms (SLAM), extended Kalman filter sampling, mobile robot localization, 284
single-input single-output state representation, stochastic control, polynomial systems, 464
singular value plots, H∞ filtering, J-spectral factorization, 412–414
smoothable states, optimal smoothing estimates, 136–140
smoothing problem, discrete-time Kalman filtering, 132–140
spacecraft reentry trajectory estimation, extended Kalman filter, 308–309
spectral density
colored process noise, 124–127
continuous-time Kalman filter, linear stochastic system, 181
statistical steady state, 68–70
stochastic control, polynomial systems, 465
Wiener filtering, 34
spectral factorization theorem
causal steady-state filtering, 42–49
continuous, 36
continuous-time Kalman filter derivation, 173–174
frequency domain solution, 181–183
discrete, 35–36
frequency domain results, 112–114
H∞ filter estimation, 410–414
Wiener filtering, 35
square root Ricatti formulation, 142–143
square transfer function matrix, H∞ filter estimation, polynomial approach, 409–410
stability margin
H∞ filtering, convergence analysis, 378
Kalman filter, exponential data weighting, 233–236
state feedback optimal control law, stochastic control, state variable systems, 431–435
state models
H∞ filtering, continuous-time systems, 353–355
radar tracking, 89–91

state models (*contd.*)
  steady-state Kalman filter, 122–123
  Wiener filtering, 147
state-space equations
  finite horizon robust Kalman *a priori* filter, 317–321
  H∞ filtering, continuous-time systems, 353–355
state space representation, H∞ filtering applications, channel equalization, 415–418
state-variable feedback, stochastic control, state variable systems, discrete-time systems, 427–435
state variable systems, stochastic control
  continuous-time linear quadratic Gaussian problem, 443–453
    complete state information, 445–449
    incomplete state information, separation principle, 449–453
  continuous-time systems, 435–443
  discrete-time linear quadratic Gaussian problem, 453–457
    complete state information, 454–455
    incomplete state information, 455–457
  discrete-time systems, 424–435
  dynamic programming, 423–443
stationary H∞ filters
  a posteriori filtering, 405–408
  dscrete-time H∞ filtering, 391–397
  Riccati equation approach, 365–373
    guaranteed and actual norm, 371–373
    linear time-invariant filters, 373
stationary Markov process, linear stochastic systems, 64–65
stationary robust Kalman filter, *a priori* properties, 321–326
statistical steady state
  continuous-time Kalman filter derivation, error dynamics, 177–180
  error dynamics, 101–102
  linear stochastic systems, 68–70
  spectral densities and, 68–70
steady-state solution
  angle-of-attack measurements, 122–123
  causal steady-state filtering, 41–49
  Chang-Letov design, 116, 147–148
  continuous-time Kalman filter derivation, 209
    alternative design method, 208–209
    Chang-Letov equation, 186–188
    damped harmonic oscillator, 165–166
    harmonic oscillator, 205
    no process noise, 210
    scalar system, 160–166
    spectral factorization, 181–183
  continuous Wiener process, suboptimal gains, 252–253
  damped harmonic oscillator, 108
  discrete Wiener filter derivation, 120–123
  infinite-delay steady-state smoothing, 38–41
  Kalman filter
    divergence, process noise injection, 229–230
    exponential data weighting, 233–236
  linear estimation, 37–38
  scalar case, 107–108
  suboptimal Kalman gains, 247–253
stochastic control
  bilinear systems, 457–458
  continuous-time Kalman filter derivation, computer simulation, 204–205
  discrete-time Kalman filtering, 141
  external fields, 457
  hyperstate propagation, discrete systems, 259–263
  polynomial systems
    linear quadratic Gaussian regulator, 473–481
    minimum variance control, 469–473
    optimal prediction, 465–469
    overview, 463–464
  state variable systems
    continuous-time linear quadratic Gaussian problem, 443–453
      complete state information, 445–449
      incomplete state information, separation principle, 449–453
    discrete-time linear quadratic Gaussian problem, 453–457
      complete state information, 454–455

incomplete state information, 455–457
dynamic programming
continuous-time systems, 435–443
discrete-time systems, 424–435
stochastic convergence, Cramér-Rao bound, 26–28
stochastic harmonic oscillator, continuous-time Kalman filter derivation, 206
stochastic linear quadratic regular, Hamilton-Jacobi-Bellman (HJB) equation derivation, 438–443
stochastic projection theorem
orthogonality principle, 13–14
Wiener filtering, 34
suboptimal gains
H∞ filtering, continuous-time systems, 354–355
finite horizon linear filters, 361–365
Kalman filters, 249–253, 257
continuous Wiener process, 250–253
filtering problems, 257
reduced-order Kalman filter decoupling, 237–244
system stability/instability, deterministic state observer, 62–64

T

tachometer vibration, colored measurement noise, 148
Taylor series
measurement update approximation, extended Kalman filter, 272–273
stochastic control, state variable systems, continuous-time systems, 437–443
time update approximation, extended Kalman filter, 271–272
time-domain function, causal steady-state filtering, 44–49
time-invariant model
continuous Wiener process, suboptimal gains, 251–253
Kalman filter, exponential data weighting, 232–236
robust stationary Kalman filtering, 322–326
time update estimates
continuous-discrete Kalman filter, 196–197
delayed process noise, 129
extended Kalman filter approximation, 271–272
continuous systems, 273–283
FM demodulation, 276–277
satellite orbit determination, 281–283
hyperstate propagation, discrete systems, 262–263
moments, 306
multiple sampling rates, 98–101
nonlinear systems, 266–268
optimal smoothing, 134–140
recursive estimation, scalar unknown, 79–81
ship navigational fixes, 76–78
time-variant plant
continuous-time Kalman filter derivation, 180
discrete-time Kalman filtering, 144
error dynamics, 110–112
time-varying stochastic system
continuous-time Kalman filter, robust filtering for, 341–342
covariance propagation, 68–70
derived measurement, 210
time-varying Wiener filter, linear estimation, 36–38
tracking systems, stochastic control, polynomial systems, linear quadratic Gaussian control, 480–481
transfer functions
causal steady-state filtering, 43–49
Wiener filtering, 35
transition probability density function, stochastic control, state variable systems, 433–435
true divergence, Kalman filter, 223–226
truth model, Kalman filters, modeling errors, 213–223
two-person zero-sum game, H∞ filtering, continuous-time systems, 356–357

U

uncertainty
discrete-time H∞ filtering, 387–390

uncertainty (*contd.*)
- extended Kalman filter, linear field closed-form estimation, 290–295
- mobile robot localization, 297–301
- norm-bounded uncertainty
  - finite horizon robust Kalman *a priori* filter, 320–321
  - robust Kalman filter modeling, 316–317
    - linear matrix inequality, 332–335
- polytopic uncertainty, robust Kalman filtering, linear matrix inequality, 335–340
- robust Kalman filter modeling, 315–317
  - continuous-time systems, 341–342
  - linear matrix inequality, 331–340
- robust stationary Kalman filter, 321–326

unique positive solution, algebraic Riccati equation, 104–106
unit rectangle definition, continuous stochastic systems, discrete-time Kalman filtering, 86–87
unknown constant parameter, H∞ filtering, 382
unmodeled dynamics, radar tracking, 90–91
unstable states
- H∞ filtering, J-spectral factorization, 412–414
- Kalman filter divergence, 223–226

upper triangular Cholesky decomposition, scalar measurement updating, Kalman filter, 253–254

V

velocity error variances, ship navigational fixes, 77–78
vibration isolation system, discrete-time Kalman filtering, 145–146
voltage estimation, continuous and discrete filters, 207–208

W

waste sewage treatment, nonlinear system estimations, 312
weakly coupled systems, reduced-order Kalman filter decoupling, suboptimal filters, 237–244
weighted covariance, Kalman filter, exponential data weighting, 231–236
weighted minimum-variance estimate, algebraic Riccati equation, 110
whitening property
- causal steady-state filtering, 42–49
- continuous-time Kalman filter derivation, spectral factorization, 183
- discrete Wiener filter, 119–123

white noise
- continuous stochastic systems, discrete-time Kalman filtering, 85–98
- finite horizon H∞ filters, *a posteriori* filtering, 402–405
- linear stochastic systems, 64
- mobile robot localization, simultaneous localization and sampling, 300–304

Wiener filtering, 34–49
- angle-of-attack measurements, 121–123
- continuous spectral factorization, 36
- continuous-time Kalman filter derivation
  - motion estimation, 174–177
  - scalar system, 164
  - spectral factorization, 183
- discrete derivation, 118–123
- discrete spectral factorization, 35–36
- Fokker-Planck equation, 305
- hyperstate propagation, continuous systems, 264–265
- Kalman filters, modeling errors, constant unknown condition, 218–223
- linear estimation, 36–38
- linear measurements, 46–49
- optimal state estimation, 147
- prediction problems, 56
- second-order Markov process, 57
- steady-state filter design, 209
- suboptimal gains, 250–253
- suboptimal Kalman gains, 247–253

Wiener-Hopf equation
- causal steady-state filtering, 41–49
- continuous-time Kalman filter, 166–167
  - optimal impulse response, 169–170

infinite-delay steady-state smoothing, 38–41
linear estimation, 37–38
prediction problems, 56

Z

zero-input response, Kalman filter divergence, 224–226
zero noise conditions
alpha-beta radar tracking, 93–95
continuous-time Kalman filter derivation, scalar system, 162–163
H∞ filtering *vs.* Kalman filtering, 373–374
zeroth stage optimization, stochastic control, state variable systems, 431–435